COMPLICATIONS IN
DIAGNOSTIC RADIOLOGY

'Never delay investigations, even surgical exploration, if a disease is suspected which demands early treatment, and if the hazards of the diagnostic procedures are less than those of the disease which is suspected.'

LORD COHEN OF BIRKENHEAD
Annual Oration, The Medical Society
of London (1966)

COMPLICATIONS IN DIAGNOSTIC RADIOLOGY

EDITED BY

G. ANSELL

MD, DMRD, FRCP, FRCR
Consultant Radiologist in Charge
Whiston Hospital
Lecturer in Radiodiagnosis
University of Liverpool

BLACKWELL SCIENTIFIC PUBLICATIONS
OXFORD LONDON EDINBURGH MELBOURNE

© 1976 Blackwell Scientific Publications
Osney Mead, Oxford,
8 John Street, London, WC 1,
9 Forrest Road, Edinburgh,
P.O. Box 9, North Balwyn, Victoria, Australia.

All rights reserved. No part of this publication
may be reproduced, stored in a retrieval system,
or transmitted, in any form or by any means,
electronic, mechanical, photocopying, recording
or otherwise without the prior permission of
the copyright owner.

First published 1976

British Library Cataloguing in Publication Data

Complications in diagnostic radiology.
  Bibl.–Index.
  ISBN 0-632-00039-2
  1. Ansell, George
  616.07'57 RC78
  Diagnosis, Radioscopic

Distributed in the United States of America by
J. B. Lippincott Company, Philadelphia,
and in Canada by
J. B. Lippincott Company of Canada Ltd, Toronto.

Printed in Great Britain at
the Alden Press, Oxford
and bound by
Mansell (Bookbinders) Ltd,
Witham, Essex.

CONTENTS

	Contributors	vii
	Preface	ix
1	Contrast media in urography G. ANSELL	1
2	Aortography and peripheral arteriography H. HERLINGER	42
3	Cardiopulmonary angiography A. J. S. SAUNDERS AND J. D. DOW	76
4	Renal complications of angiography LEE B. TALNER	111
5	Spinal cord complications of angiography G. MARGOLIS	134
6	Cerebral angiography A. LUNDERVOLD AND A. ENGESET	151
7	Pneumoencephalography and ventriculography J. V. OCCLESHAW	183
8	Myelography J. V. OCCLESHAW	199
9	Phlebography M. LEA THOMAS	215

Contents

10	Percutaneous transheptatic cholangiography, endoscopic retrograde cholangiopancreatography and spleno-portography W. B. YOUNG AND E. C. LIM	239
11	Oral and intravenous cholegraphy G. ANSELL	264
12	Bronchography G. ANSELL	278
13	Lymphography J. S. MACDONALD	301
14	Hysterosalpingography E. BARNETT	317
15	Amniography and intraperitoneal transfusion E. BARNETT	323
16	Alimentary tract G. ANSELL	333
17	Miscellaneous procedures G. ANSELL	369
18	Anaesthetic problems S. LIPTON	382
19	Radiation problems G. M. ARDRAN AND H. E. CROOKS	404
20	Isotope techniques E. S. WILLIAMS	439
21	Diagnostic ultrasound E. BARNETT	447
22	Electrical and mechanical hazards in the X-ray department G. R. HIGSON	452
23	Notes on radiological emergencies G. ANSELL	471
	Index	487

CONTRIBUTORS

GEORGE ANSELL, M.D., D.M.R.D., F.R.C.P., F.R.C.R. Consultant Radiologist in Charge, Whiston Hospital. Lecturer in Radiodiagnosis, University of Liverpool.

G. M. ARDRAN, M.D., F.R.C.P., F.R.C.R. Radiologist, Nuffield Institute for Medical Research, University of Oxford. Consultant Radiologist, Environmental and Medical Sciences Division, A.E.R.E. Harwell.

ELLIS BARNETT, F.R.C.P., D.M.R.D., F.R.C.R. Consultant Radiologist, Western Infirmary, Glasgow.

H. E. CROOKS, F.S.R. Higher Scientific Officer, Environmental and Medical Sciences Division, A.E.R.E. Harwell.

J. D. DOW, M.C., M.D., F.R.C.P., F.R.C.R. Director, X-Ray Department, Guy's Hospital, London.

*Dr Med. ARNE ENGESET Assistant Professor, Oslo University. Head of the Department of Neuroradiology, Rikshospitalet, Oslo, Norway.

H. HERLINGER, M.D., D.M.R.D., F.R.C.R. Consultant Radiologist, St James Hospital, Leeds. Senior Clinical Lecturer in Radiology, University of Leeds.

G. R. HIGSON, B.Sc., F.Inst.P., C.Eng., M.I.E.E. Assistant Director of Scientific & Technical Services, Department of Health and Social Security.

* Professor Arne Engeset died in September 1973.

E. C. LIM, M.B., D.M.R.D. University Hospital, Kuala Lumpur, Malaya.

SAMPSON LIPTON, F.F.A.R.C.S. Consultant Anaesthetist, Department of Medical and Surgical Neurology, Walton Hospital, Liverpool. Visiting Fellow, Bio-Engineering Department, University of Salford.

Dr Med. ARNE LUNDERVOLD Assistant Professor, Oslo University. Head of the Department of Clinical Neurophysiology, Rikshospitalet, Oslo, Norway.

J. S. MACDONALD, M.B., D.M.R.D., M.R.C.P.E., F.R.C.R. Consultant Radiologist, Royal Marsden Hospital and Institute of Cancer Research, London and Surrey.

GEORGE MARGOLIS, M.D. Professor of Pathology, Dartmouth Medical School, Hanover, N.H. 03755, U.S.A.

J. V. OCCLESHAW, M.D., D.M.R.D., F.R.C.R. Consultant Radiologist, Neuroradiology Department, Manchester Royal Infirmary.

A. J. S. SAUNDERS, M.R.C.P., F.R.C.R. Consultant Radiologist, Guy's Hospital, London.

L. B. TALNER, M.D. Professor of Radiology and Head, Section of Uroradiology, University of California, San Diego, University Hospital, San Diego, Calif. 92103, U.S.A.

M. LEA THOMAS, M.A., M.B., F.R.C.P., F.R.C.R., D.M.R.D. Vascular Radiologist, St Thomas' Hospital, London. Honorary Lecturer, St Thomas' Hospital Medical School.

E. S. WILLIAMS, M.D., Ph.D., B.Sc., M.R.C.P. Professor of Nuclear Medicine, University of London. Director of the Department of Nuclear Medicine, Middlesex Hospital Medical School.

W. B. YOUNG, M.B.E., F.R.C.S.E., F.R.C.R. Director, X-Ray Department, Royal Free Hospital, London.

PREFACE

In recent years there have been spectacular advances in radiology. As in all other spheres of medicine, however, the increased benefit to the patient in diagnostic accuracy must be balanced against the inevitable small but significant risk of complications which may ensue. Because of the rarity of many of these complications, an individual radiologist will fortunately witness few of them during his career. Nevertheless, every radiologist must be conversant with these possibilities so that if a complication does arise, he will be able to diagnose this promptly and institute appropriate treatment. This book provides a comprehensive and authoritative review of the subject. Particular emphasis has been placed on techniques and fundamental principles, thereby providing a rational approach to prevention and treatment.

I am grateful to my colleagues, all acknowledged experts in their fields, who so readily agreed to contribute to the book. Radiologists will mourn the sad loss of Professor Arne Engeset and the chapter written in collaboration with Professor Lundervold will be a lasting tribute to his memory.

I should also like to thank the many radiologists who participated in the U.K. National Radiological Survey for the valuable information derived from their reports. Much of this material has been incorporated into the book where it provides an important and easily accessible source of reference.

Finally, I should like to pay tribute to the publishers for their helpful co-operation and for the excellent reproduction of the illustrations.

June 1976　　　　　　　　　　　　　　　　　　　　　　　GEORGE ANSELL

CHAPTER 1. CONTRAST MEDIA IN UROGRAPHY

G. ANSELL

EXCRETION UROGRAPHY

In recent years, several major reviews of contrast media have been published [1-3]. The intravascular contrast media used for excretion urography and angiography are water-soluble derivatives of organic acids and their radio-opacity is directly related to their iodine content. The di-iodinated pyridine derivatives iodomethamate (iodoxyl) and iodopyracet (diodone) are now largely of historical interest. Modern tri-iodinated compounds are derivatives of benzoic acids. The earliest of these, acetrizoate (Diaginol, Urokon) has now been superseded for intravascular use in the United States, United Kingdom and Scandinavia because of its relative toxicity but it is still used in some other countries. The main tri-iodinated benzoic acid derivatives in current use are diatrizoate (Hypaque, Urografin, Renografin), iothalamate (Conray), metrizoate (Isopaque, Triosil) and iodamide (Uromiro).

In terms of toxicity on an equiosmolar basis, there is little difference between these anions. However, in certain specific circumstances, the methylglucamine cation decreases local endothelial toxicity. This is mainly of value in the cerebral and coronary circulations and for peripheral arteriography. For excretion urography, the evidence now tends to favour the use of pure sodium media since methylglucamine may indeed have certain disadvantages which are discussed in later sections of this chapter. Commercial preparations contain varying combinations of sodium and methylglucamine salts. This can result in some confusion when comparisons are made between individual media. In addition, these media usually contain very small amounts of sodium citrate as a buffer, and salts of the chelating agent ethylenediaminetetracetic acid (EDTA) to stabilize the solution.

If large doses of contrast media are administered to animals to determine the LD_{50}, a characteristic syndrome occurs [4]. As lethal dose levels are approached, the animals become apprehensive; vomiting (except in rodents), urination and defaecation occur, followed by muscle twitchings and convulsions. At a later stage, capillary breakdown develops in the lungs causing pulmonary haemorrhage and right heart failure. The LD_{50}, i.e. the dose causing the death of 50 per cent of the animals, is mainly of value in comparing the toxicity of different contrast media but it does not give a direct indication of the *safe* dose for clinical use in man. In any large population sample, a dose–response curve would indicate that a few individuals would be unduly susceptible to a dose which is safe for the majority. At the other extreme of the curve, a few individuals would tolerate an even larger dose than the majority.

In clinical use, the manifestations of contrast media reactions will depend partly on the dose involved and partly on the manner in which it is administered. They can be considered under three main categories:

1 Idiosyncrasy reactions in a susceptible patient from a dose of contrast medium which would be harmless to most patients. These reactions were originally considered to be allergic, but despite a voluminous literature on the subject, their aetiology is still poorly understood and there are probably several different factors involved in this group.

2 Reactions following the use of a large total dose of contrast medium in high-dose urography, angiography or, occasionally, as a result of accidental overdose.

3 Reactions occurring when a concentrated bolus of contrast medium has been delivered to a critical area such as the myocardium, brain, spinal cord or kidneys.

The reactions in the last two categories depend largely on the known chemotoxic effects of contrast media and on their hypertonicity. This chapter deals primarily with the general effects of contrast media. The specific effects, occurring as a result of their use in specialized procedures, are considered in the corresponding chapters.

IDIOSYNCRASY REACTIONS

Classification and incidence

Since the aetiology is unknown, any classification of idosyncrasy reactions must be arbitrary but a practical classification can be based on the severity of the reaction as it involves the general condition of the patient and the necessity for treatment. In the U.K. National Survey which was based on more than 300 000 urographic examinations [5], the following classification was therefore used (Table 1.1).

1 Minor reactions

Those which usually required no treatment.

2 Intermediate reactions

These usually required some form of treatment but there was no undue alarm for the patient's safety, and the response to treatment was usually rapid.

TABLE 1.1. Classification of reactions [5].

Minor	Intermediate	Severe
Nausea, retching	Faintness	Severe collapse
Slight vomiting	Severe vomiting	Loss of consciousness
Feeling of heat	Extensive urticaria	Pulmonary oedema
Limited urticaria	Oedema of face or glottis	'Cardiac arrest'
Mild pallor or sweating	Bronchospasm	Myocardial infarction syndrome
	Dyspnoea	Cardiac arrhythmias
Itchy skin rashes	Rigors	
Arm pain	Chest pain	
Sneezing	Abdominal pain	
	Headache	

3 Severe reactions

There was often fear for the patient's life and intensive treatment was required in most cases.

4 Death

In the survey, a considerable number of reactions were classified as 'minor' but, since the reporting of this type of reaction in a heterogeneous survey was likely to be variable, no attempt was made to compute their incidence. In a study at the Ochsner Clinic involving 40 000 urograms [6], the incidence of 'mild' reactions was 8 per cent and, in a Mayo Clinic series involving approximately 30 000 patients [7] the incidence of 'minor' reactions was 5.1 per cent. However, most patients in the Mayo Clinic series suffered from trivial symptoms such as mild hot flush, metallic taste in the mouth, mild nausea, cough, sneezing, tingling of the skin, etc. Similar 'trivial' symptoms were reported in 59 per cent of urograms, in a smaller recent investigation [8], and they appeared to be related to the dose and to speed of injection. Among the trivial symptoms reported in this study were unpleasant perineal sensations such as burning, a feeling of wetness, or a desire to empty the rectum or bladder, sometimes accompanied by a spurious sensation of having done so. These latter sensations apparently occurred in nearly 40 per cent of cases [8].

The incidence of intermediate reactions, severe reactions and death in the U.K. survey are shown in Table 1.2. It is difficult to compare these with other series in the literature due to differences in classification. In the Mayo Clinic series [7], the incidence of 'moderate' reactions was 1 in 112 and of

TABLE 1.2. Incidence of reactions in 318 500 excretion urograms. U.K. National Survey 1966–69 [5].

	Intermediate reactions	Severe reactions	Death
Reports received	142	24	8
Incidence	1 in 2000 (0.05%)	1 in 14 000 (0.007%)	1 in 40 000 (0.0025%)

'severe' reactions 1 in 1100, but only about one-third of this severe group appeared to be 'life-threatening' giving an incidence of 1 in 3000 to compare with the incidence of 1 in 14 000 found in the U.K. survey. In a survey of 12 419 paediatric patients, the overall incidence of reactions noted was 3.4 per cent and the incidence of severe reactions was 1 in 2500 [9].

In the classical North American survey during the years 1942 to 1958, Pendergrass et al. [10] found a mortality rate of 8.6 per million urograms (1 in 117 000). In France between the years 1955 and 1965, Wolfromm [11] found an incidence of approximately 1 in 61 000, and from Italy, Toniolo [12] reported a mortality rate of 1 in 85 000. It was widely believed that with the reduced toxicity of currently used contrast media, the mortality rate would improve. However, this has not in fact occurred. The mortality in the U.K. survey was 1 in 40 000, and this has since been confirmed by a recent survey in the U.S. where Fischer [13] found a mortality rate of 1 in 50 000.

TABLE 1.3. Urographic reactions related to age [5].

Age (years)	Unkn.	0-9	10-19	20-29	30-39	40-49	50-59	60-69	70+	Under 50	Over 50
Intermediate reactions (Total = 164)	17	4	10	20	26	32	21	27	7	92	55
Severe reactions (Total = 43)	8		1	3	5	8	5	5	8	17	18
Deaths U.K. survey (Total = 13)		2*					6	2	3	2	11
Deaths Pendergrass [10] (Total = 61)	3	3		4	5	7	10	12	17	19	39

* Overdosage.

This table also includes a number of cases reported to the U.K. survey prior to 1966 which were not used in computing the incidence rates shown in Table 1.2.

Indeed, these rates might have been higher, had it not been for the use of modern methods of resuscitation which would not have been available at the time of the earlier surveys. It seems possible, therefore, that the relative increase in mortality, as compared with the earlier surveys, may be due to the less restrictive selection of patients for urography and possibly also to the higher dosage schedules currently in use. Shehadi's most recent figures suggest a mortality rate as high as 1 in 13 000 urograms [14].

There does not appear to be any sex difference in the incidence of reactions but data derived from the U.K. survey suggests that age has a significant effect on the severity of the reactions sustained (Table 1.3) [5]. If it is assumed that the more numerous intermediate reactions are broadly representative of the age distribution of urographic examinations, then there appears to be a progressive spectrum of increasing severity of reactions with advancing age. This is more easily seen if an arbitrary dividing line is taken at the age of 50. Intermediate reactions are more frequent below this age; severe reactions show a higher age-peak with equal numbers occurring below and above the age of 50; whilst the majority of deaths occur in the older age groups. Data from Pendergrass' paper [10] have also been analysed in a similar manner and are given in Table 1.3. Although these show a wider scatter, deaths again predominate in older patients. These findings suggest that when a reaction occurs in an older patient, there may be an impaired ability of the cardiovascular system to respond to the insult. These findings in the U.K. survey are in contradistinction to those of an earlier much quoted study [15] which showed a lower incidence of side effects in the elderly as compared with younger adults, but this related mainly to minor reactions.

CLINICAL CHARACTERISTICS IN NONFATAL REACTIONS

Reactions are unpredictable and usually occur either during the injection or within the following five to ten minutes. However, occasionally, they may be delayed in onset. Symptoms of the minor reactions have already been discussed briefly. Table 1.4 shows the relative frequency of the various clinical features of intermediate and severe reactions in the U.K. survey [5]. In the individual cases, there was often more than one basic feature and these are each counted separately so that the numbers of 'symptoms' exceeds the numbers of patients with reactions.

Mucocutaneous lesions

Under this heading are included reactions involving the skin or mucous membranes, such as erythema, rashes, urticaria or angioneurotic oedema. This group most resembles the symptoms which might be expected if an allergic reaction or histamine release were a significant factor. These lesions occurred in more than one-third of the intermediate reactions but in only one-sixth of the severe reactions. Sneezing is another occasional symptom

TABLE 1.4. Clinical analysis of intermediate and severe reactions following excretion urography.

	Intermediate reactions (164 patients)	Severe reactions (43 patients)
Mucocutaneous	66 (40.2%)	7 (16.3%)
Hypotension	46 (28.0%)	35 (81.4%)
Bronchospasm	23 (14.0%)	5 (11.6%)
Myocardial	0	12 (27.9%)
Vomiting	20 (12.2%)	5 (11.6%)
Rigors	29 (17.7%)	2 (4.7%)
Headache	14 (8.5%)	0
Abdominal pain	8 (4.9%)	2 (4.7%)
Pain in chest	7 (4.3%)	1 (2.3%)
Convulsions	3 (1.8%)	4 (9.3%)
Sneezing	4 (2.4%)	4 (9.3%)
Paraesthesiae	7 (4.3%)	0

Many reactions had more than one clinical feature so that the totals for 'Symptoms' exceed the totals for the numbers of 'Reactions'.

Percentage figures indicate *comparative* frequency of symptoms in intermediate and severe reactions only. Incidence rates for reactions are shown in Table 1.2.

suggesting an allergic or histamine-like reaction. It is usually of only minor inconvenience but it may rarely presage a severe or even fatal reaction. In one patient, marked oedema of the face persisted for three days. In another patient, oedema of the face was associated with blurring of vision. Rarely, there may be delayed skin rashes and severe toxic eruptions have occurred up to two weeks after urography.

Bronchospasm

Asthmatic patients appear to form an unduly susceptible group. Severe bronchospasm sometimes occurred with cyanosis, and in one patient the asthmatic attack progressed to cardiac arrest which was fortunately reversible. As little as 0.5–1 ml of contrast medium could provoke a serious attack and one-quarter of the patients received less than 10 ml. Methylglucamine contrast media appeared to be more likely to cause bronchospasm and this is discussed in a later section. Sixteen of the twenty-eight patients had a previous history of asthma and two had an alternative allergic history. However, skin rashes appeared to be uncommon in association with bronchospasm. Although there is an increased risk in asthmatic patients, only a small proportion actually develop an attack following urography. In the Mayo Clinic series [7], this occurred in 6 per cent of patients with a history of asthma.

Vomiting

This appeared to be a relatively nonspecific symptom but it could be distressing and occasionally it presaged a severe reaction. In one patient, inhalation of vomit lead to death. There appeared to be some evidence of a dose relationship. In the intermediate and severe reactions in the U.K. survey, the relative incidence of vomiting in infusion urograms was some seven times that noted with conventional doses. In one patient, the worst vomiting was delayed until 1½ hours after the infusion. Metoclopramide may be useful in the treatment of persistent vomiting [16].

Hypotension

Hypotensive collapse is one of the most important features of contrast media reactions. Hypotension occurred in about one-quarter of the intermediate reactions and in the majority of the severe reactions. In the latter group, it was usually profound, often associated with transitory loss of consciousness, and occasionally accompanied by incontinence. In the unconscious patient there is a major risk of airways obstruction by the tongue falling back, with resultant cyanosis. This requires immediate attention. The pulse in hypotensive collapse was sometimes rapid and thready whereas in other cases there was bradycardia. The skin was often cold and clammy and occasionally there was profuse perspiration. In the Mayo Clinic series [7] apparently all the cases of severe hypotension were associated with a diffuse erythematous rash and preceded by nausea and vomiting. In the U.K. series [5] an erythematous rash was only reported infrequently in the severe cases of hypotensive collapse. The reason for this discrepancy is uncertain. The contrast medium used in the Mayo Clinic was 69 per cent sodium methylglucamine diatrizoate (Renovist). It may be a coincidence that in the few cases in the U.K. series where an erythematous rash was reported in hypotensive collapse, a number had received a similar medium (Urovison).

Syncope may also occur due to fear or nausea or it may occasionally result from decreased venous return to the heart as a result of inferior vena caval obstruction by abdominal compression.

Minor degrees of hypotension usually require no treatment but in severe cases, vigorous treatment is often required with vasopressors and steroids. Prolonged shock may also be associated with hypovolaemia and it may fail to respond to vasopressors until this has been corrected [17]. A syndrome resembling shock-lung with diffuse pulmonary infiltrates has also been described following a severe contrast medium reaction [18]. Hypotensive shock usually occurs within the first few minutes of the injection but may rarely be delayed. In two patients, hypotension developed approximately one hour after the examination [5]. In one patient, circulatory collapse with prolonged unconsciousness was followed, on recovery, by the passage of offensive bloodstained stools [19].

Cardiac disorders

Sudden cardiac arrest following the administration of contrast medium is the most dramatic of the severe reactions but, with external cardiac massage and resuscitation, there is a reasonable prognosis for full recovery. Arrhythmias may occur and there may be other transitory electrocardiographic changes. These changes are more likely to occur in older patients or in patients with a previous history of heart disease. In several cases they were initially attributed to myocardial infarction but review of the follow-up data showed that the ECG changes were nonspecific. Occasionally however, hypotensive collapse may be associated with unequivocal ECG evidence of coronary infarction and it may then sometimes be problematical whether the coronary infarction occurred incidentally, or whether it resulted from a hypotensive reaction to the contrast medium. It seems likely that in the majority of cases, ECG changes could be due to the direct toxic affect of the contrast medium on the myocardium [5]. Supporting evidence for this view has been provided by Berg *et al.* [20] who monitored electrocardiograms in ten patients undergoing excretion urography with 50 ml of 60 per cent methylglucamine diatrizoate and in twenty patients receiving infusions of 300 ml 30 per cent methylglucamine diatrizoate. The incidence of underlying heart disease was similar in both groups. Venepuncture alone produced no abnormalities but ECG changes occurred more frequently in patients undergoing infusion urography, where there were six major and three minor abnormalities. In the group receiving the standard dose, there was only one minor abnormality. The major ECG abnormalities included ischaemic changes, bigeminy or trigem-

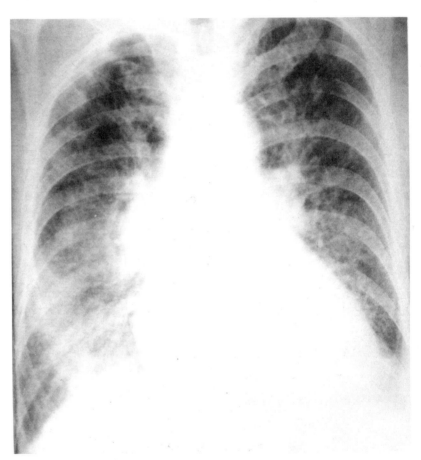

FIGURE 1.1. Male age 69 referred, with incomplete clinical history, for infusion urography to investigate renal failure (blood urea = 74 mg %). Immediately following infusion acute pulmonary oedema developed requiring emergency treatment with diuretics and oxygen. It then emerged that the patient had only recently recovered from an attack of left ventricular failure due to hypertension and that there had been ECG evidence of myocardial damage. Film taken on ward, approximately two hours after infusion, shows resolving pulmonary oedema.

iny, and accentuation of bundle-branch block with ischaemic change and chest pain. In three of the six patients showing major abnormalities, there had been recent myocardial ischaemic events. There was also a significant correlation with the older age groups. Two cases of ventricular tachycardia have also been detected by routine electrocardiographic monitoring during high-dose urography in elderly patients [21]. In the U.K. series, chest pain or retrosternal discomfort surprisingly did not usually appear to be of great significance but it is interesting to speculate whether simultaneous ECG tracings might have shown minor arrhythmias. In patients with incipient cardiac failure, infusion urography may precipitate acute pulmonary oedema [19, 8]. This may be partly due to the volume of fluid administered but the main factor is the high osmotic load of the contrast medium which causes withdrawal of fluid from the interstitial tissues into the intrasvascular compartments with resulting hypervolaemia. The cardio-toxic action of contrast media may also be a factor in the causation of pulmonary oedema. Although the risk of pulmonary oedema is now well established, unsuitable patients may be referred for examination with an inadequate clinical history (Fig. 1.1), and the radiologist would be well advised to check the patient's records when high-dose urography is requested. With sodium-containing contrast media, the high sodium load may be an aggravating factor in congestive cardiac failure and methylglucamine media may be safer but this is not certain. In animal experiments on the rabbit heart, both sodium chloride and methylglucamine chloride individually have a depressor action on myocardial contractility [22].

Rigors

In large-dose urography, using 60 per cent methylglucamine diatrizoate (Urografin), Doyle et al. [23] found that all five patients receiving a dose of 4 ml/kg developed malaise and rigors. More recently, Smith et al. [24] who were using large doses of contrast medium for parathyroid venous sampling, found that ten out of nineteen patients receiving methylglucamine salts developed a characteristic syndrome with rigors, pyrexia, and hypotension, 90 minutes to 6 hours after the examination. These reactions were confined to patients over the age of 50. Similar doses of sodium-containing contrast media did not cause this syndrome even in patients of the older age group. In the U.K. survey, the incidence of rigors appeared to be appreciably higher in association with methylglucamine media but rigors have also occurred with sodium media. Although rigors are probably commonest after large doses of contrast media, they may also occur after as little as 2 to 3 ml. Faulty batches of methylglucamine iothalamate, which contained pyrogens, caused rigors in four patients, and a faulty batch of sodium diatrizoate caused severe headache in three patients.

Abdominal pain

Severe abdominal pain may occur in anaphylactic reactions, presumably as a result of intestinal oedema or spasm and, in one of our cases, the pain was relieved by atropine. Most of the cases of abdominal pain reported in the U.K. survey were associated with hypotensive collapse and it is possible that this may sometimes result in transitory intestinal ischaemia. In some cases, abdominal pain was associated with paraesthesiae occurring elsewhere. Loin pain following excretion urography may conceivably be due to the diuretic effect of contrast medium in the presence of partial ureteric obstruction.

Arm pain and soft tissue changes

Arm pain was common with the early irritant contrast media such as iodomethamate. With most modern contrast media it is rarely more than an occasional minor symptom. However, with the highly concentrated sodium media such as sodium iothalamate (Conray 420), severe arm pain may occur. In addition, at least five cases of late thrombophlebitis have developed following use of this medium [25]. Davies *et al.* [8] made a detailed analysis of the incidence of arm pain in 755 patients and correlated this with routine straight radiographs of the injected arm. They found that in most cases, pain at the injection site was associated with a small perivenous injection which was otherwise not clinically apparent. Perivenous injections of Conray 420 were almost always associated with persistent pain at the injection site. With similar perivenous injections of Urovison, pain was less frequent, whilst perivenous injections of Hypaque rarely caused pain. Pain extending up the arm following injections of Conray 420 or Urovison was usually found to be associated with stasis of contrast medium in the arm vein. In experiments on rabbits, intravenous injection of contrast media caused minor endothelial changes in the vein which appeared to be related to the concentration of sodium in the medium [26].

In infants, contrast media are sometimes administered by subcutaneous or intramuscular injection. McAlister and Palmer [27] encountered two cases of severe sloughing of the soft tissues following this procedure and were aware of twelve similar cases. To investigate these effects, they administered Conray 60 (methylglucamine iothalamate), Conray 400 (sodium iothalamate),

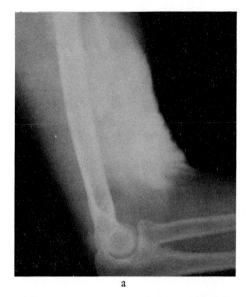

FIGURE 1.2. a 30 ml Hypaque 45 inadvertently injected outside vein into subcutaneous tissues and muscle in this fourteen-year-old girl, causing severe pain with swelling. b One hour later, contrast medium partially absorbed but oedema of the soft tissues present. At 48 hours the arm was still hot, swollen and painful. Prednisone was then administered orally and produced rapid improvement of symptoms. However, this was then followed by prolonged invalidism which appeared to be mainly psychogenic.

Hypaque 50 (sodium diatrizoate) and Renografin 60 (sodium and methylglucamine diatrizoate) by subcutaneous or intramuscular injection to rats, and they then examined the tissues histologically. All of these media produced an acute inflammatory response but the change following Hypaque 50 was slightly less than that caused by Renografin 60. The response to intramuscular injection was more localized and severe, but there was a greater risk of sloughing following subcutaneous injections. Addition of hyaluronidase, which has frequently been advocated to aid absorption, actually caused a considerable increase in the tissue damage. Dilution with water or saline, or the addition of procaine, did not diminish the inflammatory response. These authors therefore suggested that, when intravenous administration is not possible, contrast media should be injected in small quantities, and at multiple sites, to prevent tissue sloughing. In other animal experiments using a hamster-pouch preparation, no appreciable difference in tissue toxicity was found between sodium and methylglucamine media, but Biligrafin Forte (50 per cent methylglucamine iodipamide) caused more damage than Urografin 76 (methylglucamine sodium diatrizoate) [28]. In a recent case, where a large quantity of contrast medium was accidentally injected into the soft tissues of the arm, a severe inflammatory response persisted for several days and systemic steroid therapy appeared to be useful in diminishing this (Fig. 1.2).

Tetani and convulsions

Tetani may occur due to overbreathing in an anxious patient. However, in a few cases tetanic contractions or muscle spasm were reported in which overbreathing did not appear to be a factor. It has been shown [29] that contrast media such as Hypaque may cause a decrease in the blood calcium and magnesium levels. This might possibly be due to the small quantity of chelating agents (EDTA) present in these media. Although these changes in calcium and magnesium levels might perhaps be relevant to the occurrence of tetani, the rarity of the condition suggests that there may be other, as yet unknown, predisposing factors.

In patients suffering from intermediate reactions, convulsions were mainly due to exacerbation of a latent tendency to epilepsy. In severe reactions they were secondary to hypotensive collapse, cardiac arrest, or overdose.

Salivary gland enlargement and iodide effects

Parotid gland swelling has rarely been recorded following excretion urography with the former conventional doses of contrast medium [30]. However, with the introduction of high-dose urography, and particularly in those cases were there is prolonged retention of contrast medium due to delayed excretion from renal failure, salivary gland enlargement is becoming more frequent [31]. Nevertheless, this still appears to be an idiosyncrasy reaction, occurring in only a very small proportion of patients, and it may occur in the same patient with repeated examinations. The swelling usually occurs two to four days after the contrast medium has been administered and may last for several days. In these delayed cases there may be both organically bound iodine and inorganic iodide in the saliva [32]. Moreover, in uraemic patients, there is evidence that the retained contrast medium is subject to deiodination, *in vivo* [33]. It has recently been shown that, contrary to previous assumptions, small but significant quantities of free iodine are present in commercial preparations of contrast medium. The free iodine content of methylglucamine preparations (1.1–4.6 mg/100 ml) was higher than that in sodium-containing media (0.2–1.4 mg/100 ml) [34]. Evanescent salivary gland enlargement has also been reported in two cases within a few minutes of injection of methylglucamine diatrizoate and the swelling subsided over a period of a few hours [35]. In an isolated case, parotid gland enlargement was associated with paralysis of the facial nerve which developed over a period of twelve hours after the injection of 60 ml 50 per cent sodium diatrizoate. The parotid gland swelling responded to prednisone and diphenhydramine administered on the fourth day, but the facial nerve palsy persisted. Operative decompression of the facial nerve at five weeks revealed the presence of haemorrhage and oedema and was followed by partial recovery of function [36]. Nerve involvement of this nature appears to be exceptional but if a similar case occurs in the future, steroid and possibly antihistamine therapy should presumably be commenced as soon as possible.

The small amount of free iodine which has now been demonstrated in urographic contrast media is sufficient to account for the transitory depression of radioiodine uptake in the thyroid gland [34]. Although reference is usually made in contrast medium information literature to the risk of administration in patients with thyroid disease, there has hitherto been little objective evidence to support this. Recently however, several cases have been reported in which there has been clinical and laboratory evidence of hyperthyroidism developing in patients with previously nontoxic goitres, following the administration of large doses of contrast medium for urography or cardio-angiography [37]. In another case, transitory hyperthyroidism was induced; the clinical signs and laboratory values returned to normal without specific treatment after a period of approximately two months [38]. It is possible that some similar cases may have been overlooked since there may be an interval of one week or longer, after the contrast medium has been administered, before signs of hyperthyroidism develop. An unusual case of transitory swelling of the thyroid resembling an acute thyroiditis was associated with conjunctival injection and rhinorrhoea [19]. The effects of other contrast media on thyroid function are discussed on pages 267 and 327.

Haematological changes

Stemerman *et al.* [39] describe what they believe to be the first reported case of *pancytopenia* due to diatrizoate. The patient had three urographic examinations and a renal angiogram, each with diatrizoate, over a period of fifteen days, for the investigation of hypertension. One hour after angiography there was a severe anaphylactic reaction with dizziness, pruritus, dyspnoea, pyrexia and hypertension. This was followed by a decrease in haematocrit platelets and leucocytes with a hypoplastic bone marrow. Four months later the blood picture had returned to normal. This patient had also received penicillin and cephalothin but the timing of the reaction provided strong presumptive evidence that the diatrizoate was responsible.

When red blood cells are mixed with contrast media *in vitro*, there is initially a decrease in red cell diameter due to the hypertonic environment. However, as the concentration of contrast medium increases, the red cells begin to show an increase in diameter, due to damage to the cell wall, and haemolysis may then occur. Contrast media may also cause clumping of the red cells. When this occurs *in vitro* it may be reversed by agitation, but clumping which occurs *in vivo* may cause increased viscosity of the blood [40, 41]. With the more toxic media such as acetrizoate or iodipamide, scanning electron-microscopy showed severe morphological changes in the erythrocytes, with formation of cells resembling sea urchins, an appearance to which the name 'echinocytes' has been given. These changes may occur after brief exposure to the contrast medium. They are thought to be due to damage to the cell membranes by the protein-binding action of the medium. Diatrizoate and iothalamate, on the other hand, produced only a few echinocytes and this only at high concentrations of the contrast medium [42]. When blood is aspirated into a syringe during the administration of contrast medium, the red cells are subjected to a very high concentration of contrast medium. This will render them liable to subsequent haemolysis, with liberation of haemoglobin and possibly other toxic substances. It is therefore preferable to avoid re-injection of this blood with the contrast medium [43].

Three cases of *haemolysis* and *haemoglobinuria* have been reported following cardio-angiography with diatrizoate [44].

Contrast medium may also alter the haemoglobin–oxygen equilibrium. In 25 out of 29 children with congenital heart disease undergoing cardio-angiography with diatrizoate, there was a shift to the left in the oxyhaemoglobin dissociation curve. *In vitro* experiments showed that this effect was dependent on the presence of an intact red cell membrane and that the effects on free haemoglobin were more complex with methylglucamine in particular causing a shift to the right [45]. Iodipamide also causes a shift to the right [46].

There is a possibility that contrast media may occasionally cause *sickle cell crises* in patients homozygous for sickle cell haemoglobin (SSHb), and two such cases have been reported following cerebral angiography [47]. A fatality has also been reported following selective coronary arteriography in a patient with unsuspected sickle cell disease. Following the injection of 1 ml of Renografin 76 into the left coronary artery, there was immediate thrombus formation in the artery [48]. The addition of diatrizoate to samples of SSHb blood *in vitro* causes sickling and this is particularly marked when the concentration of diatrizoate exceeds 35 per cent. This appears to be partly due to the acidic nature of the contrast medium since buffering of the medium to pH 7.4 reduced the degree of sickling by approximately 50 per cent [47]. The effects of contrast media on the oxyhaemoglobin dissociation curve may also be a factor in causing sickling due to deoxygenation [49]. The risk of sickling is probably greater during angiography where relatively high concentrations of contrast medium occur in the blood. With slow intravenous injections, on the other hand, admixture with blood will tend to decrease the concentration of contrast medium in the blood. Considering the widespread prevalence of sickle cell disease, surprisingly few incidents have been reported, but it would appear sensible to use minimal concentrations and volumes of contrast medium in these patients. In high risk cases, buffering of the contrast medium to pH 7.4, and warming the medium to 37–38°C may be useful, as may oxygen therapy [47]. Blood transfusion with packed normal red cells may also be used to reduce the risk of sickling [50].

Administration of large doses of contrast medium during angiography may cause *hypocoagulability* of the blood with inhibition of clotting, and increased fibrinolytic activity. These coagulation defects, which may persist for up to 24 hours after angiography, are believed to be due to interference with the protein factors responsible for coagulation [51]. It has also been suggested that depression of the serum calcium may play a part in the hypocoagulable state [52].

Chapter 1

PREDISPOSING FACTORS IN REACTIONS

History of allergy and previous reactions

It is generally believed that reactions to contrast media are more common in the presence of an allergic history. In the U.K. survey [5] (Table 1.5), approximately one-third of the patients with intermediate reactions had a previous history of allergy or sensitivity of one type or another. The commonest specific drug sensitivity mentioned was that to penicillin. However, in patients with severe reactions and life-endangering cardiovascular collapse, a history of preceding allergy was generally lacking. In the Mayo Clinic

TABLE 1.5. Reactions following excretion urography. History of allergy and previous reactions.

Previous history (54 patients)	Intermediate reactions (164 patients)	Severe reactions (43 patients)
Penicillin sensitivity	10	0
Hay fever and other sensitivities	19	0
Asthma	15	3
Previous contrast medium reaction	8	1
Previous I.V.P. with *no* reaction (17 patients)	15	2

series [7], the average incidence of 'acute' reactions in patients with a preceding history of asthma, hay fever, hives or food allergy varied between 4 and 7 per cent, whilst a history of sensitivity to inorganic iodide was associated with a 13 per cent risk of a reaction to contrast medium. By comparison, the incidence of 'acute' contrast media reactions in nonallergic subjects was only 1.2 per cent. As in the U.K. survey, the incidence of *severe* reactions did not appear to be increased in patients with a history of allergy.

In those cases where there had been a previous reaction to contrast medium, the risk of a recurrence at a subsequent examination was in the region of 35 per cent, but even in these cases, reactions were inconstant and might be present on some occasions and not on others [7, 13]. It is also important to realize that reactions may occur *de novo* even though the patient has had previous uneventful examinations with contrast medium (Table 1.5).

Methylglucamine

Methylglucamine was originally introduced into the formulation of contrast media many years ago as a means of increasing the solubility of diatrizoate [53]. Subsequently, it was found to reduce the toxicity of contrast media in the peripheral, cerebral and coronary circulations [54, 55, 56]. The first clinical evidence to suggest that methylglucamine might have undesirable side effects emerged in the U.K. survey [5], where it was shown that the incidence of bronchospasm in patients receiving methylglucamine contrast

TABLE 1.6. Comparison of methylglucamine and sodium media (intermediate and severe reactions 1967–69).

	Methylglucamine media (44 900 urograms)	Sodium media (100 700 urograms)
Bronchospasm	11 (1 in 4000)	6 (1 in 16 800)
Mucocutaneous	9 (1 in 5000)	17 (1 in 6000)
Hypotension	12 (1 in 3700)	20 (1 in 5000)

Methylglucamine media: Urografin 60, Urografin 76, Conray 280.
Sodium media: Hypaque 45, Conray 420, Conray 480.

media was four times that found with sodium media (Table 1.6). Other side effects such as mucocutaneous changes and hypotension showed only a marginal increase with methylglucamine media. More recently, delayed reactions with rigors, pyrexia and hypotension have also been reported following large doses of methylglucamine media [24].

Route of administration

Lang [57] investigated 22 patients who had previously suffered reactions during intravenous urography and who subsequently underwent arteriography. He concluded that the incidence of reactions was lower following arteriography. However, since contrast media reactions do not invariably recur, even at subsequent urographic examinations, the significance of Lang's findings is problematical. Undoubtedly, idiosyncrasy reactions may occur following arteriography [5]. Mild allergic reactions were noted in 6 out of 167 patients (3.6 per cent) who underwent arteriography with iodamide [58]. Severe circulatory collapse has also occurred following nephro-angiography in two patients who had previously had reactions following urography [39, 59]. In another patient, severe laryngeal oedema with a woody oedema of the neck occurred, following injection of contrast medium into the ascending aorta, necessitating emergency laryngostomy [60]. It is possible, from the description in the latter case, that there may have been some acute thyroid enlargement but this was not considered by the author.

Dose and rate of injection

Major reactions may rarely occur after minute doses of contrast medium and, because of this, it has sometimes been assumed that the size of the dose does not influence the risk of a reaction occurring. Thus, in a recent paper, it is stated that 'there is no evidence that increases in dose increase the incidence of major, minor or allergic reactions but they are associated with an increase in trivial reactions' [8]. Similarly, it has been claimed that the incidence of reactions with infusion urography is no greater than that from urography with conventional doses of contrast medium [61]. On pharmacological grounds this would be most improbable and both of these papers involved relatively small numbers of patients. In the Mayo Clinic series [7], there was suggestive evidence of a higher incidence of acute reactions in patients receiving more than 50 ml of contrast medium but the number of patients receiving this larger dose was insufficient for statistical evaluation. In the U.K. survey [5] the incidence of reactions following infusion urography was appreciably higher than that from all other types of excretion urograms (Table 1.7). This was probably partly related to the dose factor and partly to the general condition of the patients selected for infusion urography. There is also evidence that in older patients with heart disease, electrocardiographic abnormalities are more likely to occur following large-dose urography [20]. This could, therefore, be a significant factor in the causation of cardiovascular collapse in severe reactions.

In animal experiments, when large doses of contrast medium are injected very rapidly, as in angiocardiography, there is a considerable increase in toxicity. Whereas the LD_{50} of sodium diatrizoate by slow intravenous drip in the dog is 13 200 mg/kg, with rapid injection of 90 per cent diatrizoate, the LD_{50} drops to 2700 mg/kg (3 ml/kg). Death usually occurred rapidly and was associated with acute hypotension, marked electrocardiographic change and haemorrhagic pulmonary oedema [62]. While these factors are not directly applicable to excretion urography, it might be expected on theoretical grounds that slower injections of contrast medium would be safer, particularly in patients with cardiovascular disease. Davies *et al.* [8] noticed an increased incidence of warmth after rapid injections but otherwise the

TABLE 1.7. Incidence of reactions related to dose (1966–69).

Total infusion urograms = 13 500
Total all other excretion urograms = 305 000

	Intermediate reactions	Severe reactions	Death
Infusion urograms	15 (0.1%)	3 (0.02%)	3 (0.02%)
All other excretion urograms	127 (0.04%)	21 (0.007%)	5 (0.002%)

incidence of reactions did not appear to vary with injection speeds. Likewise, Toniolo did not consider that the rate of injection affected the incidence of mortality from reactions to contrast media [12]. There is, however, still a dearth of clinical information on this subject and no firm conclusion can be reached at present.

Accidental overdose

The dose of contrast medium used in children is often relatively larger than that used in adults. Moreover, the recent tendency towards the use of increasing doses of contrast medium may cause confusion. During a twelve-month period, three cases of accidental overdosage in infants occurred in the U.K., with fatal results in two cases [63]. A 2.95 kg infant was given an infusion of the adult dose of 250 ml 30 per cent methylglucamine diatrizoate in error. Cardiac arrest developed at 30 minutes. The infant was resuscitated but died 19 hours later with convulsions. A second infant weighing 2.28 kg with renal and cardiac lesions died from pulmonary oedema after receiving 25 ml Urovison (sodium methylglucamine diatrizoate). In the third infant (weight 3.65 kg) 40 ml of 60 per cent methylglucamine diatrizoate caused convulsions. A fatal case has also been reported from the U.S. where a 1.6 kg infant died from haemorrhagic pulmonary oedema following an injection of 14 ml 60 per cent methylglucamine iothalamate. In subsequent animal experiments, five out of six beagle dogs receiving 9 ml/kg of 60 per cent methylglucamine iothalamate died [64]. Recently, a case of reversible cardiac arrest and pulmonary oedema has occurred in a 3.7 kg infant receiving only 12 ml of 45 per cent sodium diatrizoate (3.3 ml/kg) [25]. It seems probable that the current doses of contrast media used for high-dose urography are only just below toxic levels. The toxic effects of high doses in infants appear to be partly due to the chemotoxic action of the contrast medium. The major factor, however, appears to be the induced hyperosmolar state which causes pulmonary oedema and may also cause cerebral symptoms due to hypertonic dehydration [65]. Large doses of contrast medium may also cause acidosis [66]. Moreover, the low glomerular filtration in neonates may result in slower excretion of contrast medium [67]. Accidental overdosage in infants should be a largely avoidable hazard. Mistakes may occur when calculating doses on the basis of the infant's weight. In particular, pounds may be confused with kilograms. Mistakes may also occur when it is the practice for contrast medium to be diluted before injection—undiluted medium may be injected in error. A clearly defined simple dosage schedule should therefore be posted in the X-ray room, using a single standard concentration of contrast medium.

Renal function

In recent years it has become accepted practice to use larger doses of contrast medium for excretion urography [68]. This is particularly the case in patients with renal failure where a dose equivalent to 2.2 ml/kg of 45 per cent diatrizoate (0.6 g I/kg) may be used [69]. With these doses, the question of renal toxicity has naturally to be considered and this subject has recently been reviewed [70, 71]. A particular problem in assessing this factor is the difficulty of differentiating between natural progress of disease and the effects of urography. Bergman *et al.* (72) showed that the use of dehydration before high dose urography could result in transitory renal failure, and a number of other cases of this nature have since been reported. Until recently, however, it was generally believed that high-dose urography did not affect renal function providing that the patient was adequately hydrated. Thus, in the U.S., Davidson *et al.* [73] analysed the effect of high-dose urography in 82 patients who had impaired renal or hepatic function and they were unable to demonstrate any significant deterioration of these functions as a result of the procedure. This assumption is unfortunately no longer completely tenable in so far as renal function is concerned. Milman & Stage [74], who used doses of up to 1.3 g I/kg, have made a careful and detailed analysis of the effects of high-dose urography in a series of patients with renal failure who had not been dehydrated. They showed that there was commonly a transient small, but statistically significant, increase of serum creatinine following the urography. In three patients there was clinical deterioration; one patient had a temporary delay in improvement; the second developed a permanent but moderate decrease in creatinine clearance; whilst the third patient developed anuria requiring dialysis and did not return to his former level of renal function until 45 days after urography. In this series there also appeared to be a positive correlation between the dose of contrast medium and the increase in serum creatinine.

Kleinknecht et al. [75, 76] encountered nine cases of transitory oliguria or anuria following high dose urography in 300 patients with renal insufficiency. In six cases, preceding dehydration of the patient was a possible factor but in the other three cases this was not so. Two patients were diabetic and three patients had hyperuricaemia due to gout. These authors also point out that sodium restriction or the use of diuretics in the days preceding urography can result in dehydration of the patient; vomiting or diarrhoea may, of course, also cause dehydration. In a further study from the same clinic [77], renal biopsies were performed in 211 patients within ten days of high-dose urography using up to 1 g I/kg. In 47 of these cases, histological changes of osmotic nephrosis were present in the proximal tubular cells. The severity of the osmotic nephrosis appeared to bear some relationship to the dose of contrast medium used, to repetition of the examination after a short interval, and possibly also to pre-existing renal disease. In some of these patients there was a transitory deterioration in renal function, but this was not invariable. The clinical significance of these histological appearances is not yet clear. There appears to be a particular risk of renal failure developing from high-dose urography in azotaemic and dehydrated diabetic patients [78].

A characteristic radiological picture has been described following excretion urography in some oliguric patients. There is a brief 'flash filling' of the pelvi-calyceal system by contrast medium, followed by a prolonged nephrogram lasting for many hours or even for days after the examination. It is believed that this appearance may be due to precipitation of Tamm-Horsfall protein by the contrast medium in the renal tubules [79]. Tamm-Horsfall protein is a mucoprotein which is normally present in the urine. It is excreted in increased amounts during exercise and in oliguric patients. The effect of dehydration or of hypertonic infusions is to concentrate any Tamm-

Horsfall protein already present in the collecting tubules. It may then form aggregates that coalesce into viscid casts. Experimentally, all the currently used contrast media may cause a viscid gel precipitation of Tamm-Horsfall protein and this effect is increased by acidification [80].

Barshay et al. [81], reported five cases of temporary acute renal failure which developed following infusion urography in azotaemic diabetic patients. In three patients there had been a history of antibiotic nephrotoxicity two to five weeks before urography. The patient with the most severe degree of renal failure had also received three infusion urograms within five days and the large total dose of contrast medium could well have been an aggravating factor. It is probably unwise to repeat high-dose urograms at intervals of less than one week in such patients. In another diabetic patient, two separate episodes of renal failure have been documented following use of contrast medium, even though there was an intervening period of five weeks [82].

For some years it had become accepted that in *myelomatosis* there was a small but definite risk of causing transient or fatal renal failure as a result of either excretion or retrograde urography [83]. Subsequently, *in vitro* tests [84, 85] showed that the earlier urographic media such as diodone and acetrizoate and the cholangiographic medium iodipamide, all caused precipitation of myeloma protein, but that diatrizoate and iothalamate did not cause these precipitates. On this basis, it was suggested that myelomatosis should no longer be regarded as a contraindication to excretion urography providing that dehydration was avoided. However, it appears possible that Tamm-Horsfall protein may be a factor in causing renal failure in myelomatosis and this is, of course, susceptible to precipitation by currently used contrast media [80]. Furthermore, cases of renal failure in myelomatosis have now been reported in which the urographic contrast medium was diatrizoate, although dehydration was a possible complicating factor [86, 87]. Renal failure has also occurred in myelomatosis following renal angiography [88] and cerebral angiography [25]. It would therefore seem that contrast media should be avoided in myelomatosis unless there are compelling reasons for their use.

When excretion urography is undertaken in the presence of renal failure, large quantities of contrast medium may remain in the circulation for several days and extrarenal excretion may occur through the liver, resulting in a cholecystogram, and also through the small intestine [89]. Contrast medium can, however, be removed rapidly by dialysis and this may be necessary if a sensitivity reaction occurs. The effects of contrast media on renal function receive further consideration in Chapter 4.

URINALYSIS

Examination of the urine after intravenous urography may give rise to misleading results. The contrast medium causes an increase in the specific gravity of the urine. There may be a false positive test for protein when the sulphosalicylic acid or nitric acid ring test is used, but the bromophenyl dye test (Albutest) is not affected. There may also be a false positive black-copper reduction reaction when a Clinitest tablet is added to the urine, thereby simulating the findings in alcaptonuria. This reaction may occur with either diatrizoate, iothalamate or iodipamide [90].

AETIOLOGY OF IDIOSYNCRASY REACTIONS

Although much effort and research has been devoted to the investigation of contrast media, it must be admitted that the precise aetiology of idiosyncrasy reactions is still largely unknown. Indeed, it appears likely that a number of different factors may be involved in their causation [91] and some of these concepts will therefore be briefly discussed.

The allergic hypothesis

For many years it was assumed that these reactions were allergic in nature, but as early as 1955 Sandström [92] questioned this concept by showing that reactions may occur in patients who had not previously been exposed to a sensitizing dose of contrast medium. However, this would not exclude sensitization to one of the other trace constituents of commercial contrast media. Halpern et al. [94, 95] claimed to have demonstrated an allergic mechanism in patients with contrast media reactions by using the *in vitro* 'lymphocyte transformation test', but other investigators have had little success with this technique in assessing sensitivity to contrast media [5, 96]. Similarly, other types of pretesting have been unreliable in predicting reactions [13]. Attempts to produce circulating antibodies to contrast media in rabbits were unsuccessful [91]. In a recent isolated case report, an IgM immunoglobulin antibody was found to react with iothalamate, diatrizoate, acetrizoate and iodipamide [97], but the general significance of this is not yet clear. Occasional cases have been noted in which the history suggested the development of hypersensitivity to contrast medium [5] and, more recently, Lasser appears to have been able to induce an idiosyncrasy response in a dog, by repeated injections of contrast medium [98].

On the evidence available at present, it seems unlikely that the majority of contrast media reactions are allergic in the sense that they are immunologically induced. Nevertheless, it is possible that, in a few cases, there may be a true allergic hypersensitivity to contrast medium. In addition, since it has now been demonstrated that small quantities of free iodine may be present in contrast medium [34], it is possible that certain phenomena such as conjunctival injection, rhinorrhoea, salivary gland enlargement and delayed skin rashes might, in some cases, be due to true iodism, particularly when these have followed large doses of contrast medium.

Histamine liberation

Certain drugs can cause direct liberation of histamine from the mast cells without the intervention of an immunological reaction. Mann [99] drew attention to a number of similarities between the effects of injecting such drugs, or histamine itself, and the features of contrast media reactions. On this basis, she postulated that contrast media reactions were due to the mechanism of histamine release. She also suggested that the susceptibility of allergic subjects to contrast media reactions could be accounted for by the known fact that, in such patients, the tissues contained a higher quantity of histamine than normal, and that these patients were also more sensitive to the effects of histamine. Subsequent *in vitro* experiments showed that contrast media could act as histamine-releasing agents in mast cells [100]. Initial *in vivo* attempts to demonstrate histamine release were negative [101]. However, following the recognition of a clinical association between methylglucamine contrast media and bronchospasm [5], Lasser *et al.* [102, 103] subsequently demonstrated that methylglucamine contrast media could act as histamine-releasing agents in the animal liver and lung, and that hypertonic methylglucamine chloride itself was also a potent histamine-releasing agent. By contrast, the sodium salts of the same media were much less active. Elevation of plasma histamine has also been demonstrated in some patients following the administration of methylglucamine containing salts of contrast media. However, there did not appear to be any consistent correlation between the elevated histamine levels and the presence of either contrast media reactions, or haemodynamic changes [104, 105]. In the U.K. analysis [5] hypotensive collapse was poorly correlated with the histamine-like mucocutaneous phenomena, whilst headache, which is a characteristic symptom following the injection of histamine, was even rarer. Moreover, in animal experiments Rosenberg [106] found that antihistamines had no protective effect against the induction of hypotension by contrast media. It seems possible that histamine release might be a factor in some of the minor reactions, but there must be considerable doubt about its importance in causing hypotensive collapse, which is the major problem in many severe reactions. Other vaso-active substances such as Slow-Reacting Substance (SRS-A) and bradykinin are known to play a part in anaphylaxis and might possibly be relevant in contrast media reactions [103] but there is, as yet, no evidence either for, or against, this possibility.

Protein-binding and enzyme inhibition

Lasser *et al.* [107] demonstrated protein-binding of contrast media to bovine albumin *in vitro*. Iopanoic acid had the strongest binding effect followed in decreasing order by iodipamide, acetrizoate and diodone. On the other hand, diatrizoate appeared to have little, if any, protein-binding action to bovine albumin. However, in the human, Kutt *et al.* [29] showed that diatrizoate produced changes in the electrophoretic mobility of proteins, affecting not only albumin but also alpha, beta and gamma globulins. There also appeared to be some protein-binding of the diatrizoate. In the case of the animal sera, there appeared to be a correlation between the degree of protein-binding and the relative toxicities of contrast media as measured by their $LD_{50}s$ [107]. Lasser suggested that this may be partially due to the resulting alteration of the proteins in cell membranes of blood vessel endothelium, or to alterations in the proteins of enzymes. At relatively high concentrations of contrast media there may be partial inhibition of various enzymes such as acetyl-cholinesterase, beta-glucuronidase, lysozyme, alcohol-dehydrogenase and glucose-6-phosphate dehydrogenase [91]. It has been suggested that inhibition of acetyl-cholinesterase might produce various cholinergic side effects such as diarrhoea or vasodilatation, etc., but it is not certain whether enzyme inhibition is relevant in the clinical context in most contrast media reactions. More recently, Lasser [108] has shown that contrast media may activate the serum complement system *in vitro* and *in vivo* and it is possible that this might be a factor in some adverse reactions.

Drug interactions

Many therapeutic drugs are partially bound to serum protein and, if they were to be displaced by protein-binding of contrast media, this might potentiate their therapeutic effects by increasing the concentration of the free drug in the plasma. Thus, pentothal anaesthesia may be potentiated by the administration of acetrizoate or iodipamide [109]. In the U.K. survey [5] a possible association was noted between the induction of cardiac arrhythmias by contrast media and digitalis medication. Fisher *et al.* [110] have now shown, by experimental work in animals, that dogs premedicated with near toxic levels of ouabain developed arrhythmias following an intravenous bolus of contrast medium. However, an injection of saline of equivalent tonicity failed to induce arrhythmias in these dogs, suggesting that the contrast medium molecule itself was the toxic factor. This group also noticed a possible association between the use of tranquillizers and the induction of arrhythmias by contrast media in patients. As yet, little is known about the possible interactions of drugs and contrast media but this could be a subject of considerable importance.

PREVENTION AND TREATMENT OF REACTIONS

Since the mechanisms of contrast media reactions are still in doubt, treatment schemes are basically symptomatic and empirical. Prior to the administration of contrast medium, it is usual to enquire whether there is any history of allergy. Patients with an allergic background and, more particularly, those sensitive to drugs such as penicillin, are more likely to react to other drugs in a similar manner. This appears to have some relevance in relation to minor contrast media reactions, but in the great majority of the severe and fatal reactions, there does not appear to be any significant preceding history of allergy. However, patients with a history of asthma or angioneurotic oedema are somewhat exceptional in that they may be prone to develop quite severe reactions.

In those cases where a patient believes that he or she has previously had a reaction to contrast medium, careful personal enquiry by the radiologist will help to determine whether this was only a minor reaction such as nausea or flushing, in which case the examination can usually proceed normally. If, however, there was a significant and typical reaction to contrast medium, a decision has to be made whether the examination should be performed. When a patient has had a severe life-endangering reaction to contrast medium in the past, it would obviously be unwise to repeat the exposure unless there are imperative clinical reasons. With less severe reactions the risk of a recurrence appears to be in the region of 30 to 35 per cent, and usually the reactions which occur are no more severe than on the previous occasion [13, 7]. However, if there is a history of reactions on several occasions, and if these were of increasing severity, this may indicate that the patient is developing a true hypersensitivity to the contrast medium [5].

Where there has been a previous history of a significant reaction it will usually be appropriate to consult the clinician to determine the importance of the proposed examination for the management of the patient's condition. If, following consultation, the examination is still considered to be necessary, the patient should not be denied its benefit, but full precautions should be taken to treat any emergency which might arise. Lalli [93] has, in fact, recently described nine patients who had previously suffered severe hypotensive reactions with unconsciousness following urography. In all of these patients repeat urography, performed after adequate reassurance, caused no reaction.

A vague history of 'iodism' is another problem which occasionally arises. Here again, questioning may often reveal that there was nothing more than a skin burn resulting from tincture of iodine. True iodism is relatively rare but, in the Mayo Clinic series [7] such patients appeared to have an increased liability to contrast media reactions. Although the typical contrast medium reaction is not considered to be a manifestation of iodism, small amounts of free iodine may be present in contrast medium and de-iodination can occur if there is delayed excretion of the contrast medium. There could, therefore, be a problem if large doses are used, particularly in renal failure.

Pretesting is unreliable in predicting the risk of a contrast medium reaction [13]. A patient may have no reaction to a small dose of contrast medium and yet develop a fatal reaction to a larger dose [19]. On the other hand, the test dose itself may even cause death [13]. Nevertheless, it is a sensible precaution to wait for a short period after injection of the first millilitre or so of the contrast medium before proceeding with the injection. Where there is a reason to fear a reaction, a slow rate of injection may allow the examination to be interrupted, or abandoned, at the earliest sign of trouble. This is, however, a personal opinion and many radiologists consider that a rapid injection is preferable [8, 103]. On the other hand, in patients with cardiovascular disease, and particularly in those with a previous history of coronary artery disease, a fast injection may possibly predispose to arrhythmias and a slow rate of injection is probably preferable [111].

There is controversy concerning the value of *routine* prophylaxis with intravenous antihistamines and recent analyses have not demonstrated any significant beneficial value from their use [6, 8]. Moreover, intravenous antihistamines themselves may produce undesirable and sometimes severe reactions. In such a case it may then be difficult to know whether a reaction has been due to contrast medium or to the antihistamine. It is probable, therefore, that intravenous antihistamines could cause more harm than benefit if given prophylactically in all patients. It is suggested that they should be reserved for the treatment of *established* reactions such as urticaria or angioneurotic oedema. Moreover, it is inadvisable to mix antihistamines and contrast media in the same syringe since precipitation may occur, particularly with iodipamide [112].

There may be a case for some form of prophylaxis in patients who have previously had reactions to contrast media. Schatz *et al.* (113) recommend the use of prophylactic antihistamine for these cases. However, steroid therapy would appear to be more rational. Zweiman & Hildreth [114] used large doses of steroids in eight patients who had previously had anaphylactoid phenomena such as wheezing, dyspnoea or shock, following a contrast medium injection. In these cases, they administered 150 mg of prednisone during the eighteen hours prior to the repeat contrast medium examination and a further 75 mg of prednisone in the twelve hours after the examination. A mild reaction occurred in one patient and none in the other seven. While this was an uncontrolled trial, it compares favourably with the normally expected figure of 30–35 per cent recurrence. There is, however, no completely reliable method of preventing contrast medium reactions. Wolfromm *et al.* [11] have reported severe reactions and even death despite the use of prophylactic steroids but it is probable that the steroid doses involved were much smaller than those advocated by Zweiman.

In an interesting paper, Lalli [115] found that prior hypnotic relaxation significantly decreased the incidence of contrast media reactions. He also drew attention to the role of fear in the causation of reactions and stressed the importance of a calm reassuring approach to the patient. Most experienced radiologists would fully endorse the latter recommendation.

It is neither desirable nor even necessary to institute active treatment for every contrast medium reaction, but these should be immediately assessed by the radiologist. An alarm system should be available to summon medical assistance without the radiographer having to leave the patient. Minor reactions usually subside without treatment and even syncopal attacks may respond to such simple measures as raising the patient's legs. Providing that the patient's general condition is satisfactory or improving, it is justifiable to spend a few minutes in deciding whether the reaction is subsiding spontaneously or whether treatment is required. In the case of major reactions, however, the most important factor in reducing mortality is the immediate availability of full facilities for resuscitation [116–118]. Immediate treatment will depend partly on the clinical presentation, for example, cardiac massage may be required in cardiac arrest (see Chapters 18 and 23). Depending upon the presenting symptoms, life-saving symptomatic treatment should be

Contrast Media in Urography

Reaction	Oxygen	Steroids (I.V.)	Adrenaline * (S.C. or I.M.)	Aminophylline (I.V.)	Antihistamine (I.V.)	Additional measures which may be required
Hypotension cardiac arrest	+	+	?			Posture, aramine I.M, cardiac massage, etc.
Respiratory failure	+	+	?			Maintain airway pulmonary ventilation (? Nikethamide I.V.)
Bronchospasm	+	+	+	+	?	
Angioneurotic oedema	+	+	+		+	May require laryngotomy or tracheotomy
Pulmonary oedema	+	+	+	+		Lasix (I.V.), ? venesection sit up if possible ? morphine and atropine I.V., ? Digoxin, ? Positive pressure ventilation
Toxic convulsions	+	+				Valium I.V. (? artificial ventilation)
Cerebral oedema	+	+				20% Mannitol I.V.
Hypertensive crisis (phaeochromocytoma)	+	+				Rogitine I.V. ? Propranolol I.V.

Steroid (100–200 mg) Efcortisol 1–2 ml (I.V.).
* Adrenaline (Epinephrine) 0.5 ml (S.C. or I.M.).
Aminophylline 0.25 g in 10 ml (I.V. slowly).
* Antihistamine: Piriton (10 mg) 1 ml (I.V.).
* Aramine 0.2–1 ml (I.M.)
Nikethamide 1–4 ml (I.M. or I.V.).
Valium (10 mg) in 2 ml (slow I.V.) (Infants 0.25 mg/kg)
For other doses etc. see 'Drug Information', p. 482.

* See 'Drug interactions' p. 485.
S.C. = Subcutaneous (hypodermic).
I.M. = Intramuscular.
I.V. = Intravenous.

commenced (Table 1.8). Adrenaline (epinephrine) has a physiological antihistamine action and an additional stimulating action on the circulation. In many cases, and particularly in the treatment of asthma, it will be the most appropriate initial drug. It can be given in a dose of 0.5 ml either subcutaneously or, in shocked patients, intramuscularly. However, except when it is being used for the treatment of cardiac arrest, it is generally unwise to give adrenaline intravenously because of the risk of inducing ventricular fibrillation. In older patients with cardiovascular disease it is even possible that treatment with adrenaline may sometimes have been an aetiological factor in arrythmias which have occurred after contrast media reactions, particularly when the adrenaline has been given intravenously. Patients receiving tricyclic antidepressant drugs may also have an increased sensitivity to adrenaline (see page 485).

As soon as possible, an intravenous route should be set up either by the use of a butterfly needle or intravenous infusion. In severe bronchospasm or pulmonary oedema, intravenous aminophylline is often valuable but it should be injected very slowly since it may cause hypotension. The main value of antihistamines is in the treatment of urticarial type reactions or angioneurotic oedema. They are unlikely to be of value in the treatment of hypotensive collapse. It is therefore illogical to use antihistamines as a routine form of treatment for all types of contrast medium reactions.

On the other hand, there does appear to be an undoubted major beneficial value from the use of intravenous steroids, although the exact manner in which they exert their effect in these cases is still not certain. There is usually a delay before steroids become effective and this emphasizes the importance of the other immediate symptomatic treatment. However, intravenous steroids should be given as soon as practicable in *all severe* reactions, whatever the symptomatology. If the patient fails to respond to treatment, large doses of steroids may sometimes be required and no patient should be allowed to die from a contrast medium reaction without having had the benefit of a generous dosage of steroids. The blood levels of steroids, following an intravenous injection, decline after approximately four hours so that booster doses or a steroid infusion may be required to prevent a late relapse. In prolonged shock, failure to respond to treatment may also be due to hypovolaemia which requires correction by plasma expanders. Measurement of the haematocrit and central venous pressure are important in the diagnosis and management of these cases [119]. If a contrast medium reaction occurs in a patient with renal failure, dialysis may be required to remove the medium [68].

RETROGRADE UROGRAPHY

Retrograde pyelography

In patients with presumed sensitivity to contrast medium, retrograde pyelography may sometimes be considered as an alternative procedure to excretion urography when it is essential to delineate the pelvicalyceal system or ureter. The question therefore arises as to what extent contrast medium may be absorbed following this procedure. From animal experiments [120] it appears that contrast media may be absorbed from the intact pelvicalyceal system but that there is very little absorption when the isolated ureter is perfused. These experiments suggested that up to 17.8 per cent of I^{131} labelled diatrizoate, instilled into the renal pelvis, could be absorbed at hydrostatic pressures which may occur in hydronephrosis. In clinical measurements when I^{131} labelled diatrizoate was used for retrograde pyelography, 12.5 per cent of the medium was absorbed in a patient who had pyelo-renal intravasation but in eight out of the ten cases investigated, the absorption was less than 1 per cent [121]. In another clinical evaluation, blood levels of I^{131} labelled diatrizoate absorbed following retrograde pyelography did not exceed 6.6 per cent of the blood levels achieved by an intravenous dose of only 1 ml of contrast medium [122]. It would therefore seem that, provided pyelo-renal intravasation or prolonged stasis of contrast medium in the renal pelvis can be avoided, the risks of a reaction from systemic absorption of contrast medium will be minimal. Nevertheless, in cases of extreme hypersensitivity, there might be a theoretical possibility of a reaction developing as a result of mucosal contact with an offending allergen.

It is generally believed that retrograde pyelography may occasionally cause deterioration of renal function but there are relatively few recorded cases. In one patient, severe bilateral flank pain and anuria followed bilateral retrograde pyelography. On cystoscopy, there was inflammatory oedema of the ureteric orifices and the anuria was relieved by ureteric catheterization. In this patient there was a positive skin test to the contrast medium (Skiodan, sodium methiodal) and a sensitivity reaction was postulated [123]. However, mucosal changes can also result from a local irritative action of contrast medium [124]. Anuria, following retrograde pyelography, may also occur in the absence of ureteric obstruction, when it is presumably due to a direct nephrotoxic effect on the renal parenchyma. In one such case [125], bilateral retrograde pyelography was performed with Skiodan and an increased sense of resistance was noted during the injection. Pyelo-renal intravasation occurred on the left side only, but there was immediate severe bilateral flank pain with anuria which lasted for five days, and this was followed by a diuresis. A renal biopsy subsequently showed evidence of chronic pyelo-nephritis which may have been a predisposing factor. Fatal acute renal failure has also occurred in a patient with two functioning kidneys, after unilateral retrograde pyelography with diatrizoate [126].

Perforation of the kidney by a retrograde catheter may rarely be followed by retroperitoneal phlegmon [127] and an isolated case of miliary tuberculosis has been reported following pyelo-renal intravasation at retrograde pyelography in a tuberculous kidney [128].

An unusual complication occurred in a four-day-old infant, with bilateral hydronephrosis, following retrograde pyelography with 50 ml Retrografin. This commercial preparation contains 2.5 per cent neomycin. There was extensive intravasation of contrast medium in the kidneys and retroperitoneal space. Following this, the neomycin was absorbed, causing neuromuscular blockade with flaccid paralysis and apnoea. Positive pressure ventilation was initially required but the neuromuscular disorder subsequently responded to intravenous neostigmine [129].

Thortrast has long been discontinued as a contrast medium due to the risk of radiation damage, but delayed effects may occur many years after retrograde pyelography with this substance resulting in perinephric or periureteric fibrosis. Urinary tract tumours may also occur [130]. In these cases, opaque Thorotrast residues may frequently be visible in the radiograph.

Retrograde urethrocystography

Retrograde urethrocystography is often followed by oedema of the urethral and bladder mucosa, and intravasation of contrast medium into the corpora cavernosa may occur in 5 per cent of cases. Sorensen [131] has recently investigated the tissue toxicity of local anaesthetics and contrast media used for urethrography with his hamster pouch preparation. Diodone-diethanolamine, the contrast medium in Perjodal, has a marked local toxic effect and, amongst thickening agents, dextran has a greater local toxicity than methyl cellulose gel. 'Thesat', which was used in one of the local anaesthetic gels to lower surface tension, was extremely toxic to tissues. Xylocaine (2 per cent) (lignocaine, lidocaine) showed moderate local toxicity, whereas Carbocaine (1.5 per cent) showed little local toxicity by itself although a Carbocaine–Thesat preparation was toxic. It should also be remembered that absorption of local anaesthetics may cause systemic effects (see Chapter 12). Two cases of hypotensive collapse have been reported following urethral anaesthesia with 2 per cent lignocaine jelly (a 1 per cent preparation of lignocaine should be adequate for urethral anaesthesia) [132]. McAlister *et al.* [124] noted oedema and haemorrhage in the bladder mucosa on the day following cystography. They therefore evaluated the histological effect of 30 per cent Cystokon (sodium acetrizoate), 25 per cent Hypaque (sodium diatrizoate) or Renografin 30 (methylglucamine and sodium diatrizoate), in the rat bladder. With all these media, there was an appreciable inflammatory response which was maximal at 48 hours and still present at 1 week. Factors which lessened the inflammatory response were dilution of the medium and use of smaller volumes, instilled at low pressures. In this investigation there was little difference between the histological effects produced by the different contrast media. In clinical use, however, acetrizoate appears to have a more immediate irritant effect on the bladder than diatrizoate [133]. On the other hand, acetrizoate has bactericidal properties which may be useful in certain cases [134].

The use of barium sulphate for cystography has, in general, been discontinued. If ureteric reflux occurs, the barium may become inspissated in the calyceal system of the kidney and experimental work in animals suggests that this could cause a granulomatous reaction with fibrosis [135].

An unusual intravesical foreign body was found to result from the precipitation of sodium alginate by neomycin. The sodium alginate was a constituent of catheter lubricant, whilst the neomycin was present in Retrografin used for cystography [136].

REFERENCES

1 KNOEFEL P.K. (Section Editor) (1971) Radio contrast agents. *International Encyclopaedia of Pharmacology & Therapeutics*, Section 76, Vols 1 & 2. Pergamon Press, Oxford.
2 Symposium on contrast media toxicity of the American Association of University Radiologists. (1971) *Invest. Radiol.* 5, 373–502.
3 SHERWOOD T. (Editor) (1972) Radiology of renal disease. *Br. med. Bull.* 28, 189–262.
4 HOPPE J.O. (1959) Some pharmacological aspects of radiological compounds. *Ann. N.Y. Acad. Sci.* 78, 727–739.
5 ANSELL G. (1970) Adverse reactions to contrast agents. Scope of problem. *Invest. Radiol.* 5, 374–384.
6 OCHSNER S.F. & CALONJE A. (1971) Reactions to intravenous iodides in urography. *Sth. Med. J. (Bgham, Ala.)* 64, 907–911.
7 WITTEN D.M., HIRSCH F.D. & HARTMAN G.W. (1973) Acute reactions to urographic contrast medium. Incidence, clinical characteristics and relationship to history of hypersensitivity states. *Am. J. Roentg.* 119, 832–840.
8 DAVIES P., ROBERTS M.B. & ROYLANCE J. (1975) Acute reactions to urographic contrast media. *Br. med. J.* 2, 434–437.

9 GOODING C.A., BERDON W.E., BRODEUR A.E. & ROWEN M. (1975) Acute reactions to intravenous pyelography in children. *Am. J. Roentg.* **123**, 802–804.
10 PENDERGRASS H.P., TONDREAU L., PENDERGRASS E.P. *et al.* (1958) Reactions associated with intravenous urography: historical and statistical review. *Radiology* **71**, 1–12.
11 WOLFROMM R., DEHOUVE A., DEGAND F. *et al.* (1966) Les accidents graves par injection intraveineuse de substances iodées pour urographie. *J. Radiol. Electrol.* **47**, 346–357.
12 TONIOLO G. & BUIA L. (1966) Risultati di uno ischiesta nazionale sugli incidente mortali da iniezioni di mezzi di contrasto organo-iodata. *Radiol. Med.* (*Torino*) **7**, 625–657.
13 FISCHER H.W. & DOUST V.L. (1972) An evaluation of pretesting in the problem of serious and fatal reactions to excretory urography. *Radiology* **103**, 497–501.
14 SHEHADI W.H. (1975) Adverse reactions to intravascularly administered contrast media: a comprehensive study based on a prospective survey. *Am. J. Roentg.* **124**, 145–152.
15 MACHT S.C., WILLIAMS R.M. & LAWRENCE P.S. (1966) Study of 3 contrast agents in 2,234 intravenous pyelographies. *Am. J. Roentg.* **98**, 79–87.
16 WEBER G.N. (1972) Personal communication.
17 DELORME P. & LETENDRE J. (1972) Prolonged shock after intravenous pyelography. *Canad. med. Ass. J.* **106**, 343–344.
18 EDDE R.R. & BURTIS B.B. (1973) Lung injury in anaphylactic shock. *Chest* **63**, 636–638.
19 ANSELL G. (1968) A national survey of radiological complications: interim report. *Clin. Radiol.* **19**, 175–191.
20 BERG G.R., HUTTER A.M. & PFISTER R.C. (1973) Electrocardiographic abnormalities associated with intravenous urography. *New Engl. J. Med.* **289**, 87–88.
21 STADALNIK R., DAVIES R., ZAKAUDDIN V., HILLIARD G. & DA SILVA O. (1974) Ventricular tachycardia during intravenous urography. *J. Am. med. Ass.* **229**, 686–687.
22 SALVESEN S., NILSEN P.L. & HOLTERMAN H. (1967) Ameliorating effect of calcium and magnesium ions on the toxicity of isopaque sodium. II. Studies on the isolated heart and auricles of the rabbit. *Acta. Radiol.* Suppl. **270**, 30–43.
23 DOYLE F.H., SHERWOOD T., STEINER R.E., BRECKENRIDGE A. & DOLLERY C.T. (1967) Large dose urography. Is there an optimum dose? *Lancet* **2**, 964–966.
24 SMITH M.J.G., KENDALL B.E. & TOMLINSON S. (1974) Adverse reactions to high doses of methylglucamine-based contrast media. *Br. J. Radiol.* **47**, 566–569.
25 Reports to National Radiological Survey.
26 PENRY J.B. & LIVINGSTON A. (1972) A comparison of diagnostic effectiveness and vascular side-effects of various diatrizoate salts used for intravenous pyelography. *Clin. Radiol.* **23**, 362–369.
27 MCALISTER W.H. & PALMER K. (1971) The histologic effects of four commonly used media for excretory urography and an attempt to modify the response. *Radiology* **99**, 511–516.
28 SORENSEN S.E. (1971) Changes in vascular permeability after the application of roentgen contrast media in the hamster cheek pouch. *Acta Radiol Diagn.* **11**, 274–288.
29 KUTT H., MILHORAT T.H. & MCDOWELL F. (1963) The effect of iodinized contrast media upon blood proteins, electrolytes and red cells. *Neurology* **13**, 492–499.
30 SUSSMAN R.M. & MILLER J. (1956) Iodide mumps after intravenous urography. *New Engl. J. med.* **255**, 433–434.

31 NAKADAR A.S. & HARRIS-JONES J.N. (1971) Sialadenitis after intravenous pyelography. *Br. med. J.* 3, 351-352.
32 TALNER L.B., LANG J.H., BRASCH R.C. & LASSER E.C. (1971) Elevated salivary iodide and salivary gland enlargement due to iodinated contrast media. *Am. J. Roentg.* 112, 380-382.
33 TALNER L.B., COEL M.N. & LANG J.H. (1973) Salivary secretion of iodine after urography. *Radiology* 106, 263-268.
34 COEL M.N., TALNER L.B. & LANG J.H. (1975) Mechanism of radioactive iodine uptake depression following intravenous urography. *Br. J. Radiol.* 48, 146-147.
35 NAVANI S., TAYLOR C.E., KAUFMAN S.A. & PARLE R.H. (1972) Evanescent enlargement of salivary glands following tri-iodinated contrast media. *Br. J. Radiol.* 45, 19-20.
36 KOCH R.L., BYL F.M. & FIRPO J.J. (1969) Parotid swelling with facial paralysis: a complication of intravenous urography. *Radiology* 92, 1043-1044.
37 BLUM M., WEINBERG U., SHENKMAN L. & HOLLANDER C.S. (1974) Hyperthyroidism after iodinated contrast medium. *New Engl. J. Med.* 291, 24-25 and 682-683.
38 SHETTY S.P., MURTHY G.G., SHREEVE W.W., NWAZ A.M. & RYDER S.W. (1974) Hyperthyroidism after pyelography. *New Engl. J. Med.* 261, 682.
39 STEMERMAN M., GOLDSTEIN M.L. & SCHULMAN, P.L. (1971) Pancytopenia associated with diatrizoate. *N.Y. St. J. Med.* 71, 1220-1222.
40 CHAPLIN H. & CARLSSON E. (1961) Changes in human red blood cells during in vitro exposure to several roentgenographic contrast media. *Am. J. Roentg.* 86, 1127-1137.
41 BERNSTEIN E.F. (1970) Discussion. Symposium on contrast medium toxicity. *Invest. Radiol.* 5, 416-422.
42 SCHIANTARELLI P., PERONI F., TIRONE P. & ROSATI G. (1973) Effects of iodinated contrast media on erythrocytes. I. Effects of canine erythrocytes on morphology. *Invest. Radiol.* 8, 199-204.
43 ANSELL G. & PITCHFORD A.G. (1974) Dangers of re-sterilized disposable syringes in urography. *Br. J. Radiol.* 47, 300.
44 COHEN L.S., KOKKO J.P. & WILLIAMS W.H. (1962) Hemolysis and hemoglobinuria following angiocardiography. *Radiology* 92, 329-332.
45 ROSENTHALL A., LITWIN S.B. & LAVER M.B. (1973) Effect of contrast media used in angiocardiography on haemoglobin-oxygen dissociation. *Invest. Radiol.* 8, 191-198.
46 LASSER E.C. (1973) Contrast material-red blood cell interactions. *Invest. Radiol.* 8, 189-190.
47 RICHARDS D. & NULSEN F.E. (1971) Angiographic media and the sickling phenomenon. *Surgical Forum* 22, 403-404.
48 MCNAIR J.D. (1972) Selective coronary angiography. Report of a fatality in a patient with sickle cell haemoglobin. *Calif. Med.* 117, 71-75.
49 LASSER E.C. (1975) Personal communication.
50 PEARSON H.A., SCHIEBLER G.L., KROVETZ J., BARTLEY T.D. & DAVID J.K. (1965) Sickle cell anaemia associated with tetralogy of Fallot. *New. Engl. J. Med.* 273, 1079-1083.
51 STEIN H.L. & HILGARTNER M.W. (1968) Alteration of coagulation mechanism of blood by contrast media. *Am. J. Roentg.* 104, 458-463.
52 CHANDRA R. & ABRAHAM J. (1973) Preliminary studies on in vitro and in vivo effects of 50% Hypaque on coagulation in man. *Angiology* 24, 199-204.

53 FISCHER H.W., ROSENBERG F.J., HOEY G.B. (1970) Discussion. *Invest. Radiol.* 5, 486.
54 HILAL S.K. (1966) Haemodynamic changes associated with the intra-arterial injection of contrast media. New toxicity tests and a new experimental contrast medium. *Radiology* 86, 615-633.
55 FISCHER H.W. & REDMAN H.C. (1971) Comparison of a sodium methylglucamine diatrizoate contrast medium of minimal sodium content with a pure methylglucamine diatrizoate preparation. *Invest. Radiol.* 6, 115-118.
56 GENSINI G.G. & DI GIORGI S. (1964) Myocardial toxicity of contrast agents used in angiography. *Radiology* 82, 24-34.
57 LANG E.K. (1965) Clinical evaluation of side effects of radiopaque contrast media administered via intravenous and intra-arterial routes—a preliminary report. *Radiology* 85, 665-669.
58 KAUDE J. (1971) Angiografi med jodamid—klinisk prövning ab ett trijoderat kontrastmedel. *Lakartdningen* 68, 42-48.
59 SÖDERMARK T. & MINDUS P. (1968) A retrospective study of complications following nephroangiography, cardioangiography and coronary angiography. *Acta Med. Scandinav.* 183, 177-182.
60 SEYMOUR J. (1969) Severe laryngeal oedema during injection of sodium metrizoate ('Triosil'). Survival after emergency laryngostomy. *Br. Heart. J.* 31, 529-530.
61 SCOTT W.C. (1968) Infusion pyelography in perspective. *Clin. Radiol.* 19, 83-89.
62 BERNSTEIN E.F., PALMER J.D., AABERG T.A. & DAVIS R.L. (1961) Studies of the toxicity of Hypaque-90 per cent following rapid intravenous injection. *Radiology* 76, 88-95.
63 ANSELL G. (1970) Fatal overdose of contrast medium in infants. *Br. J. Radiol.* 43, 395-396.
64 MCCLENNAN B.L., KASSNER E.G. & BECKER J.A. (1972) Overdose at excretory urography: toxic cause of death. *Radiology* 105, 383-386.
65 GIAMMONA S.T., LURIE P.R. & SEGAR W.E. (1963) Hypertonicity following selective angiocardiography. *Circulation* 28, 1096-1101.
66 MARSHALL M. & HENDERSON G.A. (1968) Tendency to acidosis following injection of radiopaque contrast material. *Br. J. Radiol.* 41, 190-192.
67 NOGRADY M.B. & DUNBAR J.S. (1968) Delayed concentration and prolonged excretion of urographic contrast medium in the first month of life. *Am. J. Roentg.* 104, 289-295.
68 SAXTON H.M. (1969) Review Article. Urography. *Br. J. Radiol.* 42, 321-346.
69 CATTELL W.R., MCINTOSH C.S., MOSELEY I.I. & FRY I.K. (1973) Excretion urography in acute renal failure. *Br. Med. J.* 2, 275-278.
70 TALNER L.B. (1972) Urographic media in uraemia. Physiology and pharmacology. *Radiol. Clin. N. Amer.* 10, 421-432.
71 GRAINGER R.G. (1972) Renal toxicity of radiological contrast media. *Br. med. Bull.* 28, 201-215.
72 BERGMAN L.A., ELLISON M.R. & DUNEA G. (1968) Acute renal failure after drip-infusion pyelography. *New Engl. J. Med.* 279, 177.
73 DAVIDSON A.J., BECKER A.J., ROTHFIELD N. *et al.* (1970) An evaluation of the effects of high-dose urography on previously impaired renal and hepatic function in man. *Radiology* 97, 249-254.

74 Milman N. & Stage P. (1974) High dose urography in advanced renal failure II. Influence on renal and hepatic function. *Acta Radiol. Diagn.* **15**, 104–112.
75 Kleinknecht D., Jungers P., Michel J.R. et al. (1971) Les accidents de néphrose osmotiques après urographie par perfusion chez l'insuffisant rénal. *Ann. Med. Intern.* **122**, 1167–1171.
76 Kleinknecht D., Jungers P. & Michel J.R. (1972) Les accidents anuriques après urographie par perfusion chez l'insuffisant rénal en dehors du myélome. *Sem. Hôp. Paris* **48**, 3383–3387.
77 Moreau J., Droz D., Sabto J., Jungers P., Kleinknecht D., Hinglais N. & Michel J.R. (1975) Osmotic nephrosis induced by water-soluble triiodinated contrast media in man. A retrospective study of 47 cases. *Radiology* **115**, 329–336.
78 Pillay G., Robbins P.C., Schwartz F.D. & Kark R.M. (1970) Acute renal failure following intravenous urography in patients with long-standing diabetes mellitus and azotaemia. *Radiology* **95**, 633–636.
79 Berdon W.E., Schwartz R.H., Becker J. & Baker D. (1969) Tamm-Horsfall proteinuria. Its relation to prolonged nephrogram in infants and children and to renal failure in adults with multiple myeloma. *Radiology* **92**, 714–722.
80 Schwartz R.H., Berdon W.E., Wagner J. et al. (1970) Tamm-Horsfall urinary mucoprotein precipitation by urographic contrast agents: in vitro studies. *Am. J. Roentg.* **108**, 698–701.
81 Barshay M.E., Kaye J.H., Goldman R. & Coburn J.W. (1973) Acute renal failure in diabetic patients after intravenous pyelography. *Clin. Nephrol.* **1**, 35–39.
82 Feldman H.A., Goldfarb S. & McCurdy D.K. (1974) Recurrent radiographic dye-induced acute renal failure. *J. Am. med. Ass.* **229**, 72.
83 Brown M. & Battle J.D. (1964) The effect of urography on renal function in patients with multiple myeloma. *Canad. med. Ass. J.* **91**, 786–790.
84 Lasser E.C., Lang J.H. & Zawadzki Z.A. (1966) Contrast media: myeloma protein precipitates in urography. *J. Am. med. Ass.* **198**, 945–947.
85 Cwynarski M.T. & Saxton H.M. (1969) Urography in myelomatosis. *Br. Med. J.* **1**, 486.
86 Gross M., McDonald H. & Waterhouse K. (1968) Anuria following urography with meglumine diatrizoate (Renografin) in multiple myeloma. *Radiology* **90**, 780–781.
87 Myers G.H. & Witten D.M. (1971) Acute renal failure after excretion urography in multiple myeloma. *Am. J. Roentg.* **113**, 583–588.
88 McEvoy J., McGeown M.G. & Kumar R. (1970) Renal failure after radiological contrast media. *Br. Med. J.* **4**, 717–718.
89 Van Waes P.F.G.M. (1972) *High-dose urography in oliguric and anuric patients.* Excerpta Medica, Amsterdam.
90 Lee S. & Schoen I. (1966) Black-copper reduction reaction simulating alcaptonuria—occurrence after intravenous urography. *New Engl. J. Med.* **275**, 266–267.
91 Lasser E.C. (1971) Metabolic basis of contrast media toxicity. Status 1971. *Am. J. Roentg.* **113**, 415–422.
92 Sandström, C. (1955) Secondary reactions from contrast media and the allergy concept. *Acta Radiol.* **44**, 233–242.
93 Lalli A.F. (1975) Editorial. Urography, shock reaction and repeated urography. *Amer. J. Roentg.* **125**, 264–268.

94 HALPERN B., KY N.T. & AMACHE N. (1967) Diagnosis of drug allergy in vitro with the lymphocyte transformation test. *J. Allerg.* **40**, 168–181.
95 HALPERN B., KY N.T. & AMACHE N. (1969) 'In vitro' lymphoblast transformation test (L.T.T.) as a tool for the study of drug hypersensitivity. *Proc. Europ. Soc. Stud. Drug Toxicity* **10**, 27–35.
96 DI BENEDETTO G., BARABINO A. & INDIVERI F. (1973) Richerche prelimiari in tenua di ipersensibilita ai mezzi di contrasto iodati. Applicazione del test di blastizzazione dei linfociti in vitro. *Arch. E. Maragliano Pat. Clin.* **29**, 43–49.
97 KLEINKNECHT D., DELOUX J. & HOMBERG J.C. (1974) Acute renal failure after intravenous urography: detection of antibodies against contrast media. *Clin. Nephrol.* **2**, 116–119.
98 LASSER E.C., SOVAK M. & LANG J.H. (1976) Development of contrast media idiosyncrasy in the dog. *Radiology* **191**, 91–95.
99 MANN R.M. (1961) The pharmacology of contrast media. *Proc. Roy. Soc. Med.* **54**, 473–475.
100 ROCKOFF S.D., BRASCH R., KUHN C. & CHRAPLYVY M. (1970) Contrast media as histamine liberators. I. Mast-cell histamine release in vitro by sodium salts of contrast media. *Invest. Radiol.* **5**, 503–509.
101 LASSER E. (1970) Discussion. *Invest. Radiol.* **5**, 440.
102 LASSER E.C., WALTERS A., REUTERS S. & LANG J. (1971) Histamine release by contrast media. *Radiology* **100**, 638–686.
103 LASSER E.C., WALTERS A.J. & LANG J.H. (1974) An experimental basis for histamine release in contrast media reactions. *Radiology* **110**, 49–59.
104 BRASCH R.C., ROCKOFF S.D., KUHN C. & CHRAPLYVY M. (1970) Contrast media as histamine liberators. II. Histamine release into venous plasma during intravenous urography in man. *Invest. Radiol.* **5**, 510–513.
105 ROCKOFF S.D., & AKER C.T. (1972) Contrast media as histamine liberators. VI. Arterial plasma histamine and haemodynamic responses following angiocardiography in man with 75% Hypaque. *Invest. Radiol.* **7**, 403–406.
106 ROSENBERG F.J. (1970) Discussion. *Invest. Radiol.* **5**, 389.
107 LASSER E.C., FARR R.J., FUJIMAGARI T. & TRIPP W.M. (1962) Significance of protein-binding of contrast media in roentgen diagnosis. *Am. J. Roentg.* **87**, 338–360.
108 LASSER E.C. (1974) Contrast media activation of serum complement system. *Invest. Radiol.* **9**, 6A.
109 LASSER E.C., ELIZONDO-MARTEL G. & GRANKE R.C. (1964) The roentgen contrast media potentiation of nembutal anaesthesia in rats. *Am. J. Roentg.* **91**, 453–460.
110 FISCHER H.W., SCHABEL S. & THORNBURY J. (1975) Personal communication.
111 PFISTER R.C., YODER I.C., HUTTER A.M. & BERG G.R. (1974) The effect of intravenous urography on cardiac rhythm and ischemia. *Radiological Society of North America*, Chicago, Dec. 1–6.
112 MARSHALL T.R., LING J.T., FOLLIS G. & RUSSELL M. (1965) Pharmacological incompatibility of contrast media with various drugs and agents. *Radiology* **84**, 536–539.
113 SCHATZ M., PATTERSON R., O'ROURKE J. et al. (1975) The administration of radiographic contrast media to patients with a history of a previous reaction. *J. Allerg. Clin. Immunol.* **55**, 358–366.
114 ZWEIMAN B. & HILDRETH E.A. (1974) An approach to contrast studies in reactive humans. *J. Allerg. Clin. Immunol.* **53**, 97.

115 LALLI A.F. (1974) Urographic contrast media reactions and anxiety. *Radiology* **112**, 267–271.
116 ANSELL G. & ANSELL A. (1964) Medical emergencies in the X-ray department: prevention and treatment. *Br. J. Radiol.* **37**, 881–897.
117 BARNHARD F.M. & BARNHARD H.J. (1968) The emergency treatment of reactions to contrast media: updated 1968. *Radiology* **91**, 74–81.
118 ANSELL G. (1973) *Notes on Radiological Emergencies*. 2nd Edition. Blackwell Scientific Publications, Oxford.
119 DELORME P. & LETENDRE J. (1972) Prolonged shock after intravenous pyelography. *Canad. med. Ass. J.* **106**, 343–344.
120 MARSHALL W.H. & CASTELLINO R.A. (1970) The urinary mucosal barrier in retrograde pyelography. The role of the ureteric mucosa. *Radiology* **97**, 5–7.
121 LYTTON B., BROOKS M.B. & SPENCER R.P. (1968) Absorption of contrast material from urinary tract during retrograde pyelography. *J. Urol.* **100**, 779–782.
122 GANGAI M.P. (1972) Reaction to contrast media—is it a dose related phenomenon? *Milit. Med.* **137**, 388–389.
123 BURROS H.M., BORROMEO V.H.J. & SELIGSON D. (1958) Anuria following retrograde pyelography. *Ann. Intern. Med.* **48**, 674–676.
124 MCALISTER W.H., SHACKELFORD G.D. & KISSANE J. (1972) The histological effects of 30% Cystokon. Hypaque 25 and Renografin 30 in the bladder. *Radiology* **104**, 563–565.
125 ALFREY A.C., ROTTSHAFER O.W. & HUTT M.P. (1967) Acute renal failure following retrograde pyelography. *Arch. Intern. Med.* **119**, 214–217.
126 MIHALECZ K., WÖLFER E., CZÁSZÁR J. & PINTER J. (1967) Acute nierensuffisienz nach unilateraler retrograder pyelographie. *Z. Urol. Nephrol.* **60**, 783–788.
127 HOWARDS S.S. & HARRISON J.H. (1973) Retroperitoneal phlegmon: a fatal complication of retrograde pyelography. *J. Urol.* **109**, 92–93.
128 BABANOURY A.P. & BOGGS L.K. (1966) Miliary tuberculosis following retrograde pyelography. *J. Urol.* **96**, 101.
129 PERCY A.K. & SAEF E.C. (1967) An unusual complication of retrograde pyelography: neuromuscular blockade. *Pediatrics* **39**, 603–606.
130 TUCHSCHMID C. & LAHLAIDI A. (1972) La thorotrastose renale. *J. Med. Maroc.* **8**, 45–57.
131 SORENSEN S.E. (1972) Local toxic effects of anaesthetics and contrast media in urethrography. *Acta Radiol. Diagn.* **12**, 225–240.
132 DIX V.W. & TRESIDDER G.C. (1963) Collapse after use of lignocaine jelly for urethral anaesthesia. *Lancet* **1**, 890.
133 SHOPNER C.E. (1967) Clinical evaluation of cystourethrographic contrast media. *Radiology* **88**, 491–497.
134 KUHNS L.R., BAUBLIS J.V., GREGOR J. & POZNANSKI I.K. (1972) The in vitro effect of cystographic contrast media on urinary tract pathogens. *Invest. Radiol.* **7**, 112–117.
135 BRODEUR A.E., GOYER R.A. & MELLICK W. (1965) A potential hazard of barium cystography. *Radiology* **85**, 1080–1084.
136 MERTZ J.H. & BROWN J.R. (1968) Iatrogenic vesical foreign body. *J. Urol.* **99**, 439.

CHAPTER 2: AORTOGRAPHY AND PERIPHERAL ARTERIOGRAPHY

H. HERLINGER

INTRODUCTION

Transfemoral catheter aortography was first described by Farinas of Cuba [1] who regarded it a safer alternative to direct needle puncture. The modification by Seldinger [2] made it possible to catheterize arteries percutaneously, and opened the door to the development of a universally practised radio-diagnostic procedure which continues to grow in volume and scope.

The incidence and pattern of complications is no longer that reported by earlier angiographers [3-6]. Improvements of contrast media, catheters and guide wires and better image intensification have considerably reduced or even abolished some of the complications. Partial femoral occlusion by thrombus was formerly a major source of complications. The adoption of methods to reduce thrombosis has already lowered the overall incidence of complications and a further reduction should be possible.

The main purpose of this chapter is not to catalogue all possible complications of angiography but to emphasize methods designed to avoid them. It relies heavily on published material and is also based on a series of 2454 angiograms done in adult patients, in this hospital group.

Angiographic complications can be classified in the following way [7]:

Arteriographic death is directly due to angiography or to surgery undertaken to correct a major complication.

Major complications threaten life, limb or visceral integrity and require surgical intervention or prolong hospitalization.

Minor complications are either asymptomatic or clinically insignificant. They will not be discussed.

ARTERIOGRAPHIC DEATH

It is usual to attribute to angiography a death which occurs during the procedure or within the subsequent 48 hours [8]. In earlier years there has been doubtless causal relationship between angiography and fatality. McAfee [3] who surveyed 13 207 mostly translumbar aortograms in 1957 reported 37 deaths (0.28 per cent). In a study of 1706 retrograde thoracic aortograms reported in 1957 [9] there were 29 deaths (1.7%). Ansell [10] believes that the earlier, more toxic contrast media were chiefly responsible.

Deaths clearly related to angiography have become very rare. In a recent report two fatalities were associated with surgery to relieve femoral thrombosis [11]. In 1966 Baum *et al.* [13] compared the incidence and type of fatal complications during the 48-hour period preceding cancelled angiograms with those occurring during the 48 hours after the performance of angiographies. There was surprising similarity, indicating that the pre-existing disease and not the angiogram bore responsibility for most angiographic fatalities in more recent years. The growing use of vascular radiology in acute and severely ill patients—both for diagnosis and treatment—must lead to an increase in the number of deaths which show a time relationship to the procedure.

Three post-angiographic deaths have occurred in our series. They were all in patients with haematemesis and liver failure and are considered to have been due to the underlying disease process.

MAJOR COMPLICATIONS

These will be dealt with separately in relation to the following angiographic approach methods:
1 Transfemoral catheterization
2 Direct femoral arteriography
3 Translumbar aortography
4 Transaxillary catheterization

TRANSFEMORAL CATHETERIZATION

Our own experience derives from 1152 examinations performed by the author.

Femoral thrombosis

Thrombus formation

It is now generally agreed that thrombus formation commences as platelet aggregation on the foreign-body surface of the catheter within the artery. Fibrin appears in the spaces between the aggregates, platelet membranes disintegrate and irreversible agglutination results. Red blood corpuscles are entrapped by the fibrin. The deposition of fibrin normally ceases when the whole catheter surface has been encased, but an adequate blood flow is required for self-limitation.

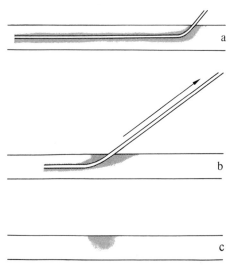

FIGURE 2.1. Concept of thrombus formation. a. Catheter surface within the artery covered by thrombus. b. During withdrawal the surface thrombus is stripped at the exit point. c. After withdrawal a plug of thrombus is left behind.

On withdrawing the catheter the encasement thrombus is readily wiped off its surface at the point where it passes through the artery wall. A plug of thrombus is thus left behind, projects into the lumen and usually adheres at the puncture site (Fig. 2.1). It can be well demonstrated by a 'pull-out angiogram' as first described by Siegelman et al. [14]. The catheter is drawn back to lie close to the entry point, contrast medium is injected and films are taken at twice magnification (Fig. 2.2). The filling defect shown in this way was believed by Siegelman to represent a 'thrombus of injury'. Amplatz [15] suggested that stripping of the catheter surface thrombus was the operative mechanism.

More rarely thrombus is formed in relation to intimal tears or damaged atherosclerotic plaques at or near the puncture point. Infrequently thrombus may become dislodged to embolize with clinical significance.

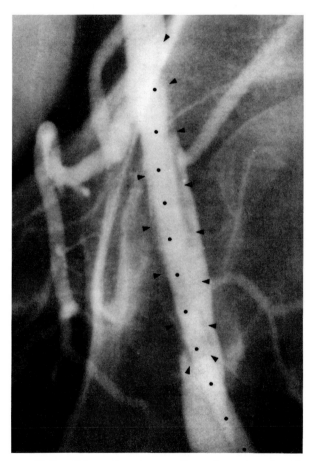

FIGURE 2.2. 'Pull-out' femoral angiogram after arch aortography using a Kifa grey catheter. Arrowheads outline non-opaque surface thrombus accumulating near catheter exit. (Position of catheter shown by dots.)

Thrombus formation is influenced by a number of circumstances:
Catheter length and width are considered relevant by several authors [16–

18]. We believe this to be the most important factor since the extent of the outer surface area of the catheter within the artery must directly relate to the size of the eventual thrombic plug. A catheter with an outer diameter of 2.8 mm (Kifa gray) if taken into the aortic arch exposes to thrombus deposition a surface which is three times that of a 2-mm (6F) catheter in the coeliac axis (Fig. 2.2).

Artery diameter. The width of the common femoral artery determines the effect of the thrombus on the blood supply to the limb. Female patients have narrower arteries and show a greater tendency to develop spasm near the puncture site. Stasis increases thrombus deposition. Cramer *et al.* [11] have found more than three times as many thrombotic complications in women than in men. *Oral contraceptives* may also be a contributory factor to thrombotic complications in women, and they should preferably be discontinued prior to arterial catheterization [12].

Time in artery. Formanek *et al.* [19] believe this to be more important than catheter diameter. Our experience has not shown that catheterization time by itself mattered, provided spasm was avoided. Even indwelling catheters have not presented problems in this respect. Experimentation on dogs should not be directly transposed into human terms; in dogs thrombus forms more rapidly and is more substantial.

Each *catheter change* will add to the bulk of the thrombus plug. *Inexperience* of the angiographer has been shown by Meaney [20] to increase the incidence of thrombus significantly. *Atheroma* is not now considered an important factor in thrombus formation, particularly as spasm does not occur in rigid and often ectatic arteries. The *patient's age* influences thrombosis inversely; younger patients are at greater risk [21, 22]. The type of *catheter surface* matters to a slight degree; the somewhat rougher surface of a polyurethane catheter may favour fibrin deposition compared with the smoother surface of a polyethylene catheter [23, 24].

Incidence and diagnosis

Incidence and diagnostic criteria are interrelated. Where diagnosis rests on purely clinical evidence of ischaemia the reported incidence ranges from nil [18] and 0.35 per cent [7] to 1.2 per cent [25]. Jacobsson *et al.* [26] rely on comparison oscillography and consider a post-catheterization reduction of pulsation by at least 50 per cent to indicate significant ischaemia; this was found in 1.6 per cent of 880 patients and confirmed by surgery or angiography in most of them. Cramer *et al.* [11] were able to show pull-out angiogram evidence of a thrombus in about half of their patients but considered thus clinically relevant in only 1.2 per cent. Siegelman et al. [14] believed a 'pull-out defect' in excess of 2.5 mm to indicate relevance; this was found in 31 of 173 patients but was associated with significant diminution of oscillometry in only 7 (4 per cent).

Clinical relevance is crucial in attempting an evaluation of the wide range of reported incidence and the *need for surgical intervention* should express this clearly. However, indications for surgery vary enormously. Surgery has been performed on a number of asymptomatic patients in order to prevent the possible development of claudication later [11]. Meaney [20] and Beven [27] treat medically and observe milder cases of ischaemia in order to differentiate arterial spasm from thrombosis and to allow an adequate collateral circulation to develop; more severe ischaemia is treated surgically. Yellin & Shore [17] counsel early surgery and warn against a diagnosis of 'vasospasm' after arteriography. Recently a nonsurgical catheter method of treatment has been suggested [28].

It is clear that the incidence of significant thrombosis following transfemoral angiography can only be given as a wide range between nil and 4 per cent. It is likely that higher incidence levels reflect a greater diagnostic effort and a preponderance of long/wide catheter studies. In our own material there has been only a single case of femoral thrombosis and this did not require surgery. This represents an incidence of less than 0.1 per cent and is due to a predominance of short/narrow catheter angiograms and a mainly clinical approach (including comparison pulses) to the diagnosis of ischaemia.

Prevention of thrombosis

Gangrene and amputation are the occasional, disastrous consequence of thrombosis [10]. All possible aspects of thrombosis prevention must be considered carefully.

Premedication. Aspirin given as a single dose the day before angiography inhibits platelet adhesion and ADP release [29]. It could be shown in cats [30] that aspirin reduced to about half the incidence of thrombus formation on catheters. Dipyridamole is one of a number of compounds which affect platelet aggregation [31]. Both drugs in combination have been found to influence platelet survival in patients with prosthetic heart valves [32]; their thrombosis-preventing function does not seem to have been studied in angiography.

Dextran reduces platelet adhesiveness with increasing effect from about two hours after administration. Jacobsson [33] compared a group of untreated and pretreated patients and found thrombus in 8 per cent of the former and none of the latter group. Disadvantages are the possible hazards of large infusion and the problem of fitting infusion, time to await its effect, and the angiogram itself, into one working day. Whereas some recent animal experiments [34] do not confirm earlier claims of protection by dextran against the cerebral toxicity of contrast agents, others [35] do suggest a protective effect on the blood–brain barrier.

Pre-sedation. All patients about to undergo angiography show anxiety to a varying degree. Increased apprehension might possibly affect the coagulation process [11]. Whenever feasible the radiologist should visit the patient in advance to explain the method and purpose of the examination and to establish a personal relationship.

We no longer employ narcotics and do not require general anaesthesia in adults. Two drugs are in use at present, diazepam and lorazepam. Wilson & Ellis [36] have found memory recall for the period under sedation to be profoundly influenced by lorazepam only. Anxiolysis is more effectively produced by lorazepam than diazepam [37]. Glanville [38] gives lorazepam 4 mg by mouth $1\frac{1}{2}$ hours before examination, prefers its soporific effect to that of diazepam and confirms its amnesic and anxiolytic value. Diazepam (10 mg) is given intravenously at the time of the angiogram in patients in whom lorazepam has been omitted.

Side effects are normally insignificant, rarely excitement occurs in place of sedation. In patients with depressed liver function all pre-sedation can be disastrous and should be omitted. We must also warn against the *intra-arterial* use of barbiturates or tranquillizing drugs, since these may cause intimal damage and thrombosis [39].

The use of hypothrombogenic catheters. The bonding of heparin to vascular prosthetic materials was reported in 1966 [40]. This was applied to arterial catheters, at first by attempting to bind a monolayer of heparin to the catheter surface after immersion in a cationic surfactant [41, 42].

Heparin can form a stable compound with benzalkonium chloride, a commonly used antiseptic and germicide. Cramer *et al.* [43] describe in detail a method of preparing the waxy B–H precipitate and applying it to catheter surfaces. The compound is insoluble in water but is slowly desorbed in blood, full effectiveness being limited to about one hour. B–H coating minimizes post-catheterization thrombosis. Hawkins & Kelley [44] used B–H coated catheters in 563 patients and did not see a single case of thrombosis; in 100 of the patients pull-out angiograms showed minor clot in only four. In the hands of Cramer *et al.* [11] the use of B–H-treated catheters reduced pull-out angiogram evidence of thrombus to 16 per cent (vs. 50 per cent with uncoated catheters) and clinically relevant thrombosis to 0.6 per cent, less than half the incidence when using uncoated catheters.

There have been no reports of increased bleeding with the use of heparinized catheters, the effects of heparin being strictly local. A major disadvantage of B–H coating is that catheter shaping becomes more difficult and curves given are inadequately retained. This has made it increasingly necessary to rely on a deflector guide wire system[1] for selective catheterization, at times even for advance into subselective positions[2].

Guide wires can also be B–H coated [45] and this is indicated in cerebral and coronary angiography. Siliconization of catheters has not been found to influence thrombogenicity [46].

Systemic heparinization. While flushing with heparined saline will prevent thrombus formation within the lumen of the catheter, the systemic effect of this procedure has been considered insufficient to prevent thrombus deposition on the outer surface. Adequate low-grade systemic heparinization can be produced by giving 45 units of heparin per kg, an average dose of 3000 units, injected through the catheter immediately after its introduction into the aorta or iliac artery [47].

There is increased danger of bleeding during the procedure and after withdrawal. Nejad *et al.* [48] suggest heparin neutralization with protamine four minutes before withdrawal. The recommended dose is 10 mg of protamine sulphate for each 1000 units of heparin injected initially.

Thrombosis prevention practised by us

For abdominal arteriography we do not employ a catheter larger than 6 French. B–H coating is then unnecessary. Instead of intermittent catheter flushing by means of syringes we use a high-pressure drip infusion system[3] via a three-way tap and connect it to the catheter as soon as it has been introduced into the aorta (Fig. 2.3). Flushing is only interrupted during guide-wire-based manoeuvres and during contrast medium injection. The infusion fluid is saline with heparin 1500 units in 500 ml, this being run in at the rate of 90 drops/minute to last about one hour. We assume that during retrograde arterial catheterization the outer surface of the catheter is thus continually rinsed with heparin solution of sufficient local concentration to reduce thrombus deposition to acceptably low levels (Fig. 2.4).

[1] Cook Deflector, obtained from W. Cook, Europe A/S, 24 Oakfield Ave., Hitchin, Herts., England.

[2] Heparin bonded catheters are now commercially available.

[3] Consists of Pressure Infuser and disposable Transfer Pack and Arterial Recipient Set. Obtained from Baxter Laboratories, Thetford, Norfolk, England.

Furthermore it is essential to test femoral pulsation immediately beyond catheter entry repeatedly throughout the procedure. Spasm would then be detected at an early stage and the examination cut short before irreversible thrombotic damage has been done. In the event of a spasm we would inject

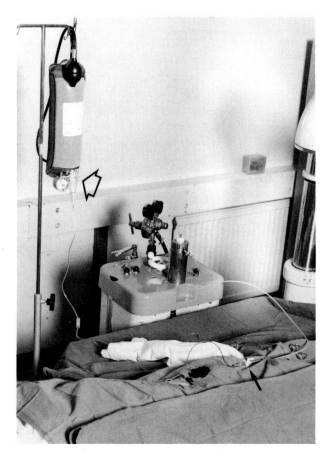

FIGURE 2.3. Continual infusion arrangement during angiography. Pressure infusor (open arrow) connects to the catheter through a three-way tap (arrow).

through the catheter 5 ml of 1 per cent procaine prior to its withdrawal. Alternatively, tolazoline (Priscol) can be injected through the catheter to relax arterial spasm [69]. We have adopted the catheter removal routine described by Jeffery [49], and designed to expel fresh clot from the puncture site; we have occasionally recovered small fragments in this way.

Aortography and Peripheral Arteriography

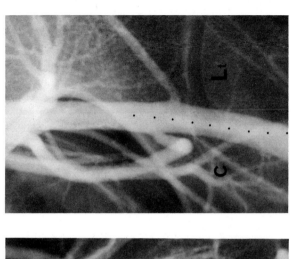
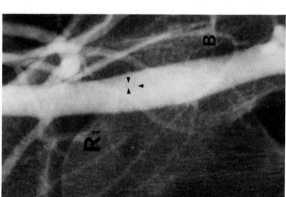
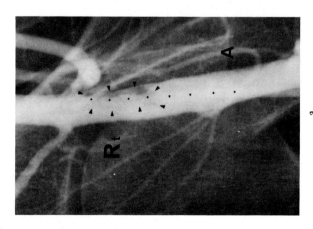

FIGURE 2.4. 'Pull-out' angiogram in a female patient with multiple hepatoma in a huge liver. For adequate contrast medium perfusion two 6F polyethylene catheters were taken into the common hepatic artery, one from each femoral. Only the left catheter was heparin coated. a. Pull-out angiogram of the non-coated catheter shows moderate thrombus aggregation (arrowheads). b. After withdrawal and haemostasis the right common femoral artery was opacified through the left catheter. Only very little thrombus remains (arrowheads). c. No thrombus is visible at 'pull-out' angiography of the heparin-coated left femoral catheter.

Embolism

Peripheral embolization

The thrombus stripped from a catheter on withdrawal from the artery usually adheres to the exit area. Out of 32 patients in whom significant thrombus occurred, peripheral embolization was found in only 7 [17]. A similar proportion of embolizing by adherent thrombus was reported by Cornier & Lagneau [50].

Small emboli will lodge in arteries supplying the toes or the skin of the foot and may produce patches of tender cyanosis, usually with early and spontaneous return to normality. We believe that thrombus initially adherent at the puncture site may become fragmented during post-withdrawal compression, to disappear into the periphery without clinical signs of embolism (see Fig. 2.4). Larger emboli will be arrested in more proximal arteries and may require surgical removal.

Athero-embolism

Occlusion of medium-sized peripheral arteries by dislodged atheromatous plaques has been reported [11]. Anderson [51] has described inferior mesenteric artery occlusion and ischaemic colitis associated with aortography in the presence of atheroma. The routine use of movable core J-type guide wires during the introduction of catheters [52] should avoid plaque traumatization during that stage of angiography.

Embolism of arteriosclerotic debris into peripheral vessels of elderly patients with already compromised distal circulation could tilt the balance towards dangerous ischaemia [22]. Numerous small athero-emboli were found in several of the organs of an elderly patient who died one week after aortography [53]. Embolization by arteriosclerotic debris constitutes the chief danger of angiography in the geriatric age group. While it is almost invariably possible to advance a thin-walled catheter into the aorta without damage, it is absolutely essential—and a hallmark of angiographic experience—to recognize overwhelming difficulty and not to persevere with attempts at selective catheterization in patients with advanced atheroma.

Renal artery embolism

This is a rare and potentially disastrous complication. The embolized clot seems to derive from the catheter itself [54], probably having formed between the distal side hole and the end hole in a space not perfused until the higher pressure injection of contrast medium. Immediate heparin anticoagulation succeeded in a case repeated by McConnell et al. [55]. Buxton & Mueller [54] used a trans-catheter method by which fresh clot was removed with success. Renal complications of angiography are discussed in greater detail in Chapter 4.

Cotton fibre embolization

Four examples of multiple small cotton fibre infarcts—renal and cerebral—have been reported by Adams et al. [56]. The fibres derived from an open bowl of heparinized saline used to fill syringes for catheter perfusion. Other particulate matter and pyrogenic material may also be infused in this way. As mentioned earlier, we advise the use of closed system arterial infusion.

Air embolism

During angiography, contrast medium passes through the narrow syringe nozzle at high velocity and this may create a negative pressure at the junction of the nozzle and the distal connecting tube. Unless there is a perfect airtight fit between the connectors, bubbles of air may be sucked into the column of contrast medium and give rise to air embolism. The result may be particularly serious if this occurs in the cerebral or coronary circulation. Malfitting of high-pressure connections appears to be more common with disposable plastic syringes and fittings but, even with metal connectors, it is essential to ensure that they fit perfectly and do not become worn [57].

Bleeding and haematoma

These are regarded as major complications if surgery is required to arrest bleeding or to evacuate a collection of blood. Of significance would also be a haematoma causing pain and impairing mobility over a few days. Lang [8] considers venospasm, phlebitis and thrombophlebitis to be a possible further complication of a haematoma. Wound healing may be impeded where an incision for vascular surgery has to involve skin suffused with extravasated blood [58].

Inadequate technique

This is the most common cause of bleeding. The femoral artery can be lacerated due to guide-wire kinking during the introduction of a catheter [20]. This may occur where undue pressure is needed to thread the catheter towards the artery, either because of an inadequate skin opening or an approach through scar tissue or through the inguinal ligament. Atheromatous arteries should be punctured with circumspection and require sharp needles and well-fitting, carefully rounded-off catheter tips; they should be entered at a more acute angle in an attempt to avoid penetration through the posterior wall.

Post-catheterization compression of the puncture site should be carried out by the radiologist until bleeding has stopped. Pressure should be applied at artery puncture level (above the skin puncture) as it is only there that arrest of bleeding can be achieved with pressures which do not totally obliterate the peripheral pulse. Moderate compression continues for ten minutes after bleeding has ceased and is followed by observation of the uncovered puncture site (except for a plastic dressing). Should a small haematoma have formed, we inject it with Hyalase 1000 units straight away.

Bleeding around the catheter during angiography is not uncommon in arteriosclerotic patients, especially after catheter change. Intermittent compression during the procedure will usually allow the examination to be carried through.

Problem patients. Hypertension as well as arteriosclerosis may render haemostasis more difficult and post-angiographic observation more important. Liver failure, uraemia, bleeding disorders, and coumarin drugs are not absolute contraindications to angiography but should induce a more careful evaluation of the diagnostic purpose. It will be wise to keep the examination as brief as possible and to retain the patient in the angiographic area for a prolonged period of compression and observation.

No patient in our series has required surgery for bleeding or a haematoma. Approximately 10 per cent of patients received an injection of Hyalase at the end of the angiogram.

Artery dissection

This is a fairly frequent complication, usually minor, which relates to catheter manipulation or to catheter whip during pressure injections. Dissection implies the entry of a guide wire, catheter or a jet of contrast medium into the subintimal space of the artery wall. It can be recognized by an increased resistance to the normally effortless movement of an instrument within the artery and by failure to aspirate pulsatile blood freely in the case of an end-hole catheter (with side holes aspiration does *not* exclude dissection). It is made clear by a small test injection of contrast medium which would then be retained at the injection site to assume an eccentric position and a line of demarcation against more diluted medium within the lumen.

Dissection against the direction of blood flow is a minor complication, the subintimal space becoming occluded by the pressure of blood after withdrawal of the catheter. The potentiating effect, however, due to chemotoxicity of older contrast media has previously been considered relevant [59]. Somewhat more dangerous are antegrade dissections, especially of arteries to organs where collateral supply is nonexistent or may be inadequate. Renal artery dissection may provoke a hypertensive crisis.

Dissection used to occur most frequently during the introduction of the guide-wire/catheter combination, an incidence of 25 per cent had been reported [60]. This has been virtually eliminated by the use of the J-type guide wire. Dissection is now most often produced in the course of efforts to catheterize selectively and subselectively. The straightening of the end curve of a catheter (e.g. in the aortic arch) during high-pressure injection may force the tip against the intima to produce dissection by the jet of contrast medium.

Much of dissection can be avoided by a combination of visual observation and a tactile ability to appreciate changes of resistance at catheter tip level. It is dangerous to attempt forcing a catheter of unsuitable shape over a tip deflector wire into a subselective position. The introduction of a floppy J-type guide ahead of the selectively placed catheter may help to advance the latter into a subselective position with safety [61]. A test injection by hand should always be done before a full series. We also test inject after changing a patient's position as this may have altered the location of the catheter tip. Aortic arch or other midstream injections are best done through a pigtail catheter [62] so as to avoid catheter whip or the inadvertent entry into a small branch.

In our series minor degrees of dissection (Fig. 2.5) have occurred during subselective catheterization in almost 10 per cent of examinations. This high incidence had been brought about by employing a deflector guide wire for subselective entry into atheromatous vessels. In such patients we now use specially shaped catheters and restrict the deflector to aid with preliminary selective catheterization. Major dissections have occurred in two patients. The films in Fig. 2.6 are those of a seriously ill patient, with jaundice, bloodstained ascites and a distended, tender abdomen. The coeliac axis was readily catheterized, and a test injection showed nothing of note. However, the television detail was poor at the time and a control injection film ought to have been done. The pressure injection of contrast medium was attended with increasing dissection which involved the aorta, coeliac axis and main branches. It was still possible to take the catheter into the common hepatic artery and demonstrate a large pancreatic carcinoma with liver secondaries. The patient died ten days later. Autopsy demonstrated areas of liver infarction related to secondary deposits but major arteries were patent.

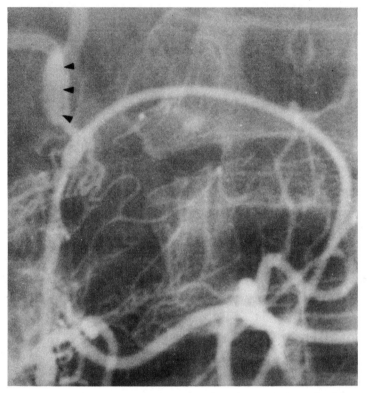

FIGURE 2.5. Small dissection in the proper hepatic artery (arrowheads) produced in the course of subselective catheterization of the gastroduodenal artery.

Spinal cord damage

The tragic complication of paraplegia may result from the entry of a high concentration or of a large bolus of contrast medium into vessels supplying the spinal cord. Two mechanisms are implicated:

1 Direct injection into a lumbar or intercostal artery which happens to give origin to the major anterior radicular artery of Adamkiewicz [63]. This vessel feeds into the anterior spinal artery and arises at any level between T8 and L4, more often on the left side.

2 Diversion of contrast medium into the spinal vascular bed in the presence of an organic outflow obstruction in the aorta or iliac arteries or in response to vasopressor drugs which constrict somatic vessels and do not affect the spinal vasculature [64]. Flow of contrast medium to the spinal cord is gravitationally increased in the supine position.

Incidence

McAfee [3] reported major neurological complications in 0.22 per cent of examinations using older, more toxic contrast media. Such complications however, do still occur at the present time [10, 65–67], and it is believed that published reports do not reflect the real incidence. We have been fortunate enough not to have encountered a single example.

Presentation

The relatively minor complication of transient spasticity of the legs was observed by Cornell [68] in 5 of 1014 aortographies. Tonic and clonic contractions and a Babinsky reflex are a manifestation of more severe cord damage and may progress to paraplegia. Paraplegia may appear without perliminary signs of spinal irritation, particularly in angiograms performed under general anaesthesia.

Treatment

The most promising emergency treatment is that described by Mishkin [66] and is intended to correct the raised iodine content of the cerebrospinal fluid found during neurological complications. An immediate lumbar puncture is performed with the patient in a head-up position, CSF is repeatedly withdrawn in 10-ml aliquots and replaced by equal volumes of normal saline. Head-high position is maintained after withdrawals. A return to near-normal iodine levels of the CSF accompanies clinical improvement.

Prevention

Unnecessary midstream aortograms should be avoided; it is not necessary to perform a 'flush aortogram' before every selective catheter study. In the presence of outflow restricting atheroma, the quantity of contrast medium injected into the aortic midstream should be kept low. Inadvertent catheter

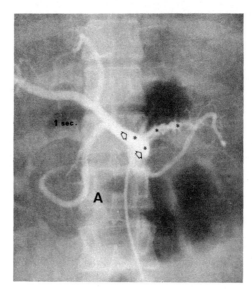

a

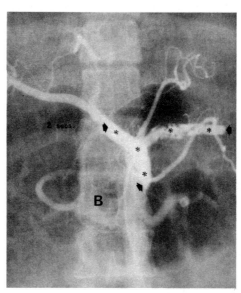

b

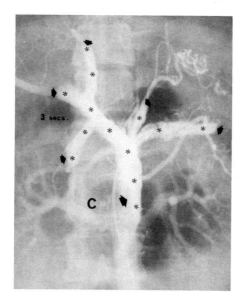

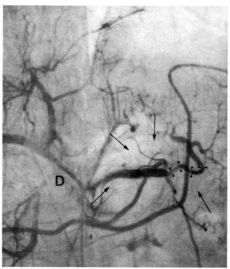

d

FIGURE 2.6. Major dissection. Coeliac catheterization, dissecting position of the catheter tip unsuspected at test injection. a–c. Show increasing dissection during pressure injection of contrast medium (asterisks over area of dissection, arrows to its margin). d. Catheter then passed into the common hepatic artery. Large carcinoma tail and body of pancreas (large arrows). Dissection into the splenic artery has been modified and limited by tumour encasement (arrowheads). Subtraction print.

entry into a lumbar or intercostal artery often occurs during attempts to catheterize visceral arteries selectively; initial test doses of contrast medium should be in the 1–2 ml range. Diagnostic injections into arteries from which radiculo-medullary branches to the cord may originate should employ the smallest quantity of contrast medium at the lowest useful concentration. Spinal cord damage is discussed in greater detail in Chapter 5.

Other complications

False aneurysm

This presents as a pulsatile mass caused by persistent communication between artery lumen and a space within an organized clot. The diagnosis is not usually made until a week or two after the angiogram. A bruit or thrill is typically present but may also be found when a simple haematoma causes artery compression. Treatment is by surgical repair. It is an unusual complication and we have never seen it.

Arteriovenous fistula

An extreme rarity at the puncture site; only one case was reported by Meaney [20].

Artery perforation

This occasionally occurs when a straight guide wire is taken up through a tortuous and atheromatous vessel. Simple guide wire perforations are not usually serious though the patient ought to be observed carefully over the next 24 hours. Where a guide wire perforation is not recognized, and a catheter is forced through the artery wall and contrast medium injected outside the artery, the consequences would be more severe and surgery may have to follow. All forms of perforation are readily avoided by the observance of basic principles of angiographic care.

Gastrointestinal

Bleeding may be provoked by the intra-arterial injection of a vasodilator substance. In two of our patients with portal hypertension tolazoline has been held responsible for a resumption of variceal bleeding. *Ileus* of small and large bowel occurs occasionally. In an azotaemic patient on long-term dialysis and warfarin, an ileus followed arteriography to show a possible bleeding site [70]. The toxic effect of prolonged elevation of the contrast medium blood level together with its vicarious excretion across the intestinal tract were believed responsible. The acute syndrome resolved after dialysis.

Infection

Angiography is carried out under aseptic conditions and this should comprise sterile covering of X-ray controls used by the radiologist. In addition to towelling, we apply a Steri-Drape square[1] round the puncture point in male patients to prevent contamination by the penis when the patient is turned into the oblique position. Puncture site infection is most unusual and it has never happened here. A haematoma in the groin or in relation to an arterial perforation may become infected via the blood stream.

Catheter knotting

During brachiocephalic catheterization rare cases of catheter knotting have been reported. Non-surgical unknotting has been described, either using a deflector guide wire [71] or another catheter introduced from the opposite femoral artery [72].

Catheter impaction

Catheter impaction may lead to chemotoxic damage if associated with overinjection into a small vessel of a sensitive organ like the kidney. Impaction occluding flow through a mesenteric branch will often produce ischaemic pain which should be relieved by immediate catheter withdrawal.

Guide wire breakage

Detachment of the flexible leader has been reported in the past and should not happen with present safety guide wires [73]. Pre-use testing is, however, still necessary.

[1] Steri-Drape No. 1020, made by the 3M company.

DIRECT FEMORAL ARTERIOGRAPHY

The common femoral artery is entered either by a Teflon sleeve needle[1] or a 6F polyethylene catheter introduced a short distance. Some 885 femoral arteriograms have been done by members of this department.

We no longer employ direct puncture in patients whose ipsilateral femoral pulse is of poor quality:
(a) atheroma usually exists above femoral level and may extend into the aorta; it is better demonstrated from above.
(b) the examination is less likely to be free of complication.

Thrombosis

The type of thrombosis found in transfemoral catheterization does not occur. Only a short length of tubing is introduced; it is of narrow gauge and the examination is brief.

Damage to atheromatous plaques

It has been claimed that plaque elevation may occur during puncture, cause subintimal bleeding and lead to thrombosis. Sutton [74] was able to inspect artery puncture sites at autopsy and found no evidence to support this belief, even in cases where needle entry had been made through plaques.

It may, however, be possible to dislodge atheromatous material and cause further impairment of an already inadequate peripheral blood supply. This complication is avoided by never advancing the catheter or Teflon sleeve until after a resistance-free insertion into the lumen of a fine J-type guide with its core wire pulled back. Contrast medium should be injected at hand pressure.

In two patients examined without this safeguard and whose pulses were of poorer quality than we would now admit to the direct procedure, the state of peripheral vascularity deteriorated rapidly, presumably because of embolization of atheromatous debris.

Bleeding and haematoma

Transfemoral catheterization is more likely to cause bleeding and haematoma than direct femoral arteriography. In the latter procedure, no catheter

---

[1] Longdwell, $18G \times 4$ inch, thin wall. Becton, Dickinson & Co., Rutherford, N.J., U.S.A.

changes are needed, duration is shorter, the arterial opening smaller. The same post-withdrawal routine is required.

Dissection

Should return of blood be unsatisfactory at first puncture, it is best to withdraw and repuncture after short compression. The test injection of a small quantity of contrast medium under vision further eliminates the risk of dissection.

In centres where high-pressure retrograde aortography from the femoral artery is practised [75], dissection may result from the contrast medium jet being delivered too close to a plaque of atheroma. Cope [76] has described a safety device which ensures the alignment of the Teflon cannula in the long axis of the artery lumen.

Major complications should not occur in direct femoral arteriography provided unsuitable patients are excluded. We have found the Teflon sleeve needle to be the safest and most convenient instrument for use in this examination.

TRANSLUMBAR AORTOGRAPHY

The technique of the examination is basically that described by Dos Santos *et al.* in 1931 [77]. It was the predominant aortographic method until mid-1950; the series of 13 207 angiograms reported by McAfee in 1957 [3] included 12 832 aortograms by the lumbar route. Subsequently, transfemoral catheterization displaced it in most centres but it soon became evident that, especially in patients with lower limb ischaemia, the translumbar aortogram provided a method which was safer and just as simple and adequate.

Reports on complications vary greatly. Lindgren in 1953 [78] seems to have encountered no major complications at all but does not mention the number of cases done. Weyde [79] performed 210 examinations and Goetz [80] 700 without serious complication. Kottke [6], in the course of 80 aortograms, found three peri-aortic injections at the test dose stage and two modest retroperitoneal haematomata at the end of the procedure; he considered it a safer method than transfemoral catheterization. In 1972 Reiss [7] reported five major complications with 427 translumbar aortograms, an incidence of 1.2 per cent; this was slightly better than the complication rate for transfemoral catheterizations given in the same paper (1.5%).

Other authors have been less optimistic. Crawford *et al.* [81] recorded nine examples of renal complication among 300 aortograms; three of them were fatal. Renal damage was the predominant cause of the fatalities and major complications in the series reviewed by McAfee [3]. Sutton [74] felt that the dangers of aortography have been underestimated but found only two major complications (dissection) in his own 1500 aortograms. Schinz et al. [82] in their text book published in 1968 state: '... the percutaneous puncture of the aorta ... makes the examination considerably simpler, but also more hazardous'.

Complications due to the puncture

Either a short, bevelled needle or a Teflon sleeve needle is used to puncture the aorta. It is important that they are of sufficient rigidity to ensure that the needle tip does not become deflected from the line of entry and is able to respond to any alterations made to the angle of entry. The Teflon sleeve should be 16 gauge (1.6 mm outer diameter) and 24 cm long.

A few examples of inadvertent puncture of other structures have been recorded. In two instances [83] excessively high needle insertion led to lung penetration and puncture of the thoracic aorta with profuse post-withdrawal bleeding. Pancreatitis, pneumothorax, chylothorax, even cardiac puncture have been reported. Entry into the spinal canal has occurred with disastrous results when contrast medium was injected as a localizing test dose [84].

Careful observance of the puncture technique clearly described by several authors [74, 80, 85] will avoid such complications.

Dissection and perithecal injection

Dissection is usually of minor significance. However, Gaylis & Laws [86] have shown that a dissecting aneurysm may follow the subintimal injection of contrast medium; the obliteration of important side branches may be a further and possibly disastrous consequence [87, 88].

The tip of the needle or Teflon sleeve is likely to have been introduced into subintimal or transmural position at the time of primary puncture in those cases in whom dissection or perithecal injection are demonstrated. A preliminary test dose of contrast medium under image intensifier control must be made even where backflow of blood is satisfactory. Only a clearly shown midstream delivery of the test dose should be accepted. A röntgenogram should be taken if there is any doubt. If necessary, the position of the instrument should be adjusted or the puncture repeated once only at a different level.

With the tip of the needle or Teflon sleeve shown to be in mid-lumen, further dislodgement is not likely in the course of the procedure. Thomas & Andress [89], who turn their patients into oblique positions to demonstrate profunda femoris origins, have never seen displacement, the needle being firmly held by the muscles of the lumbar region. We would, however, in this examination prefer to angle the tube rather than rotate the patient. It has been suggested [82] that vasoconstriction in the conscious patient could bring needle and aortic wall into closer apposition and lead to dissection. However, the tunica media of the aorta is almost entirely made up of elastic tissue and aortic constriction is not feasible.

It is important not to inject the main bolus at delivery rates which exceed hand pressure significantly. A safer type of needle, with blocked tip and with side hole [90], has distinct advantages and is widely used. Dissection can be avoided if meticulous care is taken.

Injection into an aortic branch

This was a potent source of fatal and serious complication when more toxic contrast media were in use. Irreversible damage was reported in kidney and bowel and mostly associated with high aortic puncture. Dos Santos referred to it as the 'zone dangereuse'.

Even when using newer contrast media it is possible to damage kidney or bowel by the direct injection of an excessive quantity or concentration. A test injection will prevent it.

Haematoma

Significant bleeding after completion is unlikely even in heavily calcified aortas. The considerable elastic material in its wall will help to seal the small opening.

Patients with severe hypertension, a possible aneurysm, with bleeding diathesis and those on coumarin drugs are not suitable for translumbar aortography as more severe retroperitoneal bleeding may be produced.

Neurological damage

Killen & Foster in 1960 analysed 38 cases of spinal cord damage collected from the literature and from colleagues in response to a questionnaire [91].

Direct injection into a lumbar artery is the likeliest cause of cord damage at the present time and can be prevented by pretesting. The prone position of the patient helps to protect the spinal vasculature from contrast medium entry during midstream delivery. It is commonly agreed that the quantity and concentration of contrast medium should not be greater than is required to obtain diagnostic information. While larger doses may produce more pleasing radiographs, they increase the danger of damage to the cord, the kidney and bowel, and this particularly where outflow is impeded by aorto-iliac atheroma. Yet it is just in these patients that the full extent of an occlusion and the state of the distal arteries need to be demonstrated so that surgery can be planned. The larger dose of contrast medium required for this purpose is justified by the clinical need. No complications have followed larger volume aortography in 36 patients in whom the test injection had revealed aortic occlusion [85].

Translumbar catheterization

Stocks *et al.* [92] have modified the Teflon sleeve technique to carry out retrograde catheterization of the aortic arch and selective entry into brachiocephalic branches. This has been used in fifteen patients without complication.

In our series of 353 translumbar aortograms we have seen two perithecal injections of the test dose, one fairly extensive dissection and one patient who developed a retroperitoneal haematoma. There were no serious consequences and surgery was not required. Translumbar aortography is now widely used in this country and is potentially the safest method of aortography in patients with aorto-iliac disease.

TRANSAXILLARY CATHETERIZATION

The transaxillary method has many advocates and the following advantages are claimed for it:
(a) it can be performed where transfemoral catheterization is contraindicated, in severe aorto-iliac atheroma or with an aneurysm of the abdominal aorta;
(b) the antegrade approach may facilitate the selective and subselective catheterization of the unpaired branches of the abdominal aorta. Of fourteen patients in whom the transfemoral approach failed or proved inadequate, axillary catheterization was successful in twelve [93];
(c) thrombosis of the axillary artery is a less serious complication than that

of the femoral. Usually, adequate collateral flow evolves through the humeral circumflex arteries;

(d) a short catheter can be used for the aortic arch area, giving better manipulative control [94].

Like most other centres we were concerned when Staal [95] first reported the complication of permanent brachial plexus injury in 1966. It made us reserve the axillary approach for only those patients in whom transfemoral or translumbar examinations could not be done. In consequence, only 64 transaxillary angiograms have been performed by the author.

Thrombosis

The mechanism is that described for transfemoral catheterization. Catheter length and width are important factors. Being an antegrade approach, the continual heparin infusion will be less effective in preventing deposition of thrombus on the catheter surface. Systemic heparinization may be unwise as haematoma formation could be promoted. Heparin-coated catheters should be employed.

Spasm of the axillary artery occurs not infrequently after catheter withdrawal [96]. This could promote the size and significance of the strip thrombus. It can be avoided by injecting 5 ml of 1 per cent procaine through the catheter immediately before its withdrawal. Signs of cortical irritation may follow if the injection is made too close to carotid/vertebral origins. The catheter should first be pulled back towards the puncture site and the injection given slowly. Intra-arterial tolazoline is also effective in preventing arterial spasm [69].

Of 1762 examinations reported by Molnar & Paul [97] thrombosis occurred in nine. In five of the patients brachial pulses became normal within 24 hours; in a further three patients pulsation was not re-established but no ischaemia was evident. In only one patient was it necessary to proceed to surgery. Among the 239 axillary arteriograms done by Westcott and Taylor [94] there was only one case of thrombosis in need of surgery. Brachial artery catheterization is more dangerous in this respect as collateral flow seems less adequate and spasm more likely. In almost 4000 brachial angiograms [98] there were twenty thromboses of whom eight required surgery.

None of our patients developed thrombosis or post-withdrawal spasm. We have not used catheters larger than 6F. Like Molnar & Paul we puncture the axillary artery distal to the origin of the circumflex branches in order to preserve a potential collateral flow channel.

Haematoma

The axillary artery is a fairly mobile structure and readily traumatized during puncture and catheterization. Post-withdrawal compression may also be more difficult unless the more accessible third part of the artery has been punctured. A fairly large haematoma may form in the loose tissue space of the axilla before becoming obvious, and this particularly in obese patients.

Axillary artery compression should be done above puncture level and should achieve pulse obliteration. Pressure must be relaxed periodically just sufficient to allow peripheral pulses to be felt. Compression should continue for at least fifteen minutes after bleeding has stopped, and must be followed by observation for a period of at least an hour. It is wrong to cover the area with a compression bandage.

A false aneurysm is not an uncommon further development of a haematoma and requires surgery.

Brachial plexus injury

Two mechanisms exist: (1) By far the most important is plexus compression by a haematoma or false aneurysm. This possibility gives added relevance to the prevention of haematoma formation in transaxillary catheterization. Even a small haematoma can produce nerve damage. (2) Traumatization during axillary puncture. To a large extent this can be avoided by entering the artery in its distal third where plexus overlap is usually absent [97].

The first reported neurological complications involved seven patients in only twenty-one examinations; in two patients paralysis ensued [95]. Two of 305 patients examined by Dudrick *et al.* [99] developed sensory and motor impairment and five examples of nerve damage were reported by Kerdiles *et al.* [100].

Early diagnosis is all important. Tingling and numbness in the arm frequently results from local anaesthesia, posture and post-withdrawal compression. It should clear within one to two hours. Any residual paraesthesia or its development after a free interval must be *regarded as abnormal*. Minor degrees, possibly even associated with slight motor weakness of the fingers—provided there is no progression—may be kept under observation and may clear spontaneously. Of major significance is a paresis which appears within hours or days of the examination, usually in association with a haematoma, and seems to get worse. This must be recognized without delay since surgical decompression of the axillary neurovascular sheath is then an urgent requirement and the only possible way to prevent progression to paralysis.

In the large series reported by Molnar & Paul [97] there were seven patients with progressive paresis, associated with a haematoma in five and a false aneurysm in two. All came to surgery, four recovered but three failed to respond. As functional recovery depends on early recognition of plexus damage and on prompt surgery, it is advisable to instruct the patient to report to the nursing staff any development of tingling, pain, numbness or weakness in the arm after return to the ward, as this would indicate haematoma formation. Arrangements should be made for the radiologist to be called to the patient immediately should this occur. By their preventive care during and after axillary catheterization, Molnar & Paul [97] were able to perform the last 602 angiograms of their series without brachial plexus paralysis occurring in a single patient. No nerve damage has taken place in the small series done by the author.

Embolism

Dislodgement of smaller athero-emboli or of thrombi forming on guide wires may occur. Two patients developed hemiparesis [97] and one patient hemianopia [94]. A floppy J-type guide wire introduced through the cannula should enter the aorta without encountering resistance, and its advance should always be checked fluoroscopically. Only then is it right to introduce the catheter. When guide wire passage is resisted the examination had better be abandoned. This will prevent athero-embolism and possible dissection. Guide wires used within the catheter during later manoeuvres should be withdrawn after about 30 seconds and thoroughly rinsed. It would be safer still to use heparin-coated guide wires.

Patient selection

Axillary aortography is attended with a higher complication rate in the following circumstances [101]:
(a) approach from the lower pressure side, in patients in whom significant pressure differences exist;
(b) in patients with severe hypertension;
(c) obesity (greater difficulty of catheterization and control of bleeding);
(d) with pronounced atheroma, right innominate/subclavian tortuosity, marked elongation of the aortic arch;
(e) in patients on coumarin drugs.

It would be necessary to study critically the indications and purpose of the examination in such patients and to take particular care over the prevention of haematoma formation in the department and after return to the ward.

CONCLUSION

Established ideas in Medicine take a very long time to adapt to changed circumstances. In skilled hands arteriography is not now the relatively perilous procedure it was 20 years ago.

In *transfemoral catheterization* thrombosis has been the main danger and can now be avoided. Dissection, a not infrequent and only rarely serious complication, can be held at low incidence by greater care in the technique of selective and subselective catheterization.

Direct femoral arteriography carries no dangers provided it is not done through significantly diseased arteries.

Translumbar aortography is the method of choice in aorto-iliac-femoral disease. Dissection is the only real hazard and is totally preventable.

Transaxillary catheterization demands close attention to technique. The possibility of brachial plexus damage from haematoma will continue to be with us and must be countered by well-organized post-angiographic observation and, where indicated, by early recourse to surgery. The procedure is not recommended where other approach methods are open.

In *the geriatric age group* angiography may be an attractive alternative to examinations like barium enema which can impose a considerable strain on the patient. Thrombosis is not a problem. Harm can be done where advanced atheroma impedes selective catheter entry and the radiologist perseveres for too long. Our own series of transfemoral catheter studies includes 148 patients aged 70 years and over; none developed major complications.

ACKNOWLEDGEMENT

The author records his indebtedness to his clinical colleagues and to the consultant radiologists, particularly Dr J. N. Glanville, who have carried out many of the examinations reviewed here. Sincere appreciation is due to the radiographers and nurses whose skill and zest have been invaluable.

REFERENCES

1 FARINAS P.L. (1941) A new technique for the arteriographic examination of the abdominal aorta and its branches. *Am. J. Roent.* 46, 641–645.

2 SELDINGER S.I. (1953) Catheter replacement of the needle in percutaneous arteriography. A new technique. *Acta Radiol.* **39**, 368–376.
3 MCAFEE J.G. (1957) A survey of complications of abdominal aortography. *Radiology* **68**, 825–835.
4 LANG E.K. (1963) A survey of the complications of percutaneous retrograde arteriography. *Radiology* **81**, 257–263.
5 HALPERN M. (1964) Percutaneous transfemoral arteriography. An analysis of the complications in 1,000 consecutive cases. *Am. J. Roentg.* **92**, 918–934.
6 KOTTKE B.A., FAIRBAIRN J.F. & DAVIS G.D. (1964) Complications of aortography. *Circulation* **30**, 843–847.
7 REISS M.D., BOOKSTEIN J.J. & BLEIFER K.H. (1972) Radiologic aspects of renovascular hypertension. Arteriographic complications. *J. Am. med. Ass.* **221**, 374–378.
8 LANG E.K. (1967) Prevention and treatment of complications following arteriography. *Radiology* **88**, 950–956.
9 ABRAMS H.L. (1957) Radiologic aspects of operable heart disease. The hazards of retrograde thoracic aortography. *Radiology* **68**, 812–824.
10 ANSELL G. (1968) A national survey of radiological complications; Interim report. *Clin. Radiol.* **19**, 175–191.
11 CRAMER R., MOORE R. & AMPLATZ K. (1973) Reduction of the surgical complication rate by the use of a hypothrombogenic catheter coating. *Radiology* **109**, 585–588.
12 BEHAR V.S. (1970) Symposium on contrast media toxicity. *Invest. Radiol.* **5**, 493.
13 BAUM S., STEIN G.N. & KURODA K.K. (1966) Complications of 'no arteriography'. *Radiology* **86**, 835–838.
14 SIEGELMAN S.S., CAPLAN L.H. & ANNES G.P. (1968) Complications of catheter angiography. *Radiology* **91**, 251–253.
15 AMPLATZ K. (1968) Catheter embolisation. Editorial. *Radiology* **91**, 392–393.
16 JACOBSSON B. & SCHLOSSMAN D. (1969) Thromboembolism of leg following percutaneous catheterisation of the femoral artery for angiography. Predisposing factors. *Acta Radiol. Diag.* **8**, 109–118.
17 YELLIN A.E. & SHORE E.H. (1973) Surgical management of arterial occlusion following percutaneous femoral angiography. *Surgery* **73**, 772–777.
18 HAWKINS I.F. (1972) 'Mini-catheter' for femoral run-off and abdominal arteriography. *Am. J. Roentg.* **116**, 199–203.
19 FORMANEK G., FRECH R.S. & AMPLATZ K. (1970) Arterial thrombus formation during clinical percutaneous catheterisation. *Circulation* **41**, 833–839.
20 MEANEY T.F. (1973) Percutaneous femoral angiography. In: *Complications and legal implications of radiologic special procedures* (Ed. by MEANEY T.F., LALLI A.F. & ALFIDI R.J.), ch. 2. C. V. Mosby, St. Louis.
21 WEILL F., RICATTE J-P., KRAEHENBUHL J.R., BECKER J-C., PREVOTAT N. & GILLET M. (1972) In whom is arterial catheterisation dangerous? *Nouv. Presse Med.* **1**, 2247.
22 MEANEY T.F. (1974) Complications of percutaneous femoral angiography. *Geriatrics* **29**, 61–64.
23 NACHNANI G.H., LESSIN L.S., MOTOMIYA T. & JENSEN W.N. (1972) Scanning electromicroscopy of thrombogenesis on vascular catheter surfaces. *N. Eng. J. Med.* **286**, 139–140.
24 OLBERT F., DENCK H. & WICKE L. (1973) Complications during catheter angiographies. Causation and treatment. *Wien. Med. Wchschr.* **123**, 293–298.

25 MOORE C.H., WOLMA F.J., BROWN R.W. & DERRICK J.R. (1970) Complications of cardio-vascular radiology: Review of 1,204 cases. *Am. J. Surg.* **120**, 591–593.
26 JACOBSSON B., PAULIN S. & SCHLOSSMAN D. (1969) Thromboembolism of leg following percutaneous catheterisation of the femoral artery for angiography. Symptoms and signs. *Acta Radiol. Diagn.* **8**, 97–108.
27 BEVEN E.G. (1973) Surgical treatment of angiographic complications. In: *Complications and legal implications of radiologic special procedures* (Ed. by MEANEY T.F., LALLI A.F. & ALFIDI R.J.), ch. 5. C. V. Mosby, St. Louis.
28 SNYDER C. & AMPLATZ K. (1973) Non-surgical treatment of post-catheterisation femoral artery occlusion. A new technique. *Am. J. Roentg.* **119**, 590–596.
29 O'BRIEN J.R. (1968) Effects of salicylates on human platelets. *Lancet* **1**, 779–783.
30 KRICHEFF I.I., ZUCKER M.B., TSCHOPP T.B. & KOLODJIEZ A. (1973) Inhibition of thrombosis on vascular catheters in cats. *Radiology* **106**, 49–51.
31 MUSTARD J.F., KINLOUGH-RATHBONE R.L., JENKINS C.S. & PACKHAM M.A. (1972) Modification of platelet function. *Ann. N.Y. Acad. Sci.* **201**, 343–359.
32 HARKER L.A. & SLICHTER S.J. (1972) Platelet and fibrinogen consumption in man. *New Engl. J. Med.* **287**, 999–1005.
33 JACOBSSON B. (1969) Effect of pre-treatment with dextran 70 on platelet adhesiveness and thromboembolic complications following percutaneous arterial catheterisation. *Acta Radiol. Diagn.* **8**, 289–295.
34 DOUST B.D. & FISCHER H.W. (1972) The effects of dextran infusion, glucose infusion and state of hydration on cerebral toxicity of arteriographic contrast media. *Radiology* **103**, 607–609.
35 MURPHY D.J. (1973) Cerebrovascular permeability after meglumine iothalamate administration. *Neurology (Minneap.)* **23**, 926–936.
36 WILSON J. & ELLIS F.R. (1973) Oral premedication with lorazepam (Ativan): A comparison with heptabarbitone (Medomin) and diazepam (Valium). *Br. J. Anaesth.* **45**, 738–744.
37 HEDGES A., TURNER P. & HARRY T.V.A. (1971) Preliminary studies on the central effects of lorazepam, a new benzodiazepine. *J. Clin. Pharm.* **110**, 423–427.
38 GLANVILLE J.N. (1974) Personal communication.
39 KLATTE E.C., BROOKS A.L. & RHAMY R.K. (1969) Toxicity of intra-arterial barbiturates and tranquilising drugs. *Radiology* **92**, 700–704.
40 GOTT V.L. (1966) The causes and prevention of thrombosis on prosthetic materials. *J. Surg. Res.* **6**, 274–283.
41 ERIKSON J.C., GILLBERG G. & LAGERGREN H. (1967) New method for preparing non-thrombogenic plastic surfaces. *J. Biomed. Mater. Res.* **1**, 301–312.
42 BJORK L. (1972) Heparin coating of catheters against thromboembolism in percutaneous catheterisation for angiography. *Acta Radiol. Diagn.* **12**, 576–578.
43 CRAMER R., FRECH R.S. & AMPLATZ K. (1971) Preliminary human study with simple non-thrombogenic catheters. *Radiology* **100**, 421–422.
44 HAWKINS J.F. & KELLEY M.J. (1973) Benzalkonium-heparin coated angiographic catheters. *Radiology* **109**, 589–591.
45 OVITT T.W., DURST S., MOORE R. & AMPLATZ K. (1974) Guide wire thrombogenicity and its reduction. *Radiology* **111**, 43–46.
46 DURST S., JOHNSON B.S. & AMPLATZ K. (1974) The effect of silicone coating on thrombogenicity. *Am. J. Roentg.* **120**, 904–906.

47 WALLACE S., MEDELLIN H., DEJONGH D. & GIANTURGO C. (1972) Systemic heparinisation for angiography. *Am. J. Roentg.* 116, 204–209.
48 NEJAD M.S., KLAPER M.A., STEGGERDA F.R. & GIANTURGO C. (1968) Clotting on the outer surfaces of vascular catheters. *Radiology* 91, 248–250.
49 JEFFERY R.F. (1972) A simple method of removing intra-arterial clots formed during catheterisation. *Radiology* 103, 573–575.
50 CORNIER J-M. & LAGNEAU P. (1972) Severe complications due to retrograde percutaneous arteriographies. *J. Chir. (Paris)* 104, 395–412.
51 ANDERSON P.E. (1969) Ischaemic colitis caused by angiography. *Clin. Radiol.* 20, 414–417.
52 JUDKINS M.P., KIDD H.J., FRISCHE L.H. & DOTTER C.T. (1967) Lumen-following safety J-guide for catheterisation of tortuous vessels. *Radiology* 88, 1127–1130.
53 LONNI Y.G.W., MATSUMOTO K.K. & LECKY J.W. (1969) Post-aortographic cholesterol (atheromatous) embolisation. *Radiology* 93, 63–65.
54 BUXTON D.R. & MUELLER C.F. (1974) Removal of iatrogenic clot by transcatheter embolectomy. *Radiology* 111, 39–41.
55 MCCONNELL R.W., FORE W.W. & TAYLOR A. (1973) Embolic occlusion of the renal artery following angiography. *Radiology* 107, 273–274.
56 ADAMS D.F., OLIN T.B. & KOSEK J. (1965) Cotton fibre embolisation during angiography. *Radiology* 84, 678–681.
57 BERGERON R.T. & RUMBAUGH C.L. (1971) Air embolism associated with the use of malfitting plastic connectors in angiography. *Radiology* 98, 689–690.
58 OLSON J.R., MILLER R.E. & MANDELBAUM I. (1971) Angiographic haematomas complicating surgery. *J. Ind. St. Med. Assoc.* 64, 123–124.
59 GILLANDERS L.A. (1963) Aortic dissection during translumbar and retrograde abdominal aortography. *Br. J. Radiol.* 36, 725–728.
60 Editorial. (1965) Mechanical hazards in catheter aortography. *Circulation* 32, 876–877.
61 ROESCH J. & GROLLMAN J.H. JR (1969) Superselective arteriography in the diagnosis of abdominal pathology: Technical consideration. *Radiology* 92, 1008–1013.
62 JUDKINS M.P. (1968) Percutaneous transfemoral selective coronary arteriography. *Radiol. Clin. N. Amer.* 6, 467–492.
63 ADAMKIEWICZ A. (1882) Blood vessels of the human spinal cord. *S.-B. Akad. Wiss. Wien, math.-nat. Kl.* 85, 101–130.
64 MARGOLIS G. (1970) Pathogenesis of contrast media injury: Insights provided by neurotoxicity studies. *Invest. Radiol.* 5, 392–406.
65 Editorial. (1973) Spinal cord damage after angiography. *Lancet* 2, 1067–1068.
66 MISHKIN M.M., BAUM S. & DI CHIRO G. (1973) Emergency treatment of angiography-induced paraplegua and tetraplegia. *N. Engl. J. Med.* 288, 1184–1185.
67 BRODEY P.A., DOPPMAN J.L. & BISACCIA L.J. (1974) An unusual complication of aortography with the pigtail catheter. *Radiology* 110, 711.
68 CORNELL S.H. (1969) Spasticity of the lower extremities following abdominal aortography. *Radiology* 93, 377–379.
69 PARVER M., STARGARDTER F.L. & GLICKMAN N.G. (1974) The use of tolazoline hydrochloride (Priscoline) in the treatment of complications at arteriography. *Brit. J. Radiol.* 47, 299.
70 FALCHUK C.K. & FALCHUK Z.M. (1974) A complication of angiography in chronic renal failure: the acute abdomen. *Am. J. Gastroent.* 61, 223–225.

71 HAWKINS I.F. & TOMKIN A. (1973) Deflector method for non-surgical removal of knotted catheters. *Radiology* **106**, 705.
72 CHINICHIAN A., LIEBESKIND A., ZINGESSER L.H. & SCHECHTER M.M. (1972) Knotting of an 8-French 'head hunter' catheter and its successful removal. *Radiology* **104**, 282.
73 DOTTER C.T., JUDKINS M.P. & FRISCHE L.H. (1966) Safety guide spring for percutaneous cardiovascular catheterisation. *Am. J. Roentg.* **98**, 957–960.
74 SUTTON D. (1962) Arteriography. E. & S. Livingstone Ltd., Edinburgh and London.
75 WALDHAESEN J.A. & KLATTE E.C. (1962) Direct retrograde femoral aortography. *N. Engl. J. Med.* **267**, 480–482.
76 COPE C. (1967) A new safety device for retrograde power femoral aortography. *Radiology* **88**, 797–798.
77 DOS SANTOS R., LAMAS A.C. & CALDAS P.J. (1931) L'arteriographie des membres et de l'aorte abdominale. Masson, Paris.
78 LINDGREN E. (1953) Lumbar aortography. In: *Modern Trends in Diagnostic Radiology* (2nd series) (Ed. by MCLAREN, J.W.), ch. 2. Butterworth & Co. Ltd., London.
79 WEYDE R. (1952) Abdominal aortography in renal diseases. A preliminary report. *Br. J. Radiol.* **25**, 353–359.
80 GOETZ R.H. (1964) In: *Translumbar aortography in Vascular Roentgenology* (Ed. by SCHOBINGER R.A. & RUTZICKA F.F. JR). The Macmillan Co., New York; Collier-Macmillan Ltd., London.
81 CRAWFORD E.S., BEALL A.C., MOYER J.H. & DE BAKEY M.E. (1957) Complications of aortography. *Surg. Gynec. Obstet.* **104**, 129–141.
82 WELLAUER J. (1968) In: *Roentgen Diagnosis* (Ed. by SCHINZ H.R., BAENSCH W.E., FROMMHOLD W., GLAUER R., UEHLINGER E. & WELLAUER J.), vol. 1, 2nd American edition, Grune and Stratton, New York and London.
83 POULIAS G.E. & STERGIOU L.E. (1968) Pulmonary infarction and haemothorax as a post-translumbar aortography sequel. *Br. J. Radiol.* **41**, 866–869.
84 DOROSHOW L.W., YOON H.Y. & ROBBINS M.A. (1962) Intrathecal injection, an unusual complication of translumbar aortography: case report. *J. Urol.* **88**, 438–441.
85 THOMAS M.L. & FLETCHER E.W. (1970) Large volume translumbar aortography in aortic occlusion. *Am. J. Roentg.* **109**, 541–548.
86 GAYLIS H. & LAWS J.W. (1956) Dissection of aorta as a complication of translumbar aortography. *Br. Med. J.* **2**, 1141–1146.
87 MCDOWELL R.F.C. & THOMPSON I.D. (1959) Inferior mesenteric artery occlusion following lumbar aortography. *Br. J. Radiol.* **32**, 344–346.
88 BOBLITT D.E., FIGLEY M.M. & WOLFMAN E.F. JR (1959) Roentgen signs of contrast material dissection of the aortic wall in direct aortography. *Am. J. Roentg.* **81**, 826–834.
89 THOMAS M.L. & ANDRESS M.R. (1972) Value of oblique projections in translumbar aortography. *Am. J. Roentg.* **116**, 187–193.
90 HUSNI E.A. (1958) New needle and safer method of arteriography. *Surg. Gynec. Obstet.* **106**, 627–630.
91 KILLEN A.D. & FOSTER J.H. (1960) Spinal injury as a complication of aortography. *Ann. Surg.* **152**, 211–230.
92 STOCKS L.O., HALPERN M. & TURNER A.F. (1969) Complete translumbar aortography. *Am. J. Roentg.* **107**, 835–839.

93 BOIJSEN, E. (1966) Selective visceral angiography using a percutaneous axillary technique. *Br. J. Radiol.* 39, 414–421.
94 WESTCOTT J.L. & TAYLOR P.T. (1972) Transaxillary selective 4-vessel arteriography. *Radiology* 104, 277–281.
95 STAAL A., VAN VOORTHUISEN A.E. & VAN DIJK L.M. (1966) Neurological complications following catheterisation by the axillary approach. *Br. J. Radiol.* 39, 115–116.
96 NG A.C. & MILLER R.E. (1971) Intra-arterial injection of procaine. *Am. J. Roentg.* 111, 791–793.
97 MOLNAR W. & PAUL D.J. (1972) Complications of axillary arteriotomies. *Radiology* 104, 269–276.
98 McBURNEY R.P., LEE L. & FEILD J.R. (1973) Thrombosis and aneurysm of the brachial artery secondary to brachial arteriography. *Am. Surgeon* 39, 115–117.
99 DUDRICK S., MASLAND W. & MISHKIN N. (1967) Brachial plexus injury following axillary artery puncture. *Radiology* 88, 271–273.
100 KERDILES Y., SIGNARGOUT J. & LOGEAIS Y. (1972) Nerve damage due to retrograde percutaneous axillary artery catheterisation. *J. Chir. (Paris)* 104, 323–330.
101 MOLNAR W. (1973) In: *Complications and legal implications of radiologic special procedures* (Ed. by MEANEY T.F., LALLI A.F. & ALFIDI R.J.), ch. 3. C. V. Mosby, St Louis.

CHAPTER 3: CARDIOPULMONARY ANGIOGRAPHY

A. J. S. SAUNDERS & J. D. DOW

INTRODUCTION

In this chapter the complications described in the literature in relation to cardiopulmonary investigations will be evaluated in the light of the experience gained at Guy's Hospital in which nearly 3200 of these cases were accurately documented and followed up. In the literature most of the complications and their incidence have been derived from specialized units: the department in which these investigations were performed is not in any sense specialized but undertakes all types of cardiovascular investigations, including peripheral, abdominal and cerebral. It is hoped that a description of the problems which have been encountered in the cardiopulmonary field in such a general unit will be of some value to other similar departments elsewhere. No mention will be made in this section of the complications which may arise during the introduction of catheters or of the general systemic effects of opaque media as these are dealt with elsewhere in the book.

Of the total of 3200 cases, nearly 800 comprise selective coronary arteriograms. Certain of the complications of this latter group will be dealt with separately as they appear to have a higher complication rate than the other investigations. These other investigations included right and left heart angiography, aortic root angiography and pulmonary arteriography. The prime purpose of carrying out each investigation was angiographic, but routine cardiac catheterization was also performed where appropriate. It has been reported [1] that there is no difference in the overall incidence of complications in patients in whom angiocardiography is performed compared with those in whom cardiac catheterization alone is carried out. Major complications have been described as occurring in 3.1 per cent [2] to 3.6 per cent [1] of all investigations.

PRELIMINARY CONSIDERATIONS

As a preliminary to the main sections it is, perhaps, worth while making certain comments on the general risks of cardiac catheterization and angiocardiography, with special reference to their prevention; and also to comment on certain procedures which have been described as carrying an overall increased risk.

CATHETER PROBLEMS

(1) Choice of catheter

As will become apparent in the section on perforation of the heart and great vessels, the choice of catheter is of extreme importance in any cardiopulmonary investigation. A catheter with an end hole, with or without side holes, is satisfactory for hand injections of opaque medium but is unsatisfactory for pressure injections [3]. The peak pressure at the tip of an end hole catheter often exceeds 500 mmHg at injection pressures of 100 psi and 1000 mmHg at 600–800 psi. It is the use of these catheters which leads very largely to major myocardial extravasation in right and left heart angiography (Fig. 3.1). The functional equivalent—i.e. the Ducor pigtail catheter [4], although it is much safer than an ordinary end hole catheter, is still not as safe as a catheter with side holes and a closed end (*vide infra*, perforation of the heart and great vessels). However, it is common clinical practice to catheterize the heart percutaneously from the leg and in this situation the Ducor pigtail catheter provides an acceptable alternative to a closed-end catheter for intracardiac pressure injections. Open-ended catheters may safely be used for intra-aortic injections.

(2) Manipulation of catheters

Even using closed-end catheters, care should be taken in manipulation. Some catheters (e.g. No. 8 NIH) are much more rigid than others (e.g. No. 9 NIH). In certain situations, e.g. right heart catheterization in major pulmonary embolism, when the right ventricle is partially obstructed and when its wall becomes extremely soft and easily penetrated, it is wise always to use the softest catheter which can be manipulated into the correct position. Even taking this precaution, perforation or trapping of the catheter in the right ventricular wall has occurred in 6 out of 91 cases with obstructed right

78   Chapter 3

heart outflow due to massive pulmonary embolism. Only one of these cases
had any clinical sequelae (*vide infra*).

(3) Catheter placement

No catheter should ever be allowed to remain in a vessel or in a cardiac
chamber without ascertaining its exact position. In addition to monitoring

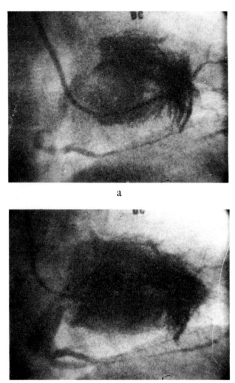

FIGURE 3.1. Gross extravasation in the left ventricle with rupture of a cardiac
vein and immediate filling of the major cardiac veins and coronary sinus, associated with a pressure injection through an open-ended catheter. The patient was
a man of 56, the investigation was left ventricular angiocardiography and
coronary arteriography using the Sones technique.

the pressure, it is desirable to check the catheter position by injection of 1 or
2 ml of opaque medium. Obstruction of a coronary orifice has occurred in this
series as a result of carelessly leaving a catheter in the region of the sinuses
of Valsalva after an aortic root injection, even although the monitor showed
a normal aortic trace.

No injection, pressure or hand, should ever be repeated without first checking the position of the catheter on fluoroscopy (Fig. 3.2). Particularly after a pressure injection, catheter recoil may occur, e.g. into a Thebesian vein. A second injection may then result in some disaster such as an extensive myocardial extravasation. This happened in one of the earliest cases in our

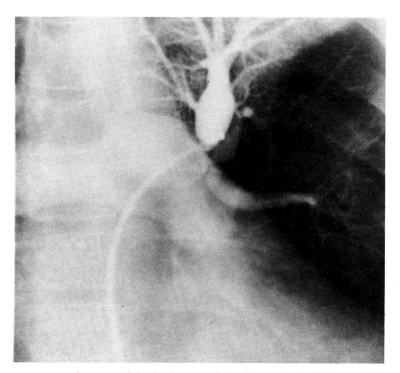

FIGURE 3.2. A pressure injection into a peripheral artery in the left upper lobe. The catheter was initially placed correctly in the main trunk of the pulmonary artery but in the middle of the series the Schonander film changer jammed. The series was then repeated without first checking that the catheter had not moved. The patient developed an upper lobe 'consolidation' but subsequently recovered completely.

series, before simultaneous antero-posterior and lateral angiocardiograms could be performed. In the antero-posterior injection catheter placement was correct; in the lateral, which followed immediately, extensive myocardial extravasation had occurred, presumably due to catheter recoil.

(4) Contrast medium

The volume and concentration of contrast medium used should be recorded throughout the examination and the total in adults should not exceed accepted limits (4 ml/kg of body weight of Renografin 76 or equivalent) provided dehydration is avoided [5]. In children the quantity injected should be very much less, particularly in the neonatal period when the hypertonicity of any opaque medium produces increased osmolality of the blood. This increase in serum osmolality leads to rapid passage of extravascular water into the intravascular space and such expansion of the plasma volume can be disastrous in infants already in heart failure. The raised serum osmolality and increased plasma volume take longer to return to normal in infants with congenital heart disease because of the combination of delayed excretion of the contrast medium resulting from impaired renal function, with the effects of cyanosis and heart failure. This problem occurred in 2 out of 67 infants investigated in the neonatal period in the Guy's series. Both suffered bradycardia and hypotension and one had convulsions. This infant died but the other recovered completely.

During prolonged catheterization and angiographic procedures, infants who are particularly at risk, but who are not in heart failure, can be protected to a certain extent by ensuring good hydration and not restricting fluids beforehand. In addition, the dose of contrast medium, including test doses, should not exceed 1.75 ml/kg of Triosil 75 or equivalent for each angiogram [6]; and the time between successive angiograms should be at least 25 minutes. Children in heart failure, who are more at risk from sudden expansion of plasma volume, should be fasted before catheterization and the smallest possible dose of contrast medium should be used.

(5) Catheter clotting

Clot may form either in or on the surface of a catheter [7] or may occur around a guide wire and be wiped off when the catheter has passed over it [8].

The occurrence of clot in a catheter can be minimized by heparinizing the patient [9, 10] and the solution used for perfusion, and by never allowing blood to remain in the lumen of the catheter. In particular, a partially occluded catheter should never be forcibly cleared; instead it should always be replaced. This is easy to do in open procedures but with percutaneous techniques the passage of a wire through a partially occluded catheter may itself dislodge clot: in these circumstances it is always better to withdraw the catheter completely and to repuncture. One of the advantages of open-ended catheters without side holes is that, providing constant forward perfusion is maintained throughout the procedure, clot is less likely to develop at the tip of the catheter. If, however, there are side holes present, clot may develop despite forward perfusion. Clot formation on guide wires can be reduced by the measures described above but particularly by using heparin-coated wires [11].

It is difficult to estimate how often some clot does develop in a catheter. Even if a pulmonary or cerebral complication arises, it is seldom possible to be certain whether it is due to dislodgement of clot in the heart or to passage of clot from the catheter. In this series, as will be described, there were three cases of cerebral emboli following left heart procedures. With right heart procedures there were two cases of sudden chest pain during catheterization, probably due to small pulmonary emboli. Subsequently one showed a small area of consolidation in the left lung, possibly embolic in origin; in the second case the lung fields remained radiologically clear. Both patients made uneventful recoveries. The problems of catheter-induced thrombus formation will be dealt with further in the section on coronary angiography.

(6) Breakage of catheter or wire

Breakages of the tip of a catheter [12] or of a wire [13] or of a transseptal needle [14, 15] have been reported, although the incidence of these occurrences is small. In this series no case of breakage of a catheter or of a wire occurred in the heart, but the tip of one wire broke off in a femoral artery and had to be retrieved surgically. Wire breakage in peripheral vessels is particularly likely to occur if a catheter is advanced over a wire whilst the tip is not lying free. All catheters should be checked prior to use and the number of times they have been used should be recorded. If they are carefully checked and discarded at the first sign of a fault, and not used too often, catheter breakage should be largely eliminated [15]. As far as guide wires are concerned, it is seldom that they should still be *in situ* when the catheter reaches the heart. Although the practice of probing a stenosed aortic valve with the flexible tip of a guide wire has been recommended in attempting to gain entry to the left ventricle [16], we and others consider this to be a dangerous practice and to carry the risk of coronary embolism and dissection.

(7) Knotting or buckling of catheter

Knotting of a catheter can occur [17], usually in cases of congenital heart disease with cardiac chambers in abnormal situations. This can be prevented in the great majority of cases by avoiding too many catheter loops in the heart and by taking care not to pull the loops too tightly. Buckling of catheters may also occur, particularly when large-bore catheters are used. On manipulation, their lack of torque tends to form a spiral. If they are then withdrawn without being carefully unwound, arterial damage may result. This possibility should always be considered when blood cannot be aspirated. Avoidance of these complications largely depends on the skill of the operator.

(8) Catheter causing orifice obstruction

In severe cases of pulmonary outflow tract obstruction, a catheter tip may be sufficient to cause total obstruction to pulmonary outflow [18] and similar near-total obstruction to outflow has been described by a catheter in the left side of the heart in aortic valvar stenosis [19]. Both these complications are uncommon and both are manifested by rapidly developing bradycardia and a falling blood pressure. It is difficult to know how frequently this occurs in small infants with congenital heart disease but certainly in our 67 cases of angiocardiograms performed in infants of 20 lb (9 kg) or less in weight, bradycardia and falling blood pressure occurred quite frequently. The immediate treatment is always the same: to withdraw the catheter and apply such resuscitative measures as are necessary.

(9) Miscellaneous

It is axiomatic that no catheter should be inserted into the region of the heart without constant ECG and pressure monitoring. Teflon-coated guide wires should be used whenever possible. Obviously, lubricated and heparinized catheters should be used as soon as they become available [11]. All catheter changes should be performed under fluoroscopic control. When Kifa-type catheters are used, the greatest care should be taken in inserting side holes. These should never be less than 1 cm apart and should always alternate on opposite sides of the catheter. Partial catheter tip breakage occurred in this series on two occasions. Once when two side holes were placed nearly opposite each other near the tip of the catheter, and secondly in a Sones catheter. In both instances the breakage was observed only when the catheter was withdrawn.

ACCIDENTAL INTRODUCTION OF AIR OR INFECTION

Air embolism has been described [12]. This complication, of course, can be lethal in the left side of the heart by causing either cerebral or coronary embolism. So far as is known, no case of air embolism in the left side of the heart has occurred in our series. In one patient, air was identified in the catheter during a right ventricular pressure injection. This was at a time when an old ram-type injector was being used and there has been no other similar episode with the more modern types of injectors. The patient suffered no apparent disability.

The control of infection depends very largely on the sterility of the equipment and the standard of theatre practice in any individual institution. Bacterial endocarditis from contamination of a pressure recording device has been described [20] and occurred in three cases out of 12 367 [21]. In our series, which was undertaken with full surgical sterility, significant localized wound skin infection occurred on seventeen occasions but no case of arterial or vein infection occurred at the site of any procedure nor was there any generalized infection following a procedure. Every open procedure had routine antibiotic cover.

Electrical Hazards

These represent a very real danger to patients in respect of ventricular fibrillation [2]. A current of 0.3 milliamps (0.0003 amp) will produce only a faint tingling sensation on dry human skin but when there is any conductor, such as a catheter or guide wire in the heart, currents as small as 20 micro amperes (0.00002 amp) may induce ventricular fibrillation. Currents of this small magnitude cannot be detected by the operator and may go completely unrecognized. It is therefore vital that all electrical apparatus used in cardiac catheterization and angiocardiography should be earthed to a common grounding system which effectively penetrates the water table of the earth. One case of a patient receiving an electric shock occurred in the Guy's series through the anaesthetist's trolley coming into contact with a faulty foot switch. The anaesthetist received a shock and current passed through the non-static anaesthetic tube to the patient. T-wave inversion and bradycardia, accompanied by hypotension, occurred immediately. The current was switched off within seconds but for the next fifteen minutes the ECG changes remained. Eventually the ECG reverted to normal and both the blood pressure and pulse rate were back to normal in about half an hour. There were no late sequelae. However, after this incident a system of regular checks by qualified personnel was instituted as the only way to ensure that defects do not develop in the apparatus and that the equipment is safe to use. Electrical hazards are considered in greater detail in Chapter 22.

Procedures Carrying Special Risks

(1) Transseptal left heart investigation

Gorlin [22] quotes an incidence of 100 complete perforations of the heart in 12 367 investigations; and of these, 47 occurred in 1 765 transseptal procedures. All instances of perforations of the aortic root (17 cases) occurred in transseptal studies, as did 30 of the 33 cases of perforation of the right atrium. In this series there were also 17 other major complications. These were mainly extravasation of opaque medium in the wall of the left atrium, and deep thrombophlebitis involving the femoral veins through which the procedure was performed. A similar high incidence of complications has been described in most series in which transseptal investigations were performed. At Guy's, very few transseptal investigations have been done. We have preferred the alternative procedure of direct left ventricular puncture.

(2) Direct left ventricular puncture

Direct percutaneous left ventricular puncture is used mainly in cases of aortic stenosis, to determine the gradient between the left ventricle and the aorta, in those cases in whom the ventricle cannot be entered by retrograde catheterization of the valve because of severe stenosis. The procedure was originally described by Fleming and by Ross [23, 24]. In the latter paper, there were 2 deaths in 184 cases. Since then a further 200 cases have been performed in this series. One death occurred from haemopericardium, which developed ten days after the investigation, following discharge from hospital. In one other case there was a pericardial reaction similar to that which may occur after mitral valvotomy but the patient suffered no long-term ill effects. The commonest complication of the procedure, occurring in about 15 per cent of cases, is pericardial pain, presumably due to some pericardial bleeding. The pain usually passes off after 24–48 hours. Two cases of pneumothorax and one of haemothorax have also been reported [25]. Since the original descriptions, the technique has been modified to use a Teflon cannula over a fine needle: this appears to be safer and easier than using a rigid needle. The cannula can usually be guided to the aorta using a guide wire and this has proved to be an easy, acceptable and relatively safe method of obtaining the gradient in cases of aortic valvar stenosis. The essential requirement for safety in this procedure is, however, that it should be performed *only in cases with definite evidence of left ventricular hypertrophy*. Percutaneous puncture of a normal left ventricle can easily lead to tamponade. Levy [26] described a series of 122 direct left ventricular punctures. In those that went to surgery soon after the procedure 8 per cent were observed to have evidence of haemopericardium of between 50 and 200 ml. One patient required pericardiocentesis for tamponade. One death occurred in a six-year-old child due to an intramural injection leading to cardiac arrest.

We have found the practice of performing angiocardiography through the Teflon catheter, as described by Grossman [27], to be dangerous. In three out of five cases in which this procedure was attempted, the Teflon catheter

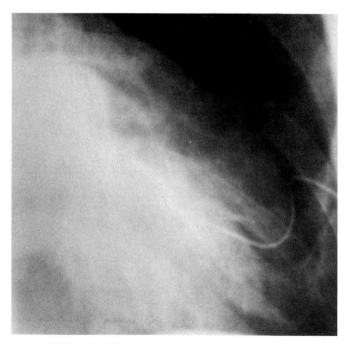

FIGURE 3.3. Left ventricular angiocardiography by the method of direct left ventricular puncture using a Beckton-Dickinson Long-dwel 18 gauge, 8 inch needle. In three of five cases with pressure injection the Teflon kinked as illustrated here and ruptured towards the end of the injection.

became kinked and ruptured on pressure injection (Fig. 3.3). Two of these cases suffered no ill effects but the third developed tamponade which was treated successfully by pericardiocentesis.

(3) Rashkind balloon septostomy

Rashkind [28] describes septostomy performed by this method in 31 infants without major complications. Transient arrhythmias occurred but no permanent conduction abnormality was induced. The procedure, however, is potentially dangerous particularly when the catheter is withdrawn through the atrial septum. Verification of the balloon position should be made by angiography, by sampling, by monitoring the atrial pressure and by observing the catheter position posteriorly at the upper left cardiac border. Further confirmation of its position can be made when entry into a pulmonary vein occurs. Complications have been reported by other authors including atrial wall perforation and tricuspid laceration [29]. Arrhythmias such as primary heart block, bradycardia, wandering pacemaker and nodal rhythm have also been recorded [30]. It has been suggested that a fast pull-back through the septum is less likely to produce an arrhythmia [29, 30].

At Guy's, atrial septostomy was performed in only eleven cases. Ten cases showed varying degrees of arrhythmia from which all recovered, but in one instance the tip of the catheter broke leaving the balloon in the left atrium. Thoracotomy was performed within an hour but the infant died.

Angiocardiography

The haemodynamic effects of contrast media injections have been well documented [31–33]. Haemodynamic effects of considerable magnitude are seen when contrast media are injected into the left ventricle or the proximal aorta. The cardiac output is increased by about 50 per cent, as is the stroke volume. The cardiac rate may remain the same or it may also be increased. The right and left atrial pressures become elevated as does the left ventricular end diastolic pressure. The blood volume expands and the peripheral blood flow increases. Initially the systemic arterial pressure increases and then falls as the systemic resistance is lowered. The haematocrit falls and venous pressure gradually rises.

Injection in the right side of the heart produces transient pulmonary hypertension, systemic hypotension and decreased cardiac output. The pulmonary hypertension is partly due to an increase in the pulmonary vascular resistance from clumping and sludging of the red cells, while the early systemic hypotensive phase may be due to the decreased cardiac output. The hypotension which may develop later is probably due to the vasodilator effects of contrast medium on the systemic vessels.

Several factors are associated with these haemodynamic changes. Among them must be included the volume, concentration and composition of the contrast medium; the hypertonicity and viscosity of the solution; and possibly also potassium release from the cells. As has already been described, the rapid elevation of plasma osmolality, produced by the hypertonicity of opaque media, causes a rapid expansion of plasma volume. Water shifts from the extravascular fluid spaces to the blood; it also moves out of the red cells, which shrink and become crenated. This increase in blood volume is particularly important in infancy (*vide supra*).

Most of the extreme haemodynamic effects described in the literature [34] have been derived from animal experimentation and although they undoubtedly occur to a greater or lesser extent in routine investigations, it is nevertheless true that angiocardiography in experienced hands is a remarkably safe procedure. As has already been mentioned, Swan [1] found no difference in the incidence of complications when angiocardiography was performed in addition to cardiac catheterization.

An incidence of 10 per cent of minor complications has been quoted [2] including nausea, vomiting, transient arrhythmias and allergic skin manifestations such as urticarial weals and rashes. The incidence in our series was much smaller than this, less than 5 per cent, although it is certainly possible that minimal complications such as transient nausea or a fleeting arrhythmia could have gone unrecorded. The great majority of these complications had disappeared before the patient left the department.

MAJOR COMPLICATIONS

1 *Death*

In considering the incidence of death as a complication, a sharp distinction should be made between infants, usually with congenital heart disease in the neonatal period, and others. In infancy the mortality rate has been quoted as high as 10–15 per cent [12] and in the first month of life as 12 per cent [7]. After about the third month, however, fatalities drop dramatically, and, after two years of age the incidence is quoted as 0.15 per cent in over 10 000 cases [21].

It is generally recognized that in infants death may occur without any specific precipitating cause, usually preceded by bradycardia, hypotension and depressed respiration. This syndrome typically follows hypoxic or cyanotic attacks during the procedure. When a specific event could be identified in infancy, this was usually due to perforation of the outflow tract of the right ventricle in attempts to reach the pulmonary artery, or to extravasation of opaque medium in this region during angiocardiography. While extravasation may have no clinical sequelae in older children and adults, it is nearly always an extremely serious complication in infancy [35]. Administration of an excess of opaque medium and orifice obstruction by a cardiac catheter are other causes which have been quoted [19]. In our series of 67 investigations in the first few days of life, four deaths occurred, two during the procedure and two more within 48 hours after return to the ward. The two deaths during the procedure both showed the syndrome of bradycardia, hypotension and depressed respiration, following a hypoxic attack and one had convulsions; of the two late deaths, one, already described, was attributed to an excess of opaque medium while in the fourth case no specific cause of death could be identified, although the symptoms suggested that it might have been due to a direct effect of opaque medium on the brain (see p. 92). In infancy, as well as following the regime already described in relation to heart failure (see p. 80) it has been recommended [19] that the procedure be carried out under 100 per cent oxygen cover, that the pH be measured frequently, and that alkali be administered rapidly if there is a reduction of the pH to below 7.25. In addition body temperature should be recorded by a rectal thermometer and maintained at a uniform level.

In adults, on the other hand, there is in the majority of cases a specific cause of death. The main causes are perforation of the heart and great vessels, severe arrhythmias or major myocardial extravasations following contrast medium injection. Death is, however, much less common. Thus in nearly 2200 of our cases over the age of 2 years, only three deaths occurred in the department and one subsequently. The three deaths in the department occurred during investigation of dissecting aneurysms involving the root of the aorta and coronary arteries. The fourth death was in a case of Fallot's Tetralogy, in whom asystolic arrest occurred suddenly 24 hours after an uneventful cardiac catheterization and angiocardiographic procedure.

2. *Perforation of the heart or great vessels and myocardial extravasation*

Braunwald [36] quotes an incidence of perforation of the heart or one of the great vessels of 0.8 per cent, i.e. on 100 occasions out of a total of 12 367 examinations. In this series there were also fourteen cases of serious myocardial extravasation of contrast medium (i.e. partial perforation). This probably underestimates the frequency of the occurrence because in many cases in adults there is an almost total absence of significant clinical sequelae, although that is certainly not the case in infants [35]. Of the 100 cases of total perforation quoted, 24 occurred under the age of fifteen and most of these were in infancy, emphasizing the particular danger of this complication in the early age groups. Of these 100 perforations 33 occurred in the right atrium, 21 in the right ventricle, 10 in the left atrium, 12 in the left ventricle,

17 in the aortic root and 7 in the extrapericardial portion of the aorta or one of its minor branches. Forty-seven of these perforations resulted from transseptal puncture—30 in the right atrium and 17 in the aortic root. Pericardial fluid was recognized in 11 of the right atrial perforations and 7 cases required immediate pericardio-paracentesis or thoracotomy. Two of these patients died. Of the 17 aortic perforations, tamponade developed in 3 patients with one death. In the remaining 53 perforations, closed pericardio-paracentesis was required in 8 patients and thoracotomy in 10. All these patients survived.

Apart from transseptal catheterization, which appears to have its own particular dangers, most cardiac perforations result from the use of open-ended catheters [2] (Fig. 3.1, p. 78). In the chambers of the heart the incidence of perforation is increased by the use of stiff, strong catheters and perhaps also by the use of too much force by the operator.

The danger of perforation of the outflow tract of the *right ventricle* in infancy and early childhood has already been mentioned. In adults, no particular clinical condition is mentioned in the literature as being particularly liable to be associated with perforation. In our series, however, perforation or partial perforation has tended to occur especially in cases of acute right ventricular failure following massive pulmonary embolism. In these patients, the ventricular wall appears to become extremely soft and can be entered with surprising ease. In this series there were six cases in which perforation of the right ventricle or trapping of the catheter in its wall occurred, whereas there was only one perforation of the left ventricle. In addition there was one case in which the *superior vena cava* was perforated.

The six cases of right ventricular perforation all occurred in a series of 91 patients who were being investigated for major pulmonary embolism. In all of these cases the perforation was recognized by the abnormal position of the catheter and was confirmed by the hand injection of 1–2 ml of opaque medium. The catheter was withdrawn immediately in all cases and in five instances the patient suffered no disability. In the sixth case, tamponade occurred and, despite initial pericardio-paracentesis followed by thoracotomy, the patient died. At autopsy it was not possible to be certain whether death was directly attributable to the myocardial perforation or to the massive pulmonary embolus which the patient had sustained.

In the *left ventricle*, perforation tends particularly to occur near the apex, especially when catheter wedging within ventricular trabeculation has occurred. One such case of ventricular perforation with pericardial injection of opaque medium occurred in the Guy's series out of 1040 left ventricular angiocardiograms. This incident occurred several years ago when open-ended catheters were used and no such occurrence has developed since the routine use of closed-end catheters or their functional equivalent. The pressure at the tip of an end-hole catheter always exceeds that at the side holes. High velocity end-hole jets can be very injurious to tissues in line with the tip and may even cause cardiac perforation on their own [37]. The 'functional equivalent' of a closed tip catheter—the Ducor pigtail catheter—for use in the left ventricle is certainly much safer than a conventional end-hole catheter but is probably still not quite as safe as a closed-end catheter. In 527 cases in which this type of open-ended catheter was used, myocardial extravasation occurred in five instances, although it was of a minor degree, and the patients suffered no residual disability. There was no case of right atrial perforation in our series.

Perforation of the heart or the great vessels should be suspected on fluoroscopy if there is difficulty in moving the catheter and particularly if there is slight resistance on attempted withdrawal. It tends also to be confirmed by an unusual position of the catheter, by the inability to withdraw blood, by a sudden alteration in pressure or by the aspiration of serous fluid. It can be definitely established by the injection of 1–2 ml of opaque medium. Clinically important perforations are recognized by falling blood pressure, bradycardia, loss of consciousness and signs of cardiac tamponade. If cardiac perforation does occur, it has been suggested [7] that the catheter should be left in position till surgical intervention, but this is not the general practice and it is certainly not ours. Trapping of a catheter in a ventricular wall has occurred on a number of occasions, including the cases of confirmed perforation mentioned above. Despite this, immediate withdrawal of the catheter has resulted in only one case, already described, in which tamponade developed.

3 *Arrhythmias*

Almost any arrhythmia may occur at almost any time during cardiac catheterization with or without angiocardiography. It may be precipitated either by manipulation of a catheter in any of the chambers of the heart or, rather less frequently, by the injection of opaque medium. Fortunately most of these arrhythmias are transient and minor, with normal rhythm being resumed spontaneously. In a proportion of cases, however, the abnormality persists and requires prompt recognition and institution of treatment. Major cardiac arrhythmias are much more common in infants with congenital heart disease than in any other age group or type of examination with the exception of coronary arteriography.

The most complete analysis available for the incidence of arrhythmias appears to be that given by McIntosh [38]. Major arrhythmias occurred in 1.2 per cent of all investigations. In 12 367 investigations, ventricular fibrillation occurred in 59 instances; ventricular tachycardia in 12; bradycardia or asystole in 37; complete heart block in 7 and supraventricular arrhythmias in 35. These were 12 deaths directly attributable to the arrhythmia.

It seems very likely, however, that the true incidence of arrhythmias during cardiac catheterization and angiocardiography is higher than this and that these figures represent only those cases in whom the arrhythmia was sufficiently severe to cause significant sequelae. Certainly it is quite common during manipulation of a catheter to produce a sudden burst of supraventricular tachycardia, or for a brief period of asystole to occur.

(a) *Ventricular fibrillation*. This is the commonest of the major arrhythmias. Of the 59 patients quoted by McIntosh [38], 36 occurred either in infancy or during selective coronary arteriography and there is general agreement in the literature [39, 35] that the main incidence of ventricular fibrillation occurs in these two categories. This is confirmed by our experience in which, excluding these two categories, only one case of ventricular fibrillation occurred in 1040 left ventricular examinations. This case was easily and successfully defibrillated using DC conversion. This compares with 31 cases of ventricular fibrillation requiring defibrillation occurring in coronary arteriography alone (4 per cent).

Ventricular tachycardia not amounting to fibrillation occurred in 2 of the 1040 examinations. One of these patients reverted to sinus rhythm spontaneously after a few minutes. The second was subsequently converted to normal rhythm by DC defibrillation. Despite the fact that in routine right and left heart examinations ventricular fibrillation is not common, it is nevertheless mandatory that a DC defibrillator should be available and that defibrillation should be carried out immediately ventricular fibrillation is confirmed, without waiting for reversion to occur spontaneously.

(b) *Asystole*. This was the second commonest major arrhythmia quoted by McIntosh [38]. Occasional brief periods of asystole may occur during catheter manipulation or in association with the injection of opaque medium. However, true significant asystolic arrest is very frequently preceded by bradycardia and a steadily falling blood pressure. The closest watch should therefore be kept on the heart rate and blood pressure in any cardiac investigation. If the blood pressure begins to fall, immediate steps such as removal of the catheter, administration of atropine, etc., should be taken to remedy the situation. Excluding infants in the neonatal period, asystolic arrest of a sufficient length of time to require external cardiac massage occurred in 11 of 1800 investigations (0.6 per cent) in whom the right or left ventricles were entered. None of these patients died but two suffered some residual brain damage.

Of 67 infants in the neonatal period, however, two died with this syndrome. Both these children had complex cardiac defects and in both the pattern was the same—progressive, irreversible bradycardia with a slowly falling blood pressure which, despite immediate withdrawal of the catheter, administration of vasopressor drugs and 100 per cent oxygen, progressed inevitably to asystolic arrest and death. The measures which should be taken in prevention have already been described (see pp. 80 and 88).

(c) *Other arrhythmias*. Complete heart block occurred in one case, a child of 15 lb with Fallot's Tetralogy. Persistent right or left bundle branch block occurred in three patients, two cases of transposition of the great vessels and one case of truncus arteriosus. Atrial fibrillation which persisted developed in three cases and transient atrial tachycardia occurred in two. Of these nine cases, eight of the arrhythmias occurred during manipulation of the catheter in the heart and one occurred following pressure injection of contrast medium in the right ventricle.

4 Neurological complications

These are quite uncommon in all the series which have been published. Swan [40] described 24 such complications in 12 367 investigations and of these 14 had simultaneous angiocardiography. Of the 24 patients, 11 sustained generalized cerebral dysfunction and 13 had central nervous system complications with focal signs. Approximately half of these cases developed right hemiparesis and the other half left hemiparesis. In all patients, the systemic ventricle, the left atrium or the ascending thoracic aorta had been entered by a catheter which could have caused embolism.

During cardiac catheterization and angiocardiography it would appear that central nervous system complications may arise in any of three ways:
1 First by a direct effect on the brain, possibly due to hypertonicity of opaque medium injected into the ascending aorta or left side of the heart. The symptoms usually consist of loss of consciousness, sometimes with periodic respiration and convulsions [7]. The incidence was low—five patients in 1859 examinations. In our series of 1320 angiocardiograms in which the left side of the heart or ascending aorta was injected, two patients suffered convulsions without focal signs. One of these recovered completely but the second, an infant of 6 lb with truncus arteriosus, died 40 hours after the investigation.
2 Secondly, manipulation of a catheter in the left side of the heart may dis-

Cardiopulmonary Angiography

lodge an embolus, either atheroma from the ascending aorta, a vegetation or calcium from a diseased valve, or clot from the left ventricle; and, of course, there is always the possibility of embolism from clot in the catheter or on the Seldinger wire if a percutaneous puncture has been performed. Three cases with focal signs considered to be due to emboli occurred in the 1320 cases in our series in which aortic root investigations or left ventricular angiography was performed. In one patient with aortic valvar stenosis, considerable difficulty was experienced in negotiating the aortic valve to obtain the left ventricular pressure. During the procedure the patient lost consciousness and subsequently developed a right hemiplegia, from which he recovered only partially. The second patient, a case of Marfan's syndrome, was being investigated for aortic incompetence and aneurysm of the ascending thoracic aorta, and he also subsequently developed a right hemiplegia. The third patient, a case of combined aortic and mitral valve disease, had a left ventricular angiocardiogram and a retrograde aortogram. She developed a headache after the left ventricular injection and had an epileptic fit after the aortic injection. These last two patients recovered completely without any residual disability. It is nevertheless surprising that embolism does not occur more frequently. In our series of 1040 left ventricular angiocardiograms, clot was demonstrated in the left ventricle in 84 cases but despite manipulation of a catheter in the ventricle and injection of opaque medium, none of these patients developed any embolic signs.

3 The third cause of central nervous system complication is said to result from the hyperosmolality of opaque media, possibly combined with dehydration in a long angiocardiac procedure, producing focal cerebral thrombosis. Swan [40] reported three such cases. As has been mentioned, in two out of 67 neonatal infants in the Guy's series, hyperosmolality of the opaque medium was considered responsible for symptoms; one infant suffered convulsions and subsequently died.

Wise et al. [73] described focal brain ischaemia occurring in ten normotensive patients during arteriography, without systemic hypotension, cardiac arrythmias or failure. In eight of these patients, cerebral function improved when the blood pressure was increased by one or more carefully controlled brief courses of vasopressor drugs. They recommended that treatment should begin within 15 minutes after the onset of brain ischaemia.

Another neurological complication which occurred in our series involved a patient aged 11 with pulmonary valvar stenosis. Combined with cardiac catheterization, antero-posterior and lateral right ventricular angiocardiograms were performed, using Schonander cut film apparatus. To obtain lateral films the arms, of course, have to be positioned above the head and, although many cases of this type were investigated before and after, and although the possibility of nerve damage was clearly appreciated, this patient subsequently developed *brachial palsy*, from which, after a very long period of time, she eventually recovered completely. One other patient suffered brachial trunk damage, from attempted axillary artery puncture. Similar peripheral nerve damage has been described elsewhere [40].

5   *Pulmonary complications*

(a) With normal pulmonary artery pressure

The following complications have been described: pulmonary embolism and infarction; pulmonary oedema; pneumothroax and haemothorax.

Pulmonary embolism with or without infarction may occur at the time of the procedure or subsequently. At the time of the procedure, the embolus presumably arises from the catheter or guide wire on the one hand or from dislodgement of pre-existing clot, e.g. in the right atrium, on the other. Both are uncommon. Ross [25] quotes only four cases out of 12 367 examinations in whom symptoms and signs of embolism developed on the same day as the procedure and other references in the literature are scanty. It may well be that embolization from small clots in or around a catheter is quite common [41] but it appears likely that they are seldom large enough to cause signs or symptoms. In the Guy's series of 185 pulmonary arteriograms, including 91 cases with massive embolism, two patients complained of sudden chest pain during a procedure. In one the pain subsided with no further symptoms or signs, and whether or not it was due to a small infarct is not known; in the other case a very small patch of consolidation, which could have been the result of an embolus, was seen subsequently at the left base.

Embolization occurring from seven to fourteen days after a procedure is more common (7 out of 12 367 cases) [25] but is still very rare. So far as is known, no significant late pulmonary embolism occurred in our series.

Pulmonary ischaemic lesions have also been described in 9 out of 125 patients (7.2 per cent) in whom flow-directed balloon-tip catheters were used [42, 43]. The mechanism of ischaemia included persistent wedging of the catheter tip in a peripheral artery by an inflated balloon (one case), and pulmonary embolism from venous thrombosis developing around the catheter (two cases). Awareness of the tendency for spontaneous wedging of the catheter to occur, and of the possibility of air remaining in the balloon after use, should tend to reduce the frequency of these complications [44, 45].

The incidence of *pulmonary oedema* must depend, to a certain extent, on the clinical condition of patients submitted for angiographic investigation or cardiac catheterization. Ross [25] quotes only two instances of this occurring in 12 367 examinations but in 1320 cases in whom the left heart or aortic root was catheterized in our series it has occurred nine times. These patients were all in varying degrees of left heart failure when the examination was started. In four the procedure was completed, but in five others it had to be abandoned. The reason for the onset of left heart failure is not far to seek: that the patients could not tolerate the recumbent position. All these patients responded to routine measures.

(b) With elevated pulmonary artery pressure

Any of the complications listed above may, of course, occur but there is one which appears to be peculiar to cases in whom the pulmonary artery pressure is elevated. Injection of opaque medium into the pulmonary circulation can result in sludging and sedimentation of the red cells. This mechanism has been invoked to account for deaths which have been reported in pulmonary arteriography with an elevated pulmonary artery pressure [46, 47]. In most of our 91 cases of pulmonary arteriography for massive embolism, the pulmonary artery pressures were raised but no death occurred from this cause and this complication appears to be rare.

6 *Rupture of the coronary sinus*

A catheter in the coronary sinus may appear on fluoroscopy to be in the right ventricle (Fig. 3.4a) and the pressure recording may be very similar to a true right ventricular trace. If a pressure injection is performed in this situation the coronary sinus may be ruptured and this has been reported [18, 48]. In one of the cases described death occurred.

It is difficult to believe that this complication should ever occur. A hand injection of a few ml of opaque medium prior to the pressure injection will immediately reveal the true situation of the catheter (Fig. 3.4b).

7 *Systemic arterial embolism, excluding cerebral embolism*

Ross [49] describes seven such cases out of a series of 12 367. Of these cases four had rheumatic heart disease, one had arterio-sclerotic heart disease and two had congenital heart disease. The presumed source of embolism was from the aortic valve in two patients, the heart in three patients and the femoral artery at the site of arterial entry in two. The point of final lodgement of the embolus was the coronary artery in two patients, the aortic bifurcation in one and the leg in four. None of the patients died.

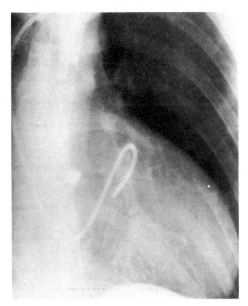

FIGURE 3.4a. To show the position of a catheter coiled in the heart. This catheter could be either in the right ventricle or in the coronary sinus.

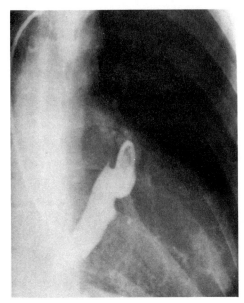

FIGURE 3.4b. A hand injection of a few ml of opaque medium demonstrating that the catheter is in the coronary sinus.

Other than cerebral emboli, systemic embolism is obviously quite uncommon. In our series of 1320 patients in whom the ascending aorta or left side of the heart were entered, systemic embolism occurred in only one case, a patient with combined mitral and aortic rheumatic valve disease. In this instance the embolus lodged at the aortic bifurcation and had to be removed surgically. Obviously, the most potentially lethal site of impaction of a systemic embolus is in the coronary arteries, but in fact neither of the two cases described by Swan died.

8 *Miscellaneous problems*

Although these are not complications in the accepted sense of the word, in most investigative units a number of situations occur which prevent a satisfactory examination being performed. Among these are the following:

(a) Congenital abnormalities

Most of the causes of failure in this group occurred on the venous side and were due to variations in the insertion of the systemic veins. In seven infants with complex heart lesions, both the inferior vena cava and the superior vena cava drained into the coronary sinus so that the right side of the heart could not be entered. In two cases the inferior vena cava was absent, drainage from the lower half of the body being by the azygos system, while at the same time the superior vena cava drained to the coronary sinus so that again the right side of the heart could not be reached. In one final case, the inferior vena cava was absent while the superior vena cava was so grossly dilated and tortuous that it was impossible to manipulate a catheter through it.

On the arterial side only one similar situation occurred—a patient had a localized aortic atresia in the normal situation of a coarctation, so that the left heart could not be reached from the leg; in addition there was atresia of the proximal parts of each subclavian artery so that equally the heart could not be entered from either arm. In the end, the anatomy in this patient was demonstrated by left ventricular angiography carried out by direct percutaneous puncture, using a hand injection.

(b) Malfunction of apparatus

Occasional malfunction of apparatus occurs in any department; there is only one instance which is worth recording. During investigation of an 11-year-old child with suspected Fallot's tetralogy, a lateral tube failed to function.

The angiocardiogram taken in the antero-posterior position was satisfactory with the catheter tip lying free in the ventricle. As a lateral projection was considered to be essential, the child was turned on his side with the catheter in position to use the antero-posterior tube for the lateral projection. After injection of opaque medium the heart rate slowed and the blood pressure began to fall. On fluoroscopy in the antero-posterior position it was found that the catheter had penetrated the right ventricle (Figs 3.5a and b).

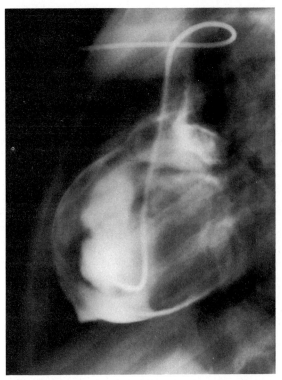

FIGURE 3.5a. Perforation of the anterior wall of the right ventricle with opaque medium in the pericardium.

It was withdrawn immediately and after a few minutes the pulse rate and blood pressure returned to normal. At operation fourteen days later the site of perforation was identified on the anterior wall of the right ventricle.

This patient is included in the section on perforation of the heart and is mentioned here only to make the point that it can be dangerous to move a patient with a catheter in the heart.

CORONARY ARTERIOGRAPHY

Techniques

Before considering the complications which accompany selective injection of the coronary arteries a résumé of the two methods commonly used is in order, since the approach used does have some bearing on the complication rate.

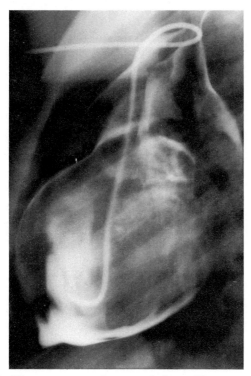

FIGURE 3.5b. The same case a few seconds later. The opaque medium has escaped through a congenital defect in the pericardium into the pleura.

The perforation resulted from moving the patient from the supine to a right lateral position after failure of the lateral X-ray tube. The patient suffered no ill effects. At operation fourteen days later for correction of Fallot's tetralogy, the scar of the perforation was identified on the anterior wall of the right ventricle.

(a) Catheters

The method described by Sones *et al.* [50] requires the placement of a flexible woven Dacron catheter into the ascending aorta via a right brachial arteriotomy. The catheter has a slight taper and there are both end and side holes.

There is a slight preformed angle towards the end of the catheter and this, combined with manipulation and the catheter 'memory' for the curve formed across the sinuses of Valsalva, enables the tip to be placed at, or in, the coronary ostium.

An alternative technique was developed later and described by Judkins [51] using preformed Ducor catheters of different size and shape for the right and left coronary arteries. An appropriate size of right or left catheter is chosen and introduced via a Seldinger puncture of the femoral artery. In order to reduce the risk of intimal dissection of the coronary artery, the tip of the catheter has only a slight taper and the use of a vessel dilator is necessary prior to introduction into the femoral artery. Due to the steel mesh embedded in the catheter wall a high degree of torque control is possible. The tip can be placed into the appropriate coronary artery with minimal manipulation. There are no side holes and the tip must be placed within the artery rather than simply at the ostium.

Other types of torque-control preformed catheters have since been developed. That described by Amplatz [52] is suitable for use from both the leg and the arm and this catheter is used by many operators as an alternative to both the Sones and Judkins catheters, when difficulty in selecting one or other coronary artery is encountered. The right coronary artery is generally found to be more difficult to enter with the Judkins catheter and vice versa with the Sones technique.

Each of the catheters has advantages and disadvantages. The preformed Ducor catheters are much easier to use than the Sones catheters in that injections can easily be made in standard projections. This is an advantage in serial examinations of the same patient. The single end hole (without side holes) also reflects the true pressure in a coronary artery more accurately than the Sones catheter in which the pressure trace may be reflecting the aortic pressure through its side holes. On the other hand, the in-built 'spring' in the left Judkins catheter is said to be a predisposing factor in causing coronary artery dissection, especially when the main left coronary artery trunk is small. The use of all the above catheters has been described using a percutaneous axillary approach. Catheters specifically designed for this method have also been described [53].

(b) Contrast medium

The choice of contrast medium is particularly important if arrhythmias are to be avoided. Ischaemia due to blood replacement and alteration of normal haemodynamics by hypertonic fluid are likely to affect myocardial function.

Further, the cation constituent is of considerable importance, particularly the amount of sodium. Media containing a proportion of methylglucamine are less toxic than pure sodium salts [54]. In an attempt to reduce toxicity still further, a medium containing only 0.04 mEq/litre of sodium was introduced [55]. This consisted almost entirely of methylglucamine salt. However, far from being less toxic, a marked increase in the incidence of ventricular fibrillation was observed. Although no specific amount of sodium has been proven to be ideal, amounts near to physiological levels are desirable. We have found media with a combination of sodium and methylglucamine diatrizoate in the proportions 25 per cent:50 per cent W/V (Hypaque 65 per cent) and 10 per cent:66 per cent W/V (Urografin 76 per cent) to be satisfactory.

Alteration of myocardial function occurs normally following the injection of contrast medium into a coronary artery and is reflected in the electrocardiogram. Transient T-wave changes, prolongation of the QT interval and sinus bradycardia are almost the rule and rarely last longer than one minute [56]. If these changes do not occur, it is possible that insufficient opaque medium has been injected. The occurrence of sinus bradycardia, which probably arises through a vagal mechanism, requires close attention. If the bradycardia is more than transient, intravenous administration of 1 mg of atropine will usually restore the normal rate. It is in fact unwise to start the procedure in the first place, if the resting heart rate is below 70 beats per minute or the systolic blood pressure below 100 mmHg. Failure to raise the heart rate with atropine necessitates the placement of a temporary pacemaker in the right ventricle. Routine premedication with atropine is advocated by Green [57]. Chest pain is not unusual and can occur in the normal subject, presumably due to ischaemia induced by replacement of blood by contrast medium.

Complications of coronary arteriography

It should be understood that any of the complications which may develop following the passage of a catheter to the aorta and left side of the heart, described previously, may occur during selective coronary arteriography. The purpose of this additional section is merely to highlight the specific dangers which may occur during this examination.

Complications may be precipitated by complete or incomplete dissection of a coronary artery by the catheter (Figs 3.6 and 3.7), by embolism with clot or air, or they may occur apparently spontaneously after injection of opaque medium, without any known cause being discovered. They may occur either during the procedure or subsequently. To date our series of selective coronary arteriograms consists of 765 cases, 238 having been performed by the Sones technique and 527 by the Judkins.

Death

Nine deaths have occurred in this series (an incidence of 1.17 per cent), four from irreversible ventricular fibrillation, four from asystolic arrest

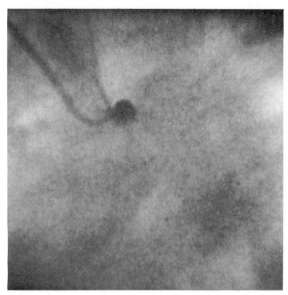

FIGURE 3.6. Dissection of the main left coronary artery following injection of a test dose of 2 ml of contrast medium, using a left Judkins catheter. This episode was followed by progressive bradycardia and hypotension. Death occurred within 5 minutes.

following bradycardia and hypotension and one from myocardial infarction occurring suddenly 24 hours after an uneventful procedure. No specific cause of death was discovered at autopsy in any of the cases of ventricular fibrillation. In one case of asystolic arrest, dissection of the left coronary artery occurred after a test injection, using a Judkins catheter. Death occurred within minutes and the dissection was confirmed at autopsy. In the second of these cases, using a Sones catheter, death occurred within minutes of a right coronary artery injection. No dissection was observed or confirmed at

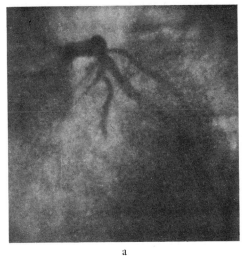

a

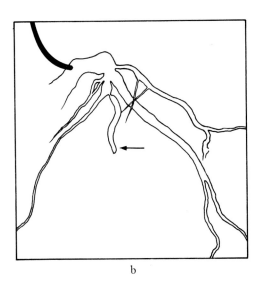

b

FIGURE 3.7a and b. Dissection of an accessory obtuse marginal branch of the circumflex. This occurred during the injection of 8 ml of contrast medium into the left coronary artery. It is difficult to explain because at no time could the catheter possibly have been near the origin of this artery. However, it was without doubt a dissection since it persisted for nearly ten minutes after contrast was washed out of the other arteries. There were no ECG changes other than the usual ones associated with contrast medium injection and the patient suffered no apparent disability.

autopsy but when the catheter was withdrawn, it was found to be fractured at its tip. The edges of the fracture were irregular and sharp and it is certainly possible that the intima of the artery could have been damaged.

Mortality figures published in the literature vary between 0.05 per cent [58] and 8 per cent [59] with an overall mortality of about 1 per cent. Sones, however, is now achieving mortality rates as low as 0.017 per cent [60].

Complications not causing death

(i) *Severe arrhythmias*. Ventricular fibrillation requiring DC shock occurred in 31 cases in this series (an incidence of 4 per cent). This is considerably higher than the 1 per cent expected [22]. Most cases of ventricular fibrillation revert promptly to normal rhythm with DC shock but in this series, as has been mentioned, four deaths occurred.

Severe sinus bradycardia and hypotension requiring external cardiac massage occurred on eleven occasions (1.4 per cent). Takaro [8] describes eight of sixty-six deaths resulting from acute arrhythmia. Other major hazards include total heart block and right and left bundle-branch block.

(ii) *Myocardial infarction*. This occurred in four patients in our series (an incidence of 0.5 per cent), all after the procedure had been terminated. Adams [58] in a survey of over 40 000 coronary arteriograms from 173 hospitals quotes an incidence varying between 0 and 2.5 per cent. Takaro [8] in his survey of 66 deaths arising from 3044 arteriograms, states that at least 60 per cent were due to myocardial infarction. It is of interest that he included deaths up to ten days after the examination. Twenty-six per cent occurred more than 24 hours after the procedure.

Prevention of complications

It is generally agreed that coronary angiography is a much more dangerous procedure (a) where patients are in the pre-infarction state with crescendo angina and (b) in patients within one month of a previous infarct. If it is clinically possible these times should be avoided but, of course, this is often not possible. Similarly, the risk is increased in patients with a poor cardiac reserve whether or not they are in overt left heart failure. It is unwise to undertake the investigation in patients with incipient pulmonary oedema. Injection of opaque medium into the left ventricle or a coronary artery may raise the left ventricular end diastolic pressure very rapidly [61] particularly if the myocardium is damaged, and so induce severe left heart failure. If the patient has an arrhythmia it is always preferable to delay the investigation until it has been stabilized. This is particularly true of digitalis intoxication. Abnormal serum potassium levels should be corrected so that the blood level is 4 mg per cent or above. As already indicated, the incidence of severe bradycardia and hypotension can be reduced by the prophylactic administration of atropine. If this fails to raise the pulse rate and blood pressure to a satisfactory level a temporary pacemaker should be inserted.

When patients have a poor cardiac reserve, the incidence of complications can be diminished by limiting the number of injections from the routine five to two or three. It is always advisable to inject the left coronary artery first for two reasons: (i) because in this way it can be determined whether or not the left coronary artery is dominant; if so, the right artery is likely to be hypoplastic and an appropriately small amount of opaque medium should be injected into it; (ii) because arrhythmias are more common following injection of the right coronary artery and if the left artery has already been visualized, the procedure can be terminated without much loss of information. The first injection in each artery, or into a graft, should always be of a small amount of opaque medium to try to evaluate the position, because if there is a major vessel or graft occlusion, or if the vessel is supplying a small territory, overloading of the area may easily occur with consequent morbidity.

Mortality figures can vary considerably between departments [58, 62, 63, 57, 64, 59]. These figures will obviously be influenced by the number of patients with normal coronary arteries whose symptoms were originally suspected as being due to angina. Similarly, a large number of patients studied with only mild disease will also favourably affect a department's mortality rate. Those departments, however, with a higher than average mortality tend to have two factors in common. They are likely to perform a smaller than average number of examinations and they will tend to use the transfemoral route rather than the brachial.

A number of surveys have been published putting the transfemoral route in an unfavourable light [63, 8, 64]. This method is technically much easier to perform than the Sones and consequently good-quality images can be produced by the relatively inexperienced. It seems likely that this is a factor in the higher mortality which has been reported with the Judkins technique, compared with the Sones. It is of interest that of the seven deaths in our series using the Judkins technique, four occurred in the first hundred cases, when we were inexperienced in the use of these catheters, although we had previously performed nearly 150 selective coronary arteriograms with the Sones method, without a fatality. This represents a mortality of 4 per cent in the first 100 cases and 0.7 per cent in the subsequent 427. One cannot help sympathizing with Judkins [65] that his catheters were being condemned by people almost totally inexperienced in coronary arteriography despite the fact that his figures compare very favourably with those of any other techniques. Despite this, however, it has long been recognized that the less flexible, preformed Judkins femoral catheters are potentially hazardous and that great care must be exercised when introducing the tip into a coronary ostium. The catheter is sensitive to slight manipulation and the tip can easily be guided to its destination. It is unnecessary to use the relatively vigorous manipulation required with the Dacron NIH and Sones catheters. Constant monitoring of the catheter tip pressure is essential; a damped tracing will indicate that the end hole is in contact with the vessel wall and, in this situation, there is a danger of intimal dissection or of dislodgement of an atheromatous plaque if an injection is performed.

An acceptable mortality can be achieved using the femoral approach but this tends to be seen in units performing over 100 examinations per year. In his survey, Adams [58] quotes a department performing 1183 examinations over two years with a mortality rate of 0.08 per cent. The equivalent figure for the brachial approach in a department with 1831 coronary arteriograms over two years was a mortality of 0.05 per cent. Morbidity is also lower in those departments performing a large number of examinations. A recent recommendation from the Royal College of Physicians [66] has stated that no new department should be set up to perform coronary angiography unless it is likely to perform more than 100 examinations per annum. A mortality rate of under 1 per cent should then be attainable.

Takaro [8] in his analysis of 66 deaths attempts to explain why the femoral route can be hazardous. Thirty-five (53 per cent) of these deaths were due to acute coronary occlusion and in fifteen there was evidence of a long, coiled platelet thrombus similar to that described by Price [67]. It is well known that the surfaces of catheters and guide wires are thrombogenic and Formanek [41] showed by pull-out angiograms that thrombus can be wiped off the wall of the catheter at the puncture site. It has been postulated that this thrombus can be transferred to the tip of a new catheter subsequently inserted and then propelled to a distant site. The adherent thrombus may be blown off the catheter tip by contrast injection or forced flushing, or simply detached by the change of direction of blood flow relative to the catheter tip. It is difficult to believe that this should be a potent cause of risk if the catheter is routinely force-flushed in the descending aorta before being passed to the aortic root, and it can be almost entirely eliminated by adequate heparinization of the patient [68]. However, it is of interest that in Takaro's series [8] only two deaths in 720 examinations (0.3 per cent) occurred with the brachial approach, whilst 51 deaths in 2500 examinations (2.2 per cent) followed the femoral approach. In almost all deaths suspected of being embolic in origin, at least one catheter change had taken place. Although significant disease was present in all but 2 of the 51 deaths, fatal thromboembolism has been reported in the presence of normal coronary arteries [69].

Conclusion

A department should be able to achieve a mortality rate of under 1 per cent and a low morbidity provided it is performing well over 100 investigations per annum; and the higher the number of investigations performed, the lower is the morbidity likely to be. Clearly, however, there is a higher risk implicit in coronary arteriography than there is in angiographic procedures applied to less vital organs. Further improvement in the mortality and the morbidity figures can be expected with more physiological forms of contrast media [70], and perhaps particularly with the use of metrizamide [71, 72]. The routine use of heparin by systemic administration [9, 10], or coated on catheters and guide wires [11] should further reduce the risk of thromboembolism.

REFERENCES

1 SWAN H.J.C. (1968) Complications associated with angiocardiography. *Circulation* 37, No. 5, Supp. No. 3, III.81.
2 BECKMANN C.H. & DOOLEY B. (1970) Complications of left heart angiography—A study of 1,000 consecutive cases. *Circulation* 41, 825–832.
3 LIM T.P.K. & CADWALLADER J.A. (1967) Delivery pressure at the catheter tip in selective angiocardiography. *J. Am. med. Ass.* 199, 11–14.
4 BOIJSEN E. & JUDKINS M.P. (1966) A hook-tail 'closed-end' catheter for percutaneous selective cardioangiography. *Radiology* 87, 872–877.
5 DOUST B.D. & REDMAN H.C. (1972) The myth of 1 ml/kg in angiography. *Radiology* 104, 557–560.
6 LACHLAN H. (1970) Biochemical and other changes occurring in infants during angiocardiography. *Proc. R. Soc. Med.* 63, 46–48.
7 HO C.S., KROVETZ L.J. & ROWE R.D. (1972) Major complications of cardiac catheterisation and angiocardiography in infants and children. *Johns Hopkins med. J.* 131, 247–258.
8 TAKARO T., HULTGREN H.N., LITTMANN, D. *et al.* (1973) An analysis of deaths occurring in association with coronary arteriography. *Am. Heart J.* 86, 587–597.
9 NELSON R.M. & OSBORN A.G. (1971) Systemic heparinisation for percutaneous catheter arteriography. *Circulation* 44, No. 4, Supp. II, 205.
10 WALKER W.J. *et al.* (1973) Systemic heparinisation for femoral percutaneous coronary arteriography. *New Engl. J. Med.* 288, 826–828.
11 AMPLATZ K. (1971) A simple non-thrombogenic coating. *Invest. Radiol.* 6, 280–289.
12 KROVETZ L.J., SHANKLIN D.R. & SCHIEBLER G.L. (1968) Serious and fatal complica-

tions of catheterization and angiocardiography in infants and children. *Am. Heart J.* **76**, 39–47.
13 COPE C. (1962) Intravascular breakage of Seldinger spring guide wires. *J. Am. med. Ass.* **180**, 1061–1063.
14 PARKER J.O., WEST R.O. & FAY J.E. (1964) The Brockenbrough transseptal catheterisation. An unusual complication. *Circulation* **30**, 743–744.
15 SUSMANO A. & CARLETON R.A. (1964) Transseptal catheterisation of the left atrium. *New Engl. J. Med.* **270**, 897–898.
16 ARNOLD J.R., FRASER D.J. & NOCERO M.A. (1974) New technique for retrograde catheterisation of stenosed aortic valve. *Amer. Heart J.* **88**, 128.
17 JOHANSSON L., MALMSTRÖM G. & UGGLA L.G. (1954) Intracardiac knotting of the catheter in heart catheterisation. *J. thorac. Surg.* **27**, 605–607.
18 ZIMMERMAN H.A. (1966) *Intravascular catheterization*, Second edition. Charles C. Thomas, Springfield, Illinois.
19 RUDOLPH A.M. (1968) Complications occurring in infants and children. *Circulation* **37**, No. 5, Supp. No. 3, III.59.
20 WINCHELL P. (1953) Infectious endocarditis as a result of contamination during cardiac catheterisation. *New Engl. J. Med.* **248**, 245–246.
21 BRAUNWALD E. et al. (1968) Summary. *Circulation* **37**, No. 5, Supp. No. 3, III.93.
22 GORLIN R. (1968) Perforations and other cardiac complications. *Circulation* **37**, No. 5, Supp. No. 3, III.36.
23 FLEMING H.A. et al. (1958) Percutaneous left ventricular puncture with catheterisation of the aorta. *Thorax* **13**, 97.
24 ROSS D.N. (1959) Percutaneous left ventricular puncture in the assessment of the obstructed left ventricle. *Guy's Hospital Reports* **108**, 159.
25 ROSS R.S. (1968) Pulmonary complications. *Circulation* **37**, No. 5, Supp. No. 3, III.46.
26 LEVY M.J. & LILLEHEI C.W. (1964) Percutaneous direct cardiac catheterisation. A new method, with results in 122 patients. *New Engl. J. Med.* **271**, 273–280.
27 GROSSMAN W. (1974) *Cardiac catheterisation and angiography*, p. 23. Lea & Febiger, Philadelphia.
28 RASHKIND W.J. & MILLER W.W. (1968) Results of palliation by balloon atrioseptostomy in thirty-one infants. *Circulation* **38**, 453–462.
29 VENABLES A.W. (1970) Balloon atrial septostomy in complete transposition of great arteries in infancy. *Br. Heart J.* **32**, 61–65.
30 SINGH S.P., ASTLEY R. & BURROWS F.G.O. (1969) Balloon septostomy for transposition of the great arteries. *Br. Heart J.* **31**, 722–726.
31 FISCHER H.W. (1968) Haemodynamic reactions to angiographic media. A survey and commentary. *Radiology* **91**, 66–73.
32 SIMON A.L., SHABETAI R., LANG J.H. & LASSER E.C. (1972) The mechanism of production of ventricular fibrillation in coronary angiography. *Am. J. Roentg.* **114**, 810–816.
33 GRAINGER R.G. (1970) Radiological contrast media. *Modern Trends in Diagnostic Radiology* (Ed. by MCLAREN J.W., BUTTERWORTH'S, chapter 15, 254–265.
34 GIAMMONA S.T., LURIE P.R. & SEGAR W.E. (1963) Hypertonicity following selective angiocardiography. *Circulation* **28**, 1096–1101.
35 GRAINGER R.G. (1965) Complications of cardiovascular radiological investigations. *Brit. J. Radiol.* **38**, 201–215.

36 BRAUNWALD E. (1968) Deaths related to cardiac catheterisation. *Circulation* 37, No. 5, Supp. No. 3, III.17.
37 SELLERS R.D., LEVY M.J., AMPLATZ K. & LILLEHEI C.W. (1964) Left retrograde cardioangiography in acquired heart disease; technique, indications and interpretations in 700 cases. *Am. J. Cardiol.* 14, 437–447.
38 MCINTOSH H.D. (1968) Arrhythmias. *Circulation* 37, No. 5, Supp. No. 3, III.27.
39 MILLER G.A.H. (1974) Congenital heart disease in the first week of life. *Brit. Heart J.* 36, 1160–1166.
40 SWAN H.J.C. (1968) Complications associated with angiocardiography. *Circulation* 37, No. 5, Supp. No. 3, III.42.
41 FORMANEK G., FRECH R.S. & AMPLATZ K. (1970) Arterial thrombus formation during clinical percutaneous catheterisation. *Circulation* 41, 833–839.
42 FOOTE G.A., SCHABEL S.I. & HODGES M. (1974) Pulmonary complications of the flow-directed balloon-tipped catheter. *New Engl. J. Med.* 290, 927–931.
43 MCLOUD T. & PUTMAN C.E. (1975) Radiology of the Swan–Ganz catheter and associated pulmonary complications. *Radiology* 116, No. 1, 19–22.
44 SWAN H.J.C., GANZ W., FORRESTER T. *et al.* (1970) Catheterisation of the heart in man with use of a flow-directed balloon-tipped catheter. *New Engl. J. Med.* 283, 447–451.
45 LIPP H., O'DONOGHUE K. & RESNEKOV L. (1971) Intracardiac knotting of a flow-directed balloon catheter. *New Engl. J. Med.* 284, 220.
46 WATSON H. (1964) Severe pulmonary hypertensive episodes following angiocardiography with sodium metrizoate. *Lancet* 2, 732.
47 SNIDER G.L. *et al.* (1973) Primary pulmonary hypertension: A fatality during pulmonary angiography. *Chest* 64, 628–635.
48 HAMMERMEISTER K.E., MURRAY J.A. & BLACKMON J.R. (1973) Revision of Gorlin constant for calculation of mitral valve area from left heart pressures. *Brit. Heart J.* 35, 392.
49 ROSS R.S. (1968) Systemic arterial embolism. *Circulation* 37, No. 5, Supp. No. 3, III.48.
50 SONES F.M., SHIREY E.K., PROUDFIT W.L. & WESTCOTT R.N. (1959) Cine-coronary arteriography. *Circulation* 20, 773.
51 JUDKINS M.P. (1967) Selective coronary arteriography. Part 1: A percutaneous transfemoral technique. *Radiology* 89, 815–824.
52 AMPLATZ K. *et al.* (1967) Mechanics of selective coronary artery catheterisation via femoral approach. *Radiology* 89, 1040–1047.
53 BROWN O.L. (1975) Preshaped catheters for percutaneous transaxillary selective coronary arteriography. *Radiology* 114, 732–733.
54 GENSINI G.G. & DI GIORGI S. (1964) Myocardial toxicity of contrast agents used in angiography. *Radiology* 82, 24–34.
55 SNYDER C.F., FORMANEK A., FRECH R.S. & AMPLATZ K. (1971) The role of sodium in promoting ventricular arrhythmia during selective coronary arteriography. *Am. J. Roentg.* 113, 567–571.
56 BANKS D.C., RAFTERY E.B. & ORAM S. (1970) Evaluation of contrast media used in man for coronary arteriography. *Br. Heart J.* 32, 317–319.
57 GREEN G.S., MCKINNON C.M., RÖSCH J. & JUDKINS M.P. (1972) Complications of selective percutaneous transfemoral coronary arteriography and their prevention. A review of 445 consecutive examinations. *Circulation* 45, 552–557.

58 ADAMS D.F., FRASER D.B. & ABRAMS H.L. (1973) The complications of coronary arteriography. *Circulation* **48**, 609–618.
59 SELZER A., ANDERSON W.L. & MARCH H.W. (1971) Indications for coronary arteriography. Risks vs. benefits. *Calif. Med.* **115**, 1–6.
60 SONES F.M. (1975) Paper read at the Congress of the European Society of Cardiology.
61 LEVIN D.C. & BALTAXE H.A. (1971) The effect of intracoronary and intraventricular injection of radioopaque contrast material upon left ventricular end diatolic pressure (LVEDP). Exhibited at the annual meeting of The American Roentgen Ray Society, Sept. 20, 1971. *N.Y. St. J. Med.* **46**, 589.
62 BOURASSA M.G., LESPERANCE J. & CAMPEAU L. (1969) Selective coronary arteriography by the percutaneous femoral artery approach. *Am. J. Roentg.* **107**, 377–383.
63 PETCH M.C., SUTTON R. & JEFFERSON K.E. (1973) Safety of coronary arteriography. *Brit Heart J.* **35**, 377–380.
64 CHAHINE R.A., HERMAN M.V. & GORLIN R. (1972) Complications of coronary arteriography: Comparison of the brachial to the femoral approach. *Ann. int. Med.* **76**, 862.
65 JUDKINS M.P. (1973) Lecture at the International Congress of Radiologists, Madrid.
66 EMANUEL R. (1975) Proceedings of the British Cardiac Society. *Br. Heart J.* **37**, No. 5, 549 Para. 10.
67 PRICE H.P. *et al.* (1973) Unusual coronary emboli associated with coronary arteriography. *Chest.* **63**, 698–700.
68 JUDKINS M.P. & GANDER M.P. (1974) Prevention of complications of coronary arteriography. *Circulation* **49**, 599–602.
69 CHENG T.O. (1972) Fatal thromboembolism following selective coronary arteriography. *Chest* **62**, 1–2.
70 SPINDOLA-FRANCO, H. *et al.* (1975) Coronary vascular patterns during occlusion arteriography. *Radiology* **114**, 59–63.
71 TRAGARDH B., LYNCH P.R. & VINCIGUERRA T. (1975) Effects of metrizamide, a new non-ionic contrast medium, on cardiac function during coronary angiography in the dog. *Radiology* **115**, 59.
72 TRAGARDH B., ALMEN T. & LYNCH P. (1975) Addition of calcium or other cations and of oxygen to ionic and non-ionic contrast media. Effects on cardiac function during coronary arteriography. *Invest. Radiol.* **10**, 231–238.
73 WISE G., FONTANA M.E., BURKHOLDER J. *et al.* (1974) The treatment of brain ischaemia following arteriography. *Radiology* **110**, 383–390.

CHAPTER 4: RENAL COMPLICATIONS OF ANGIOGRAPHY

LEE B. TALNER

INTRODUCTION

The kidneys are uniquely vulnerable to damage from angiography. In the case of selective arteriography the kidneys are similar to other organs as they are subject to both the toxic effects of locally high concentrations of contrast media and to vascular accidents related to catheter and guide wire manipulation. Unlike other organs, however, kidneys may be significantly damaged by the presence in the circulation of relatively low concentrations of angiographic contrast media as, for example, after arteriography or venography of remote organs. In this respect the resulting renal insult is analogous to that occasionally observed after intravenous urography. For the sake of discussion these two varieties of renal complications of angiography will be considered separately, although there is, in fact, considerable overlap.

COMPLICATIONS OF NONRENAL ANGIOGRAPHY

The types of renal damage to be described here may occur with any of the currently acceptable tri-iodinated contrast agents. This group consists of the sodium, methylglucamine or sodium plus methylglucamine salts of diatrizoate, iothalamate, and metrizoate in concentrations ranging from 50 to 80 per cent. As yet there is no evidence to suggest that one particular contrast material is more at fault than any other. The use of acetrizoate or other older agents is no longer justified, since their high nephrotoxicity has been documented repeatedly in both renal and nonrenal angiography.

Acute renal failure

The most serious and clinically significant renal complication of nonrenal angiography is acute renal failure. This syndrome has been reported following mesenteric and coeliac angiography [1, 2], angiocardiography [2, 3], aortography [2, 4, 5], peripheral arteriography [6], and spinal arteriography [7]. Although it is perhaps better known as a complication of intravenous urography, particularly in diabetic and dehydrated patients, there is little doubt now that it may occur after *any* intravascular administration of contrast media.

In the typical patient with this syndrome there is a decrease in urine output within the first 24 hours following the procedure, often progressing to anuria. Unless urine output is recorded and determinations of blood urea

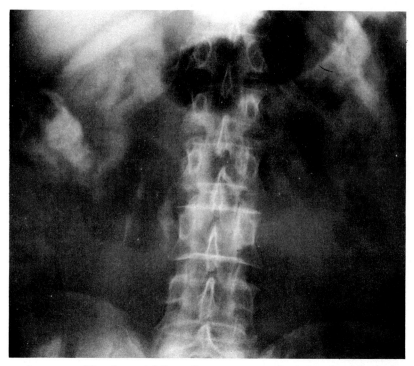

FIGURE 4.1. Oliguric renal failure after coronary arteriography. An abdominal radiograph taken 36 hours after the procedure shows bilateral, dense nephrograms with faint filling of nondilated calyces. Extrarenal excretion of contrast media is manifested by opacification of the gallbladder and ascending colon. The findings are similar to those in acute tubular necrosis.

nitrogen (BUN) or serum creatinine are done within 24–48 hours after angiography, the patient may be the first to appreciate the existence of a problem by noting a diminished urine output. Uraemia will not be apparent clinically until several days have passed.

The presence of bilateral, dense nephrograms on the abdominal radiograph taken after an angiographic procedure signals the evolution of this type of acute renal failure (Fig. 4.1). In the full-blown syndrome dense nephrograms may persist for several days, as occurs with acute tubular necrosis [8].

Once the diagnosis is established the patient should be treated as he would be for acute renal failure of any toxic aetiology. The administration of intravenous mannitol or a diuretic such as furosemide, in an attempt to stimulate a diuresis, has not proved successful in most instances, but it is still practised and with some justification. Fortunately, in almost every case the syndrome spontaneously remits, heralded by increasing urine output and clearing of azotaemia. The average duration of oliguria is five to seven days, with a few patients having gone as long as three to four weeks before diuresis ensued.

It is difficult to establish the incidence of acute renal failure after angiography. In Lang's review of over 11 000 percutaneous retrograde arteriograms only two patients developed postangiographic acute renal failure [9]. The syndrome was not observed at all in 1750 angiographies reviewed by Robertson et al. [10]. In a retrospective analysis of the Mayo Clinic experience between 1968 and 1972, 8 patients were found who developed acute renal failure after angiography, an incidence of approximately one case in 1000 angiographic procedures [2]. The actual incidence of the syndrome is probably close to this figure or even slightly higher, since mild cases remain undetected unless careful postangiography follow-up is practised [84]. Five additional unpublished cases of this syndrome have been brought to the attention of this writer, and all were similar to the cases already published.

A number of investigators have attempted to sort out the factors predisposing to this nephrologic complication of intravascular contrast material. The following conditions appear to increase the likelihood of developing acute renal failure: (1) Severe underlying medical disorders such as low output heart failure, hypertensive atherosclerotic cardiovascular disease, diabetes mellitus, hyperuricaemia, and multiple myeloma. (2) Dehydration and/or blood volume depletion leading to diminished renal perfusion. (3) Mild to moderate renal disease with elevation of preangiography serum creatinine. (4) Use of large volumes of contrast media, near or over 5 ml/kg total for the procedure. (5) Recent administration of drugs capable of causing acute renal failure.

It is perhaps unnecessary to point out that renal failure occurring after the intravascular administration of contrast media does not necessarily imply a cause–effect relationship. A variety of drugs and procedures may result in acute renal damage with variable degrees of renal insufficiency, and patients undergoing angiography often have received one or more of these agents in the pre-angiographic period. A complete list of nephrotoxic drugs is not intended, but the following commonly used agents at least should be considered in this setting: methoxyflurane (Penthrane) anaesthesia, a number of antibiotics [11] including tetracyclines, sodium cephalothin, kanamycin, trimethoprim, gentamycin, rifampin, and oral and intravenous biliary contrast media [12–14].

In view of the rarity of postangiographic renal failure, it is evident that no one of the above preconditions is very likely to result in a problem. However, the combination of two or more clearly puts the patient at greater risk. The interested reader is referred to recent discussions of these points [2, 84].

The mechanism for renal damage under these circumstances remains controversial. Given in bolus intravascular injections, all the currently used contrast media are capable of causing red blood cell crenation and sludging and a transient reduction in renal blood flow and glomerular filtration rate [15]. While the normal kidney can tolerate this insult without difficulty, the mildly diseased and/or hypoperfused kidney may respond unfavourably with the development of acute renal failure. Although this seems to be an attractive hypothesis it has not so far been supported by the experience with selective renal arteriography in which one would expect an even higher incidence of acute renal failure after bilateral studies. In over 2700 bilateral selective renal arteriograms at the Mayo Clinic not one patient developed acute renal failure [2]. Similarly, Folin did not mention renal failure in his series of selective renal arteriographies [16]. These observations suggest that the renal artery contrast medium concentration alone is not the predominant factor in the genesis of acute renal failure. Precipitation in the renal tubules of uric acid crystals or Tamm–Horsfall mucoprotein have been repeatedly mentioned as possible aetiologic factors but as yet they are unproven in this circumstance. Acute renal failure after oral or intravenous biliary contrast media is particularly suspect for a uric acid aetiology since these agents are highly uricosuric [17].

Although patients with mild renal disease and/or hypoperfusion are at greater risk to develop postangiographic renal failure, it has been difficult to detect a deleterious effect of intravascular contrast media on the kidneys of patients with well established, frank renal insufficiency. Except for rare instances neither intravenous urography [18–20] nor aortography [21] have caused permanent worsening of glomerular filtration rate, renal plasma flow, urinary protein excretion, or serum creatinine in patients with chronic renal insufficiency. In patients with acute renal failure, nonrenal angiography has been performed without noticeable adverse effects on renal function but the experience is small. In animals with acute renal failure the data are conflicting. McLachlan et al. [22] created acute renal failure in rats and found increased mortality in those who had ultra-high-dose urography (7–8 ml/kg) with either sodium or meglumine diatrizoate. On the other hand, Walsh et al. [23] performed aortography on rats with acute renal failure and could detect no worsening of the renal disease. It is now generally accepted that nonrenal angiography can be performed in patients with acute or chronic renal failure providing the volumes of contrast material are kept reasonable (less than 3.0 ml/kg total).

What can be done to avert the syndrome of postangiographic renal failure in patients at presumed greater risk? It should be widely recognized now that *no* patient should be dehydrated before angiography and that patients with any type of renal disease should be carefully hydrated prior to and during the examination [12]. Of course patients who are already oliguric or anuric from pre-existing renal disease or patients with congestive heart failure cannot be superhydrated in this manner. From a practical standpoint, the simplest and probably most effective manoeuvre in nonoliguric patients is to induce a diuresis *before* the injection of contrast material by administering intravenous saline [24]. While dextrose in water has also been recommended for this purpose [25], there is little justification for its use. In one patient at least, preparation with intravenous dextrose and water before and during the procedure failed to prevent postangiographic oliguria [26]. The intra-aortic administration of low molecular weight dextran was once suggested as an alternative prophylactic measure but dextran itself has been shown to cause osmotic nephrosis and, at times, anuria under conditions of renal hypoperfusion [27, 28].

Rare renal reactions

Two odd sequelae have been described with postangiography oliguric renal failure. Bora et al. [3] reported a 49-year-old man who developed a hypersensitivity reaction from angiocardiography with meglumine diatrizoate. The patient had a skin rash and eosinophilia and was anuric for 72 hours. Oliguria ensued which necessitated dialysis for an additional five weeks. The oliguric phase was replaced by a full-blown *nephrotic syndrome* which lasted a further two weeks before ceasing spontaneously. Renal biopsy during the oliguric phase showed proliferative glomerulonephritis. Kovnat et al. [4] described a 70-year-old woman who suffered five days of oliguric renal failure after aortography and unilateral selective renal arteriography with meglumine–sodium diatrizoate. A satisfactory diuresis followed treatment with mannitol and dextrose in saline, but serum creatinine remained elevated. The patient continued to pass large volumes of dilute urine which failed to respond to pitressin administration. This picture of *nephrogenic diabetes insipidus* gradually resolved over a three-month period and renal function returned to normal.

Infants and children at risk

There is evidence to suggest that infants undergoing angiocardiography are particularly susceptible to renal damage from the circulating contrast material. In these severely ill children hypoxia and reduced renal perfusion contribute to renal ischaemia, and contrast material may lower renal blood flow even further leading to severe renal compromise. Gruskin *et al.* [29] examined the kidneys of 34 infants who died after cardiac studies before the age of four months. In seven there were histologic renal lesions which could be related to the contrast material, including vacuolization of the proximal tubule cells and medullary necrosis. Almost all these infants had received 3 ml/kg or more during the study. The authors concluded that the renal damage was dose related, and subsequently they have limited the total volume of contrast material to less than 3 ml/kg. Since the dose limitation was instituted no histologic lesions have been found in children dying after angiocardiography. Similar tubular lesions of osmotic nephrosis have recently been found after high dose eurography in adults but the significance of these findings is not yet clear [85].

In a few cases angiography has caused clinically obvious haemolysis and haemoglobinuria, especially in children with cyanotic congenital heart disease and polycythaemia. Once this sequence is recognized, measures should be instituted to achieve a diuresis and alkalinize the urine in order to prevent renal tubular damage and anuria [30].

Complications related to technique

For many years translumbar needle aortography was the only method for studying the abdominal aorta and its branches. Instances of renal complications were not rare, reaching 5 per cent of the examinations in some series. Although mechanical problems such as subintimal injection with dissection caused renal artery compromise occasionally, most renal insults from the translumbar technique were related to contrast medium nephrotoxicity (acetrizoate, diodrast, sodium iodide) or to accidental overinjection of one renal artery (usually the right) [31, 32]. Today a correctly performed translumbar needle or catheter aortogram, with fluoroscopic monitoring and test injection to assure proper needle tip placement, is a safe procedure and should have no higher incidence of renal complications than catheter aortography with equal volumes of contrast media. Indeed it is difficult to find reported examples of renal failure occurring after translumbar aortography with the current contrast media [9]. Thus, the procedure is not only acceptable for studying peripheral vascular disease in older patients, but it is actually preferred by some. This is due to the slightly increased incidence of vascular thrombosis accompanying retrograde femoral artery catheterization in the older age group, and the greater risks attendant to axillary artery catheterization, the two alternative approaches.

Atherosclerotic plaques can be dislodged by translumbar or catheter aortography resulting in renal embolization with infarction. Thrombi forming on the exterior of aortic catheters may separate from the catheter and similarly cause renal emboli. Although all of these are unusual with nonrenal angiography, accidents related to angiographic technique are not so uncommon with selective renal studies, and as such must be considered later in more detail.

COMPLICATIONS OF SELECTIVE RENAL ARTERIOGRAPHY

Contrast media nephrotoxicity

A wide variety of contrast media are in current use for selective renal arteriography (Table 4.1), indicating that an 'agent of choice' has not yet been declared. Why this is so can be explained by several phenomena: (1) These contrast agents are relatively nontoxic to the kidney at ordinary doses, and the safety factor even at double the volume of injection is considerable. (2) Minor damage to one kidney is easily overlooked and is rarely apparent by the usual clinical criteria. (3) There is no easily measurable, sensitive index of renal damage akin to seizures in the nervous system or arrhythmias and electrocardiographic changes in the heart. In these two organs the distressing clinical consequences of contrast-media-induced damage have stimulated the search for agents of choice and these have now been defined [33, 34]. (4) Assessments of contrast medium nephrotoxicity have shown no clear-cut differences between the agents in the same concentration range.

What indices of contrast media nephrotoxicity have been utilized to select

TABLE 4.1. Contrast media in use for selective renal arteriography in the United States.

Agent	Brand name	Concentration (gm %)	Per cent iodine	Osmolality (mosm/l)
Sodium diatrizoate	Hypaque sodium	50	30	1470
Meglumine diatrizoate	Renografin-M	60	28	1400
	Hypaque–Meglumine	60	28	1350
Meglumine–sodium diatrizoate	Renografin	60	29	1420
	Renografin	76	37	1690
	Renovist	69	37	1720
	Hypaque-M	75	39	1780
Meglumine iothalamate	Conray	60	28	1400

the best nephroangiographic agent? Examination of kidney histology was formerly the standard technique for this purpose, but it has now become clear that conventional histologic examination is too insensitive as a measure of renal damage from contrast media [35]. Transient changes (usually for the worse) in several kidney parameters have been documented in the first few minutes after experimental and clinical renal arteriography, but it is not certain what relationship these transient events have to nephrotoxicity. At least they are useful as rough guides for comparing the renal 'reactivity' of the various agents. The observed rise, then fall in renal blood flow [36], depression of E_{PAH} [37], and increase in urine enzyme excretion [38] have been similar for all agents tested in the 50–60 per cent concentration range. A few authors have claimed an advantage of iothalamate or diatrizoate, but their data are not convincing [39–41]. Other work similarly has failed to establish a difference in nephrotoxicity between iothalamate and diatrizoate [42, 43]. Further improvements in contrast media formulations can be expected to reduce toxicity. Iocarmate, the dimer of iothalamate, may offer significant reduction in nephrotoxicity because of its lower osmolality. It was predicted that the dimer would be less vasoactive than the monomer and this has now been confirmed for the kidney [44]. Of course it is premature to predict the eventual role of dimers in renal arteriography but they clearly merit further trials.

In cerebral and spinal arteriography it has been clearly established that the meglumine salts are less neurotoxic than the sodium salts. Other experiments, for example, *in vitro* studies of red blood cell clumping caused by contrast media [45] have also shown meglumine to be less reactive than sodium, perhaps related to the higher electrical conductivity of the sodium compounds. Although there are some hints that meglumine salts may be preferable for renal arteriography, this has not yet been confirmed. Indeed, recent experiments suggested a possible advantage for the sodium salts for renal arteriography, since sodium iothalamate and sodium iocarmate caused less vasodilatation in the dog's renal circulation than the corresponding meglumine salts [44]. If one takes low vasoactivity as a desirable property, i.e. low vasoactivity equals low nephrotoxicity, by this criterion the sodium salts would be advantageous. At present it must be stated that insufficient information is available to make a firm choice between the sodium and meglumine contrast media for renal arteriography and that a combination appears to be a reasonable compromise for now.

A recent survey revealed that many angiographers are successfully using contrast media more concentrated than 50–60 per cent for renal arteriography, in order to more consistently demonstrate the smaller intrarenal vessels [46]. While some have warned against using these agents of higher concentration [47], there is yet no experimental or clinical evidence that media such as 76 per cent meglumine–sodium diatrizoate have increased clinical nephrotoxicity over the 50–60 per cent group [46, 48]. Of course the angiographer should choose an agent combining adequate radiodensity and viscosity with low toxicity, and in general nephrotoxicity is related to concentration, osmolality, volume, contact time, and number of injections [49, 50]. The least concentrated material adequate to the diagnostic task should be used [47].

Overinjection

The volume of contrast material injected for selective renal arteriography is best governed by the renal blood flow. In the average adult with a flow of 600 ml/min for one kidney, a bolus of 8–10 ml in one second can be injected without significant reflux to the aorta. This dose level yields consistently satisfactory visualization of the intrarenal vessels and a good nephrogram without apparent nephrotoxicity. For a small or a diseased kidney the volume and speed of injection should be reduced according to the estimated renal blood flow; for example, in severe renal disease an injection of 1–2 ml/sec for several seconds may be sufficient [51]. In other circumstances it is desirable to inject considerably larger volumes of contrast media, e.g. in the range 20–40 ml, in order to demonstrate clearly the renal veins. This has been done for the most part in tumorous kidneys scheduled to be removed or in the presence of large arteriovenous fistulae. At these elevated dose levels, histological evidence of renal damage has not been found, but it must be recognized that most of the contrast material is shunted through the tumour or fistula.

What can be expected in the case of an accidental overinjection of 40–60 ml of contrast media into the renal artery of a normal kidney without associated catheter obturation of the renal artery? When this occurred with acetrizoate and other older contrast media, permanent renal damage with shrinkage was common. On the other hand, in the few published reports of this event with modern contrast media, the injected kidneys have done amazingly well with little evidence of permanent damage [52–54]. Flank pain, haematuria, and proteinuria may develop and persist for a few days with a variable but usually temporary interference with renal function. In one case maximal renal repair took several months [52]. Excretory urography, if performed within several days of the accident, will usually reveal some decrease in density and excretion of contrast material on the affected side, but follow-up urograms return to normal. Clearly, such coarse measurements of renal function as the urogram, serum creatinine, and even glomerular filtration rate are not sensitive enough to detect minor degrees of unilateral renal compromise.

It appears that serial radionuclide studies will provide the best non-invasive measure of unilateral renal damage under these circumstances. Either conventional isotope renography [55] or quantitative renal scintigraphy [56] with I^{131} hippurate can demonstrate unilateral renal impairment. The latter technique has been able to detect a 5–10 per cent loss in unilateral renal function. Our own experience with quantitative renal scintigraphy using 99m technetium–penicillamine has also proved reliable in detecting minor degrees of unilateral renal damage [57].

Treatment of an accidental overinjection should be aimed at flushing out the contrast material and establishing a diuresis. Pepper *et al.* [54] advocated an immediate infusion of 500 ml normal saline into the renal artery, followed by intravenous fluids. Their patient received 50 ml of 76 per cent meglumine–sodium diatrizoate into an artery supplying approximately one-half of the left kidney. No long-term renal damage could be demonstrated.

Obturation arteriography

While normal kidneys may tolerate excessively large volumes of contrast media if injected through properly placed catheters, this is clearly not the case if catheter obturation of the renal artery is present during the injection. A number of investigators have even demonstrated severe renal damage after the injections of *normal* volumes of contrast media when the catheter was wedged in the renal artery [58, 59]. The combination of prolonged contact between the hypertonic contrast material and the renal tissues and the resulting renal ischaemia and chemical damage presumably were responsible for irreversible kidney damage [58, 60, 61]. Injection of contrast media through a severe renal artery stenosis can, in a similar fashion, lead to undesirable prolonged contact between the opaque material and kidney tissues. To safeguard against obturation arteriography, free backflow from the catheter should be confirmed immediately before the injection and filming sequence. Catheters within a tortuous aorta may develop considerable slack which, during injection, can result in elongation and advancement of the catheter tip; hence, free backflow of blood from the catheter and a test injection should be checked immediately before the injection and filming sequence but after the slack has been reduced by gently pulling back the catheter from its most advanced position. We agree with Meaney *et al.* [31] that it is a sound practice to withdraw catheters from small main or accessory renal arteries immediately after the injection, and not wait until films have been processed.

Emboli

Embolic occlusion of a renal artery by clot is one of the most feared complications encountered during arteriography. Infarction is likely to follow because the renal arteries function as end arteries in this type of acute obstruction. Judging by the sparsity of case reports [62–64] such embolism might be considered to be so rare as to be hardly worth discussion, but this is not the case. For reasons that are not entirely clear, renal emboli during arteriography are simply not reported in most instances. Indeed most angiographers will admit to one or more intra-arteriographic renal emboli, and this writer is aware of at least six at three nearby teaching hospitals. Factors predisposing to this complication are (1) prolonged catheter time, (2) catheter exchanges, (3) multiple side-hole cathers, including the use of a tip-occluder for aortography, (4) insufficient or inefficient catheter flushing, and (5) hypercoagulable state. Catheter exchange is probably involved in most of the larger emboli. During withdrawal of the initial catheter, small platelet and/or fibrin thrombi adherent to the external catheter surface are 'wiped off' and build up at the puncture site. The new catheter picks up the clot material during insertion and carries it upstream. If the second catheter is used for selective renal arteriography, the thrombotic material can become dislodged and enter the renal circulation (Fig. 4.2). Should the catheter be used for aortography, embolization can occur to any of the aortic branches downstream,

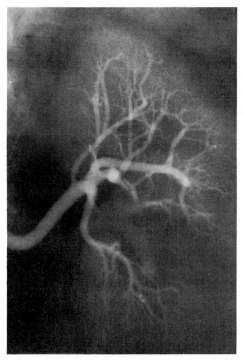

FIGURE 4.2a

including the renal arteries. The routine use of heparin-coated catheters [65] or short-term systemic heparinization virtually eliminates thrombotic complications of arteriography, but the individual angiographer will have to determine whether the potential benefits justify the risks of systemic heparinization [66]. An increasing number of angiographers are adopting this routine of short-term systemic heparinization for arteriography.

If renal embolization occurs, what can be done? Emboli to segmental or more peripheral renal arteries are best handled by immediate systemic heparinization for several days [64]. Serial radionucleide scans will aid in documenting the volume of kidney involved. If renal infarction is present arise in serum lactic dehydrogenase (LDH) will occur and the magnitude of the LDH elevation is a rough guide to the extent of the infarction [67]. Emboli to the main renal artery create a much more critical situation. Although these larger clots may lyse and move along the renal arterial tree, the possibility of total renal infarction should make one reluctant to offer expectant treatment only. Recently, transcatheter embolectomy has been tried successfully and it appears reasonable to attempt this technique initially

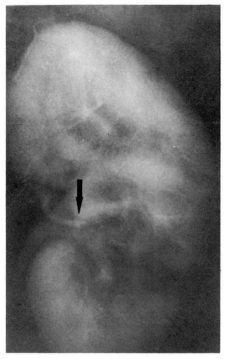

b

FIGURE 4.2. Embolus occurring during selective renal arteriography. A catheter remained in the aorta for 45 minutes while renal vein sampling for renin determinations was carried out. The subsequent aortogram was normal. Following catheter exchange, selective left renal arteriography (a) showed a fresh occlusion of several branch arteries to the lower pole. A localized perfusion defect was present on the nephrogram phase (b), and one of the obstructed arteries was still opacified with contrast material proximal to the embolus (arrow).

in conjunction with systemic heparinization [62]. In the event of an unsuccessful transcatheter embolectomy, main renal artery embolism should be treated surgically (59) if preservation of renal tissue is deemed essential and the patient can tolerate an operation.

Cholesterol emboli

Atheromatous or cholesterol embolization to the kidney is a specific complication of arteriography in patients with advanced atherosclerosis [68–70]. Release of atheromatous material from the diseased aortic wall is related to catheter, needle, or guide wire manipulation, or to the injection jet itself. Showers of emboli may reach any downstream organ, but when the kidney is involved renal failure may result. Cholesterol embolization to the kidney has also been described after abdominal aortic aneurysm surgery, but probably occurs most often in multiple spontaneous episodes. That it is a common cause of renal scarring in the elderly is supported by the autopsy incidence of 3–4 per cent [70]. Since most atheromatous emboli are small, they obstruct small muscular arteries. Thus infarcts, if produced at all, are subclinical. It is interesting to note that a majority of the patients with cholesterol embolization as a complication of arteriography had pre-existing mild azotaemia, further evidence that repeated episodes are common. The development of postangiographic renal failure in an elderly patient with advanced atherosclerosis should suggest the possibility of cholesterol embolization, especially if back pain, *livedo reticularis* of the lower trunk and/or legs, or gastrointestinal complaints are also present [71].

Small air emboli may occur during arteriography due to loose connections in the catheter system or failure to expel air completely from the system prior to injection, or due to 'cavitation' at the catheter tip. While the angiographic demonstration of air embolism may cause embarrassment to the angiographer and technologist, there is generally no problem to the patient when this happens in abdominal or renal arteriography. Clearly this is not true in the coronary or cerebral circulation. As an extreme example we have witnessed the accidental mechanical injection of 30 ml of air *instead* of contrast material for renal aortography. The resulting renal air arteriogram was distressing to say the least, but a positive contrast aortogram done 30 minutes later revealed very little perfusion abnormality and the patient suffered no adverse sequelae.

Foreign material may embolize to the kidneys if allowed access to the catheter-syringe system. Instances of cotton fibre or surgical glove powder embolization were not rare when open bowls for flushing solutions were in use, but the institution of closed reservoir-tubing systems for catheter flushing has largely eliminated this problem [72–74]. Still, wiping wound guide wires with dry or wet cotton gauze does deposit small fibres in the interstices, which can lead to fibre embolization [72]. Obviously the smoother wires such as ones coated with Teflon will be less at fault in this respect.

Dissection

Fortunately renal artery dissection as a complication of selective renal arteriography is a rare event, since the consequences are likely to be severe [31, 32, 75, 76, 83]. Narrowing or even complete occlusion of the renal artery can result from such an iatrogenic intramural haematoma. The initial tear is created either by direct trauma from a misplaced guide wire, the tip of the angiographic catheter itself, or by the jet of contrast material during injection. Careful fluoroscopic observation of the test injection should reveal the presence of this complication as contrast material will remain localized near the catheter tip in a crescentic, linear, or sleevelike accumulation. If this is recognized further test injection of contrast media into the renal artery should not be carried out as each additional intramural injection increases the likelihood of ultimate vascular occlusion. The catheter should be immediately withdrawn to the aorta for aortography to determine the extent of vascular compromise, since treatment depends on the location and extent of the dissection. Rarely, subintimal dissection will occur during the injection for filming in spite of preliminary testing having shown a satisfactory catheter position. This can be explained by forward migration of the catheter tip during patient positioning or during the injection, an undesirable consequence of too much catheter slack in a patient with tortuous arteries.

If the dissection has caused segmental renal artery occlusion with interruption of flow to a portion of the kidney only, surgical therapy is unlikely to be of benefit. As in branch artery emboli, systemic heparinization for a few days is sufficient as long as serial radionucleide studies show no deterioration of the perfusion pattern. It is likely that some degree of segmental renal infarction will result (Fig. 4.3).

If the post-dissection aortogram demonstrates total main renal artery occlusion, it would appear that immediate surgery is indicated; however, we are aware of at least two such cases where at surgery the involved renal artery was found to be patent with satisfactory pulsations distally [77, 78]. At present we believe it is desirable to perform a radionucleide blood flow study just prior to surgery to detect these spontaneous 'cures' of total renal artery occlusion from dissection, and thereby avoid unrewarding surgery.

Two patients have developed acute, severe, renovascular hypertension as a result of iatrogenic renal artery dissection [31, 76]. In one, split ureteral catheterization studies revealed characteristic findings of a functionally significant renal artery stenosis and the patient indeed became normotensive after nephrectomy [76].

While it might be assumed that patients with atherosclerosis are more liable than others to incur an iatrogenic renal artery dissection, such is not necessarily the case. Several young patients have suffered this problem, apparently without predisposing factors. In addition, renal artery dissection has been identified as a specific, spontaneous complication of fibromuscular dysplasia, particularly with intimal fibroplasia or fibromuscular hyperplasia

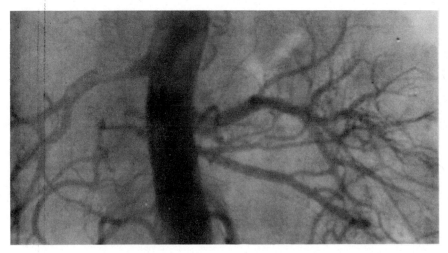

a

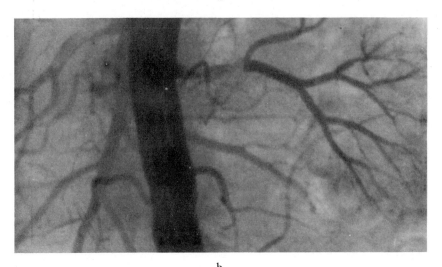

b

FIGURE 4.3a and b

[75, 79]. Since these younger patients are likely to undergo renal arteriography it is important that aortography precedes selective studies in order to document the presence or absence of a dissection before the catheter is manipulated within the renal artery.

Risks of renal angiography in diseased kidneys

While considerable experience has accumulated with intravenous urography

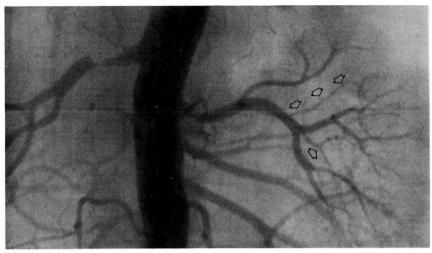

c

FIGURE 4.3. Iatrogenic left renal artery dissection. (a) Initial aortogram done for hypertension shows severe right renal artery stenosis with thrombus distal to the stenosis. Left renal artery narrowing is partially obscured by overlapping vessels. During catheterization of the left renal artery, a test injection showed the presence of a dissection. A repeat aortogram (b) demonstrates marked compromise of the renal artery lumen with complete occlusion of the dorsal division. The patient was treated with systemic heparinization only, and the aortogram done seven weeks later (c) shows re-establishment of the left renal artery lumen. However, the dorsal division is attenuated (arrows) and the kidney has lost 1.5 cm in length. [78] Courtesy *Radiology*.

and nonrenal angiography in patients with a variety of renal diseases, there is much less known about the risks of selective renal arteriography in this population. On theoretical grounds one would predict increased contrast media nephrotoxicity because diseased kidneys often exhibit reduced and slowed blood flow (increased vascular resistance) leading to a prolonged contact between contrast agent and the kidney vasculature [51]. Experimental studies support this contention of increased risk in abnormal kidneys. Chou et al. [80] found a much greater renal vasoconstriction from arteriography in kidneys with a high baseline vascular resistance, and recommended that renal arteriography be done with caution in patients whose renal resistance is likely to be high. Another problem with arteriography in kidneys with chronic disease relates to the size of the main renal artery. Since these are often small there is an increased chance of catheter-related complications and obturation arteriography.

In one series of patients undergoing selective renal arteriography, pre-

and postangiography I^{131} hippuran renograms were done to assess renal damage from the angiographic procedure [55]. Twelve per cent of diseased kidneys developed an abnormal postangiography renogram compared to only 5 per cent of normal kidneys. From the data and pictures presented it appeared that the increased ineidence of abnormal renograms in diseased kidneys could best be explained by overinjection of contrast media, either with or without catheter obturation of the renal artery, rather than by increased contrast medium nephrotoxicity *per se*.

It is extremely difficult to document a further deleterious effect of arteriography in patients with irreversible, end-stage kidney disease undergoing chronic dialysis. In this group renal function is already so severely compromised that it is hard to imagine angiography making it any worse. On the other hand, in patients with renal insufficiency who are not yet requiring dialysis, it is extremely important to protect the remaining renal function. Therefore if renal arteriography is indicated, adequate hydration should be insured and only unilateral arteriography with a single injection is recommended. Using these guidelines and careful catheterization technique, no worsening of renal failure should be expected [51].

Risks of pharmacoangiography

Both vasoconstricting and vasodilating drugs have been used in selected circumstances for enhanced renal angiography. Based on the considerations given above, any drug which increases renal vascular resistance should potentiate the nephrotoxicity of contrast media. This has been confirmed in rabbits, where the combination of high doses of intra-arterial epinephrine followed by a reasonable dose of 76 per cent meglumine–sodium diatrizoate proved more toxic to the kidney than the administration of either substance alone [81]. No damage could be demonstrated, however, when clinically comparable levels of epinephrine were used together with a contrast material. Although documented cases of renal damage due to pharmacoangiography have not been reported in patients, this may be more a reflection of our insensitive renal function tests than a true lack of toxicity. At any rate, the potential for increased toxicity is real and more than one injection of a vasoconstricting drug is not generally recommended.

Ex-vivo *renal arteriography*

This new and unusual application of arteriography may be expected to find increased application as refinement of and enthusiasm for 'bench surgery' continues. In bench surgery the kidney is removed, perfused under hypothermic conditions, appropriate surgery performed with the kidney still outside the body, and finally autotransplanted. Special indications for the technique are still being formulated, but the following conditions have already been cited in this regard: renal trauma, bilateral renal malignancies, malignancy in a solitary kidney, and advanced stag-horn calculus disease. An arteriographic study of the perfused kidney is often essential for a successful dissection and repair, and the techniques of *ex vivo* arteriography have been described [82]. A severe complication of *ex vivo* arteriography has already been reported resulting in irreversible damage to the kidney [82]. Apparently, diatrizoate crystallizes or precipitates at the low temperatures used for hypothermic perfusion, thereby causing occlusion of the small renal vessels and ischaemic injury. Fortunately, meglumine iothalamate is stable even below 0°C, making it the agent of choice under these circumstances. The authors carefully point out that the contrast media must be immediately washed out with electrolyte solution.

Risks of arteriography in transplanted kidneys

Recently, a retrospective investigation of 173 kidney transplants concluded that allograft arteriography may contribute to, or elicit, acute rejection [86]. While this was not a controlled study, the data are impressive enough to warrant a more conservative approach to arteriography in recently-transplanted patients. Specifically, it does not appear wise to perform such arteriography for a 'baseline' examination in a patient who otherwise has normal renal function and no signs of rejection. There have been other suggestions in the recent literature that even excretory urography may hasten rejection on occasion, but the data at this time are not nearly so convincing.

REFERENCES

1 DOUST B.D. & REDMAN H.C. (1972) The myth of 1 ml/kg in angiography: a study to determine the relationship of contrast medium dosage to complications. *Radiology* **104**, 557–560.
2 PORT F.K., WAGONER R.D. & FULTON R.K. (1974) Acute renal failure following angiography. *Am. J. Roentg.* **121**, 544–550.
3 BORRA S., HAWKINS D., DUGUID W. & KAYE M. (1971) Acute renal failure and nephrotic syndrome after angiocardiography with meglumine diatrizoate. *New Engl. J. Med.* **284**, 592–593.
4 KOVNAT, P.J., LIN K.Y. & POPKY G. (1973) Azotemia and nephrogenic diabetes insipidus after arteriography. *Radiology* **108**, 541–542.

5 STARK F.R. & COBURN J.W. (1966) Renal failure following methylglucamine diatrizoate (Renografin) aortography: report of a case with unilateral renal artery stenosis. *J. Urol.* 96, 848–851.
6 KHAN M.A., PILLAY V.K.G. & WANG F. (1972) Peripheral arteriography causing acute renal failure. *South Afr. Med. J.* 46, 522.
7 TEAL J.S., RUMBAUGH C.L., SEGAL H.D., BERGERON R.T. & SCANLAN R.L. (1972) Acute renal failure following spinal angiography with methylglucamine iothalamate. *Radiology* 104, 561–562.
8 CATTELL W.R., MCINTOSH C.S., MOSELEY I.F. & FRY I.K. (1973) Excretion urography in acute renal failure. *Br. Med. J.* 2, 575–578.
9 LANG, E.G. (1963) A survey of complications of percutaneous retrograde arteriography: Seldinger technique. *Radiology* 81, 257–263.
10 ROBERTSON P.W., DYSON M.L. & SUTTON P.D. (1969) Renal angiography. A review of 1750 cases. *Clin. Radiol.* 20, 401–409.
11 KOVNAT P., LABOVITZ E. & LEVISON S.P. (1973) Antibiotics and the kidney. *Med. Clin. N.A.* 57, 1045–1063.
12 Editorial. (1971) Contrast media and the kidney. *Br. Med. J.* 2, 4–5.
13 MCEVOY J., MCGEOWN M.G. & KUMAR R. (1970) Renal failure after radiological contrast media. *Br. Med. J.* 4, 717–718.
14 DUGGAN F.J., JR., ROHNER T.J., JR. (1973) Acute renal insufficiency following oral cholecystography. *J. Urol.* 109, 156–159.
15 MUDGE G.H., COOKE W.J. & BERNDT W.O. (1974) Renal excretion of urea in the dog during onset and subsidence of diuresis. *Am. J. Physiol.* 227, 369–376.
16 FOLIN J. (1968) Complications of percutaneous femoral catheterization for renal angiography. *Radiologé* 8, 190–195.
17 MUDGE G.H. (1971) Uricosuric action of cholecystographic agents: a possible factor in nephrotoxicity. *New Engl. J. Med.* 284, 929–933.
18 TALNER L.B. (1972) Urographic contrast media in uremia. *Radiol. Clin. N.A.* 10, 421–432.
19 DAVIDSON A.J., BECKER J., ROTHFIELD N. et al. (1970) An evaluation of the effect of high dose urography on previously impaired renal and hepatic function in man. *Radiology* 97, 249–254.
20 MILMAN N. & STAGE P. (1974) High dose urography in advanced renal failure: II. Influence on renal and hepatic function. *Acta Radiol. Diagn.* 15, 104–112.
21 METYS R., HORNYCH A., BURIÁNOVÁ & JIRKA J. (1971) Influence of tri-iodinated contrast media on renal function. *Nephron* 8, 559–565.
22 MCLACHLAN M.S.F., CHICK S., ROBERTS E.E. & ASSCHER A.W. (1972) Intravenous urography in experimental acute renal failure in the rat. *Invest. Radiol.* 7, 466–473.
23 WALSH P.C., GITTES R.F. & LECKY J.W. (1970) Aortography in experimental renal failure: evaluation of contrast media toxicity. *Radiology* 97, 33–37.
24 TASKER P.R.W., MACGREGOR G.A. & DE WARDENER H.E. (1974) Prophylactic use of intravenous saline in patients with chronic renal failure undergoing major surgery. *Lancet* 2, 911.
25 RHEA W.G., JR., KILLEN D.A. & FOSTER J.H. (1964) Protection against nephrotoxicity of iothalamic acid. *Arch. Surg.* 89, 294–298.
26 FELDMAN H.A., GOLDFARB S. & MCCURDY D.K. (1974) Recurrent radiographic dye-induced acute renal failure. *J. Am. med. Ass.* 229, 72.

27 MAILLOUX L., SWARTZ C.D., CAPIZZI R., KIM K.E., ORESTI G., RAMIREZ O. & BREST A.N. (1967) Acute renal failure after administration of low-molecular-weight dextran. *New Engl. J. Med.* 277, 1113–1118.
28 JANSSEN C.W. (1968) Osmotic nephrosis: a clinical and experimental investigation. *Acta Chir. Scand.* 134, 481–487.
29 GRUSKIN A.B., OETLIKER O.H., WOLFISH N.M., GOOTMAN N.L., BERNSTEIN J. & EDELMANN C.M., JR. (1970) Effects of angiography on renal function and histology in infants and piglets. *J. Pediatrics* 76, 41–48.
30 COHEN L.S., KOKKO J.P. & WILLIAMS W.H. (1969) Hemolysis and hemoglobinuria following angiography. *Radiology* 92, 329–332.
31 MEANEY T.F., LALLI A.F. & ALFIDI R.J. (1973) *Complications and Legal Implications of Radiologic Special Procedures*, ch. 2. C. V. Mosby Co., St Louis.
32 WEIGEN J.F. & THOMAS S.F. (1973) *Complications of Diagnostic Radiology*, ch. 5. Charles C. Thomas, Springfield, Illinois.
33 HILAL S.K. (1966) Hemodynamic changes associated with the intra-arterial injection of contrast media. *Radiology* 86, 615–633.
34 PAULIN S. & ADAMS D.F. (1971) Increased ventricular fibrillation during coronary arteriography with a new contrast medium preparation. *Radiology* 101, 45–50.
35 EVENSON A. & SKALPE I.O. (1971) Cell injury and cell regeneration in selective renal arteriography in rabbits. *Invest. Radiol.* 6, 299–303.
36 SHERWOOD T. & LAVENDER J.P. (1969) Does renal blood flow rise or fall in response to diatrizoate? *Invest. Radiol.* 4, 327–329.
37 SORBY W.A. & HOY R.J. (1968) Renal arteriography and renal function: the short-term effect of aortography, selective renal arteriography, adrenalin suppression and renal venography on PAH extraction ratio. *Austral. Radiol.* 12, 252–255.
38 TALNER L.B., RUSHMER H.N. & COEL M.N. (1972) The effect of renal artery injection of contrast material on urinary enzyme excretion. *Invest. Radiol.* 7, 311–322.
39 LÉLEK I. (1970) Experimental study of tubular lesions and chronic renal failure caused by selective renal arteriography in the dog. *Acta chir. Acad. Sci. hung.* 11, 289–301.
40 LÉLEK I. & POKORNY L. (1967) Untersuchungen über die nephrotoxizität der jothalamathaltigen kontrastmittel mittels experimenteller renaler angiographien. *Fortschr. Röntgenstr.* 106, 247–255.
41 HAYES C.W., FOSTER J.H., SEWELL R. & KILLEN D.A. (1966) Experimental evaluation of concentrated solutions of iothalamic acid derivatives as angiographic contrast media. *Am. J. Roentg.* 97, 755–761.
42 STEWART B.H., DIMOND R.L., FERGUSON C.F. & SHEPARD P.B. (1965) Experimental renal arteriography: comparison of spinal cord and renal toxicity from iothalamate and diatrizoate compounds. *J. Urol.* 94, 695.
43 RHEA W.G., JR., KILLEN D.A. & FOSTER J.H. (1965) Relative nephrotoxicity of Hypaque and Angio-Conray. *Surgery* 57, 554–558.
44 RUSSELL S.B. & SHERWOOD T. (1974) Monomer/dimer contrast media in the renal circulation: experimental angiography. *Br. J. Radiol.* 47, 268–271.
45 KUTT H., VEREBELY K., BANG N., STRUELI F. & McDOWELL F. (1966) Possible mechanisms of complications of angiography. *Acta Radiol.* 5, 276–289.
46 TALNER L.B. & SALTZSTEIN S. (1975) Renal arteriography: the choice of contrast material. *Invest. Radiol.* 10, 91.

47 OLSSON O. (1971) Technique and hazards of renal angiography. In: *Angiography* (Ed. by H. Abrams), 2nd edn, ch. 46. Little, Brown & Co., Boston.
48 KNOX F.G., BUNNELL I.L., ELWOOD C.M. & SIGMAN E.N. (1965) The effect of selective renal angiography on glomerular filtration rate and renal plasma flow in man. *The Physiologist* 8, 210.
49 IDBOHRN H. & BERG N. (1954) On the tolerance of the rabbit's kidney to contrast media in renal angiography: a roentgenologic and histologic investigation. *Acta Radiol.* 42, 121–140.
50 MUDGE G.H. (1970) Some questions on nephrotoxicity. In: The Symposium on Contrast Media Toxicity of the Association of University Radiologists. *Invest. Radiol.* 5, 407–423.
51 MENA E., BOOKSTEIN J.J. & GIKAS P.W. (1973) Angiographic diagnosis of renal parenchymal disease. *Radiology* 108, 523–532.
52 SIDD J.J. & DECTER A. (1967) Unilateral renal damage due to massive contrast dye injection with recovery. *J. Urol.* 97, 30–32.
53 LAUBSCHER W.M.L. & RAPER F.P. (1960) A report of a case of the injection of a massive dose of Urografin into the renal artery. *Br. J. Urol.* 32, 160–164.
54 PEPPER H.W., KOROBKIN M.T. & PALUBINSKAS A.J. (1974) Massive injection of contrast medium into a renal artery segment: a case report. *Radiology* 112, 273–274.
55 KAUDE J. & NORDENFELT I. (1973) Influence of nephroangiography on 131-I-hippuran nephrography. *Acta Radiol.* 14, 69–81.
56 HAYES M., BROSMAN S. & TAPLIN G.V. (1974) Determination of differential renal function by sequential renal scintigraphy. *J. Urol.* 111, 556–559.
57 TALNER L.B. & SOKOLOFF J. Quantitative renal scintigraphy with ^{99m}Tc-penicillamine: a noninvasive split renal function test. (In preparation.)
58 OBREZ I. & ABRAMS H.L. (1972) Temporary occlusion of the renal artery: effects and significance. *Radiology* 104, 545–556.
59 HARTMANN H.R., NEWCOMB A.W., BARNES A. & LOWMAN R.M. (1966) Renal infarction following selective renal arteriography. *Radiology* 86, 52–56.
60 SMIDDY F.G. & ANDERSON G.K. (1960) Tolerance of the kidneys to the contrast medium Urografin. *Br. J. Urol.* 32, 156–159.
61 FARRY P.J., BEALE L.R. & MACBETH W.A.A.G. (1970) Intrarenal extravasation complicating selective vessel angiography. *N. Zealand Med. J.* 72, 17–18.
62 BUXTON D.R., JR. & MUELLER C.F. (1974) Removal of iatrogenic clot by transcatheter embolectomy. *Radiology* 111, 39–41.
63 MORROW I. & AMPLATZ K. (1966) Embolic occlusion of the renal artery during aortography. *Radiology* 86, 57–59.
64 MCCONNEL R.W., FORE W.W. & TAYLOR A. (1973) Embolic occlusion of the renal artery following arteriography: successful management. *Radiology* 107, 273–274.
65 CRAMER R., FRECH R.S. & AMPLATZ K. (1971) A preliminary human study with a simple nonthrombogenic catheter. *Radiology* 100, 421–422.
66 WALLACE S., MEDELLIN H., DEJONGH D. & GIZNTURCO C. (1972) Systemic heparinization for angiography. *Am. J. Roent.* 116, 204–209.
67 SAKATI I.A., DEVINE P.C., DEVINE C.J., FIVEASH J.G. & POUTASSE E.F. (1968) Serum lactic dehydrogenase in acute renal infarction and ischemia. *New Engl. J. Med.* 278, 721–722.

68 HARRINGTON J.T., SOMMERS S.C. & KASSIVER J.P. (1968) Atheromatous emboli with progressive renal failure. *Ann. Int. Med.* 68, 152–160.
69 LONNI Y.G.W., MATSUMOTO K.K. & LECKY J.W. (1969) Postaortographic cholesterol (atheromatous) embolization. *Radiology* 93, 63–65.
70 SIENIEWICZ D.J., MOORE S., MOIR F.D. & MCDADE D.F. (1969) Atheromatous emboli to the kidneys. *Radiology* 92, 1231–1240.
71 SCHWARTZ S. & WATERS L. (1973) Cholesterol embolization. *Radiology* 106, 37–41.
72 KAY J.M. & WILKINS R.A. (1969) Cotton fibre embolism during angiography. *Clin. Radiol.* 20, 410–413.
73 ADAMS D.F., OLIN T.B. & KOSEK J. (1965) Cotton fiber embolization during angiography. *Radiology* 84, 678–681.
74 YUNIS E.J. & LANDES R.R. (1965) Hazards of glove powder in renal angiography. *J. Am. med. Ass.* 193, 304–305.
75 HARE W.S.C. & KINCAID-SMITH P. (1970) Dissecting aneurysm of the renal artery. *Radiology* 97, 255–263.
76 GILL W.B., COLE A.T. & WONG R.J. (1972) Renovascular hypertension developing as a complication of selective renal arteriography. *J. Urol.* 107, 922–924.
77 REISS M.D., BOOKSTEIN J.J. & BLEIFER K.H. (1972) Radiologic aspects of renovascular hypertension. Part 4. Arteriographic complications. *J. Am. med. Ass.* 221, 374–378.
78 TALNER L.B., MCLAUGHLIN A.P. & BOOKSTEIN J.J. (1975) Renal artery dissection: a complication of catheter arteriography. *Radiology* 117, 291–295.
79 KINCAID O.W., DAVIS G.D., HALLERMANN F.J. & HUNT J.C. (1968) Fibromuscular dysplasia of the renal arteries: arteriographic features, classification, and observations on natural history of the disease. *Am. J. Roentg.* 104, 271–282.
80 CHOU C.C., HOOK J.B., HSIEH C.P., BURNS T.D. & DABNEY J.M. (1971) Effects of radiopaque dye on renal vascular resistance. *J. Lab. Clin. Med.* 78, 705–712.
81 REDMAN H., OLIN T.B., SALDEEN T. & REUTER S.R. (1966) Nephrotoxicity of some vasoactive drugs following selective intra-arterial injection. *Invest. Radiol.* 1, 458–464.
82 CORMAN J.L., GIRARD R., FIALA M., GALLOT D., STONINGTON O., STABLES D.P., TAUBMAN J. & STARZL T.E. (1973) Arteriography during *ex vivo* renal perfusion: a complication. *Urology* 2, 222–226.
83 ENGBERG A., ERIKSON U., KILLANDER A. *et al.* (1974) An unusual complication of selective renal angiography: a case presentation. *Austral. Radiol.* 18, 304–307.
84 OLDER R.A. (1976) Angiographically induced renal failure and its radiographic detection. *Am. J. Roentg.* 126, 1039–1045.
85 MOREAU J.F., DROZ D., SABTO J. *et al.* (1975) Osmotic nephrosis induced by water-soluble tri-iodinated contrast media in man. *Radiology* 115, 329–336.
86 HEIDEMAN M., CLAES G. & NILSON A.E. (1976) The risk of renal allograft rejection following angiography. *Transplantation* 21, 289–293.

CHAPTER 5: SPINAL CORD COMPLICATIONS OF ANGIOGRAPHY

G. MARGOLIS

In May 1970 this writer presented an overview of the problem of contrast media toxicity [1]. The concepts presented therein were based upon observations derived from an extended series of experimental studies of neurotoxic sequelae of retrograde aortography in the dog. That paper opened with the following image:

> Let us behold the angiographer as he chooses his favorite radiopaque medium, transfers it into his syringe and activates the plunger. His aim is to expel this agent into an arterial tree as a concentrated bolus with such rapidity that its radiopacity is not lost through dilution. Has he been adequately warned that in its packaged form, i.e., in his injection syringe, this medium is histotoxic? Does he know his gamble, that he is dependent upon a degree of dilution sufficient to reduce the concentration below the histotoxic level? I sincerely doubt it.... Stated in another way, angiography may be compared to ice skating, with the thickness of the ice representing the margin of safety. Through much of the past the ice has been perilously thin. Remarkably, angiographic accidents have been infrequent, but here again the simile holds. When accidents have occurred they have been characterized by their unpredictability and their severity, comparable to the skater breaking a hole in the ice. With the development of newer less toxic angiographic media the ice has become thicker. But with the growth of diagnostic angiology the pond is more crowded, and with the new acrobatics being used—pharmaco-angiographic approaches—the ice is under ever greater stress.

Six years later that word picture still remains valid. Witness the report of

The extended studies on which this paper is based were supported by Grants B-1102, NB-2265, NB-5160, FR-5392, HD-03298, and 5T1-GM-207 from the National Institutes of Health.

Kardjiev et al. [2] of five examples of spinal cord injury following selective bronchial or intercostal arteriography with exceedingly small volume doses of contrast media—3 to 5 ml of Jodamid-380 (70 per cent methylglucamine and 9.7 per cent sodium salt of 3-acetamide-5-(acetamide-methyl)-2,4,6-triodobenzoic acid) or 1 to 5 ml of Uropolinum (sodium 3,5-diacetamido-2,4,6-triodobenzoate). There is still no room for complacency. The concepts presented in that earlier report, therefore, will serve as an appropriate background and point of departure for the present chapter.

The advantages of this experimental model are multiple. First, the cord is a sensitive functional indicator of immediate and residual toxic effects. Second, it is a nonvital test site which allows survival of the subject and full evolution of the anatomic lesion. Third, the cord lesion serves as a prototype for contrast medium injury of the nervous system and possibly for all tissue sites. Fourth, the frequency of cord injury from unintentional injection of its vascular bed in angiography indicates the need for study of factors underlying this phenomenon. Fifth, the sensitivity of this experimental model to the modulating influences of pharmacologic and physiologic nature facilitates investigation of the role of haemodynamic factors in the injury. Sixth, with the use of fluorescein cineangiography sensitive *in vivo* observations of the relationships of vasotoxic and neurotoxic effects may be made.

A brief summary of known facts concerning contrast medium toxicity derived from studies made with this experimental model is presented below. Fig. 5.1 presents a capsule summary of the evolution of this injury as viewed in fluorescein angiography [1, 3, 4]. Features illustrated include: (1) the concentration-dependent toxic action of sodium acetrizoate (Urokon); (2) the striking alterations of circulatory dynamics (a) even following a 35 per cent concentration nontoxic injection mass (accelerated flow, vasodilatation) and (b) with a 70% concentration toxic injection mass (temporary arrest of flow, break in blood–brain barriers, local sludging, resistance to perfusion); (3) the functional disturbances (spinal convulsions, succeeded by hyperreflexia, progressing to areflexia and paralysis); (4) the anatomical lesion (myelomalacia of the injected zone). The series of events charted here were initially induced by Urokon but subsequently have been induced with other contrast agents tested in this model, including Hypaque (sodium diatrizoate), Renografin (methylglucamine diatrizoate), Angioconray (sodium iothalamate), and Conray (methylglucamine iothalamate).

The dynamics of this chain of events can be appreciated best by firsthand observations, such as described in earlier reports [3, 4] and by cine recordings, details of which are yet unpublished [5]. Notably, the onset of these reactions—seizures, arrest of flow, and disruption of the blood–brain

136   *Chapter 5*

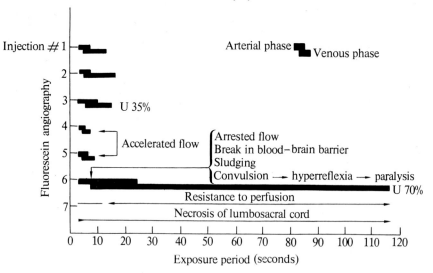

FIGURE 5.1. This chart is a composite of the major neurotoxic actions of angiographic agents, depicted by fluorescein angiography and correlated with functional and pathologic effects. These events were observed in an experimental model of midstream retrograde aortography in the dog, with the spinal cord being the site of study. Each injection indicates the flow of an aortic injection mass through the regional vascular bed of the cord which has been surgically exposed by laminectomy. Injections 1, 2, 4, 5 and 7 are of physiological saline in volume doses of 1 ml per kg. Injection 3 is of 35 per cent sodium acetrizoate (Urokon) and injection 6 is of 70 per cent acetrizoate in the same volume doses. These successive injections were made at ten to fifteen minute intervals. For further details see Margolis [1], Margolis et al. [4, 11], and Abraham et al. [14].

barrier—is timed precisely with the arrival of the injection mass in the arterial bed of the spinal cord. The instantaneous nature of the permeability break is emphasized by Fig. 5.2, a frame excerpted from a fluorescein cineangiographic record fifteen seconds after entry of a bolus of 70 per cent Hypaque into the cord circulation. Build-up of fluorescence in the cord parenchyma is so rapid that within 90 seconds the cord parenchyma is diffusely permeated (Fig. 5.3) and fluorescence of the cerebrospinal fluid is already manifest.

These observations of disruption of blood–brain–cerebrospinal fluid permeability barriers have certain practical implications. First, they are consistent with ultrastructural observations of changes induced by contrast agents in cerebrovascular endothelium. Waldron et al. [6] have recently demonstrated that repeated angiography may break down the blood–brain

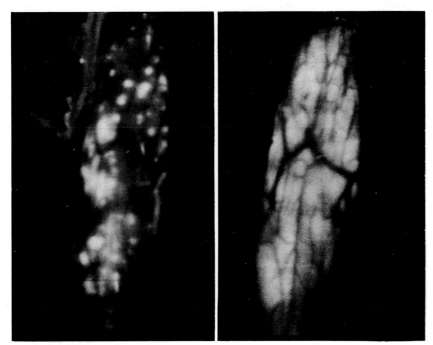

FIGURE 5.2.   FIGURE 5.3.

FIGURE 5.2. Frame from a fluorescein angiography record enlarged from a 16-mm film of the flow of a toxic dose of contrast agent (70 per cent Hypaque) through the vascular bed of the cord. The mass had been injected via retrograde aortography and rendered neurotoxic by the neuripetal shunt induced by levarterenol. This photograph, made fifteen seconds after the arrival of the fluorescent injection mass in the cord illustrates the instantaneous onset of fluorescence due to the disruptive action upon the blood–brain barrier.

FIGURE 5.3. Frame from the same film sequence 90 seconds after entry of the fluorescein 'labelled' contrast agent into the vascular bed of the cord. The parenchyma is now diffusely fluorescent, and the surface vessels appear in dark contrast against the light background. The rapidity and severity of breakdown of the blood–brain barrier clearly demonstrated that the contrast medium injury was a direct histotoxic effect, rather than a consequence of ischaemia. It was this observation that pointed to the need for a search for prophylactic rather than remedial approaches.

barrier (1) by alteration of tight junctions and allowing passage between cells at these zones and (2) by transport across endothelial cells by pinocytosis. Rapoport *et al.* [7–9] described the operation of two factors in the opening of the blood–brain barrier. One is an anion specific factor, defined as the lipid solubility of the contrast medium, which determines passage *through* cerebrovascular endothelial cell membranes of the unaltered barrier. The other is a nonspecific factor, the osmolality of the contrast agent, which can open tight junctions and permit passage *between* cells. Their analysis indicated that neurotoxicity parallels lipid solubility rather than osmolality, an interpretation which favours a direct cytotoxic action.

Observations of a therapeutic regimen used by Mishkin et al. [10] for relief of neurotoxic manifestations of clinical angiography offer further insights into the nature of contrast medium injury. They have instituted

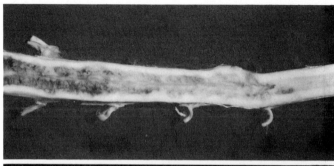

FIGURE 5.4.

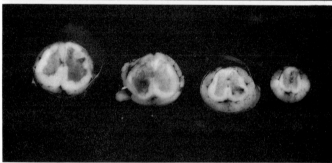

FIGURE 5.5.

FIGURES 5.4. and 5.5. Demonstration in longitudinally and transversely sectioned specimens, of the characteristic macroscopic features of contrast medium injury of the spinal cord in the experimental model of canine aortography. The striking selectivity of the noxious action upon the highly vascularized grey matter is clearly illustrated. For further details see Margolis et al. [11].

repeated cerebrospinal lavage and drainage in several subjects with neurotoxic symptoms, using cerebrospinal fluid iodine levels as indications of the relative amount of contrast agent in the cord parenchyma. Early clinical tests of this procedure have resulted in improvement of the post-injection neurological status. These results are consistent with the foregoing observations of an immediate break in permeability barriers induced by toxic injection masses. Moreover, relief of symptoms by lavage indicates that extravasated contrast media may exert not only immediate neurotoxic effects (seizures) but that their ultimate noxious action (necrotizing effect) is in large part the result of prolonged contact with neural parenchyma. Further, the ready exchange between parenchyma and cerebrospinal fluid provides an avenue for removal of extravasated contrast media.

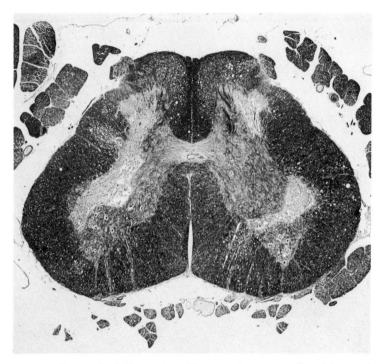

FIGURES 5.6, 5.7 and 5.8. Illustration of features of contrast medium injury to the cord at a microscopic level. Myelin appears dark grey in these sections which are stained with luxol fast blue–periodic acid Schiff–haematoxylin.

FIGURE 5.6. Sharply marginated foci of necrosis in grey matter are demarcated by their pallor in this example of moderate focal injury. Mild involvement of white matter is indicated by the vacuolar areas.

Prior to the use of the experimental aortography model the contrast media lesion had been interpreted as a transient disturbance of the blood–brain barrier, accompanied in the more severe reactions by oedema, stasis, and punctate haemorrhages. It has long been established, through study of the full evolution of the injury [11] that contrast media are capable of exerting a severe necrotizing action, associated in the early phases with development of fulminant oedema and/or haemorrhage, and eventuating in myelomalacia

(Figs. 5.4–5.8). Central grey matter is particularly involved, with extension of the injury throughout white matter and into spinal nerve roots in the most severe examples of injury. Extension of the lesion over several segments is commonly observed in the animal model.

Our observations of this fulminant necrotizing action, and of the severe disruption of the circulation of the spinal cord at the outset of the injury led us to conclude that the need was for establishment of a means of prevention,

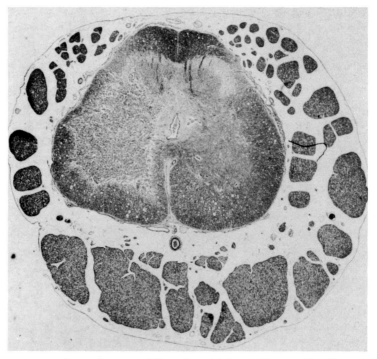

FIGURE 5.7. Severe, but essentially unilateral injury is illustrated in this section which has the distribution of an anterior sulcal artery. Almost all grey matter is destroyed. Surrounding white matter is severely involved. Scattered vacuoles appear in other areas of white matter. The necrotic zone is heavily infiltrated with phagocytic cells.

rather than treatment after the fact, of these manifestations. Manifestly, the increased vascular resistance of the spinal cord appeared to constitute a formidable obstacle to agents intended to act locally upon the vascular bed or parenchyma of the cord, certainly preventing their access via the circulatory route. It was these considerations that led us to studies of prophylactic rather than remedial approaches.

Early in the course of studies of the pathophysiology we observed that the mere act of changing the posture of the subject animal from supine to prone resulted in the conversion of a neurotoxic test dose into an innocuous one [11]. The increased vulnerability of the subject in the supine position was demonstrated to be unrelated to alterations in the circulatory dynamics of the spinal cord [3, 11]. We interpreted, by exclusion, that this augmented neurotoxic potential was a gravitational effect, with the outflow of the aortic contrast medium injection mass, heavier than blood, being directed into the

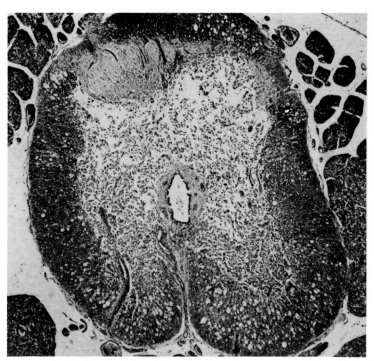

FIGURE 5.8. Myelomalacia involving virtually all grey matter bilaterally and scattered injury of surrounding white matter. The necrotic zone is disintegrated and occupied with macrophages.

vascular bed of the inferiorly situated spinal cord. This observation was of importance for several reasons. First, it demonstrated the exquisite sensitivity and flexibility of the experimental aortography model. Second, it attested to the narrow margins of safety in aortography. Third, it illustrated the 'all or none' characteristics of the injury, i.e. the abrupt transition between an innocuous and a strongly noxious effect. Fourth, this observation was made in the 'Urokon era', the period when the highest frequency of spinal cord complications of aortography was being encountered. It is more than an academic question to speculate how these statistics might have been altered, had this information been promptly applied to clinical aortography, resulting in the performance of this procedure in pronely positioned subjects.

A crucial early observation in our studies of the prophylaxis of contrast media injury was that *vasodepressor* agents exerted a strong *protective* action [12]. Theoretically, this effect was explicable, indeed, predictable, on the basis of the well-documented superiority of the extraneural over the neural

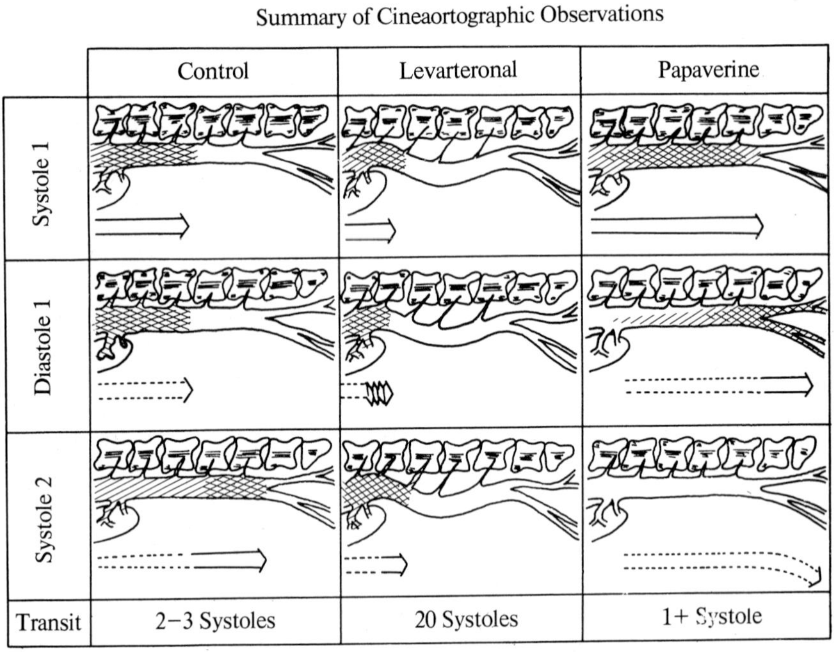

Summary of Cineaortographic Observations

	Control	Levarteronal	Papaverine
Systole 1			
Diastole 1			
Systole 2			
Transit	2–3 Systoles	20 Systoles	1+ Systole

FIGURE 5.9. Summary of cineaortographic observations under control conditions (left), following levarterenol (middle sequence) and after papaverine HCl (right sequence). Each vertical sequence represents flow of an injection mass during two cardiac systoles. Normally two to three systoles are required for transit of the injection mass through the aorta and out of its major branches. In levarterenol-induced hypertension a striking tortuosity of the aorta, minimal systolic advancement, diastolic reflux and delayed emptying, which required 20 to 30 systoles, is observed. The neuripetal diversion of aortic outflow is demonstrated in the original angiograms from which these diagrams have been constructed. In papaverine-induced hypotension an extremely rapid passage occurs, with forward flow being continued through both phases of the cardiac cycle; the injection mass requires less than two systoles to complete its transit through the aortoiliac tree. The original angiograms demonstrate the neurifugal distribution of aortic outflow. For further details see Abraham *et al.* [14].

vascular beds in levels of intrinsic tonus and in reactivity to baroreceptor, chemoreceptor and neurogenic influences. Succinctly, the greater response of the somatic vasculature induced neurifugal outflow of an aortic injection mass. A corollary of this reasoning was that *vasopressor* stimuli, by inducing a neuripetal aortic outflow, would *augment* the neurotoxic potential of aortography. As predicted, a striking augmentation of neurotoxicity was demonstrated with the use of the vasopressor, norepinephrine [13]. The potency of these haemodynamic factors is indicated by the fact that a ten- to twentyfold difference in neurotoxicity between the extremes of the protective action of vasodepressor influences and the potentiating effect of vasopressor stimuli has been demonstrated [12, 13]. The strong effects of vasoactive agents upon flow dynamics in the aorta and its major branches, and upon the distribution

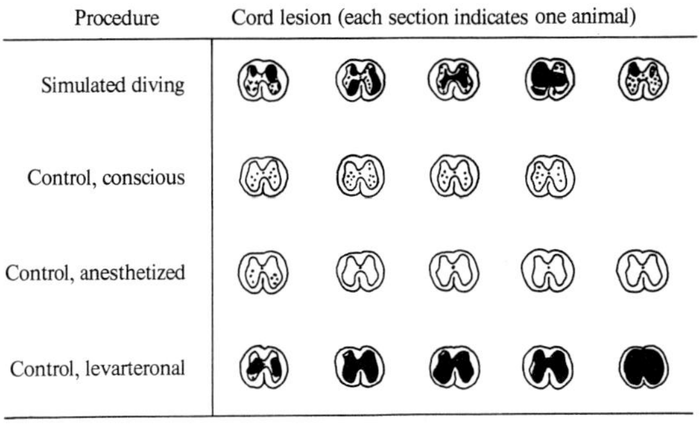

FIGURE 5.10. Cross-sections of the spinal cord showing augmentation of the neurotoxic effect of aortography (2 ml/kg 90 per cent Hypaque) by diving reflex and vasopressor shunt. The blackened area in each cross-section shows the extent of the destructive effect of the contrast agent at the lumbosacral level of the cord of individual animals. For further details see Margolis *et al.* [15] and Margolis & Yerasimides [13].

of aortic outflow have also been vividly demonstrated by cineangiographic studies [1, 4, 14], the major features of which have been diagrammatically presented in Fig. 5.9.

Observations to that point had all involved high doses of pharmacological agents. They left unanswered the question whether haemodynamic responses evoked by physiological mechanisms are as great and whether they may also play a role in the pathogenesis of neurotoxic complications of aortography. This question was resolved by the demonstration that the induction of a simulated diving experience [15] augmented neurotoxic consequences of aortography fully as strongly as did pretreatment with Levarterenol[1] (Fig.

[1] Levarterenol = noradrenaline

5.10). Cardiovascular adaptations to diving are characterized by pronounced bradycardia and forceful vasoconstrictive responses in major vascular beds throughout the body, exclusive of those in the heart and central nervous system [16]. The report by Bron et al. [17] of the diversion of contrast media to the brain with fatal consequences from an arch aortogram performed in a diving seal furnishes a striking parallel. Comparable observations have been made in still another animal under the influence of another physiologic reflex mechanism. *Apnoea*, induced by inhalation of tobacco smoke or formalin vapour, has been observed to produce in rabbits a striking vasoconstriction, manifested by transient arrest of angiographic material in the aorta, neuripetal diversion of aortic outflow and resultant severe or lethal spinal convulsions (18). Other procedures, such as *Valsalva* manoeuvre might possibly also impair the aortic outflow.

Other implications for diagnostic angiology emerge from the foregoing studies. In clinical medicine, cardiovascular pathology is the common denominator for diagnostic angiographic procedures. Generally, the more severe the disease the greater is the need for angiography. Yet, paradoxically, it has been stated frequently that in this situation arteriography carries its greatest risk. To be weighed against this viewpoint is the consideration that, because of recent advances in vascular surgery, the complications of undiagnosed, untreated and potentially remediable vascular disease far outweigh the hazards of nontreatment [19]. Indeed, for the brain, pre-existing vascular disease may decrease, rather than increase, the hazards of arteriography. Stenosis of the cervico-vertebral arteries, by raising cranial vascular resistance, could act as a protective factor. Conversely, the spinal cord, in which the march of degenerative vascular disease lags far behind that of the aorta and its major outflow tracts, may, in the absence of intrinsic vascular disease, be imperilled by functional or anatomical neuripetal shunts. This possibility can readily be envisaged by hypertension where there is a sustained elevation in systemic vascular tonus, and in occlusive aortic atherosclerosis where there is a strong collateral circulation via the important lumbar spinal radicles. Coarctation of the aorta with its multiple major collateral routes, including the spinal branches of the intercostal arteries, would appear to pose significant hazards in aortography. The operation of both of these haemodynamic factors is illustrated by two reports of unilateral renal damage occurring from aortography of patients with hypertension secondary to unilateral renal vascular disease [20, 21]. In both of these cases the better functioning kidney was severely injured, and the kidney with decreased function was protected, presumably because of its decreased blood flow.

These observations raise the question of the possible role in the production of neurotoxic complications of aortography of (1) pharmacologic adjuvants administered by the angiographer, and (2) naturally acquired neuripetal shunts of aortic outflow operating as compensatory adjustments for vascular disease. The complex interaction of these factors is illustrated by four examples of spinal cord injury complicating aortography reported by Efsen [22]. Here the following data must be analysed:

1 Aortography was performed in all four cases with the patient in the supine posture.
2 Two patients had severe longstanding hypertension.
3 One patient, in whom multiple (five) injections were made, exhibited a labile blood pressure, usually ranging from 150/90 to 90/60, but maintained at 200/100 throughout most of the procedure.
4 One patient had a history of intermittent claudication and arteriographically demonstrated narrowing of both iliac arteries and occlusion of both femoral arteries. Induction of the anaesthesia was complicated by a fall in blood pressure which required administration of a vasopressor drug before aortography.

In view of the above, the use of pharmacologic adjuvants to angiography calls for considered judgement. A classic example of this approach has been the use of epinephrine to induce constriction of normal renal vessels and enhance visualization of neovascularized renal carcinomas [23, 24]. The failure of epinephrine to induce consistent vasoconstriction of the superior mesenteric arterial tree has led to the use of a potent drug combination in selective arteriography of this vascular bed [25]. This consists of beta-adrenergic blockade by propanolol prior to epinephrine. This not only induces more consistent and sustained vasoconstriction in the mesenteric tree, but also causes gross reflux of contrast agent into the aorta.

Theoretically, angiography performed in the presence of shock might impose an increased risk to the spinal cord because of the haemodynamic adjustments to this state. Because aortography and selective arteriography have been advocated as aids in the search for sources of bleeding from gastrointestinal lesions and from splanchnic organs following abdominal trauma [26] we tested this possibility in our experimental model, under conditions of haemorrhagic shock [27]. These tests demonstrated a borderline effect, with the convulsive threshold being exceeded, but without production of structural lesions or functional defects. When we consider the fact that 'the great majority of postangiographic cord complications have occurred, and are continuing to occur at a worrisome rate, following unintentional introduction of contrast material into the spinal cord vessels' [28]—i.e. when the spinal cord is outside the target zone of the study, we might expect that angiographers would have been deterred from deliberate arteriography of the spinal cord. Yet selective arteriography has been demonstrated to be a safer and diagnostically superior approach to study of spinal cord lesions and has supplanted conventional midstream aortography [29, 30]. The experiences of the two leading proponents of this diagnostic technique attest to a low incidence of complications. DiChiro [28], as late as mid 1974, reported knowledge of but two instances of cord complications. Djindjian [30] encountered only one fatal accident in his first 240 selective cord arteriographies. This low incidence of adverse effects is accountable, in large part, by the precision of the method which allows great reduction in the volume dose. Improved safety of this procedure, as well as of selective arteriography of extraneural vascular beds from which radiculomedullary arteries originate, is unlikely to be achieved by further reduction of the low volume dose now used. It is difficult to envisage use of injection volumes lower than the 1–5 ml doses used by Kardjiev et al. [2] in their reported five cases of postbronchial arteriography cord damage. Until the development of an 'ideal' angiographic medium, this exclusion leaves open only the possibility of reduction of concentration—an all-important consideration, since contrast medium injury is a concentration-dependent effect. On the one hand, injection of an arterial tree proximate to its radiculomedullary branches will reduce the opportunity for dilution of the perfusate by admixture with blood en route to the cord. By the same token, maintenance of radiographically acceptable contrast might be attained in these circumstances with an initially lower concentration of the radio-opaque material. Experimental evidence [13] indicates the requirement for reduced concentrations of various contrast agents in order to attain a safe level for the spinal cord—Urokon below 20 per cent, Angio-conray below 30 per cent, Hypaque below 35 per cent, and Conray (methylglucamine iothalamate) below 50 per cent—if the factor of admixture with blood is discounted. These data attest to the improved safety margins of some of the newer agents.

As of this date the angiographer has to be aware of certain anatomical peril points. The greater anterior medullary artery of Adamkiewicz still remains the major concern in conventional aortography. With selective arteriography of non-neural vascular beds the possibility of unintentional injection of other important cord feeders originating from particular vessels must be recognized. The implicated vessels include (1) the vertebral arteries; (2) the fifth right intercostal artery, which might be injected in bronchial and intercostal arteriography, and from which a branch to cord segments D4–D6 emerges; (3) the costocervical trunk, from which a deep cervical branch emerges supplying the cervical enlargement, which might be injected during parathyroid arteriography; (4) the ascending branch of the thyrocervical trunk; and (5) intercostal and lumbar arteries which may be injected during visceral arteriography, from which radiculomedullary arteries supplying the lower thoracic and abdominal aorta arise.

The need for information about the status of the cord circulation in cases of post-traumatic paraplegia is urgent. DiChiro et al. [31, 32] have approached this problem with the use of selective arteriography of the spinal cord, and have been able to demonstrate a relative integrity of the major vessels after injury. They were, however, unable to obtain information on the status of the intrinsic vasculature of the cord. The information obtained from these studies provided valuable leads regarding therapeutic management of these cases. Because of the abundance of literature warning about the high toxicity of contrast agents to spinal vessels Gonzales & Osterholm [33] were concerned about the possibility that the acutely wounded cord vasculature might be especially vulnerable. They investigated this possibility in a model of spinal cord trauma in the cat, by injecting massive doses of Renografin-76 midstream into the upper thoracic aorta fifteen minutes following cord injury. Despite the high dose, the equivalent of 1050 ml singly injected in a 70-kg man, they observed no acceleration or enlargement of traumatic or haemorrhagic necrosis during a 24-hour period. They interpreted these negative observations as a possible indication 'that modern contrast agents do not cause a deleterious effect upon tissue structure as was feared in the past'. This is at variance with the observation that Renografin paralleled Hypaque in its neurotoxicity [34]. A haemodynamically oriented interpretation of the findings of Gonzales and Osterholm is that the altered haemodynamics of the injured spinal cord (contusion, haemorrhage, oedema, elevated tissue pressure) resulted in an increased vascular resistance which caused the injection mass to bypass the spinal cord. Support for such an interpretation is found in microangiographic studies by Fairholm & Turnbull [35], of the rabbit spinal cord following trauma. They described a central zone of microvascular breakdown in which capillaries progressively lost their ability to conduct blood and perfusate over a four-hour period post-trauma. A comparable phenomenon is the high perfusion resistance of the spinal cord after contrast medium injury, illustrated in Fig. 5.1 and described in detail elsewhere [1, 3]. In contrast to the findings of Gonzales & Osterholm, Fox et al. [37] have recently shown that Renografin-76 *enhances* the degree of injury in a spinal cord recovering from trauma.

Recognition and treatment of spinal cord damage

If the examination has taken place under local anaesthesia, the offending injection of contrast medium usually gives rise to severe pain in the back and or clonic movements in the legs. Objective neurological evidence of cord damage may however be delayed for several hours [2, 36]. In cases where such warning symptoms occur, it is therefore important to undertake repeated neurological examinations to detect the earliest signs of cord involvement. At the same time, it would also be rational to commence immediate systemic steroid therapy in an attempt to curtail oedema in the spinal cord [38-40].

Mishkin *et al.* [10] and DiChiro [28] have described an emergency treatment which appears to produce valuable improvement in the neurological status in these patients. Lumbar puncture is performed and the cerebrospinal fluid is withdrawn in 10-ml aliquots with replacement by isotonic saline. The patient is also placed in a head-up position so that contrast medium gravitates away from the brain. Although some spinal cord complications may resolve spontaneously, the risk of permanent paraplegia is so serious that cerebrospinal fluid replacement should be undertaken immediately there is any objective evidence of neurological involvement.

ADDENDUM

A case of paraplegia following lumber aortography for the investigation of hypertension, has recently come to notice. The patient had been instructed that, during the injection of contrast medium, he should hold his breath as if he were sitting on a commode. It seems possible that, in this case, *forced breath-holding* may have been an aetiological factor in causing a neuripetal diversion of contrast medium to the spinal cord. *Editor*

REFERENCES

1 MARGOLIS G. (1970) Pathogenesis of contrast media injury: insights provided by neurotoxicity studies. *Invest. Radiol.* 5, 392-406.
2 KARDJIEV V., SYMEONOV A. & CHANKOV I. (1974) Etiology, pathogenesis, and prevention of spinal cord lesions in selective angiography of the bronchial and intercostal arteries. *Radiology* 112, 81-83.
3 MARGOLIS G., GRIFFIN A.T., KENAN P.D., TINDALL G., LAUGHLIN E.H. & PHILLIPS R.L. (1957) Circulatory dynamics of the canine spinal cord. Temporal phases of blood flow measured by fluorescein and serioroentgenographic methods. *J. Neurosurg.* 14, 506-514.
4 MARGOLIS G., GRIFFIN A.T. & KENAN P.D. *et al.* (1959) Contrast-medium injury to the spinal cord: the role of altered circulatory dynamics. *J. Neurosurg.* 16, 390-406.

5 MARGOLIS G., BOURNE J.M., BOGDANOWICZ W. & ELLINGTON E.E. (1969) Pathogenesis of contrast medium injury—in vivo studies with fluorescein angiography. *J. Neuropath. exp. Neurol.* **28**, 155.
6 WALDRON R.L., BRIDENBAUGH R., PURKERSON M. & DEMPSEY E. (1973) The effect of angiographic media at the cellular level in the brain. *Radiology* **108**, 187–189.
7 RAPOPORT S.I. (1973) Evidence for reversible opening of blood–brain barrier by osmotic shrinkage of cerebrovascular endothelium and opening of tight junctions: relation to carotid arteriography. In: *Small vessel angiography: imaging, morphology, physiology and clinical applications* (Ed. by HILAL S.K.), p. 137. C. V. Mosby Company, St. Louis.
8 RAPOPORT S.I. & LEVITAN H. (1974) Neurotoxicity of x-ray contrast media. Relation to lipid solubility and blood–brain barrier permeability. *Radiology* **122**, 186–193.
9 RAPOPORT S.I., THOMPSON H.K. & BIDINGER J.M. (1974) Equi-osmolal opening of blood–brain barrier in rabbit by different x-ray contrast agents. *Acta Radiol. Diagn.* **15**, 21–32.
10 MISHKIN M.M., BAUM S. & DICHIRO G. (1973) Emergency treatment of angiography-induced paraplegia and tetraplegia. *New Engl. J. Med.* **288**, 1184–1185.
11 MARGOLIS G., TARAZI A.K., GRIMSON K.S. (1956) Contrast medium injury to the spinal cord produced by aortography: pathologic anatomy of the experimental lesion. *J. Neurosurg.* **13**, 349–365.
12 YERASIMIDES T.G., MARGOLIS G. & PONTON H. (1963) Prophylaxis of experimental contrast medium injury to the spinal cord by vasodepressor drugs. *Angiology* **14**, 394–403.
13 MARGOLIS G. & YERASIMIDES T.G. (1966) Vasopressor potentiation of neurotoxicity in experimental aortography: implications regarding pathogenesis of contrast medium injury. *Acta Radiol. Diagn.* **5**, 388–412.
14 ABRAHAM J., MARGOLIS G., O'LOUGHLIN J.C. & MACCARTY W.C. JR. (1966) Differential reactivity of neural and extraneural vasculature—I. Role in the pathogenesis of spinal cord damage from contrast media in experimental aortography. *J. Neurosurg.* **25**, 257–269.
15 MARGOLIS G., MILLS L.R. & NAITOVE A. (1972) Augmentation of neurotoxic effects of experimental aortography by the vascular reflexes induced during simulated diving. *J. Neurosurg.* **37**, 332–338.
16 ANDERSEN H.T. (1966) Physiological adaptations in diving vertebrates. *Physiol. Rev.* **46**, 212–243.
17 BRON K.M., MURDAUGH H.V. JR., MILLEN J.E. et al. (1966) Arterial constrictor response in a diving mammal. *Science* **152**, 540–543.
18 ANSCHUETZ R.A., SPIEGEL P.K. & FORSTER R.P. (1971) Angiographic demonstration of total renal shutdown during apneic 'diving' in the rabbit. *Comp. Biochem. Physiol.* **40**, 107–112.
19 BAUM S., STEIN G.N. & KURODA K.K. (1966) Complications of 'No Arteriography.' *Radiology* **86**, 835–838.
20 MCCALLISTER B.D., HUNT J.C. & KINCAID O.W. (1962) Unilateral renal atrophy subsequent to renal arteriography. *Proc. Mayo Clin.* **37**, 323–332.
21 SAAVENDA J.A., ABBOT J.P. & NUNNALLY R.M. (1961) Unilateral malignant nephrosclerosis with superimposed unilateral cortical necrosis following an aortogram. *Amer. J. Clin. Path.* **35**, 147–154.

22 EFSEN F. (1966) Spinal cord lesion as a complication of abdominal aortography—report of four cases. *Acta Radiol. Diagn.* **4**, 47–61.
23 ABRAMS H.L. (1964) Response of neoplastic renal vessels to epinephrine in man. *Radiology* **82**, 217–224.
24 ABRAMS H.L., BOIJSEN E. and BORGSTROM K. (1962) Effect of Epinephrine on the renal circulation. *Radiology* **79**, 911–922.
25 STECKEL R.J., ROSS G. & GROTHMAN J.H. (1968) A potent drug combination for producing constriction of the superior mesenteric artery and its branches. *Radiology* **91**, 579–581.
26 KITTREDGE R.D., COLAIACE W.M., KANICK V. & FINBY N. (1969) The angiography of hemorrhage. *Am. J. Roentg.* **107**, 181–190.
27 SWARTZ W.M., NAITOVE A. & MARGOLIS G. (1974) Arteriography in hemorrhagic shock: Potential hazards for the spinal cord. *J. Neuropath exper. Neurol.* (manuscript in preparation).
28 DICHIRO G. (1974) Unintentional spinal cord arteriography: A warning. *Radiology* **112**, 231–233.
29 DOPPMAN J.L., DICHIRO G. & OMMOYA A.K. (1969) *Selective arteriography of the spinal cord.* Green, St. Louis.
30 DJINDJIAN R., HURTH M., HOUDART R., LABOVIT G., JULIAN H. & MAMO H. (1970) *L'Angiographie de al Moelle Epiniere.* Masson & Cie, Paris.
31 GANGOUR G.W., WENER L. & DICHIRO G. (1972) Selective arteriography of the spinal cord in posttraumatic paraplegia. *Neurology* **22**, 131–134.
32 WENER L., DICHIRO G. & GANGOUR G.W. (1974) Angiography of cervical cord injuries. *Radiology* **112**, 597–604.
33 GONZALES C.F. & OSTERHOLM J.L. (1974) The lack of Renografin-76 toxicity in acute severe spinal cord injury. *Radiology* **112**, 605–608.
34 TINDALL G.T., KENAN P.D., PHILLIPS R.L., MARGOLIS G. & GRIMSON K.S. (1958) Evaluation of roentgen contrast agents used in cerebral arteriography. II. Application of a new method. *J. Neurosurg.* **15**, 37–44.
35 FAIRHOLM D.J. & TURNBULL I.M. (1971) Microangiographic study of experimental spinal cord injuries. *J. Neurosurg.* **35**, 277–286.
36 KILLEN D.A. & FOSTER J.H. (1960) Spinal cord injury as a complication of aortography. *Ann. Surg.* **152**, 211–230.
37 FOX A.L., KRICHEFF I.I., GOODGOLD J. *et al.* (1976) The effect of angiography on the electrophysiological state of the spinal cord. A study in control and traumatised cats. *Radiology* **118**, 343–350.
38 HAMMARGREN L.L., GEISE A.W. & FRENCH L.A. (1965) Protection against cerebral damage from intracarotid injection of Hypaque in animals. *J. Neurosurg.* **23**, 418–424.
39 MURPHY D.J. (1973) Cerebrovascular permeability after meglumine iothalamate administration *Neurology*, **23**, 926–936.
40 CAMPBELL J.B., DECRESCITO V., TOMASULA J.J. *et al.* (1973) Experimental treatment of spinal cord contusion in the cat. *Surg. Neurol.* **1**, 102–106.

CHAPTER 6: CEREBRAL ANGIOGRAPHY

A. LUNDERVOLD AND A. ENGESET[1]

From the patient's viewpoint and from that of many physicians, a complication from a diagnostic procedure is not to be expected and is difficult to accept. On the other hand, with present techniques and the new contrast media of low toxicity, the decision not to perform or not to complete a necessary X-ray examination may be a greater threat to the patient's health than the procedure itself. There is, therefore, almost no absolute contraindication to any radiological cerebral examination if the neurosurgeon is of the opinion that it is essential for the assessment of the patient. However, one must always remember that cerebral angiography may occasionally produce serious complications. These include a whole range of central nervous disorders from transient motor or sensory deficit to decerebration and death. Their incidence varies among reviews because of differences in definition of complications, contrast agents, techniques and patient selections [1–6]. Since the pioneers in this field started [7–14] there has been much improvement in technique and contrast media. With current contrast media and techniques, permanent complications, including death, occur with an incidence of about 1–3 per cent [15–24]. Surveys indicate a death occurring in an average of one in every 250 cerebral angiograms, compared with only one in every 40–50 000 urograms [25–27]. Sudden death can of course be expected in any large population sample, and patients attending X-ray departments are no exceptions. On the contrary, as they usually have pre-existing disease, sometimes a very serious one, they have an increased risk. If the death occurs before the examination in the X-ray department commences, the matter is in no doubt, but if the patient dies after it has started the evaluation will be difficult: is it caused by the original disease, the X-ray examination, or both? Often the complications are but an accentuation of the original signs and symptoms, so that it is impossible or very difficult to dis-

[1] Dr Arne Engeset died September 1973.

tinguish a natural progression of disease from illness enhanced by angiography or produced by the study alone. It is therefore extremely important that cerebral angiography is performed in a hospital where neurosurgical and neurological resources are present to cope with serious complications. The neuroradiologists ought also to have some clinical background so they know the indication for procedures, the relative complication rates, etc. During the actual study, unless the radiologist is aware of the signs to watch for, he may overlook subtle physical findings such as slight drooping of the mouth or a complaint of blurred vision, which may be an indication of impending neurological disaster. The physician must also talk to the patient intermittently in order to assess the patient's mental status [5].

A useful classification of reactions would be based on the severity of the reaction as it involves the general condition of the patient, and the necessity for treatment [26].

Most deaths and other serious complications that have been attributed to cerebral angiography in the past were mainly due to one of two causes. The first was the use of excessive doses and/or concentration of contrast media. The second cause was acute thrombosis following damage to the arterial wall and in particular subintimal injections leading to local dissecting aneurysm [28, 29]. The proper choice of contrast medium is consequently essential to avoid cerebral complications, and therefore it is necessary to use a medium that has undergone a thorough pharmacological and clinical trial. Media of metrizoate and diatrizoate types may therefore be used. The major Scandinavian clinics use for example 60 per cent Isopaque Cerebral (methylglucamine and calcium salts of metrizoate) both for carotid and vertebral angiography [2, 20, 30, 31]. The capacity of the internal carotid artery and its main branches does not normally exceed 6 ml. This quantity is therefore sufficient for internal carotid angiography. As the capacity of the vertebral and basilar arteries is smaller, 4 ml in vertebral angiography is usually enough [30]. Some centres use 5-8 ml for injection into the internal carotid artery [30, 31], but an increase to 9 or 10 ml may enhance the complications [2, 18]. With the low toxicity of modern contrast media an interval of five to ten minutes between each injection, to avoid any cumulative effect, is no longer necessary in uncomplicated cases [30]. However, the perfect contrast medium has not yet been developed, and therefore the total amount of a contrast medium given to a patient must always be restricted [32-37]. Many workers have also recommended that the patient should be observed after each injection and that an electrocardiogram should be taken simultaneously so that the examination may be interrupted if any alarming symptoms develop. Changes in the cerebral circulation and the cardiovascular reactions have been described by many authors [38-46]. Some also recommend simultaneously recording the electroencephalogram which may give much additional and valuable information [47-50]. If the patient is in poor condition the examination should be postponed unless it is considered absolutely necessary. When cerebral angiography is controlled in these ways, complications due solely to the direct toxicity of the contrast medium are probably extremely rare [30, 31, 51, 53-55]. Thus complications in angiography depend mainly upon the technique [52, 56-61]. The risk at the hands of an inexperienced doctor is many times that with a skilled neuroradiologist, and there is thus a strong case for limiting the procedure to special centres where it can be undertaken by workers with adequate training and skill in the technique. The methods of arterial catheterization have in this way been considerably improved both with regard to equipment and to training of personnel. In many centres the catheter technique has been used for vertebral angiography and also for direct catheterization of the carotid artery. The method of cerebral angiography using catheterization of the femoral artery has not been commonly accepted although it has often been suggested as the ideal approach, particularly when it is desirable to examine successively all brain vessels, or when the extracranial parts of the carotid and vertebral arteries should be included in the radiological investigation. Puncture of the brachial and axillary arteries carries a greater risk than femoral artery puncture, especially in patients with arteriosclerosis, and in addition, selective filling of the individual cerebral vessels is not feasible by these methods [2, 30, 31, 62, 63]. The risk of cerebral angiography is also increased in patients with cerebrovascular symptoms or serious head injury, especially if they have a major neurological deficit or an unstable cardiovascular system [64-69]. In subarachnoid haemorrhage due to ruptured intracranial aneurysm, focal signs, particularly in a conscious patient, may be due to arterial spasm, and this may be seriously aggravated by early angiography [70].

In order to prevent untoward reactions in cerebral angiography, special precautions can be taken. Patients expected to be more susceptible to complications ought particularly to be selected for special monitoring of electrocardiogram and electroencephalogram. This emphasizes the need to integrate all findings both clinically and polygraphically to arrive at a successful result, with the fewest possible complications and the lowest rate of serious side effects. Since this type of recording started in our laboratory, the incidence of complications recorded from the brain and the vascular system has steadily been diminishing. Polygraphic recordings now show changes in less than 1 per cent. A more detailed clinical examination of the patients revealed slight transitory effects, not all reflected in the recordings. Because of such changes in a patient's clinical condition and/or in the polygraphic tracings, the X-ray examination has sometimes been interrupted after discussion with or consent of the referring doctor. The improvement in this period seems to be due to different factors, but is probably mainly attributable to the availability of more advanced contrast media. Technique has also been improved in this period; for example by catheterization of the femoral artery, and the patient now always receives atropine (0.6 mg) as premedication. Increased intracranial pressure has been reduced before the X-ray examination. All these factors will be discussed in another part of this chapter.

GENERAL CAUSES FOR COMPLICATIONS

Contrast media

The intravascular injection of contrast in cerebral angiography introduces a substance differing from body fluids in viscosity, osmolarity and chemistry; it can therefore cause cellular damage in higher concentration. The new improved agents have reduced the number and severity of the complications considerably, but the perfect and completely harmless contrast medium has not yet been found. The influence of contrast media on the vessels and the brain has been examined in animal experiments as well as in clinical practice [71–86]. A general discussion of the different contrast media including the problem of anaphylaxis or allergy, is beyond the scope of this chapter and will be dealt with elsewhere in this book. As the greatest risk probably is associated with angiographic examinations of the cerebral circulation, some important aspects will be discussed and partly illustrated. The essential thing in this connection is the blood–brain barrier. The substances now in clinical use produce little alteration of the blood–brain barrier in normal concentrations, amount and rate of injection, but if given in too high a dosage or concentration, or too fast, serious brain damage and even death may occur. This has been demonstrated in several animal experiments. Given in a proper way, complications due to the direct toxic action of the new media probably only appear when the blood flow is impaired, for instance by pre-existing disease of the brain. Cellular damage then results possibly from a breakdown of the blood–brain barrier through injury of the endothelium. The usual effect of current contrast media is vasodilatation which results in an increased blood flow [40, 87–92]. But they may produce vasoconstriction when injected into an intracranial vessel, causing a reflex fall of blood pressure, bradycardia and sometimes asystole [2, 31, 93–95]. This reaction is elicited from vasomotor centres of the brain, and its efferent pathways are through the vagus nerve. In animal experiments it can be abolished by cutting the vagus nerve and in humans by giving atropine as a premedication, but not by destruction or inhibition of the carotid sinus mechanism [15, 20, 31]. One must always remember that this reflex also can be elicited in many other ways, for instance by pre-existing brain disease. The magnitude of these responses differ with the contrast medium used [15, 31]. They occur also with greater incidence when the posterior cerebral arteries are visualized. All these reactions can be demonstrated by a simultaneous registration of the electrocardiogram and the blood pressure. The use of this and other parameters in patients, started in our department of neuroradiology in 1950. The first results of the electroencephalograms and electromyograms recorded during vertebral angiography were published in 1954 [9]. Since then, continuous measurements of the intra-arterial blood pressure, respiration and the electrocardiograms have been added to the two previously mentioned electrical recordings. Not all these parameters are used in every patient, but the EEG and ECG are routinely recorded from those patients most likely to develop complications. More than 3000 patients have now been examined in this way in the laboratory. In addition, similar polygraphic recordings have been done in numerous animal experiments and in other neuroradiological examinations [18, 96, 97]. Fig. 6.1 is an example of the recordings in cerebral angiography, and shows normal tracings during carotid angiography in man. Figs 6.2, 6.3 and 6.4 illustrate different durations of asytole during cerebral angiography and their influence upon the blood pressure, the electroencephalogram and thereby on the state of consciousness. The reactions in these examples are probably caused by the contrast media used. One must remember that similar changes can be elicited in many other ways as well, for instance by the pre-existing brain lesion. In these cases, neither the X-ray examination nor the contrast has anything or very little to do with the asystole or fall in blood pressure. Fig. 6.5 shows such an example. In this case the heart stopped for about 30 seconds when the patient was moved from a supine to a sitting position in the department of neuroradiology, but before the X-ray examination had started. The patient became unconscious with a marked slowing of the EEG pattern. Such polygraphic recordings together with thorough clinical observation are of particular interest and importance in very ill patients, especially when repeated injections are necessary. In order to reduce the amount of contrast media, which may prevent such complications, it is recommended that whenever possible, films are taken in two planes simultaneously, to minimize the number of

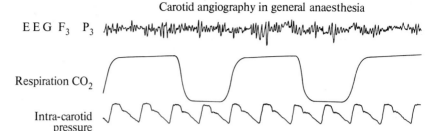

FIGURE 6.1. Left carotid angiography in general anaesthesia without any changes in the EEG, respiration, intracarotid pressure or ECG. The tracings are recorded two seconds after the injection of 7 ml Isopaque.

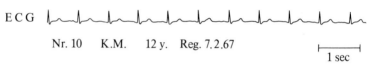

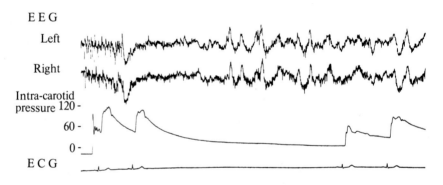

FIGURE 6.2. Asystole for six seconds and distinct fall of blood pressure with bilateral EEG changes starting two seconds after the injection.

injections needed. Similar polygraphic and clinical control is suggested if the agent itself or its concentration is changed.

Besides influencing the vasomotor centre of the brain, contrast media may also sometimes interfere unfavourably with other parts and centres of the central nervous system. The cause may differ from patient to patient,

Cerebral Angiography

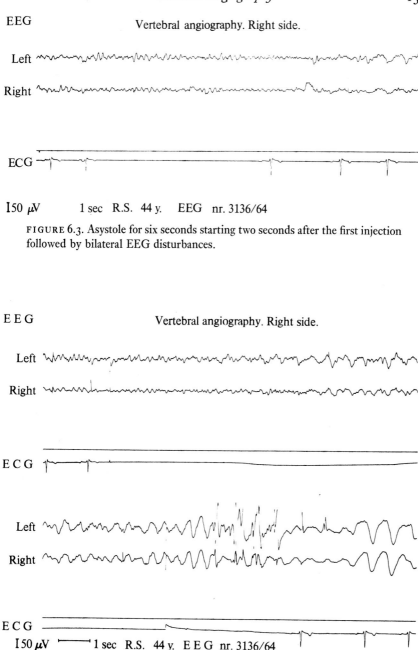

FIGURE 6.3. Asystole for six seconds starting two seconds after the first injection followed by bilateral EEG disturbances.

FIGURE 6.4. The second injection in the same patient as in Fig. 6.3 resulted in a nineteen seconds asystole with marked bilateral EEG disturbances which terminated in epileptic discharges. The patient's consciousness was disturbed.

but is usually a combination of the pre-existing disease and the X-ray examination. Most commonly, transitory hemiparesis is seen with carotid angiography whilst disturbances of vision and consciousness occur with vertebral angiography [98–103]. Previously, such temporary complications were more common, and sometimes also permanent lesions were seen. In the nineteen-fifties, with the contrast media then used, we had several patients with impaired consciousness caused by vertebral angiography. Some were even in coma and did not react to usual stimuli. Because of these and similar

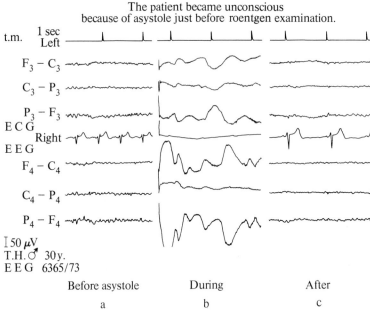

FIGURE 6.5. a Normal EEG and ECG tracings prior to an unexpected asystole before röntgen examination.

b Marked EEG pathology during a 30 seconds asystole, starting when the patient was moved from a supine to a sitting position. The patient was unconscious in this period.

c Normal tracings with some bradycardia after the heart starts beating again, and the patient had regained consciousness.

findings, and the results of polygraphic records, it has been possible to postulate some facts about, and the importance of, some minor cerebral vessels. As a rule, the EEG is the best indicator of the state of consciousness. Changes in the degree of abnormality may therefore give some indication of improvement or deterioration in the patient's condition. Some exceptions to the rule that changes in the level of consciousness are reflected in characteristic alterations of the electrocortical activity are of clinical significance. Such observations may suggest that the level of responsiveness and electrocortical activity may be affected independently [104–106]. Examples of such dissociation may occur after vertebral angiography. Some interesting X-ray and EEG findings illustrating these phenomena and demonstrating the importance of the minor vessels, can be seen in composite Fig. 6.6. Coma as a complication, during or after vertebral angiography, is commonly associated with pronounced EEG slowing. The arteriograms in these cases show contrast filling of the vertebral, basilar and both posterior cerebral arteries. This means that the brain area supplied by the posterior cerebral arteries had also been exposed to the contrast medium (Fig. 6.6a). One exception to this study was a patient who became unconscious after vertebral angiography without appreciable EEG changes. The arteriogram showed no filling of the posterior cerebral artery because of an obliterated saccular aneurysm at the top of the basilar artery (Fig. 6.6b). The patient remained unconscious for more than a year. Results of another patient in which right and left vertebral angiograms showed that this case was one with an intracranial separation of the vertebral circulation, both vertebral arteries continuing directly as two separated basilar and posterior cerebral arteries. After the left-sided injection with contrast filling only of the left part of the vertebral system, including the left posterior cerebral artery, this patient became unconscious. The electroencephalogram revealed focal disturbances in the form of slow waves over the left side of the head, during the period of altered consciousness (Fig. 6.6c). Prior to the examination the EEG was quite normal. It seems reasonable therefore to conclude that altered consciousness (responsiveness) is related to the structures supplied by the basilar artery, possibly by its pontine branches, and that electroencephalographic disturbance, on the other hand, is related to the structures supplied by the posterior cerebral artery, probably by its central or basal branches. This stresses the importance of looking for these small branches in the angiograms, as the signs and symptoms of neurologic injury will correspond to the functional components of the brain involved. Focal injury relates to the specific artery being studied. This is also the case in visual disturbances which may sometimes occur during vertebral angiography. The EEG will then show a distortion or complete disappearance of the typical alpha activity over the posterior region of the head, but there are some exceptions [9]. With half-sided visual disturbances the electroencephalograms may show theta and delta waves and a reduction of the alpha activity in the contralateral posterior hemisphere. This is especially the case when only one of the posterior cerebral arteries is filled with contrast. In some similar cases, however, electroencephalographical

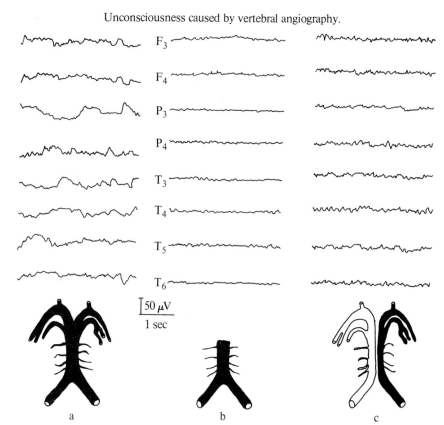

FIGURE 6.6. EEGs recorded during unconsciousness caused by vertebral angiography. The diagrammatic drawings show the vessels filled with contrast medium.

a Normal arteriogram with contrast filling of the vertebral and the basilar arteries as well as both posterior cerebral arteries, and generalized disturbances as theta and delta waves in the electroencephalogram.

b Contrast filling only of the vertebral and basilar artery, but no filling of the posterior cerebral arteries, and an electroencephalogram with normal appearance. This patient was unconscious and did not react to pain or acoustic stimuli.

c Contrast filling only of the left part of the vertebral system including the left posterior cerebral artery, in a case with intracranial separation of the vertebral circulation. The electroencephalogram showed only left-sided disturbances. The electroencephalograms recorded before the examination were normal in all these cases.

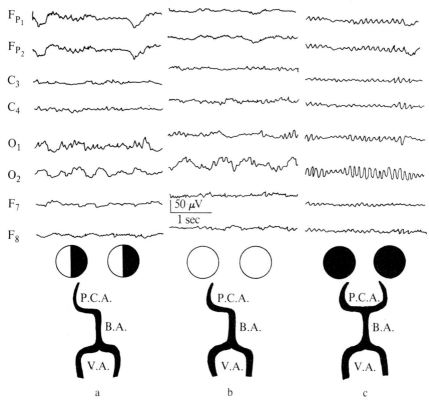

FIGURE 6.7. Electroencephalograms recorded after vertebral angiography, with diagrammatic drawings of the vessels filled with the contrast medium.

a Contrast filling of the right posterior cerebral artery while the patient had scintillation in the left visual field and electroencephalographical disturbances in the right occipital region as delta waves.

b Similar contrast filling and electroencephalographical findings, but the patient had no visual disturbances.

c Contrast filling of both posterior cerebral arteries with amblyopia, but with normal electroencephalogram.

The electroencephalograms recorded before the examination were normal in all these cases.

disturbances occur without any associated visual defect. Finally, some patients have visual defects but no electroencephalographical disturbances (Fig. 6.7).

These findings suggest that visual disturbances and electroencephalographical disturbances are not necessarily related to the same structures. The electroencephalographical disturbances may possibly originate from structures supplied by the central branches, and the visual disturbances from structures supplied by the cortical branches of the posterior cerebral artery.

With the new contrast media such complications as coma and cortical blindness are very rare, but impaired consciousness and visual disturbances can still occur during vertebral angiography. If such phenomena occur or

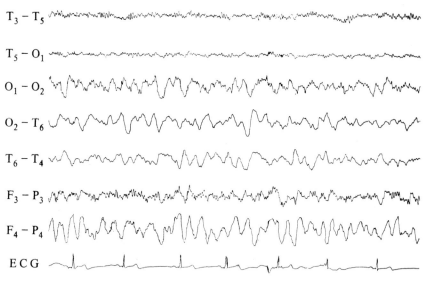

FIGURE 6.8. Ten seconds after right-sided carotid angiography there are focal right-sided EEG changes with accompanying transient paresis of the extremities contralaterally. The upper two tracings are recorded from the left temporal and occipital region and the sixth from the left frontoparietal region: these show normal or nearly normal activity. The other tracings from the corresponding regions of the right hemisphere, reveal abnormal slow wave activity.

abnormal tracings can be seen in the EEG and ECG, they should be taken as an obvious warning that further injection may cause permanent sequelae [31, 107].

In carotid angiography other parts of the brain will be flushed with the contrast media and different symptoms and signs such as motor and sensory deficits can occur. Fig. 6.8 illustrates such a case. Initially there were minor focal EEG changes on the right side after right-sided carotid angiography. After some seconds there was a marked build up of the right-sided abnormal EEG pattern, and when the patient was examined clinically it was discovered that he had a left hemiparesis which fortunately disappeared together with the EEG changes after some minutes. With the new agents, similar and other complications are, as a rule, temporary and very shortlived. Because side effects had fallen to a rate of less than 1 per cent, we tried some years ago to increase the dose from 7 to 9 ml in the hope that better contrast filling of the vessels would improve the diagnostic possibilities of carotid angiography. This seemed to result in a slight deterioration in the polygraphic parameters and also in an apparent increase in reactions seen clinically. In order to examine this more thoroughly, an intensive clinical control of all patients was instituted for about one year, both whilst in the neuroradiological department and also after the patients returned to the wards. The Seldinger technique was used for transfemoral introduction of the catheter in 554 patients (323 men and 231 women). Of these, 42 were simple arch aortographies, and the remaining 512 were selective angiographies, where more than one vessel was usually examined at the same session. A dose of 9 ml was used in the internal carotid in 419 patients and in the following 135 patients it was reduced to 7 ml.

Twenty-five of the 554 patients in the series had temporary and usually short-lived reactions. Four of these were allergic reactions to the contrast medium (urticaria), two were reactions to the local anaesthetic, and another patient had an attack of angina pectoris during the preparations for the radiographic examination. In four of the remaining eighteen cases, the reactions occurred before injection of the contrast media, but after the catheterization. It is therefore reasonable to assume that they were due either to manipulation of the catheter and possibly dislodgement of emboli, or to the triggering of some form of vasomotor effect. One of the patients had one of his recurrent epileptic fits which presumably had no connection with the examination itself. A probable cerebrovascular defect was present before the examination in three of the patients and all were in poor condition.

One of the fourteen reactions that arose after the procedure was due to hemiplegia caused by subintimal injection of the medium and partial occlusion of the carotid artery; the paralysis disappeared after about a day.

There thus remained thirteen patients, i.e. 2.3 per cent, in whom it was possible that the reactions may have been produced by the contrast medium itself in addition to any other factors. Only two of these reactions occurred after the reversion of the dose to 7 ml and were only trivial (nausea and vomiting). The duration of the effects was short, from minutes to a few days, except in a patient with mild dysphasia and facial paresis in whom they lasted three weeks; thus no patients had demonstrable permanent sequelae. Eight of these patients had complications that were so slight and of such short duration that they would almost certainly have been missed if fastidious clinical observations had not been made. This emphasizes that close observation of the patient during and immediately after the roentgen examination is essential. Polygraphic recordings were made in eleven of the patients, and in seven of them, the EEG temporarily deteriorated during the X-ray examinations. All these side effects were very shortlived and in most patients they only lasted one or two minutes, but they stress the importance of recognizing minor changes both in the clinical condition and the polygraphic tracings, because more injections may produce permanent sequelae, and therefore curtailment of the examination must be considered. In patients receiving 9 ml contrast medium, the incidence of reactions was 2.9 per cent; whereas in those receiving 7 ml, the incidence was only 1.5 per cent.

COMPLICATIONS RELATED TO GENERAL ANAESTHESIA

In patients requiring cerebral angiography the effects of a general anaesthetic on a diseased brain may be less favourable than in patients without cerebral disease. Some such complications will be illustrated and discussed here. Anaesthetic problems are discussed in greater detail in Chapter 18. There are reports stating that the incidence of complications is greater with general than with local anaesthesia, but on the other hand the use of general anaesthesia is supported in special cases by others [67, 108, 109]. In our department of neuroradiology, general anaesthesia is used whenever the patient's condition indicates its need, for instance in very sick and restless patients and in uncooperative children. This applies in some 40–60 cases each year, i.e. in about 10 per cent of the patients referred for cerebral angiography. Because of the pre-existing brain lesion and because these patients usually are very ill, we are especially afraid of complications. In order to prevent them, we routinely use polygraphic recordings in such cases. Fig. 6.1 illustrates normal tracings during carotid angiography in a patient with general anaesthesia. In these and some other special cases, we also record the electroencephalogram, the electrocardiogram, the intracarotid blood pressure and the respiration rate, and in some instances also electromyograms. As a rule we usually record the EEG and ECG, and these parameters can be very helpful to the anaesthetist in evaluating the patient's clinical condition. In a patient with increased intracranial pressure, intermittent bilateral slowing in the EEG will indicate the fluctuating origin of tentorial herniation; this can result from the original disease, the general anaesthesia, hypoventilation or cerebral angiography. Herniation is usually reversible in this early stage, but necessary precautions must immediately be taken to prevent an irreversible process. This can be achieved by simple hyperventilation or by giving pressure-decreasing medication. In order to prevent tentorial herniation, we make it a practice to administer mannitol to all patients who are suspected of having raised intracranial pressure. This applies to approximately 10 per cent of patients referred for cerebral angiography. Since

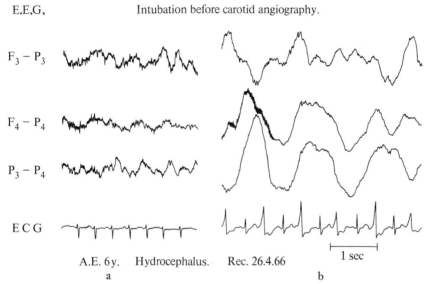

FIGURE 6.9. a Recordings of EEG and ECG with normal findings during general anaesthesia.

b Abnormal tracings during intubation of the patient before carotid angiography.

The upper tracings represent the left fronto-parietal region, the others are from the right fronto-parietal region.

Jefferson's paper in 1938 [110] it has generally been accepted that the dividing line between survival and death in cases with supratentorial masses is whether or not an irreversible tentorial herniation can be prevented. Tonsillar herniation can also be recognized from the EEG and is equally dangerous. These complications can of course also occur without general anaesthesia, and we therefore always monitor polygraphic recordings in such cases. The pre-existing brain lesion sometimes makes the central nervous system react more easily and vigorously to minor insults. For example, the simple act of intubation may, for a short period, influence the oxygen supply to the brain to such an extent that it can be recorded electroencephalographically. Fig. 6.9 shows such a case with marked changes in the EEG and some minor ones in the ECG. They were all normalized after seconds. More dramatic events may also occur during cerebral angiography under general anaesthesia. Fig. 6.10 illustrates a flat EEG, indicating brain anoxia caused

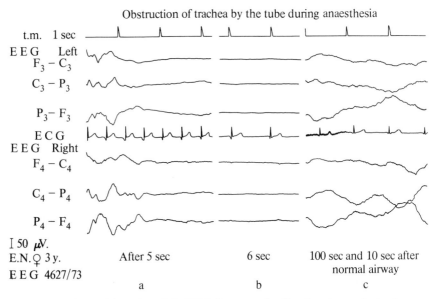

FIGURE 6.10. a Flattening of the EEG five seconds after the air supply to the lung was accidently interrupted.

b Flat EEG indicating severe brain abnormalities accompanying transient marked clinical signs of dilated pupils. A bradycardia can also be seen in the ECG.

c The improved EEG recorded ten seconds after normal airways were restored.

by a temporary interruption of the air supply to the lung. For some reason the endotracheal tube was bent during the period in which X-rays were taken. It was recognized immediately in the tracings, the patient was given oxygen and recovered completely. Even during that short period of anoxia, lasting but a few minutes, the patient developed dilated pupils: a sign of a serious brain lesion.

The influence upon the brain of the agents used in general anaesthesia can also be seen from the tracings, and give the anaesthetists warning even before any clinical signs have developed. The electrocortical alterations, characteristic of a given level of depression, appear three or four minutes before the clinical signs of the associated anaesthetic planes. The EEG may therefore be an excellent aid in managing this critical time for the patients. Thus electroencephalographic evidence of anaesthetic overdosage may allow remedial measures to be taken several minutes before clinical signs of severe

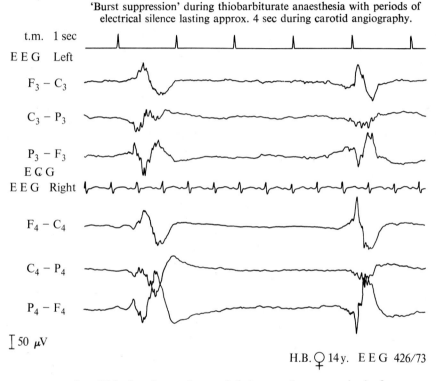

FIGURE 6.11. If the inactive or silent periods between bursts seen in the figure are greater than ten seconds, it is a grave prognostic sign and appropriate measures have to be taken.

depression occur. One must remember that different anaesthetic agents vary markedly in their effects on electrocortical activity, and that the patterns seen with one agent have little or no resemblance to those which may be recorded with a different anaesthetic. Fig. 6.11 illustrates typical 'burst suppression' which may be seen particularly when some barbiturates are used. The period of electrical silence between bursts is of special importance.

According to Courtin's classifications, in Level 6 these inactive periods are greater than 10 seconds. Level 7 is complete electrical silence as demonstrated in Fig. 6.10. As the duration of the electrical silence increases there is a progressive decrease in amplitude of the wave form comprising each burst of activity [111].

Convulsions are a complication, common to all forms of cerebral angiography and occur more often in epileptic and highly apprehensive patients (33). In our experience they may even occur before the examination has started or before the contrast injection has been given. Such an example can

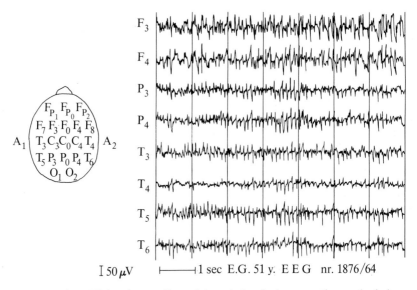

FIGURE 6.12. High-voltage spike activity starting during general anaesthesia in a curarized patient before contrast injection. The X-ray examination was postponed, and the patient had a generalized seizure when the effect of the curare subsided.

be seen in Fig. 6.12. The seizure which could be detected in the EEG commenced during general anaesthesia, but because the patient was curarized, the clinical seizure first started when the effect of curare subsided. In another patient with seizure pattern in the EEG, also curarized during anaesthesia, the roentgenological examination was continued and the epileptic activity disappeared from the electrocortical tracing. No clinical manifestation was seen in this case, probably because of the curare, and no complications could be detected afterwards. The polygraphic records are illustrated in Fig. 6.13. In other instances, convulsions represent an indication for

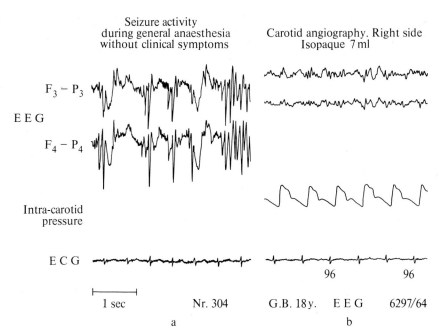

FIGURE 6.13. a The seizure pattern was recorded from a curarized patient without clinical symptoms before angiography.

b The roentgenological examination was continued, and the spike activity disappeared after injection of 7 ml Isopaque in the right carotid artery.

interruption of the roentgenological examinations. As seizures occur more commonly in epileptic patients, they can sometimes be prevented by anticonvulsant premedication.

COMPLICATIONS RELATED TO MECHANICAL OR TECHNICAL FACTORS

As previously stated, complications due solely to the direct toxic action of the contrast medium are probably extremely rare [31]. They result therefore either from the technique or because of pre-existing disease. Because only the former can be influenced to any extent, great effort should be made to minimize them by developing a safe technique.

Carotid angiography may be performed either by direct injection through a percutaneously inserted needle or by catheter technique. For a percutaneous puncture of the carotid artery, the patient lies supine with his head on the skull table, and by lowering the head, the neck is extended. The puncture should be made with a needle having an internal diameter of about 1 mm. This needle can be replaced by a catheter with the same internal diameter, a method that gives a more stable position. It is not always realized that a good blood flow 'jet' out of the needle does *not* exclude a subintimal tear

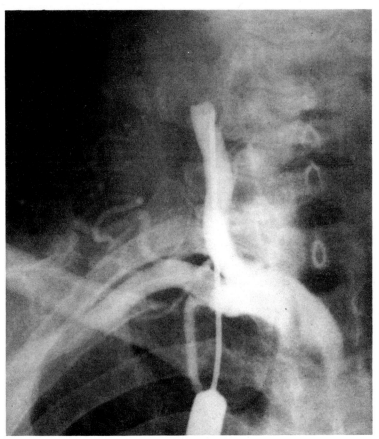

FIGURE 6.14. Subintimal dissection in a 54-year-old woman with a meningeoma in the right parietal region. The patient developed a transient hemiparesis after right-sided cerebral angiography. The tumour was subsequently removed successfully.

which may occur on the posterior wall of the carotid artery and this may produce an obstruction with a marked decrease in cerebral blood flow (Fig. 6.14). This can be avoided if a scout film of the neck is taken just after the needle has been inserted and 4 ml of contrast medium has been injected [5]. Some prefer to inject selectively into the internal or external carotid artery.

Cerebral Angiography

The patient is usually able to convey sufficient information regarding the subjective sensations to indicate clearly which vessel has been injected. A burning sensation is felt in those regions through which contrast material passes. If the puncture is made into the internal carotid artery, the heat will be felt behind the eye. If the external carotid is punctured, the heat will be noted in the cheek and in the scalp.

Puncture of the carotid sinus should be avoided because this may elicit the sinus reflex and produce pallor and a cold sweat in the patient. Atheromatous plaques are also common at the carotid sinus, and puncture at this level should therefore be avoided.

The basilar artery and its intracranial branches may be examined either through direct puncture of the vertebral artery or through introduction of a catheter. Several different techniques, both for puncture and the introduction of a catheter, have been described. The catheter may be introduced after percutaneous puncture of the femoral, axillary, or brachial arteries. A test injection should always be done first during fluoroscopy with 1 to 2 ml of contrast medium. If the patient experiences any symptoms such as dizziness or if the test injection demonstrates impairment of the flow, the catheter should be withdrawn immediately. When puncturing the vertebral artery, one must be aware that the needle is entering the vessel in a direction that is far from parallel to the artery. Furthermore, the vertebral artery is frequently very narrow. It is therefore difficult to avoid puncture of both walls, and the catheter or its guide may slide along the outer side of the posterior wall. Because the nerve roots lie in close contact with the artery, this may also cause complications. The needle may be introduced either anterior or posterior to the carotid artery in the lower part of the neck or even suboccipitally, where the vertebral artery bends around the atlas to pass into the cranial cavity. In order to avoid extravascular injection, it is preferable to puncture with a short-bevelled needle. To reduce the frequency of faulty injections, special small-bore needles have been constructed. Despite all precautions an intramural injection or haemorrhage may occur. Therefore, no attempt to puncture the opposite vertebral artery should be done in the same sitting. Utilizing this technique, the percutaneous puncture of the vertebral artery is successful on the first attempt in more than 80 per cent of cases [31].

As stressed and illustrated in the previous section, the patient's airway must be carefully maintained during cerebral angiography performed under general or local anaesthesia. Cerebral anoxia can easily be produced in a diseased brain in many different ways. The compression of the carotid artery after puncture, for instance, can produce marked changes in the electro-

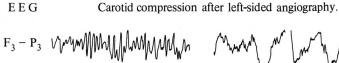
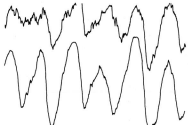
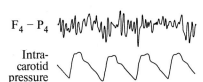

FIGURE 6.15. a Normal EEG, ECG and intracarotid pressure after left-sided carotid angiography.
  b The carotid artery was compressed after the catheter was removed in order to prevent bleeding. This resulted in partial interruption of blood supply to the brain, shown in the EEG as high-voltage slow waves. The patient's consciousness was impaired during this period.

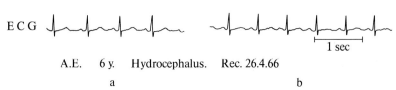
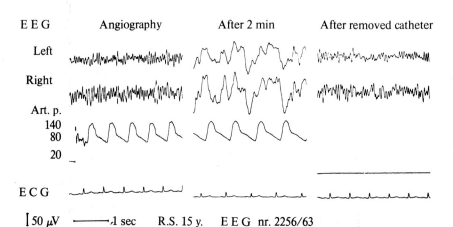

FIGURE 6.16. Sudden appearance of high voltage slow waves in the EEG, after two minutes' obstruction of the vertebral artery by the catheter. The patient became unconscious and stayed so until the catheter was removed.

cortical activity as shown in Fig. 6.15. They are as a rule accompanied by clinical signs and symptoms. The effects of catheter obstructions in the vertebral or carotid artery can be very well illustrated in polygraphic tracings. This situation occurs most commonly in the vertebral artery. Fig. 6.16 shows the EEG changes in such a case. The patient became unconscious, but recovered when the catheter was removed. Similar obstructions may also

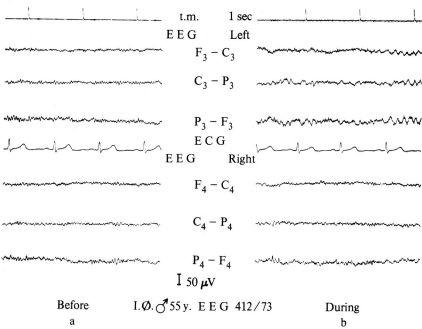

FIGURE 6.17. a Normal EEG and ECG tracings recorded from the beginning of an X-ray examination.

b Slow waves recorded from the left hemisphere after obstruction of the left carotid artery by the catheter before contrast injection. The slow waves were associated with transient paresis of the extremities contralaterally and some dysphasia.

occur, though more rarely in the carotid artery when selective internal carotid catheterization is performed. Blocking of the internal carotid artery will naturally give other polygraphic and clinical signs. This is illustrated in Fig. 6.17. The doctor obtained a warning from the unilateral focal slow waves recorded in the EEG. Some seconds later the patient became hemiparetic which could easily be detected clinically after the warning in the

EEG. Fortunately, the catheter was removed so quickly that both the clinical signs and the EEG abnormalities disappeared completely after some minutes. These cases illustrate again the importance of using polygraphic supervision, especially in selected cases with great potential risk of complications and especially when patients are under general anaesthesia and so unable to express themselves.

The hazard of broken guide wires has decreased considerably with the improvements in their manufacture and the failure rate is now only 0.0007 per cent. Catheter breakage is also very uncommon [112].

A method of foreign body retrieval that enables the angiographer to cope with breakage accidents has been described. The treatment of other complications of cerebral angiography is usually the same as for that of other neurological lesions of similar cause. Immediate treatment consists of efforts to maintain oxygenated cerebral blood flow. This may include surgical treatment of accessible intramural dissections and retrieval of foreign bodies, medical treatment for arteriospasm, cardiac or respiratory arrest, hypo- or hypertension, convulsions, etc. Oxygen inhalation in the X-ray department is often necessary in many of the emergency cases [113–117]. Accidental injection of concentrated contrast medium into the cerebral circulation may cause cerebral damage [25]. Systemic steroid therapy may be useful in diminishing damage to the blood–brain barrier and cerebral oedema in these cases and hypertonic mannitol may also be valuable [118]. This therapy should also be considered if neurological function deteriorates after routine cerebral angiography.

CONCLUSION

Cerebral angiography has become an essential part of the practice of medicine. With present technique and the new contrast media available, there is almost no absolute contraindication and the side effects are few and as a rule negligible. In order to take advantage of this improvement and their own advancement, neurosurgeons now have a more active approach to aneurysm surgery and to operations on older patients and to those in poor clinical condition. Not only in these poor risk patients, but almost in any study, cerebral angiography may occasionally cause serious complications. They occur in less than 1 per cent of cases and are, as a rule, directly related to the pre-existing disease, to errors of technique or more rarely to the contrast media used. It is therefore essential that neuroradiologists are well trained in cerebral angiography and have an adequate clinical background so they are aware of the signs to watch for in case of complications. If any such warning symptoms should appear, and the question of interruption or discontinuing the examination arises, it should be easy for the radiologists to communicate with the referring doctors. It is therefore important in these situations to have neurosurgical and neurological resources present, and even more so in cases of radiological emergency. In addition to the clinical control of the patients, we have in the department of neuroradiology now been using polygraphic recordings of mainly EEG and ECG for about 25 years, particularly in patients considered most likely to suffer complications. Sometimes the first warnings come from the polygraphic tracings, and at other times from careful clinical observation. In our opinion both methods should therefore be used, especially in patients susceptible to complications because of their pre-existing disease and clinical condition. Some general precautions should also be taken both in these and other patients in order to minimize the risk of complications.

The contrast media used for cerebral angiography have been gradually improved, but they may still occasionally influence the vasomotor centres of the brain. To avoid a possible fall of blood pressure, bradycardia or asystole, it is recommended that atropine be given as a premedication because the vagus nerve is the efferent pathway in this reflex mechanism. We use the following dose schedule:

0–½ year	0.2 ml of a 0.1 per cent atropine sulphate solution
½–1 year	0.3 ml of a 0.1 per cent atropine sulphate solution
1–3 years	0.4 ml of a 0.1 per cent atropine sulphate solution
3–7 years	0.5 ml of a 0.1 per cent atropine sulphate solution
7–18 years	0.5 ml of a 0.1 per cent atropine sulphate solution
adults	0.6 ml of a 0.1 per cent atropine sulphate solution

All cerebral angiography patients are given this dosage parenterally one hour before the examination. In addition they are given a short-acting barbiturate and/or analgesic orally.

The new contrast media have reduced the risk of complications considerably, side effects due solely to their direct toxic action being therefore extremely rare. Because of this, more comprehensive studies are now possible and desirable examinations, such as complete examination of all cerebral vessels during one stage, are now done routinely in many centres. As multiple injections are necessary within a reasonable time for completion of study, only the best contrast media available should be used. At this time the Scandinavian centres are using Isopaque Cerebral 280 (methylglucamine and calcium salts of metrizoate). This has a satisfactory concentration of 280 mg of iodine per ml and does not cause changes in the blood–brain barrier in such concentration, and when used in moderate but sufficient individual and total dosage. Urografin (meglumine diatrizoate 60 per cent), Conray 280 (meglumine iothalamate 60 per cent) also fulfil these criteria. The amount of all these contrast media necessary is 8–10 ml for injection in the common carotid artery, 5–8 ml for the internal carotid artery and 4–5 ml for the vertebral artery. Because of their low toxicity, it is no longer necessary to avoid cumulative effects by delaying many minutes between injections, but it is still advisable to restrict the total dose given to a patient, because the perfect contrast medium has not yet been developed. Eight or ten full injections, and some added contrast used during screening of catheter positions, are permissible and often unavoidable in examining all cerebral vessels selectively. To reduce the total amount it is also desirable, whenever possible, to film in two planes simultaneously.

Technical problems are a more common cause of complications in cerebral angiography than the new contrast media used. A meticulous technique by a well-trained neuroradiologist is therefore essential. The most common causes of serious complications are intramural injection which may completely occlude the injected vessel, and embolization from an atheromatous plaque after puncturing a diseased vessel wall. These problems are covered elsewhere in this book. The method of catheterization of the femoral artery has been suggested as the ideal approach particularly when a complete selective examination of all cerebral vessels is desirable. This method has been adopted with great success in our department of neuroradiology, and is also used in many other Scandinavian centres for all types of cerebral angiography. The number of complications and their severity have therefore been reduced, and at the same time cerebral angiography has become more applicable particularly to patients with cerebrovascular symptoms and also other patients in poor clinical condition. In catheter angiography, embolization due to thrombus formation on the catheter is probably one of the major sources of complication. To avoid this, the catheter should be flushed at frequent intervals, and the examination should be completed as quickly as possible, preferably within one hour.

Pre-existing disease is as a rule the main cause of complications during cerebral angiography, but very little can be done to improve it. In general, patients should be in as good a clinical condition as possible. If the patient has, or probably will develop, increased intracranial pressure, this should be reduced before the roentgenological examination. This can be done in many different ways, but we usually give hypertonic mannitol intravenously over a period of one hour. These patients, as well as those under general anaesthesia and others in poor clinical condition, must be carefully observed clinically before, during and after each injection. In those patients most susceptible to complications we strongly recommend also that some type of polygraphic recording, preferably EEG and ECG should be used. In our experience the first warnings are sometimes seen in the polygraphic tracings, at other times they are recognized clinically, but if they occur, and especially if the patient's clinical condition deteriorates, the examination should be postponed or curtailed unless it is considered absolutely imperative to continue.

REFERENCES

1 WEIGEN J.F. & THOMAS S.F. (1973) Neuroangiography. In: *Complications of diagnostic radiology*, p. 250. Charles C. Thomas. Springfield. Illinois.
2 TBLLE K. & LUNDERVOLD A. (1947) EEG and ECG and other recordings in cerebral angiography. *Acta radiol. Diag.* 15, 250–256.
3 McDOWELL F. & KUTT H. (1964) Complications of angiography. In: *Cerebral vascular diseases* (Ed. by MILLIKAN et al.), p. 18. Grune & Stratton, New York.
4 ABBOTT K.H., GAY J.R. & GOODALL R.J. (1952) Clinical complications of cerebral angiography. *J. Neurosurg.* 9, 258–274.
5 HEINZ R.E. (1973) Neuroradiologic special procedures. In: *Complications and legal implications of radiologic special procedures* (Ed. by MEANEY T.F., LALLI A.F. & ALFIDI R.J.), p. 47. The C. V. Mosby Co., St Louis.
6 BANG N.V. (1964) Angiography: Complications. *Cerebr. vasc. dis.* p. 23, editorial.
7 MONIZ E. (1940) Die cerebrale Arteriographie und Phlebographie. In: *Handbuch der Neurologie*. Springer, Berlin. Suppl. 2.
8 TORKILDSEN A. (1949) Carotid angiography. *Acta psychiat. neurol. scand.* Suppl. 55.
9 HAUGE T. (1954) Catheter vertebral angiography. *Acta radiol.* Suppl. 109.
10 RADNER S. (1951) Vertebral angiography by catherization. A new method employed in 221 cases. *Acta radiol.* Suppl. 87.
11 KAUTZKY R. & ZÜLCH K.J. (1955) *Gefahren der Angiographie in Neurologisch-Neurochirurgische Röntgendiagnostik und andere Methoden zur Erkennung intrakranieller Erkrankungen*. Springer-Verlag, Berlin–Göttingen–Heidelberg.
12 LINDGREN E. (1950) Percutaneous angiography of vertebral artery. *Acta radiol. Diag.* 33, 389–404.
13 LINDGREN E. (1954) Röntgenologie. Vol. II in *Handbuch der Neurochirurgie* (Olivecrona–Tönnis). Springer-Verlag, Berlin–Göttingen–Heidelberg.
14 DUNSMORE R., SCOVILLE W.B. & WHITCOMB B.B. (1951) Complications of angiography. *J. Neurosurg.* 8, 110–118.
15 LUNDERVOLD A. & ENGESET A. (1966) Polygraphic recordings during cerebral angiography. *Acta radiol. Diag.* 5, 368–380.
16 BINNIE C.D., BERNSTEIN D.C., BOTH A.E., McCAUL I.R., MARGERISON J.H. & SCOTT J.F. (1971) Clinical and electroencephalographic sequelae of carotid angiography. *Acta radiol. Diag.* 11, 625–640.
17 FIELD J.R., ROBERTSON J.T. & DESAUSSURE R.L. (1962) Complications of cerebral angiography in 2000 consecutive cases. *J. Neurosurg.* 19, 775–781.

18 LUNDERVOLD A., ENGESET A. & PRESTHUS J. (1964) Cerebral angiography and pneumoencephalography. A polygraphic investigation. *J. Neurol. Neurosurg. Psychiat.* **5**, 129–144.
19 LUNDERVOLD A. & ONGRE A. (1968) Polygraphic recordings in cerebral angiography with meglumine metrizoate and meglumine iothalamate. *Farmakoterapi.* **24**, 3–12.
20 LUNDERVOLD A. & ENGESET A. (1969) Electroencephalographic and electrocardiographic studies of complications in cerebral angiography. *Acta radiol. Diag.* **9**, 399–406.
21 PATTERSON R.H., GOODELL H. & DUNNING H.S. (1964) Complications of carotid arteriography. *Arch. Neurol.* **10**, 513–520.
22 PERRET G. & NISHIOKA H. (1966) Report on the cooperative study of intracranial aneurysms and subarachnoid hemorrhage. *J. Neurosurg.* **25**, 98–114.
23 SILVERSTEIN A. (1966) Arteriography of stroke. III. Complications. *Arch. Neurol.* **15**, 206–210.
24 TAVERAS J.M. & WOOD E.H. (1964) *Diagnostic neuroradiology.* Williams & Wilkins, Baltimore.
25 ANSELL G. (1968) A national survey of radiological complications: interim report. *Clin. Radiol.* **19**, 175–191.
26 ANSELL G. (1970) Adverse reactions to contrast agents. Scope of problem. *Invest. Radiol.* **5**, 374–384.
27 FISCHER H.W. & DOUST V.L. (1972) An evaluation of pretesting in the problem of serious and fatal reactions to excretory urography. *Radiology* **103**, 497–501.
28 AMUNDSEN P. *et al.* (1963) Cerebral angiography by catheterization—complications and side effects. *Acta radiol. Diag.* **1**, 164–172.
29 SUTTON D. (1962) *Arteriography.* E. S. Livingstone Ltd., Edinburgh and London.
30 AMUNDSEN P., ENGESET A. & KRISTIANSEN K. (1970) Cerebral angiography. In: *Modern trends in diagnostic radiology* 4 (Ed. by MCLAREN I.W.), pp. 155–174. Butterworths, London.
31 GREITZ T. & LINDGREN E. (1971) Cerebral angiography: Technique and hazards. In: *Angiography* (Ed. by ABRAMS H.L.), 2 edn, pp. 155–168. Little, Brown & Co., Boston.
32 AUSTEN W.G., WILCOX B.R. & BENDER H.W. (1964) Experimental studies of the cardiovascular responses secondary to the injection of angiographic agents. *J. thorac. Surg.* **47**, 356–366.
33 PERRET G. & NISHIOKA H. (1966) Cerebral angiography. An analysis of the diagnostic value and complications of carotid and vertebral angiography in 5484 patients. *J. Neurosurg.* **25**, 98–114.
34 PRIBRAM H.F.W. (1965) Complications of cerebral arteriography. In: Twelfth annual scientific meeting of the Houston Neurological Society. Charles C. Thomas, Springfield, Illinois.
35 LANG E.K. (1966) Complications of direct and indirect angiography of brachiocephalic vessels. *Acta radiol. Diag.* **5**, 296–307.
36 TAVERAS J.F. & WOOD E.H. (1964) Morbidity and complications. In: *Diagnostic Neuroradiology.* Williams & Wilkins, Baltimore.
37 TIWISINA T. (1964) *Die Vertebralis-Angiographie.* Alfred Hüthing Verlag, Heidelberg.
38 BROMAN T. & OLSSON O. (1948) The tolerance of cerebral blood-vessels to a contrast

medium of the Diodrast group. An experimental study of the effect on the blood–brain barrier. *Acta radiol.* 30, 326–342.

39 GREITZ T. (1956) A radiologic study of the brain circulation by rapid serial angiography of the carotid artery. *Acta radiol.* Suppl. No. 140. p. 123.

40 GREITZ T. (1966) Dilatation of cerebral veins during cerebral angiography with water-soluble contrast media. *Acta radiol. Diag.* 4, 625–631.

41 LUNDERVOLD A. (1957) Discussion after: Electroencephalographical expression of altered consciousness. (Paper by H. Fischgold and R. Jung). *Proc. Premier Congrès International des Sciences Neurologiques*, pp. 214–218. Bruxelles.

42 LINDGREN P. & TÖRNELL G. (1958) Blood pressure and heart rate responses in carotid angiography with sodium acetrizoate (Triurol). *Acta radiol.* 50, 160–174.

43 KÅGSTRÖM E., LINDGREN P. & TÖRNELL G. (1958) Changes in the cerebral circulation during carotid angiography with sodium acetrizoate (Triurol) and sodium diatrizoate (Hypaque). *Acta radiol.* 50, 151–159.

44 KÅGSTRÖM E., LINDGREN P. & TÖRNELL G. (1960) Circulatory disturbances during cerebral angiography. *Acta radiol.* 54, 3–16.

45 AMUNDSEN D. (1965) Double-blind test in a comparison of angiographic contrast media for intravascular use. *Acta radiol. Diag.* 3, 335–343.

46 FISHER H.W. & ECKSTEIN J.W. (1961) Comparison of cerebral angiographic contrast media by their circulatory effects. *Am. J. Roentg.* 86, 166–177.

47 INGVAR D.H. & SÖDERBERG U. (1957) Cerebral vasomotor tone and EEG during injections of Umbradil: experimental study with a new method. *Acta radiol.* 47, 185–191.

48 GREITZ T. & WEISS S. (1959) EEG and ECG in cerebral angiography with sodium diatrizoate (Hypaque) and methylglucamine diprotrizoate (Miokon). *Acta radiol.* 52, 145–148.

49 LUNDERVOLD A., SKRAASTAD E. & BORGERSEN A. (1963) Cerebral angiography with simultaneous electroencephalographic, electrocardiographic and electromyographic recordings and measurement of intraarterial pressure. *Nord. Med.* 70, 853–859.

50 INGVAR D.H. (1957) EEG during cerebral angiography. *Acta radiol.* 47, 181.

51 CHASE N.E. & KRICHEFF I.I. (1965) Comparison of complication rates of meglumine iothalamate and sodium diatrizoate in cerebral angiography. *Am. J. Roentg., Rad. Therapy & Nuclear Med.* 95, 852–856.

52 CRONQVIST S., EFSING H.O. & PALACIOS E. (1970) Embolic complicationsi n cerebral angiography with catheter technique. *Acta radiol. Diag.* 10, 97–107.

53 KUTT H., VEROBOLY K., BANG N., STROULI F. & MC DOWELL F. (1966) Possible mechanisms of complications of angiography. *Acta radiol. Diag.* 5, 276–289.

54 TÖRMÄ T. & FOGELHOLM R. (1967) Complications of cerebral angiography with Urografin. *Acta Neurol. Scandinav.* 43, 616–629.

55 SKALPE I.O. (1973) Animal experimental and clinical investigations of a non-ionic water-soluble contrast medium (metrizamide) *Universitetsforlagets trykningssentral*, Oslo.

56 ALMEN T. (1966) A steering device for selective angiography and some vascular and enzymatic reactions observed in its clinical application. *Acta radiol.* Suppl. No. 260.

57 HINCK V.C. & DOTTER C.T. (1969) Appraisal of current techniques for cerebral angiography. *Am. J. Roentg., Rad. Therapy & Nuclear Med.* 107, 626–630.

58 KRAYENBÜHL H. & YAZARGIL M.G. (1965) Komplikationen in Die cerebrale Angio-

graphie. In: *Lehrbuch für Klinik und Praxis*, 2nd edn. Georg Thieme Verlag, Stuttgart.
59 LESTER J. & KLEEN A. (1965) Complications of 337 percutaneous vertebral angiographies. *Acta Neurol. Scandinav.* **41**, 301.
60 SCHEINBERG P. (1965) Practical aspects of complications of angiography. In: *Cerebral vascular diseases* (Ed. by MILLIKAN C.H., SICKERT R.G. & WHISNANT J.P.). Grune & Stratton, New York and London.
61 SCHEINBERG P. & ZUNKER E. (1963) Complications in direct percutaneous carotid arteriography. *Archs Neurol. Psychiat., Chicago* **8**, 106–114.
62 CHYNN K.-Y. (1969) Transfemoral carotid and vertebral angiography. *Acta radiol. Diag.* **9**, 244–250.
63 KRAYENBÜHL H. & YASARGIL M. (1969) Cerebral angiography. J. B. Lippincott Company, Philadelphia.
64 WISHART D.L. (1971) Complications in vertebral angiography as compared to non-vertebral cerebral angiography in 447 studies. *Am. J. Roentg.* **113**, 527–537.
65 RIISHEDE J. (1957) Cerebral apoplexy. An arteriographical and clinical study of 100 cases. *Acta psychiat. neurol. scand.* Suppl. 118.
66 ROWE S.N. & ARDITTI J. (1961) Carotid angiograms—value and hazards. *J. Am. med. Ass.* **178**, 163–164.
67 WENDE S. & SCHULZE A. (1961) Die cerebrale Angiographie und ihre Komplikationen. *Fortschr. Geb. Röntgenstrahlen.* **96**, 494–505.
68 ALLEN J.H., PARERA C. & POTTS D.G. (1965) Relation of arterial trauma to complications of cerebral angiography. *Am. J. Roentg., Rad. Therapy & Nuclear Med.* **95**, 845–851.
69 PATTERSON R.H. JR, GOODELL H. & DUNNING H.S. (1964) Complications of carotid arteriography. *Archs Neurol. Psychiat., Chicago.* **10**, 513–520.
70 HENDERSON W.R. & MEHTA D.S. (1973) Intracranial aneurysms. *Br. med. J.* **2**, 30–35.
71 BLOOR B.M., WRENN F.R. & MARGOLIS G. (1951) An experimental evaluation of certain contrast media used for cerebral angiography. *J. Neurosurg.* **8**, 585–594.
72 BLOOR B.M., WRENN F.R. & MARGOLIS G. (1954) Effect of intracarotid diodrast upon cerebral blood flow. *Archs Neurol. Psychiat.* **71**, 358.
73 BROMAN T. & OLLSON O. (1949) Experimental study of contrast media for cerebral angiography with reference to possible injurious effect on the cerebral blood vessels. *Acta radiol.* **31**, 321–334.
74 BROMAN T., FORSSMAN B. & OLLSON O. (1950) Further experimental investigations of injuries from contrast media in cerebral angiography. *Acta radiol.* **34**, 135–142.
75 BROWNE K.M. & STERN W.E. (1954) Experimental observations concerning cerebral angiography. *A.M.A. Archs Neurol. Psychiat.* **71**, 477–487.
76 HARRINGTON G. & WIEDEMAN M.P. (1965) The effect of contrast media on endothelial permeability. *Radiology* **84**, 1108–1111.
77 JENKNER F.L. & VOGLER E. (1965) Beobachtungen über die Veränderungen der Hirndurchblutung während der Karotisangiographie. *Neurochirurgia* **8**, 60–67.
78 FISCHER H.W. & CORNELL S.H. (1965) The toxicity of the sodium and methylglucamine salt of diatrizoate, iothalamate, and metrizoate. An experimental study of their circulatory effects following intracarotid injection. *Radiology* **85**, 1013–1021.
79 FISCHER H.W. & REDMAN H.C. (1971) Comparison of a sodium methylglucamine

diatrizoate contrast medium of minimal sodium content with a pure methylglucamine diatrizoate preparation. *Invest Radiol.* 6, 115–118.
80 HILAL S.K. (1966) Hemodynamic changes associated with the intra-arterial injection of contrast media. New toxicity tests and a new experimental contrast medium. *Radiology* 86, 615–633.
81 LANCE E.M. & KILLEN D.A. (1959) Experimental appraisal of the agents employed as angiocardiographic and aortographic media. I. Neurotoxicity. *Surgery* 46, 1107–1117.
82 MARGOLIS G. & YERASIMIDES T.G. (1966) Vasopressor potentiation of neurotoxicity in experimental aortography. Implications regarding pathogenesis of contrast medium injury. *Acta radiol. Diag.* 5, 388–412.
83 ALBERTSON K.W., DOPPMAN J.L. & RAMSEY R. (1973) Spinal seizures induced by contrast media. *Radiology* 107, 349–351.
84 GABEL A.A. & KOESTNER A. (1963) The effects of intracarotid artery injection of drugs in domestic animals. *J. Am. vet. med. Ass.* 142, 1397–1403.
85 ECHLIN F.A. (1969) Vasospasm, acute and recurrent, due to experimental subarachnoid hemorrhage. *Excerpta Med.* (Int. Congress Series) 193, 49–50.
86 LINDGREN P. (1959) Carotid angiography with triiodobenzoic acid derivatives. A comparative experimental study of the effects on the systemic circulation in cats. *Acta radiol.* 51, 353–362.
87 SÖRENSEN S.E. (1971) Microvascular effects of röntgen contrast media. *Elanders Boktrykkeri Aktiebolag*, Göteborg.
88 ENGESET A. (1948) Cerebral angiography with Perabrodil. *Acta radiol.* Suppl. 29.
89 HUBER P. & HANDA J. (1967) Effects of contrast material, hypercapnia, hyperventilation, hypertonic glucose and papaverine on the diameter of the cerebral arteries: Angiographic determination in man. *Invest. Radiol.* 2, 17–21.
90 LINDGREN P., SALTZMAN G.F. & TÖRNELL G. (1968) Circulatory effects of iothalamate compounds (Conray) and contrast media of the benzoic acid type. *Acta radiol. Diag.* 7, 48–64.
91 LINDGREN P., SALTZMAN G.F. & TÖRNELL G. (1968) Vascular reaction to water-soluble contrast media: Significance of concentration and total amount of iodine. *Acta radiol. Diag.* 7, 152–160.
92 GARDNER W.J. (1969) The blood–brain barrier—why the brain dies first. *J. Am. med. Ass.* 208, 1907–1909.
93 JÖRGENSEN P.B., KARLE A. & ROSENKLINK A. (1973) Changes in blood pressure and cardiac rhythm induced by arterial contrast injection. *Neuroradiology* 5, 215–219.
94 FISHER H.W. & CORNELL S.H. (1967) Toxicity study of sodium metrizoate containing calcium and magnesium. *Acta radiol. Diag.* 6, 126–137.
95 FISHER H.W. (1970) Sodium content of contrast media. Letter to the editor. *Invest. Radiol.* 4, 266–267.
96 LUNDERVOLD A. (1967) Bevissthetsforstyrrelser ved cerebrale kontrastundersökelser. *T. norske Lægeforen.* 87, 1784–1790.
97 LUNDERVOLD A., ENGESET A. & PRESTHUS J. (1965) Forandringer i elektro-encefalogram, -cardiogram, -myogram, intraarterielt og intraspinalt trykk ved pneumoencefalografi. *Nord. Med.* 73, 609–612.
98 KRICHEFF I.I. & CHASE N.E. (1967) Evaluation of complication rates of meglumine diatrizoate and meglumine iothalamate in cerebral angiography. *Am. J. Roentg.* 101, 220–223.

99 LINDNER D.W. & GURDJIAN E.S. (1966) Complications of angiography in patients with vascular and anaplastic and other disease of the nervous system. *Acta radiol. Diag.* **5**, 352.

100 SILVERMAN S.M., BERGMAN P.S. & BENDER M.B. (1961) Dynamics of transient cerebral blindness: report of nine episodes following vertebral angiography. *A.M.A. Archs Neurol.* **4**, 333–348.

101 HOWLAND W.J. & CURRY J.L. (1964) Transient cerebral blindness: hazard of vertebral artery catheterization: report of four cases. *Radiology* **83**, 428–432.

102 LUNDERVOLD A., HAUGE T. & LÖKEN AA.C. (1956) Unusual EEG in unconscious patient with brain stem atrophy. *EEG Clin. Neurophysiol.* **8**, 665–670.

103 LUNDERVOLD A. (1966) EEG in 'Experimental' vascular brain stem disorders. In: *Clinical experiences in brain stem disorders*, pp. 81–82. Acta 25. Conventus neuropsychiatrici et EEG Hungarici, Budapest.

104 BERGER H. (1932) On the electroencephalogram of man. *Arch. Psychiat. NervKrankh.* **97**, 6–26.

105 BERGER H. (1938) On the electroencephalogram of man. *Arch. Psychiat. NervKrankh.* **108**, 407–431.

106 LUNDERVOLD A. (1974) EEG in patients with coma due to localized brain lesions. In: *Handbook of Electroencephalography and Clinical Neurophysiology*, Vol. 12. (In press.)

107 LUNDERVOLD A., SKRAASTAD E. & BORGERSEN A. (1963) Cerebral angiografi med samtidig EEG-, EKG- og EMG-registrering samt intra-arteriell trykkmåling. *Nord. Med.* **70**, 853–856.

108 LANG E.K. (1963) Survey of complications of percutaneous retrograde arteriography: Seldinger technique. *Radiology* **61**, 257–263.

109 PERRET G. (1966) Diagnostic value and complications of carotid and vertebral angiography. *Acta radiol. Diag.* **5**, 453–459.

110 JEFFERSON G. (1938) The tentorial pressure cone. *Archs Neurol. Psychiat.* **40**, 857–876.

111 BRECHNER U.L., WALTER R.D. & DILLON J.B. (1962) *Practical electroencephalography for the anesthesiologist.* Charles C. Thomas, Springfield, Illinois.

112 COOK W. (1973) Problems in manufacture of guide wires and legal implications. In: *Camplications and legal implications of radiologic special procedures* (Ed. by MEANEY T.F., LALLI A.F. & ALFIDI R.C.), pp. 137–143. C. V. Mosby Co., St Louis.

113 BARNHARD H.J. & BARNHARD F.M. (1968) The emergency treatment of reactions to contrast media. *Radiology* **91**, 74–84.

114 ANSELL G. & ANSELL A. (1964) Medical emergencies in the X-ray department: Prevention and treatment. *Br. J. Radiol.* **37**, 881–897.

115 GILSTON A. & RESENEKOV L. (1971) *Cardio-respiratory resuscitation.* William Heinemann Medical Books, London.

116 SAXTON H.M. & STRICKLAND B. (1972) *Practical procedures in diagnostic radiology*, 2nd edn. H. K. Lewis, London.

117 ANSELL G. (1973) *Notes on radiological emergencies.* 2nd edn. Blackwell Scientific Publications, Oxford, London.

118 BOUZARTH W.F., GOLDFEDDER P. & SHENKIN H.A. (1968) Hypertonic mannitol as a treatment for complications of cerebral arteriography. *Am. J. Roentg.* **104**, 119–122.

CHAPTER 7: PNEUMOENCEPHALOGRAPHY AND VENTRICULOGRAPHY

J. V. OCCLESHAW

It is the aim of any neuroradiological examination to provide the maximum information about the patient's pathological process, but to cause the minimum distress to the patient and to be as free as possible from complications. Although computer tomography, cerebral angiography and the use of radioactive isotopes for brain scanning have made great contributions to the diagnosis of brain tumours and other pathological processes, the radiological examination of the cerebral ventricles and extracerebral cerebrospinal fluid pathways still retains a place in neuroradiology.

Many patients undergoing ventriculography or pneumoencephalography already have some elevation of the intracranial pressure and are unable to withstand further alterations in the pressure relations between the ventricles and the extracerebral spaces. Such pressure changes may require the urgent intervention of a neurosurgeon as a life-preserving measure. Subclinical intracranial hypertension may cause the patient to react adversely to drugs used for premedication prior to pneumoencephalography, especially if drugs with respiratory depressing properties are employed. Many such complications can be anticipated and avoided by close cooperation between the neurologist or neurosurgeon, the anaesthetist and the neuroradiologist.

PNEUMOENCEPHALOGRAPHY

The major complications of pneumoencephalography can be divided into three groups:
(a) Complications resulting from the drugs used in the premedication of the patient.
(b) Complications occurring during the performance of the investigation.
(c) Post-examination complications.

Reaction to premedication drugs

Patients with raised intracranial pressure are often drowsy and drugs such as morphine, pethidine and phenoperidine (Operidine) may provoke respiratory depression even when given in normal dosage. Under such circumstances the dosage of these drugs should be reduced or their use avoided if possible.

If respiratory depression occurs the airway should be cleared, maintained if necessary by tracheal intubation and the patient ventilated with oxygen. The intravenous administration of 10 mg of nalorphine hydrobromide (Lethidrone) will frequently counteract the depressant effects of the offending drug, but nalorphine should not be given in a dosage greater than 30 mg, as this drug then becomes a respiratory depressant.

In extreme cases of respiratory depression which do not respond to this treatment, it may be necessary to maintain ventilation by using a mechanical respirator for a few hours until the effects of the drug have worn off.

Postural hypotension may occur on sitting some patients in the encephalogram chair and many neuroradiologists give 30 mg of ephedrine intramuscularly before starting to position the patient. The use of elastic stockings to prevent pooling of blood in the dependent legs will often prevent postural hypotension in the elderly patient. Hypertensive patients being treated with hypotensive agents and the nonhypertensive patient who has been premedicated with promethazine hydrochloride (Phenergan) or chlorpromazine hydrochloride (Largactil) are very prone to postural hypotension and it is therefore advisable to avoid premedication with these two agents.

If the patient becomes hypotensive when erect in the encephalogram chair, rotation into the horizontal position may be sufficient to restore the blood pressure. If there is no response to this simple measure, the administration of 20–40 mg of methoxamine hydrobromide (Vasoxine) intravenously is advised. If this fails to produce any response then it may be necessary to restore and maintain the blood pressure by giving the patient l-noradrenaline in an intravenous saline drip. The rate of flow of the saline-noradrenaline infusion should be carefully adjusted to control the blood pressure. Great care should be observed when giving any pressor agent to patients who are being given halothane anaesthetics as there is a risk of producing cardiac irregularities and even ventricular fibrillation. If a patient requires a noradrenaline infusion it is advisable to abandon the examination.

Collapse at the start of or during the course of a pneumoencephalogram may result from adrenal insufficiency in patients who have been on treatment with adrenocortical hormone preparations. The intravenous administration of 100 mg of hydrocortisone sodium phosphate (Efcortesol) is usually adequate to control the situation. Many patients may be protected from this complication by being given 100 mg of hydrocortisone intravenously at the commencement of the examination if they are already on steroid therapy for their neurological disorder.

The introduction of the technique of neuroleptanalgesia for neuroradiological procedures [1] has made pneumoencephalography less distressing for the patient and also reduces the incidence of complications. Phenoperidine has greater analgesic effects than its parent pethidine, but occasionally it may produce some respiratory depression in the small or frail patient. On some occasions haloperidol (Serenace) may provoke a state of agitation or a marked tremor of the limbs and trunk as a result of its effects of the extrapyramidal system. This effect is self-limiting in patients not suffering from Parkinsonism and only lasts for a few hours, but it may be necessary to control the tremor by the intravenous administration of 2-4 mg of benztropine mesylate (Cogentin) before the examination can start. In patients with Parkinsonism it is claimed that the tremor can be reduced by giving the patient 5 mg tablets of benzhexol (Artane), or 400 mg of cycloheptenyl barbituric acid by mouth before the examination [1].

Complications occurring during the performance of the investigation

Four *minor complications* can occur during the performance of a pneumoencephalogram. (a) The apprehensive or inadequately premedicated patient may become hypotensive as a result of a vaso-vagal reaction to the injection of the local anaesthetic agent or the lumbar puncture needle. Rotation of the encephalogram chair into the horizontal position for a few minutes usually corrects this reaction and allows the examination to continue. (b) Patients with a history of epilepsy are prone to have epileptic attacks during the introduction of the air. Once an epileptic reaction starts it is necessary to restrain the patient to avoid self injury, to maintain the airway, and to oxygenate the patient if there are signs of cyanosis. To continue the examination in the presence of repeated epileptic attacks, it may be necessary to control these with intravenous diazepam (Valium) given in repeated small doses of 2-3 mg via a 'butterfly' type of intravenous needle. The diazepam may be repeated up to a total dosage of 20 mg in fifteen to twenty minutes. (c) Vomiting and severe headache may occur during the introduction of the air if large amounts enter the basal subarachnoid cisterns. This usually results when the air is injected too quickly or when the patient's head is inadequately flexed. Fluroscopic control during the injection of the air helps to ensure optimum position of the head for the air to enter the ventricles. The use of a water

syphonage system to deliver the air slowly to the lumbar puncture needle is claimed to reduce the incidence of headache in pneumoencephalography [2].
(d) Changes in the heart rate or rhythm have been observed in two-thirds of patients during pneumoencephalography [3]. These take many forms such as sinus tachycardia, sinus bradycardia, sinus arrhythmia, premature auricular contractions or alterations in the electrocardiogram of the QRS complex or the S–T segment. Such reactions usually occur when over 40 ml of air have been injected, but may occur with smaller amounts of air when the intracranial pressure is elevated or in the presence of a mid-brain lesion. In most instances the changes are short lived and do not require any special treatment.

Foraminal crowding

This major complication that can occur during the performance of the pneumoencephalogram is partly due to an elevation of the intraventricular pressure. Air entering the basal subarachnoid cisterns removes the incompressible fluid cushion around the base of the brain. This may allow the distended brain to move and descend through the semi-rigid tentorial hiatus and foramen magnum. The mechanical effects of the distended brain passing through the 'narrows' of the tentorial hiatus are well recognized [4]. They are in part due to interference in the blood flow through the perforating branches of the major cerebral vessels producing ischaemic changes in the diencephalon, mid-brain and hypothalamus, and in part due to compression of adjacent cranial nerves.

The first sign of this complication is usually a fall in the pulse rate and slight elevation in the blood pressure of 20–30 mmHg. At this stage the patient usually appears pale due to peripheral vasoconstriction. Further changes soon occur with a greater fall in the pulse rate and some depression of the respiratory rate. Early compression of the third cranial nerve produces constriction of the pupils, but this does not persist for long and the pupils then start to dilate with loss of the reflex response to light.

At this stage, the patient is in need of neurosurgical intervention to reduce the intraventricular pressure as a life-preserving measure. While awaiting neurosurgical assistance emergency treatment should be commenced. The airway should be cleared and the patient oxygenated after tracheal intubation. It may be necessary to use a mechanical ventilator to maintain adequate respiratory function.

Intravenous hypertonic infusions of urea [5] or mannitol [6], may reduce the brain volume and do much to overcome the effects of elevated intracranial pressure. Urea produces a more rapid initial effect than mannitol, but as urea is able to diffuse into the intercellular space, it will produce a later rebound elevation in pressure. As these hypertonic solutions produce renal diuresis, it is essential to catheterize the patient's bladder after setting up the intravenous infusion. In many instances this regime will produce a dramatic response in the patient's condition with a return to spontaneous respiration, an increase in the pulse rate and a return to normal pupillary size and reaction. However despite recovery it is always wise to have the patient under the care of a neurosurgeon for fear of a relapse or for treatment of the underlying pathology.

Failure to respond to medical treatment requires drainage of the ventricular system, by the insertion of a catheter into the lateral ventricle via a burr hole, and slow decompression using a 'wash-bottle' drainage system. In small children it is often possible to pass a needle, through the unclosed anterior fontanelle, into the lateral ventricle and quickly reduce the intraventricular pressure by draining off some cerebrospinal fluid.

Dangerous pressure changes are more likely to occur when performing a pneumoencephalogram on a patient with an intracranial space-occupying lesion. In many neuroradiological departments where pneumoencephalography is performed on such patients in preference to ventriculography, it is considered advisable that a burr hole be made in the cranial vault prior to the examination, so that the ventricles can be readily drained if an emergency should arise.

The effects of general anaesthesia on intracranial pressure are well recognized. Nitrous oxide, being very soluble in blood and body fluids, can diffuse into closed body cavities containing air, faster than other gases can diffuse out of the cavities. This results in a rise in the pressure within air-containing cavities, such as the cerebral ventricles, during pneumoencephalography [7, 8]. There has been considerable debate, in both the anaesthetic and radiological literature, on the hazards of performing pneumoencephalography under general anaesthesia. This has resulted in most neuroradiologists performing these investigations under sedation and local anaesthesia. More recent work [9] has shown that nitrous oxide anaesthesia will produce a very gradual increase in the intraventricular pressure, but this is well below the peak pressure rises that immediately follows each fractional injection of air during the filling stages of the examination.

Post-examination complications

The sequelae to pneumoencephalography have in the last 35 years been the subject of a series of papers too numerous to list. A recent comprehensive review [10] lists the effects in decreasing order of frequency as: headache, tachycardia, pyrexia, neck stiffness, vomiting and blood pressure changes.

Headache is an inevitable reaction to pneumoencephalography and in the occasional patient may persist for up to seven days or more. The inhalation of 95 per cent oxygen following the examination is claimed to reduce the severity of this symptom [11]. Similar claims have been made for the intrathecal injection of 40 mg of methylprednisolone acetate (Depo-Medrol)* in 10 ml of saline, following the introduction of the air [12]. Desoxycortisone Acetate (DOCA) orally in a dosage of 5 mg three times per day, may also give some relief in cases of severe headache [13].

Fever following pneumoencephalography is well recognized and a careful study of this reaction noted its occurrence in 70 per cent of cases [14]. The mechanism of the pyrexia is unknown as pyrexia does not follow diagnostic lumbar puncture, and no correlation could be established between pyrexia and the underlying pathology which called for the investigation. Half the patients studied developed fever in the first twelve hours and the remainder in the next twelve hours, but all patients became apyrexial within five days.

Vomiting frequently accompanies the headache, especially if the latter is severe, but usually resolves without treatment within 36 hours. If vomiting is very severe it is often controlled by giving the patient metoclopramide monohydrochloride (Maxolon), 10 mg intravenously. Persistent vomiting may lead to a state of acidosis and it may be necessary to resort to intravenous hydration and correction of acid-base balance in extreme cases.

Persistence of the headache, vomiting, neck stiffness and pyrexia after pneumoencephalography always raises the question of meningitis following the lumbar puncture. To exclude meningitis further lumbar puncture and examination of the cerebrospinal fluid is necessary, but the results may cause some confusion, as the presence of air in the ventricles produces changes in the cerebrospinal fluid.

The effects of ventricular air on the composition of the cerebrospinal fluid have been carefully studied [15, 16]. A rise in the cell count within a few minutes of the introduction of air has been observed due to the appearance of mononuclear (M type) cells, together with a fall in the cerebrospinal fluid protein level. The features to distinguish between the pneumoencephalogram response and an inflammatory reaction are listed in Table 7.1.

Although *septic meningitis* is a rare complication of pneumoencephalography, it may occur despite careful aseptic lumbar puncture technique, but few cases are recorded in the radiological literature [17].

* Intrathecal injections of hydrocortisone produce intense pain.

Subdural haematoma formation following pneumoencephalography has been recorded in children and in adults [18–21]. The mechanism for this complication in children has been carefully studied [21]. Air is noted to be present in the subdural spaces in one-third of children during the examination and this allows the brain to fall away from the cranial vault. Tension occurs in the veins passing to the venous sinuses and tears may occur in the sheaths of vessels passing to the pachionian granulations. Damage to these structures may result in collections of blood or fluid forming in the subdural spaces.

In children the effects of these collections are to cause a prolongation of headache, vomiting and fever beyond the first 36 hours or a recurrence of

TABLE 7.1. Differentiation of pneumoencephalogram and inflammatory reactions.

	Encephalogram reaction	Inflammatory reaction
Time of onset	5 min	2–3 hrs
CSF protein	Falls	Rises
CSF glucose	No change	Falls
CSF cell count	Rises	Rises
Type of cell reaction	'M' cells	Polymorphs

these symptoms about the fifth day after the examination. Repeated aspiration of the subdural spaces via the anterior fontanelle is usually all that is required to treat this type of complication in small children, but older children may require surgical aspiration of the collections via burr holes.

In adults, subdural haematoma usually occurs in patients with cerebral atrophy and symptoms appear from two weeks to three months after the pneumoencephalogram. The general condition of the patient deteriorates with increasing unsteadiness, headache and drowsiness. Computer tomography, cerebral angiography or burr hole exploration will confirm the presence of the haematoma which can then be treated surgically.

Death following pneumoencephalography is a rare complication, but occurred in 9 of the first 4000 cases examined by the pioneer radiologists, Davidoff & Dyke [22]. In three cases, subarachnoid haemorrhage following the rupture of a cerebral artery aneurysm led to death, and four patients died as a result of haemorrhage into brain tumours. Of the remaining two patients, one died from coronary artery thrombosis, but no cause for death could be found in the last case, a child with advanced central sclerosis.

Other quoted incidences of death range from 0.22 per cent to 8.17 per cent and the most frequent cause of death found at autopsy is either the effect of an intracranial space-occupying lesion or air embolism [23]. Cardiorespiratory collapse and eventual death of a child with a subdural haematoma is reported in detail [8]. Pneumoencephalography was performed under nitrous oxide anaesthesia and further elevation of the intracranial pressure occurred following the passage of the air into the ventricles and subsequent diffusion of nitrous oxide into the intraventricular gas.

Complications occurring in children during pneumoencephalography have been the subject of study in recent years [17]. A significantly lower rate of reactions was claimed when the examination was performed under heavy sedation in preference to general anaesthesia. Most complications occurred in children under the age of five years, whilst under the age of one year, the rate of complication was four times greater if the child was examined under general anaesthesia. However, most British anaesthetists who undertake this type of work would probably not agree with these conclusions. Fever occurred in 50 per cent of children following pneumoencephalography and four patients developed meningitis (one staphylococcus albus meningitis and three aseptic meningitis). Two children were noted to have developed hemiplegia after the examination. Also recorded by these authors is the death of a three-month-old child who had been premedicated with rectal tribromethanol and later given 20 mg of pentobarbital intramuscularly. Cardiac and respiratory arrest occurred and the child could not be revived.

Further support for the use of heavy sedation in preference to general anaesthesia is found in a recent study of the reactions occurring in 100 adults and 100 children undergoing pneumoencephalography [24]. Eighty-eight of the children had been given ketamine hydrochloride (Ketalar) and the more severe pyrexial reactions occurred in the children who had been examined under general anaesthesia as a result of postanaesthetic respiratory infections. One child became cyanotic and pulseless after withdrawal of the endotracheal tube and subsequently developed 'atelectasis and pulmonary infiltration'.

Pneumomediastinum as a complication to pneumoencephalography in children has been reported [25], but the mechanism is as yet not known. Air appeared in the soft tissues of the neck during the examination and later spread to the mediastinum. Four of the children were under the age of one year and radiographs of the lumbar region did not reveal any air in the tissues at the site of the lumbar puncture. One possible mechanism put forward is that as the examinations were performed under heavy sedation and were often accompanied by crying and straining, air might have entered the tissues from ruptured alveoli at the lung bases and subsequent dissection into the mediastinum. This complication has also recently been recorded in adult patients but the evidence in these cases favoured a direct escape of air from the spinal canal into the surrounding soft tissues [26]. Discomfort in the chest should be regarded as a warning sign.

VENTRICULOGRAPHY

Except in the case of small children, when the anterior fontanelle is patent, it is necessary to have a burr hole made in the cranial vault for access to the ventricles. This surgical procedure is on rare occasions followed by infection of the scalp, skull vault or meninges. The passage of a needle or cannula through the overlying brain substance into the cavity of the lateral ventricle may provoke a haemorrhage into the brain substance, ventricle or the spaces between the brain and its coverings. When ventricular puncture is performed via an occipital burr hole, oedema in the region of the calcarine cortex may occur as a reaction to the surgical trauma with resultant transient amaurosis. With improvements in surgical techniques and the use of antibiotic cover, these complications have almost completely disappeared.

Two sequelae to needle puncture of the lateral ventricles that are of radiological interest have been recorded. (a) *Calcification* may subsequently occur along the needle track and be demonstrated in later radiographs [27]. (b) Puncture of the lateral ventricles for diagnostic or therapeutic purposes may result in the subsequent development of *cerebral cavities* at the site of the needle puncture [28, 29]. These cavities usually follow puncture of the lateral ventricles via the anterior fontanelle in hydrocephalic children. Unless the intracranial hypertension is corrected shortly after the ventricular puncture, the cavities will develop and may be seen at later ventriculography. An example of this complication is shown in Fig. 7.1. This child had a ventriculogram performed within the first week of life by puncture of the right lateral ventricle via the anterior fontanelle, but the degree of hydrocephalus present at that stage was not considered severe enough for ventriculo-atrial shunt. When a second ventriculogram was performed at the age of five weeks, the intraventricular pressure was 250 mmH_2O and a cavity was demonstrated which extended from the roof of the right lateral ventricle towards the anterior fontanelle.

As the majority of patients undergoing air ventriculography have elevated intracranial pressure, the injection of air should be compensated by the removal of a similar volume of cerebrospinal fluid. The removal of excessive amounts of cerebrospinal fluid is known to produce cardiovascular collapse

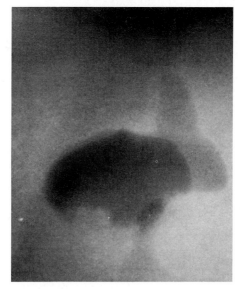

a

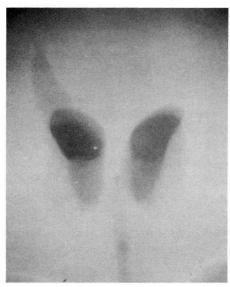

b

FIGURE 7.1. a and b. Antero-posterior and lateral films of the second ventriculogram on a five-week-old child. A cavity is seen extending upwards from the lateral aspect of the right lateral ventricle, along the needle track of the first ventricular puncture. Intraventricular pressure at the second examination was 250 mmH$_2$O.

and, on occasion, it may be necessary to inject saline into the ventricles to restore normal circulation [30].

If the procedure is performed under nitrous oxide anaesthesia, nitrous oxide diffusion into the ventricular air may, as already discussed, produce further elevation of intracranial pressure and require neurosurgical intervention.

Air embolism occurring eight hours after air ventriculography has been recorded [31]. The investigation was not followed by surgical treatment and air entered the circulation through small holes in ependymal veins, possible during a fit of coughing. In order to avoid this complication, ventriculography should, when possible, be performed immediately before neurosurgery. If no operation follows the investigation, the patient should be nursed in a horizontal position to prevent intravenous pressure in the cerebral veins becoming subatmospheric.

Positive contrast ventriculography can be performed using either the oily contrast agent iophendylate (Myodil, Pantopaque), or the water-soluble agents meglumine iothalamate (Conray 280) or meglumine iocarmate (Dimer X). Complications and reactions resulting from the use of these agents have been reported.

Although Myodil has been extensively used for myelography for many years and has come to be regarded as a 'safe' contrast agent, experimental studies in animals have shown that its intrathecal injection is followed by an inflammatory reaction in the meninges and a cellular response in the cerebrospinal fluid [32]. Further studies have shown that Myodil in the spinal subarachnoid space in the presence of blood produces an arachnoid reaction which leads to adhesion formation [33, 34]. The presence of blood in the ventricular cerebrospinal fluid at the time of the introduction of myodil for ventriculography theoretically produces the ideal situation for this type of reaction to occur. However, I know of only one patient, out of approximately 400 who have undergone Myodil ventriculography in our radiology department, who developed adhesions which obstructed the outflow paths of the fourth ventricle and produced an obstructive hydrocephalus [35]. Some claims have been made for the prevention of such reactions in the spine by the addition of methylprednislone to the Myodil at the time of its introduction, if there is any blood in the cerebrospinal fluid. The use of methylprednisolone in association with Myodil at the time of ventriculography has yet to be reported.

Minor complications that occurred in a small series of 90 examinations have recently been recorded [36]. Three patients developed pyrexia which persisted for over 48 hours and one patient developed epileptic fits as a result of an accidental subependymal injection. One patient obstructed a ventricular decompression valve with Myodil and removal of the contrast at the end of the examination is advised in patients with decompression valves. Pulmonary oil embolism following Pantopaque ventriculography in a patient with a ventriculovenous shunt has also been recorded [37]. The patient was a six-year-old hydrocephalic child and only 1 ml of contrast was introduced into the ventricles via the ventriculostomy tube. The examination was then followed by a revision of the Holter valve and reservoir. Subsequently the child became pyrexial and showed signs of meningeal irritation. Chest radiographs revealed the presence of globules of contrast in the atrial end of the catheter and a fine granular pattern was present in the lung fields.

Though *death* has been recorded following Myodil ventriculography in 1.5 per cent of cases [38], this mortality rate is similar to that of ventricular puncture via a burr hole in patients with intracranial tumours. Therefore this complication is not really attributable to the Myodil [39].

WATER-SOLUBLE MEDIA

Reactions and complications resulting from the use of meglumine iothalamate (Conray 280) for ventriculography have been recorded in three series of patients [40–42] and the results are listed in Table 7.2.

TABLE 7.2. Reactions following ventriculography with Conray 280.

	Heimburger et al. [40]	Guthkelch et al. [41]	Panda et al. [42]
Number of cases	90	91	70
Headache	32	26	28
Nausea and vomiting	38	13	26
Fever	47	8	18
Seizures	4	1	3
Meningeal reactions	2	—	—
Confusion	4	—	—
Deaths	2	—	3

Headache is not usually severe and can be controlled by simple analgesics. *Nausea and vomiting* are common reactions to all types of ventriculography and premedication of patients with barbiturate and aspirin is said to reduce the severity of the reaction. *Epileptic seizures* are recorded in most series of Conray ventriculograms and frequently follow the accidental injection of the contrast agent into the interhemispheric spaces. In order to avoid this accident most authors advise a test injection of a small amount of air down the intraventricular needle, to ensure that its tip lies within the ventricle before the introduction of the contrast agent. The seizures can often be controlled by the intravenous administration of diazepam (Valium) in repeated dosage if necessary. If it is not possible to control the seizures by this treatment, then it may be necessary to curarize the patient for some hours and maintain respiration by a mechanical ventilator [42].

The two cases of *meningeal reactions* quoted in the first series [40] were mild and only persisted for a few hours. The two *deaths* were in patients who arrived at the hospital in extremis and ventriculography was performed as a heroic measure in the presence of gross intracranial hypertension. Death could not be attributed to the contrast agent. The autopsy findings on five patients who died within 21 months of ventriculography are recorded [40] and only one patient was found to have any evidence of an ependymal reaction.

Dimer X has only been employed for ventriculography in a small number of neuroradiology departments in the last few years. The first recorded series [43] was of 120 cases, and the contrast agent was found to be more readily tolerated than Conray when used in volumes up to 5 ml. By nursing the patient flat and monitoring the EEG at six hours, epileptogenic activity could be detected and treated by intravenous Valium before the onset of seizures.

A more recent series of twenty examinations with Dimer X in the absence of EEG monitoring has been recorded [44]. Mild headache (one case), vomiting (one case) and pyrexia (two cases) were noted. Two patients received over 5 ml of Dimer X and subsequently developed clinical signs of meningitis. The first patient responded to treatment with sulphadiazine and chloromycetin and the second patient to treatment with anti-tuberculous therapy. It is very doubtful that either of these inflammatory reactions is in any way attributable to the Dimer X.

Most radiologists using Dimer X for ventriculography are finding that seizures may occur in up to 25 per cent of cases, but these fits can usually be controlled with intravenous Valium. A history of epilepsy in the past should be regarded as a contraindication to water-soluble contrast ventriculography or an indication for premedication with Valium.

The use of *metrizamide*, a new non-ionic water-soluble contrast agent, for contrast ventriculography has been recorded in five patients [45]. Headache occurred in four patients and nausea in one patient. Extracerebral injection of metrizamide was not found to cause any clinical complications and further studies could well establish this contrast agent as being more suitable for ventricular injection than either Myodil, Conray or Dimer X.

REFERENCES

1. BROWN A.S. (1964) Neuroleptanalgesia for the surgical treatment of Parkinsonism. *Anaesthesia* 19, 70–74.
2. McCALLUM A.H. (1971) A simple automatic gas dispenser for encephalography and myelography. *Br. J. Radiol.* 44, 396–398.
3. CLAYTON DAVIE J. (1963) Electrocardiographic alterations observed during fractional pneumoencephalography. *J. Neurosurg.* 20, 321–328.
4. DOTT N.M. (1960) Brain movement and time. *Br. Med. J.* 2, 12–16.
5. STUBBS J. & PENNYBACKER J. (1960) Reduction of intracranial pressure with hypertonic urea. *Lancet* 1, 1094–1097.
6. WISE B.L. & CHATER N. (1962) The value of hypertonic mannitol solutions in decreasing brain mass and lowering cerebrospinal fluid pressure. *J. Neurosurg.* 19, 1038–1043.
7. EGER E.I. & SAIDMAN L.J. (1965) Hazards of nitrous oxide anaesthesia in bowel obstruction and pneumothorax. *Anaesthesiology* 26, 61–66.
8. EGER E.I. & SAIDMAN L.J. (1965) Changes in cerebrospinal fluid pressure during pneumoencephalography under nitrous oxide anaesthesia. *Anaesthesiology* 26, 67–72.
9. GORDON E. & GREITZ T. (1970) The effects of nitrous oxide on the cerebrospinal pressure during encephalography. *Brit. J. Anaesth.* 42, 2–8.
10. WHITE Y.S., BELL D.S. & MELLICK R. (1973) Sequelae to pneumoencephalography. *J. Neurol. Neurosurg. & Psychiat.* 36, 146–151.
11. KORNREICH C.J. (1948) Relief of symptoms following encephalography by combined premedication and use of oxygen. *Arch. Neurol. & Psychiat.* 60, 512–519.
12. KULICK S.A. (1966) The clinical use of intrathecal methylprednislone acetate following pneumoencephalography and myelography. *J. Mt. Sinai Hosp.* 33, 152–157.
13. WOLFSON B., SICHER E.S. & GRAY G.H. (1970) Post-pneumoencephalography headache. *Anaesthesia* 25, 328–333.
14. LIPTON M.J. & CROWTHER M.A. (1968) Fever after air encephalography. *Br. J. Radiol.* 41, 672–673.
15. BICKERSTAFF E.R. (1950) Cerebrospinal fluid changes during pneumoencephalography. *Lancet* 2, 683–685.
16. MARRACK D., MARKS V. & COUCH R.S.C. (1961) Changes in the lumbar cerebrospinal fluid during air encephalography. *Br. J. Radiol.* 34, 635–639.
17. GROOVER R.V., CHUTORIAN A.M. & NELLHAUS G. (1966) Neuroradiologic procedures in children—comparison of heavy sedation and general anaesthesia. *Acta Radiol. Diag.* 5, 180–191.
18. BUCY P. (1942) Subdural haematoma. *Illinois Med. J.* 82, 300–310.
19. ROBINSON R.G. (1957) Subdural haematoma in an adult after air encephalography. *J. Neurol. Neurosurg. & Psychiat.* 20, 131–132.
20. CALKINS R.A., VAN ALLEN M.W. & SAHS A.L. (1967) Subdural haematoma following pneumoencephalography—Case Report. *J. Neurosurg.* 27, 56–59.

21 SMITH H.V. & CROTHERS B. (1950) Subdural fluid as a consequence of pneumoencephalography. *Paediatrics* 5, 375–389.
22 DAVIDOFF L.M. & DYKE C.G. (1946) *Normal Encephalography*, 2nd edn. Lea & Febiger, Philadelphia and London.
23 MARTINEZ NIOCHET A. & LARA G.R. (1965) Air embolism during pneumoencephalography—case report. *Acta Neurol. Lat. Amer.* 11, 215–218.
24 CLARK R.A., OBENCHAIN T.G., HANAFEE W.N. & WILSON G.H. (1970) Pneumoencephalography—comparison of complications in 100 paediatric and 100 adult cases. *Radiology* 95, 675–678.
25 BEATY W.R. & GERALD B. (1968) Pneumomediastinum as a complication of pneumoencephalography in children. *Radiology* 91, 956–958.
26 TSAI F.Y. & LEE K.F. (1974) Pneumopericardium and pneumomediastinum: Complications of pneumoencephalography. *Radiology* 112, 95–97.
27 FALK B. (1951) Calcification in the track of the needle following ventricular puncture. *Acta Radiol.* 35, 304–308.
28 LORBER J. & GRAINGER R.G. (1963) Cerebral cavities following ventricular punctures in infants. *Clin. Radiol.* 14, 98–109.
29 SALMON J.H. (1967) Puncture porencephalography—pathogenesis and prevention. *Am. J. Dis. Child.* 114, 72–79.
30 BAHL C.P. & WANDA S. (1967) Cardiovascular collapse after rapid ventricular decompression. *Br. J. Anaesthesia* 39, 657–658.
31 HEPPNER F. (1952) Air embolism eight hours after ventriculography. *Acta Radiol.* 38, 294–298.
32 FISHER R.L. (1965) An experimental evaluation of Pantopaque and other recently developed contrast media. *Radiology* 85, 537–545.
33 HOWLAND W.J., CURRY J.L. & BUTLER A.K. (1963) Pantopaque arachnoiditis. *Radiology* 80, 489–491.
34 HOWLAND W.J. & CURRY J.L. (1966) Experimental studies of Pantopaque arachnoiditis. *Radiology* 87, 253–261.
35 SEDZIMIR C.B. & IWAN S.R. (1962) Simplified contrast ventriculography. *J. Neurosurg.* 19, 657–660.
36 LANG E.K. & RUSSELL J.R. (1970) Demonstration and assessment of lesions of the third ventricle and posterior fossa. *J. Neurosurg.* 32, 5–15.
37 ALLEN W.E. & D'ANGELO C.M. (1971) Pulmonary oil embolism following Pantopaque ventriculography in a patient with a ventriculovenous shunt. *J. Neurosurg.* 35, 623–627.
38 GONSETTE R.E., DEREMACKER A., HOU H. & CORNELIS G.L. (1958) L'Iodoventriculographie. I. Technique, indications, images normales. *Acta Neurol., Psychiatr. Belg.* 58, 797–809.
39 JEFFERSON A. & OCCLESHAW J. (1960) The identification of pathological processes in the posterior cranial fossa by Myodil ventriculography. *Acta Neurochirurgica* 8, 468–494.
40 HEIMBURGER R.F., KALSBECK J.E., CAMPBELL R.L. & MEALEY J. (1966) Positive contrast ventriculography using watersoluble media. *J. Neurol. Neurosurg. & Psychiat.* 29, 281–290.
41 GUTHKELCH A.N., ZIERSKI J., FERNANDEZ-SERRATS A.A. & CHATTERJEE S.P. (1973) Ventriculography with meglumine iothalamate. *Neuroradiology* 6, 32–38.
42 PANDA D.K., DAS B.S., RATH S. & MOHANTY G.B. (1974) Conray ventriculography in neurosurgical practice. *Clin. Radiol.* 25, 145–151.

43 GONSETTE R. (1971) An experimental and clinical assessment of watersoluble contrast media in neuroradiology: a new medium: Dimer-X. *Clin. Radiol.* **22**, 44–56.
44 VAN DELLEN J.R. & LIPSCHITZ R. (1973) Meglumine iocarmate (Dimer-X) ventriculography. *Clin. Radiol.* **24**, 449–452.
45 GONSETTE R.E. (1973) Metrizamide as contrast medium for myelography and ventriculography, preliminary clinical experiences. *Acta Radiol. Suppl.* **335**, 346–358.

CHAPTER 8: MYELOGRAPHY

J. V. OCCLESHAW

Myelographic examination of the spine is carried out by both general radiologists and the specialist neuroradiologist and has now become one of the more common special investigations in diagnostic radiology. Most patients experience little discomfort from the procedure and despite the large numbers of myelographic examinations carried out, complications are not frequently recorded in the radiological literature.

The original method of myelography devised by Dandy [1] employed air introduced by lumbar puncture as the radiographic contrast agent. The investigation was subsequently made safer and improved technically by Lindgren [2] who used warmed oxygen as the contrast agent. In most Scandinavian countries this technique is the usual method of myelography for the dorsal and cervical regions.

Sicard & Forestier [3] injected Lipidol into the subarachnoid space for positive contrast myelography and later Nosik [4] used thorium dioxide (Thorotrast) as a myelographic contrast agent. These materials have now been replaced by iodophenylundecylate (iophendylate, Myodil, Pantopaque) which has now become the most frequently used contrast agent for myelography [5, 6].

Arnell & Lidstrom [7] developed the use of water-soluble contrast agents such as sodium monoidomethonesulphonate (Abrodil) for the examination of the lumbar nerve roots, but this substance being a sodium salt was irritant to the nerve roots and spinal cord, making preliminary spinal anaesthesia essential. With the introduction of the less irritant meglumine salts, spinal anaesthesia is no longer required and there has therefore been a greater acceptance of this type of examination for patients with lumbo-sacral nerve root compression.

GENERAL COMPLICATIONS

All methods of myelography require either lumbar or cisternal puncture for the introduction of the contrast agent and either of these procedures may give rise to complications.

Intrathecal injection of the local anaesthetic agent can occur in a very thin patient or if an excessively long needle is used for this part of the procedure. The patient may then be given a spinal anaesthetic with loss of power in the legs and anaesthesia in the lower part of the body. This effect usually passes off after a few hours without any consequence. A similar occurrence is recorded [8] when a radiologist accidentally interchanged the syringes of contrast and local anaesthetic agents, with the result that the soft tissues were infiltrated with Myodil and the anaesthetic agent was injected into the subarachnoid space. I have personal knowledge of a patient receiving intrathecal injections of both the anaesthetic agent and Myodil. When the patient was placed in the head-down position to examine the cranio-cervical junction, respiratory difficulty occurred. The patient required mechanical ventilation for the next twelve hours, but complete recovery ensued.

Permanent damage as a result of degenerative changes in the dorsal and ventral nerve roots, degenerative myelitis and chronic arachnoiditis are well-recorded reactions to intrathecal anaesthetic agents [9], but they have never been specifically reported after accidental intrathecal anaesthesia at myelography.

Bleeding into the subarachnoid space due to the lumbar puncture needle perforating one of the intraspinal veins may occur even with the most experienced radiologist and in itself is not a serious complication, but as detailed later, it may predispose to Myodil embolism or the production of a chronic arachnoiditis. Haemorrhage occurring during cisternal puncture may be the result of the needle damaging the posterior inferior cerebellar artery. This is more likely to happen when this artery has a low origin or if the caudal loop descends into the upper cervical canal as in cerebellar ectopia. To avoid this hazard, some radiologists advise that the puncture be made between the first and second cervical vertebrae.

Poor aseptic technique may result in the introduction of bacteria into the subarachnoid space with the subsequent development of a bacterial meningitis. This is an infrequent complication of myelography but, when it does occur, the Myodil is more likely to be blamed for causing the meningitis than any bacteria.

Malposition of the needle in the subarachnoid space may result in the contrast agent being introduced into the subdural or extradural spaces.

With Myodil, little or no discomfort is experienced by the patient. With some water-soluble contrast agents, root pains may occur, but usually resolve in a short time. Recent lumbar puncture predisposes to this complication [10] and myelography is better postponed in such circumstances, unless clinically urgent. Lumbar puncture with the patient sitting erect, and contrast injection under screening control, both help to reduce the incidence of extra-arachnoid injections, but when this occurs it may be difficult to aspirate the contrast agent.

If the lumbar puncture needle perforates a spinal vein, contrast may be introduced into the venous system. Water-soluble agents are not likely to have any untoward effects on the patient, but Myodil will give rise to pulmonary oil embolism [11, 12]. As the Myodil is being introduced it may be seen on the screen escaping into the veins and within a few seconds the patient will start to cough and may produce blood-stained sputum. Chest radiographs will often show a fine mottled pattern. A pyrexial reaction usually follows and lasts for two to three days, but recovery with resolution of the changes on the subsequent chest film is the usual outcome. This complication has also been reported when attempts to aspirate the contrast are being made at the end of the examination [13–15] and also when the patient has a bout of violent coughing during the examination [16]. Only 2–4 ml of oily contrast needs to enter the venous system to produce radiologically demonstrable oil embolism, as shown by Sicard & Forestier in their classical work on Lipiodol [3]. A recent case report [17] cites two patients who were observed to have Myodil in the vertebral veins during fluoroscopy. The lumbar puncture needle was thought to have damaged the epidural veins with the development of a fistulous communication between the cerebrospinal fluid and the veins. The pressure in the subarachnoid space of a recumbent patient is in the range of 160–200 mmH$_2$O. This is greater than the pressure in the veins during certain phases of respiration so that contrast can be drawn into these veins and then reach the lungs.

SPECIFIC COMPLICATIONS OF INTRATHECAL MYODIL

It was only one year after the introduction of Myodil before irritant effects on the meninges and turbidity of the cerebrospinal fluid was noted at operation following previous myelography [18]. In the following decade many similar reactions were reported [19–22]. The two common types of reaction are:

(a) An acute meningeal reaction which comes on a few hours after the examination.
(b) A chronic reaction which may follow the acute reaction, or may arise independently, some weeks or months later.

The *acute reaction* resembles a meningitis with headache, neck stiffness, backache, nausea, vomiting and pyrexia in the 37.2–38.9°C range. In the majority of patients this reaction will have resolved by the third day. If symptoms persist or are very severe, lumbar puncture may be necessary to exclude bacterial meningitis. If doubt of infective meningitis still exists following lumbar puncture, 1 mega-unit of benzylpenicillin *intramuscularly* every six hours may be given until an organism is isolated and its antibiotic sensitivity determined. Even in the absence of bacterial meningitis the cerebrospinal fluid protein level is usually elevated and the cell count raised [23]. Cranial nerve palsies and nystagmus have been noted as part of the acute reaction following cisternal myelography [22].

Chronic symptoms of headache, neck and low back pain may persist for a year or more and films of the spine and skull often show fixed collections of Myodil in these areas [22]. Examination of the cerebrospinal fluid may show persistent elevation of the cell count and protein level. Many patients with severe backache have undergone exploratory laminectomy and the nerve roots of the cauda equina found to be matted together by adhesions [21]. Fig. 8.1. shows extensive adhesions producing a myelographic block in the mid-lumbar region as a result of previous Myodil myelography followed by disc surgery. Histological examination of the adhesions in these cases shows fibrous tissue infiltrated by mononuclear cells and lipoid containing giant cells whilst cystic collections of Myodil may also be seen.

Myodil may reach the basal cisterns during the examination of the cranio-cervical junction or, if the contrast is incompletely removed at the end of the examination, it may ascend to the cranial cavity when the patient is recumbent. An acute irritation of the meninges and ventricles may result [24] and on occasions progress to a chronic adhesive arachnoiditis with obstruction to the outflow of the fourth ventricle and obstructive hydrocephalus [25–27].

The mechanism of these inflammatory reactions has been the subject of many experimental studies [28–31] and it is now well recognized that in many animals, retained Myodil produces an inflammatory reaction in the meninges which may progress to a chronic arachnoiditis. Dispersion of the oily contrast into fine particles in the cerebrospinal fluid has been observed to produce severe meningeal reactions in animals [28] and in man [32]. In order to keep the oily contrast agents in a stable suspension in cerebrospinal fluid, a stabilizing substance such as agar was used. Blood was found to be an excellent stabilizing agent but this suspension was fatal when injected cisternally in the dog [28]. In the spine a severe chronic adhesive

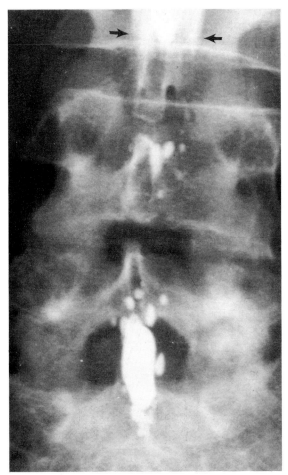

FIGURE 8.1. Previous Myodil myelography followed by surgical exploration. At subsequent water-soluble contrast myelography there were fixed pockets of Myodil in the lower lumbar canal and all water-soluble contrast was held up at the level of the L3–4 disc space due to adhesions.

arachnoiditis resulted. Other studies have confirmed the effect of a Myodil–blood combination in producing intrathecal reactions [29–31].

Hypersensitivity to Myodil has been claimed as a possible cause for the acute reactions [20], but sensitivity testing of all patients prior to myelography could well sensitize a previously nonsensitive patient.

Treatment of Myodil reactions

In the experimental animal, intrathecal hydrocortisone has been shown to reduce the effects of Myodil in producing both the acute and chronic types of reaction [29, 30, 33]. In clinical practice, intrathecal hydrocortisone produces severe pain, but methylprednisolone (Depo Medrone, Depo Medrol) is painless. Satisfactory treatment of the acute reaction has been recorded with either systemic steroids or intrathecal methylprednisolone [23, 24].

Chronic arachnoiditis involving the spinal nerve roots may be so severe that patients have been subjected to exploratory laminectomy with variable results. The application of deep X-ray therapy to the spine is alleged to produce some improvement [21].

Chronic arachnoiditis involving the basal cisterns of the brain may manifest initially by the development of cranial nerve palsies, but it may also proceed to obstruct the outflow paths of the fourth ventricle with the development of papilloedema and raised intracranial pressure. Surgical exploration of the posterior cranial fossa, after ventriculographic studies to confirm the diagnosis, have been reported. However, opening up the fourth ventricle will only have a short-term effect. The development of ventriculo-atrial and ventriculo-pleural shunts has offered a means of diverting the flow of cerebrospinal fluid and reducing the intracranial pressure, but despite these procedures death has been recorded [27].

Perhaps the best form of treatment is to avoid these complications by the complete removal of the contrast at the end of the examination and the syphonage technique is probably the most effective method available [34]. In view of a recent Appeal Court judgment in the U.S., in which a physician was held negligent due to non-removal of Pantopaque [35], complete removal of contrast is advised for the good of both patient and radiologist.

Miscellaneous complications of Myodil myelography

The formation of a *lipoid granuloma* which simulated a tumour has been recorded following Myodil myelography [36].

Serial studies of the levels of *serum protein-bound iodine* have demonstrated elevation even twelve years after Myodil myelography [37]. Careful biochemical investigations had excluded hyperthyroidism in these patients, but such elevation of the protein-bound iodine level is known to be capable of inducing hyperthyroidism in adults and causing pregnant patients to produce children with congenital goitre or hypothyroidism.

Cerebral vasospasm leading to cerebral infarction has been recorded in a patient following Myodil myelography [38]. Hemiparesis occurred three days later and carotid angiography showed minimal narrowing of the middle cerebral vessels. Repeat angiography one week later showed severe changes in the middle cerebral vessels and infarctive swelling of the cerebral hemisphere producing displacement of the mid-line vessels. Gradual improvement occurred with steroid therapy.

Visual loss 35 days after myelography has been recorded in a patient who had large amounts of Myodil retained in the region of the optic nerve [39]. Periorbital pain and neck stiffness were also present and the patient recovered after treatment with systemic corticosteroids. The mechanism in this case was considered to be an inflammatory reaction to the Myodil.

Some patients, usually those with intraspinal tumours or massive lumbar disc protrusions, may develop *dysuria or retention of urine* following myelography. This probably results from interference with the cerebrospinal fluid pressures above and below the lesion so that there is increased compression of the spinal cord or nerve roots. The development of retention of urine is usually an indication for urgent surgical decompression of the spinal cord and nerve roots. Some patients may complain of some degree of *impotence* after myelography which may be a manifestation of the primary pathology or a psychological reaction to the examination. In the latter case, this effect is usually short lived.

Headache, in the absence of any inflammatory reaction, may occur following the examination as a result of low cerebrospinal fluid pressure. Difficulty in removal of the contrast may result in the removal of a large volume of cerebrospinal fluid with the Myodil. Cerebrospinal fluid–venous fistulae as described earlier in this chapter have also been claimed as the cause of severe low-pressure headache [17]. Analgesic drugs will usually produce relief of this symptom.

The *late effects* of Myodil on the cerebrospinal fluid chemistry are well recognized. Elevation of the protein level and the cell count may persist for many weeks especially if the Myodil is retained. A recent study [40] has shown that the cell count may rise to 13 ± 4 cells per mm^3 in the first ten days, as a result of a lymphocytic response, even if the Myodil has been aspirated. Protein elevation may persist for up to 80 days and electrophoretic studies show that the pre-albumin, albumin, alpha and beta globulin fractions are all elevated with wide variations. There was also noted to be elevation of the gamma-globulin fractions and this type of response usually indicated a reaction associated with an increase of capillary permeability.

GAS MYELOGRAPHY

The general complications of lumbar puncture and cisternal puncture, as already cited, can occur in gas myelography. Too rapid removal of the cerebrospinal fluid or too rapid introduction of the gas can cause collapse with bradycardia and hypotension. The use of oxygen warmed to body temperature [41] and the slow introduction of the gas are claimed to produce only a dull sensation of pressure in the head during the examination [42]. Cardiovascular collapse can often be prevented by the intramuscular injection of 30 mg of ephedrine at the start of the examination.

During the procedure, gas has been noted to escape into the soft tissues at the site of the lumbar puncture, producing surgical emphysema locally. This may, however, extend and on occasion produce mediastinal emphysema [42]. Air embolism during cisternal gas myelography in a child has recently been recorded [43]. The initial puncture produced blood-stained cerebrospinal fluid and, when the patient collapsed, air was seen within the heart shadow. As the child was in the head-down position no air entered the cerebral circulation and this was claimed to be the reason that the child survived the examination. The use of nitrous oxide as the intraspinal gas was suggested as a means of increasing the safety of gas myelography.

As in pneumoencephalography, the presence of gas in the subarachnoid space produces an increase in the cerebrospinal fluid protein level and the cell count. At the end of the examination, when the patient is placed in the horizontal position, the gas will enter the cranial cavity and fill the basal cisterns and cortical subarachnoid spaces. This causes severe headache which is frequently accompanied by vomiting. Dihydro-ergotamine mesylate is known to reduce the severity of headache and vomiting during and after pneumoencephalography and may be of value in gas myelography [44]. It should be given intramuscularly or subcutaneously in a dosage of 1–2 mg prior to the introduction of the gas, and, if necessary, repeated twelve hours later.

Following the examination the patient should be nursed lying head down in bed without pillows and given adequate analgesics. Pyrexia as in pneumoencephalography is a frequent reaction. Over the next few days, as the headache diminishes, the patient can be given pillows and returned to a general nursing regime.

WATER-SOLUBLE CONTRAST MYELOGRAPHY

The contrast agents initially used in this type of examination were sodium salts and were irritant to the spinal cord and nerve roots, so that spinal anaesthesia was necessary before introduction of the contrast medium. Spinal anaesthesia is not devoid of complications and a large number of the reactions were the direct result of the anaesthesia. In a comprehensive review of 721 cases of Sodium Methiodol (Kontrastum U) myelography, reactions were divided into three groups [45]:
(a) *Lumbar puncture complications*—headache and neck stiffness (6 cases).
(b) *Spinal anaesthetic reactions*—hypotension leading to shock (32 cases).
(c) *Contrast media reactions*—Lumbar pain, hyperparesthesia, leg spasms and sphincter paralysis (12 cases).

In a more recent review of complications following Kontrastum U myelography [46], hypotension accounted for 20 per cent of reactions which proceeded to shock in 4 per cent, and usually occurred within the first hour. Nerve root irritation was found in 28 per cent of cases and manifest two to three hours after the examination as the effects of the spinal anaesthetic were subsiding. Other complications included paralysis of the legs and sphincters which could persist for up to three weeks [45].

The accidental use of sodium diatrizoate (Hypaque) for myelography has given rise to myoclonic spasms [47]. The spasms were so severe that muscle damage occurred with resultant myoglobinaemia producing anuria and azotaemia. Renal dialysis was necessary on four occasions, but the patient made a full recovery in a few weeks.

The successful treatment of severe spasms following accidental subarachnoid injection of diatrizoate (Urografin) at vertebral angiography has been reported [48]. When the spasms started, lumbar and cisternal punctures were performed and the contrast was washed out with normal saline. The patient was curarized and respiration was maintained by a respirator. Complete recovery occurred by the fourth day.

Meglumine contrast (Conray and Dimer X) Myelography

In an early series of twelve Conray myelograms, complications were seen in four cases [49]. One patient developed fever, tachycardia and hypotension 24 hours after the examination and three other patients had severe headache and vomiting after being placed in the head-down position during the examination. One of these three patients showed weakness of the thigh muscles and another patient had increased tonicity of the lower limbs after the examination.

In another series of 50 cases [50], 40–80 mg of methylprednisolone was added to the 5 ml of Conray, but side effects still occurred. Paraesthesia in the legs was experienced by 16 per cent of the patients, 26 per cent had leg spasms and rigidity four to five hours after the examination which persisted for two to three hours.

Many other papers have been published on the side effects of Conray and Dimer X myelography in recent years [51–56]. The most common discomfort is *headache* which occurs in 25–35 per cent of cases irrespective of the choice of contrast medium and is often accompanied by nausea and vomiting. Most radiologists have their patients nursed in the sitting position for up to six hours after the examination, so that the contrast is confined below the level of the spinal cord until absorption has occurred. It is possible, however, that this erect position may well account for the high incidence of headache. Removal of the contrast at the end of the examination followed by nursing the patients in a more horizontal position is said to reduce the incidence of severe headache.

Back pain resulting from nerve root irritation is quoted as occurring in from 2 to 21.5 per cent of patients, but this rarely requires more than simple analgesics for adequate relief. *Retention of urine* may occur in 1 per cent of patients after Conray myelography and in 2 per cent of patients after Dimer X examinations [46].

Some degree of *hypotension* either during the examination or within the next four hours has been observed by most writers, but this rarely gives rise to concern. If necessary, hypotension may be treated by the intravenous injection of 20–40 mg of methoxamine hydrochloride (Vasoxine).

The most serious effects of Conray and Dimer X in the subarachnoid space are their irritant action on the spinal cord and nerve roots with the production of *muscle spasms or myoclonic contractions*. These effects have been observed in 5 per cent of Conray examinations and in 2 per cent of Dimer X studies. The spasms or contractions do not occur for about two to four hours after the examination and are more prone to occur when contrast remains in contact with the spinal cord after the examination. If it is not possible to keep the contrast below the level of the spinal cord with the patient in the sitting position, then lumbar puncture with withdrawal of 10–15 ml of cerebrospinal fluid and contrast is advised. Spasms may also result if the patient does not follow the instruction to sit up after the examination, but lies flat in bed, thus allowing contrast to affect the spinal cord. Once spasms occur it is usually necessary to give the patient large intravenous doses of diazepam (Valium) up to 40 mg in the first fifteen minutes. This may have to be repeated over the next twelve hours to keep the patient spasm free. At this dosage the patient should be carefully observed for any respiratory depression. Many Scandinavian radiologists premedicate their patients with oral Valium for twelve hours and then give some intravenous Valium before the start of the examination. In practice I have not found such measures necessary as only one patient out of 500 Dimer X myelograms performed in my department developed myoclonic spasms.

The effects of Conray myelography on the bladder muscle and the electrical activity in the leg muscles has been studied [52]. Immediately following the injection, increased electrical activity in the leg muscles is seen on electromyography and the intravesical pressure rises indicating increased activity in the bladder musculature.

Epilepsy has been regarded as a contraindication to water-soluble contrast myelography by many radiologists because of the possibility of the contrast reaching the cerebral cortex and provoking further seizures [53]. Such complications have been studied with electroencephalographic recordings in a child of fourteen years, and convulsions occurred with both Conray and Dimer X myelography [57].

Violent myoclonic spasms may result in self-injury to the patient: fractures of the femoral necks [52, 58], and compression fractures of vertebral bodies [59] have been recorded following Conray and Dimer X myelography. Myoglobinuria as a result of severe and prolonged leg spasms following Conray myelography has also been recorded [60].

Aseptic meningitis following Dimer X myelography with pyrexia, neck stiffness, and a cellular reaction in the cerebrospinal fluid has been noted and recovery without any specific treatment occurred by the third day [46].

Chronic adhesive arachnoiditis following lumbar myelography with water-soluble contrast agents has been studied by many investigators in the last three years [61–64]. Subsequent examinations may demonstrate alterations in the size and shape of the terminal segments of the subarachnoid space, non-filling of a previously well-filled nerve root sheath, or cystic dilatations of the nerve root sheaths, all of which indicate the presence of an arachnoid reaction. This complication is more likely to occur if the patient has undergone surgical exploration of the spine between the two examinations (Fig. 8.2.), but may also occur in the absence of any surgery. In some cases the reaction produces severe symptoms of lumbar nerve root compression and may even lead to further surgical exploration.

Metrizamide myelography

The perfect contrast for myelography has yet to be developed, but the introduction of metrizamide, a water-soluble agent that does not ionize in solution, is another step towards this goal. It is well tolerated in clinical studies, but over 50 per cent of patients had headache after the examinations [65–67]. Many patients had an increase in the severity of their back pain following the

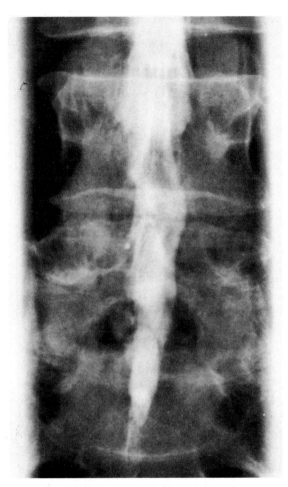

FIGURE 8.2. Arachnoiditis producing irregular narrowing of the terminal segment of the subarachnoid space. Patient has previously had a Conray myelogram followed by surgical exploration of the spine.

examination, but both types of discomfort usually resolved in the first 48 hours. Studies of the cerebrospinal fluid after the introduction of metrizamide into the subarachnoid space showed an increase in the cell count which reached a maximum at 24 hours averaging 115 WBC and similarly occurring peak for protein levels averaging 73 mg per 100 ml [66]. Electroencephalographic studies [65, 67, 68] revealed some abnormal electrical activity of the cerebral cortex, but this did not give rise to any epileptic seizures in the patients. Very extensive biochemical studies on the blood after metrizamide myelography did not reveal any disturbances in the patients' metabolic functions [67].

A recent paper on the use of Metrizamide for the examination of the cervical region after suboccipital injection has shown that reactions are less likely to occur than in lumbar or dorsal myelography [69].

REFERENCES

1 DANDY W.E. (1925) The diagnosis and localization of spinal cord tumours. *Ann. Surg.* 81, 223–254.
2 LINDGREN E. (1939) On the diagnosis of tumours of the spinal cord by the aid of gas myelography. *Acta Chir. Scandinav.* 82, 303–318.
3 SICARD J.A. & FORESTIER J. (1922/23) Méthode general d'exploration radiologique par l'huile iodée (Lipiodol). *Bull. Soc. Med. Hôp. Paris* 46, 463–468.
4 NOISK W.A. (1943) Intraspinal Thorotrast. *Am. J. Roentg.* 49, 214–218.
5 RAMSEY G.H., FRENCH J.D. & STRAIN W.H. (1944) Iodinated organic compounds as contrast media for radiographic diagnoses. IV. Pantopaque myelography. *Radiology* 43, 236–240.
6 STEINHAUSEN T.B., DUNGAN C.E., FURST J.B. et al. (1944) Iodinated organic compounds as contrast media for radiographic diagnoses. III. Experimental and clinical myelography with ethyl iodophenylundecylate (Pantopaque). *Radiology* 43, 230–234.
7 ARNELL S. & LIDSTROM F. (1931) Myelography with Skiodan (abrodil). *Acta Radiol.* 12, 287–288.
8 ANSELL G. (1968) A national survey of radiological complications: interim report. *Clin. Radiol.* 19, 175–191.
9 WOLMAN L. (1966/67) The neuropathological effects resulting from the intrathecal injection of chemical substances. *Paraplegia* 4, 97–115.
10 JONES M.D. & NEWTON T.H. (1963) Inadvertent extra-arachnoid injection in myelography. *Radiology* 80, 818–822.
11 GINSBURG L.B. & SKORNECK A.B. (1955) Pantopaque pulmonary embolism—a complication of myelography. *Am. J. Roentg.* 73, 27–31.
12 TODD E.M. & GARDENER W.J. (1957) Pantopaque intravasation (embolisation) during myelography. *J. Neurosurg.* 14, 230–234.
13 FULLENLOVE T.M. (1949) Venous intravasation during myelography. *Radiology* 53, 410–412.
14 STEINBACH H.L. & HILL W.B. (1951) Pantopaque pulmonary embolism during myelography. *Radiology* 56, 735–738.
15 KEATS T. (1956) Pantopaque pulmonary embolism. *Radiology* 67, 748–750.
16 HINKEL C.L. (1945) The entrance of pantopaque into the venous system during myelography. *Am. J. Roentg.* 54, 230–233.
17 LIN P.M. & CLARKE J. (1974) Spinal fluid–venous fistula: a mechanism for intravascular pantopaque infusion during myelography. *J. Neurosurg.* 41, 773–776.

18 Tarlov I.M. (1945) Pantopaque meningitis disclosed at operation. *J. Am. med. Ass.* **129**, 1014–1016.
19 Bering E.A. (1950) Notes on the retention of Pantopaque in the subarachnoid space. *Am. J. Surg.* N.S. **80**, 455–458.
20 Luce J.O., Leith W. & Burrage W.S. (1951) Pantopaque meningitis due to hypersensitivity. *Radiology* **57**, 878–881.
21 Hurteau E.F., Baird W.C. & Sinclair E. (1945) Arachnoiditis following the use of iodized oil. *J. Bone Joint Surg.* **36A**, 393–400.
22 Davies F.L. (1956) Effects of unabsorbed radiographic contrast media on the central nervous system. *Lancet* **2**, 747–748.
23 Mayher W.E., Daniel E.F. & Allen M.B. (1971) Acute meningeal reaction following pantopaque myelography. *J. Neurosurg.* **34**, 396–404.
24 Taren J.A. (1960) Unusual complication following pantopaque myelography. *J. Neurosurg.* **17**, 323–326.
25 Erickson T.C. & Van Baaren H.J. (1953) Late meningeal reaction to ethyliodophenylundecylate used in myelography. *J. Am. med. Ass.* **153**, 636–639.
26 Sedzimir C.B. & Iwan S.R. (1962) Simplified contrast ventriculography. *J. Neurosurg.* **19**, 657–660.
27 Mason M.S. & Raff J. (1962) Complications of Pantopaque myelography. *J. Neurosurg.* **19**, 302–311.
28 Jaeger R. (1950) Irritant effects of iodated vegetable oils on the brain and spinal cord when divided into small particles. *Arch. Neurol. & Psychiat.* **64**, 715–719.
29 Fisher R.L. (1965) An experimental evaluation of Pantopaque and other recently developed contrast media. *Radiology* **85**, 537–545.
30 Howland W.J. & Curry J.L. (1966) Experimental studies of Pantopaque arachnoiditis. *Radiology* **87**, 253–261.
31 Bergeron R.T., Rumbaugh C.L., Fang H. & Cravioto H. (1971) Experimental Pantopaque arachnoiditis in the monkey. *Radiology* **99**, 95–101.
32 Bleasel K. (1961) Nerve root radiculography. *Br. J. Radiol.* **34**, 396–401.
33 Smith J.K. & Ross L. (1957) Steroid suppression of meningeal inflammation caused by Pantopaque. *Neurology* **9**, 48–52.
34 Epstein B.S. & Epstein J.A. (1972) The syphonage technique for removal of Pantopaque following myelography. *Radiology* **103**, 353–358.
35 U.S. Court of Appeals for the Second Circuit: James F. Toal: Decided February 10, 1971. *N.Y. Law J.* March 4, 1971.
36 Sarkisian S.S. (1956) Spinal cord pseudotumour. A complication of myelography. *U.S. Armed Forces Med. J.* **7**, 1683–1686.
37 White A.G. (1972) Prolonged elevation of serum protein-bound iodine following myelography with Myodil. *Br. J. Radiol.* **45**, 21–23.
38 Smith R.A., Collier H.F. & Underwood F.C. (1973) Cerebral vasospasm following myelography. *Surg. Neurol.* **1**, 87–90.
39 Tabaddor K. (1973) Unusual complications of iodophendylate injection myelography. *Arch. Neurol.* **29**, 435–436.
40 Ferry D.J., Gooding R., Standefer J.C. & Wiese G.M. (1973) Effects of Pantopaque myelography on the cerebrospinal fluid fractions. *J. Neurosurg.* **38**, 167–171.
41 Oden S. (1953) Diagnosis of spinal tumours by means of gas myelography. *Acta Radiol.* **40**, 301–313.

42 ROTH M. (1963) Gas myelography by the lumbar route. *Acta Radiol. Diag.* 1, 53–65.
43 GARDNER L.G. (1971) Air embolism during cisternal air myelography. Case report. *Br. J. Anaesthes.* 43, 807–810.
44 BORIES J., MERLAND J.J., FREDY D. & BERNARD S. (1971) Dihydro-ergotamine and gas encephalography. *Neuroradiology* 2, 35–36.
45 LINDBLOM K. (1947) Complications of myelography by Abrodil. *Acta Radiol.* 28, 70–73.
46 IRSTAM L. (1973) Side effects of watersoluble contrast media in lumbar myelography. *Acta Radiol. Diag.* 14, 647–656.
47 MARTIN C.M. (1971) Myelography with sodium diatrizoate (Hypaque). *Calif. Med.* 115, 57–59.
48 McCLEERY W.N.C. & LEWTAS N.A. (1966) Subarachnoid injection of contrast medium. *Br. J. Radiol.* 39, 112–114.
49 CAMPBELL R.L., CAMPBELL J.A., HEIMBURGER R.F., KALSBECK J.E. & MEALEY J. (1964) Ventriculography and myelography with absorbable radiopaque medium. *Radiology* 82, 286–289.
50 DAVIES F.M., LLEWELLYN R.C. & KIRGIS H.D. (1968) Watersoluble contrast myelography using meglumine iothalamate (Conray) with methylprednisolone acetate (Depro-Medrol). *Radiology* 90, 705–709.
51 PRAESTHOLM J. & LESTER J. (1970) Watersoluble contrast myelography with meglumine iothalamate (Conray). *Br. J. Radiol.* 43, 303–308.
52 SKALPE I.O. (1971) Lumbar myelography with Conray Meglumine 282. *Acta Neurol. Scandinav.* 47, 569–578.
53 GONSETTE R. (1971) Watersoluble contrast media in neuroradiology. *Clin. Radiol.* 22, 44–56.
54 GRAINGER R.G., GUMPERT J., SHARPE D.M. & CARSON J. (1971) Watersoluble lumbar radiculography. A clinical trial of Dimer X—a new contrast medium. *Clin. Radiol.* 22, 57–62.
55 OCCLESHAW J.V. & HOLYLAND J.N. (1971) Comparative study of the effects of Conray 280 and Dimer X in lumbar radiculography. *Br. J. Radiol.* 44, 946–948.
56 LEHTINEN E. & SEPÄNEN S. (1972) Side effects of Conray Meglumine 282 and Dimer X in lumbar myelography. *Acta Radiol. Diag.* 12, 12–16.
57 DE GRAAF A.S. & KAYED K.S. (1973) Epileptic seizures and EEG changes after radiculography with meglumine iothalamate (Conray) and meglumine iocarmate (Dimer X)—a case report. *Psychiatria, Neurologia, & Neurochirugia* 76, 77–82.
58 SOMMELET J., LAMY G., VAILIANT G., SCHMITT D. & FROMENT J. (1972) Fracture spontanée et simultanée des 2 têtes fémorales—complication exceptionnelle de la radiculographie au Contrix. *Ann. Med. Nancy* 11, 459–465.
59 HASSE J., LEPSEN B.V., BECH H. & LANGEBACK E. (1973) Spinal fracture following radiculography using meglumine iothalamate (Conray). *Neuroradiology* 6, 65–70.
60 NIELSEN J.L. & NEILSEN H.O. (1974) Myoglobinuria after Conray myelography. *Ugesk. Laeg.* 136, 1581–1582.
61 AUTIO E., SUOLANEN J., NORRBACK S. & SLATIS P. (1972) Adhesive arachnoiditis after myelography with meglumine iothalamate (Conray). (1972) *Acta Radiol. Diag.* 12, 17–24.
62 AHLGREN P. (1973) Long term effects after myelography with watersoluble contrast media: Conturex, Conray Meglumine 282 and Dimer X. *Neuroradiology* 6, 206–211.

63 IRSTAM L. & ROSENCRANTZ M. (1974) Watersoluble contrast media and adhesive arachnoiditis. II. Reinvestigation of operated cases. *Acta Radiol. Diag.* **15**, 1-14.
64 LILIEQUIST B. & LUNDSTROM B. (1974) Lumbar myelography and arachnoiditis. *Neuroradiology* **7**, 91-94.
65 GONSETTE R.E. (1973) Metrizamide as contrast medium for myelography and ventriculography—preliminary clinical experiences. *Acta Radiol.* Supp. **335**, 346-358.
66 HINDMARSH T. (1973) Methiodal sodium and metrizamide in lumbar myelography. *Acta Radiol.* Supp. **335**, 359-366.
67 SKALPE I.O., TORBERGSEN T., AMUNDSEN P. & PRESTHUS J. (1973) Lumbar myelography with Metrizamide. *Acta Radiol.* Supp. **335**, 367-379.
68 KAADA B. (1973) Transient EEG abnormalities following lumbar myelography with Metrizamide. *Acta Radiol.* Supp. **335**, 380-386.
69 AMUNDSEN P. & SKALPE I.O. (1975) Cervical myelography with a watersoluble contrast medium (Metrizamide). *Neuroradiology* **8**, 209-212.

CHAPTER 9: PHLEBOGRAPHY

M. LEA THOMAS

INTRODUCTION

Demonstration of the veins after injection of radio-opaque substances into them was first described by Berberish & Hirsch in 1923 [1], using strontium bromide, and by McPheeters & Rice in 1929 [2], using iophendylate (Myodil), but the method was little used until the safer iodinated solutions became available for intravenous pyelography. In 1934 the use of diodone was reported [3] and this remained the most commonly employed medium until the advent of diatrizoates and iothalamates in the last twenty years. These more recent contrast media represent an advance not only because of their increased radiodensity but also because of their lower toxicity (see Chapter 1).

Largely as a result of these improved contrast media, phlebography is now recognized as a generally safe technique. The author favours sodium iothalamate 70 per cent (Conray 420) for the thicker parts of the body, such as the pelvis and abdomen, and where pressure injections can be safely used. The less viscous meglumine iothalamate 60 per cent (Conray 280) is preferred where hand injections are employed, and for selective studies of the smaller veins. It should be emphasized that this is a personal choice based on the author's experience not only of phlebography but other contrast examinations. Media of similar composition would produce equally good results, but experience suggests that these two media give rise to less local discomfort and to fewer minor constitutional disturbances.

This chapter is not concerned with the complications common to many contrast examinations which are fully discussed elsewhere, but solely with those associated with phlebography. The techniques used for phlebography are not described except where a particular technique in itself may give rise to a complication.

Chapter 9

LOCAL COMPLICATIONS

(A) THROMBOSIS

May [4] has demonstrated by silver staining that every contrast medium damages the intima, and that contrast medium at any concentration, if applied to the vein wall for long enough, will cause thrombosis. In one series [5] the incidence of phlebitis and thrombosis was 4.2 per cent.

It is therefore important to minimize the time with which the veins are in contact with the contrast by carrying out examinations as rapidly as possible. At the end of the examination the contrast medium should be flushed out of the veins with physiological saline. At appropriate sites this process can be assisted by gentle massage, elevation of the limb and passive and active exercise. Since introducing these measures the incidence of thrombosis attributable to phlebography has been reduced to about 0.5 per cent in the author's series of over 3600 examinations.

The routine use of heparin, before or after the examination, in doses of 2500–5000 units, to minimize the risk of phlebothrombosis has been suggested [4, 6, 7]. The writer, however, reserves this only for patients with a high risk of venous thrombosis such as those with known malignant disease. As the individual response to heparin is so variable, any dose could either be insufficient or too great. In the latter case the added complication of haemorrhage at the puncture site is possible.

Generally, it is wiser not to carry out phlebography on patients who are anticoagulated except in emergency situations such as suspected or known pulmonary embolism when anticoagulation has already been started. Patients on anticoagulants tend to bleed more freely from the puncture site whether intravenous or intraosseous. Greater care should be taken to minimize haematoma formation; thus longer pressure at the puncture site after the examination may be necessary.

The author has had no significant complications in anticoagulated patients from either intravenous or intraosseous phlebography and these examinations are not contraindicated in the presence of a clear clinical indication.

(B) EXTRAVASATION OF CONTRAST INTO THE TISSUES

It is well known that hypertonic fluids may cause necrosis of soft tissue, for instance 30 per cent glucose solution. Winzer, Langecker & Junkmann [8] tested the cellular respiration of slices of rat liver influenced by contrast media, and demonstrated a decrease in respiration after all the media tested.

If a significant amount of contrast does extravasate into the tissues considerable damage can result, as described later.

CARDIAC COMPLICATIONS

(A) PATIENTS WITH PULMONARY HYPERTENSION

Phlebography is often requested in patients with pulmonary hypertension because this may be caused by massive or repeated pulmonary emboli. Also, since patients with cardiac lesions are more prone to venous thrombosis, patients with unrelated pulmonary hypertension may require investigation.

Such patients are particularly liable to develop cardiac arrhythmias and arrest which are often impossible to reverse. These result from reduction in cardiac output due to the vasodilatation caused by the contrast medium. In such patients the pulmonary artery pressure should be continuously monitored throughout the contrast injection. The complication is more likely to occur when the contrast is delivered as a bolus as with intra-osseous phlebography with a pump injection, or when large volumes of contrast are injected rapidly through large cannulae or catheters, but the risk is present whenever contrast is injected. The vasodilatation can be counteracted by producing vasoconstriction before injection of contrast medium. Phenylephrine 0.3 mg is given intravenously and repeated if necessary, until the systolic pressure is about 20 mmHg above its original level. Volume overload by the injection of 200 ml of plasma also helps to maintain the cardiac output.

Even taking these precautions, phlebography carries a significant morbidity in patients with pulmonary hypertension.

(B) AIR EMBOLISM

This is a theoretical complication of phlebography which has not been encountered by the author. To avoid it, syringes and connecting tubing should be free of air, filled either with physiological saline or contrast medium throughout the examinations. This applies whether hand injections are being used, or injections using a mechanical pump. Should air embolism occur, the injection should be immediately stopped and the patient turned on to his left side in the head-down position, so that air in the right ventricle floats away from the pulmonary outflow tract.

(C) PULMONARY EMBOLISM

The author has not encountered a single case in which pulmonary embolism could be attributed directly to a phlebogram, even when recent thrombosis has been demonstrated. This is also the experience of others [9, 10]. On the other hand there have been a few reports attributing pulmonary embolism to the examination [5].

The author's technique [11] of calf compression to produce a bolus of contrast medium to show the iliac veins and inferior vena cava may be criticized as likely to dislodge nonadherent thrombus. Calf compression is, however, a standard part of the Doppler ultrasonic technique [12] and the compression employed is probably no more than that produced by the patient's own muscular contractions or that used clinically to assess calf tenderness. The benefit of the bolus technique, which often makes a separate pelvic phlebogram unnecessary, outweighs this theoretical complication.

The routine use of the Valsalva manoeuvre to demonstrate the internal iliac and profunda veins from foot injections, recommended by the author [13], can also be criticized on the theoretical grounds that dilatation of the pelvic veins could loosen thrombus, and the subsequent deep inspiration, by increasing venous return, suck thrombus into the lungs [14]. Again, this theoretical risk has to be weighed against the advantage of avoiding separate pelvic phlebography.

COMPLICATIONS RELATED TO PARTICULAR SITES AND TECHNIQUES OF PHLEBOGRAPHY

(A) THE LOWER LIMB

Ascending phlebography is usually performed by injection of contrast medium into a foot vein. Other sites for phlebography, both ascending and descending, are the popliteal and femoral veins. Intraosseous injections may also be employed as described later.

Minor complications include bleeding from the puncture site which is readily controlled by firm pressure with the limb elevated, and infection if a venous cut down has been used. A cut down is very rarely required and should be carried out with full sterile precautions.

Following percutaneous puncture, either by needle or cannula, a little contrast frequently leaks from the puncture site. This is probably the cause of the minimal pain, tenderness and local swelling which may be observed the day following the phlebogram in a few patients. To minimize this, the needle or cannula should be carefully placed within the lumen of the vein and its position checked by a test injection before the main injection.

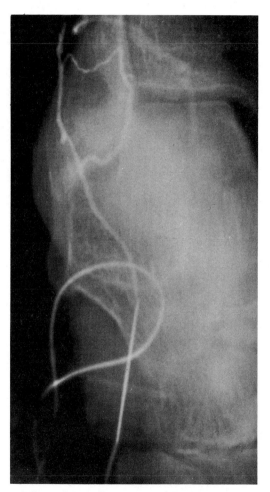

FIGURE 9.1. 21G 'butterfly' needle correctly positioned in a foot vein, i.e. there is no extravasation.

Fluoroscopy with image intensification and television monitoring is usually sufficient, but if doubt exists a film should be taken (Fig. 9.1). It has been shown (15) that extravasation occurs less frequently if a flexible catheter with a central trocar is used instead of a needle. Such a combination is, however, not as sharp as a needle and makes venepuncture less easy, especially with small veins in inexperienced hands. Multiple venepunctures in communication may allow contrast to track out of earlier venepunctures into the tissues. Thus a single venepuncture should be the aim.

Extravasation of contrast medium results in pain at the puncture site. If the patient complains of this, the injection should be stopped immediately

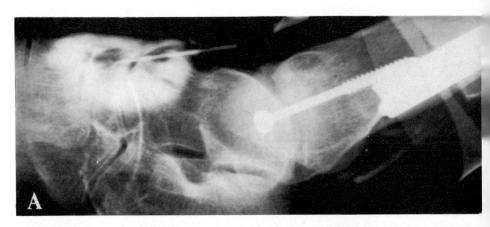

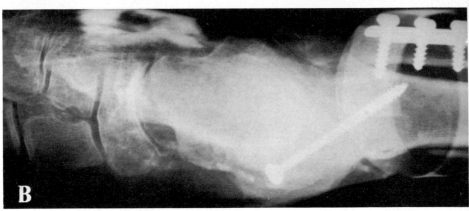

FIGURE 9.2. Extravasation due to dislodgement of needle. a Lateral projection. b P.A. projection. The contrast medium lies in the soft tissues immediately beneath the skin.

and the site of puncture examined for swelling and for extravasation, by fluoroscopy, before proceeding with the examination. The ability to screen the puncture site is one of the reasons why the author and others recommend a fluoroscopic phlebographic technique [11, 18–20] rather than an overcouch tube method [16, 17, 19, 21].

Phlebography

In the author's method, a 21G thin-walled scalp vein needle is used routinely. A smaller one may be necessary in children and adults with small veins but these have the disadvantage that injection is difficult, although venepuncture is easier. The 'butterfly' flaps enable the needle to be strapped in position to avoid dislodgement by patient movement during the examination.

Should extravasation of contrast medium occur (Fig. 9.2) it is best dispersed by gentle local massage. However injection of hyaluronidase has been advocated [22]. On the other hand, it has been suggested that hyaluronidase

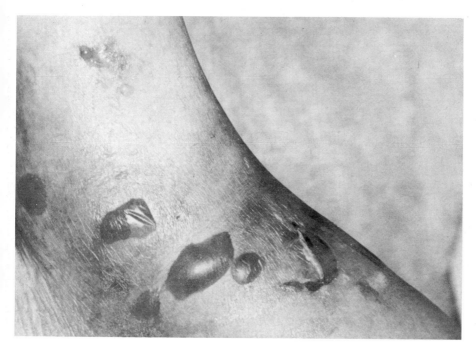

FIGURE 9.3. Clinical photograph showing cutaneous blisters twelve hours after extravasation of contrast medium into the soft tissues of the foot. One of the blisters has been punctured. The skin necrosis was treated by leaving the foot uncovered and elevating the leg. The patient had severe arterial insufficiency.

may actually increase tissue damage due to contrast media [23]. There may be a case for systemic steroid therapy (see Chapter 1, page 11).

Significant extravasation of contrast medium beneath the skin or soft tissues may lead to necrosis with sloughing of skin and even gangrene [24], sometimes sufficiently severe to warrant amputation. Five patients with these more serious complications have occurred in the author's experience. One patient had sloughing of the dorsum of the foot which required skin grafting. Two others developed superficial necrosis of the skin of the dorsum of the foot requiring hospitalization (Fig. 9.3). One of these had severe arterial insufficiency. Another patient, with congenital deficiency of the deep venous

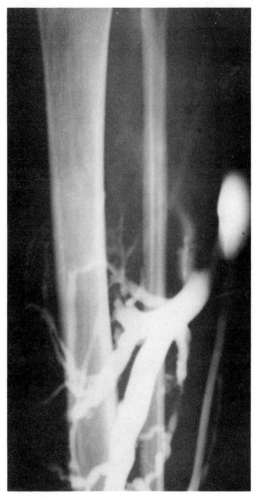

FIGURE 9.4. Phlebogram of a patient with Klippel–Trenaunay syndrome. The deep venous system is absent. The contrast medium passes into a superficial venous malformation.

system (Fig. 9.4) developed gangrene of the toes (Fig. 9.5). The fifth patient developed gangrene of the whole lower limb but died from an unrelated cerebral catastrophe. The phlebogram showed almost total occlusion of the deep venous system (Fig. 9.6). As the patient was in coma with a very low blood pressure, the circulation of the limb was greatly impaired [24].

The milder form of skin necrosis is best treated by pricking any blisters, leaving the skin uncovered, and elevating the leg.

Extravasation of this extent is often due to failure to observe that the needle has become dislodged. Other causes, as in the cases mentioned above, are congenital absence of the deep veins (Fig. 9.4) or complete blockage of the venous system due to thrombosis (Fig. 9.6). Under these circumstances

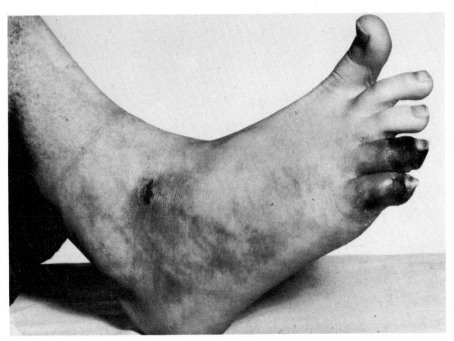

FIGURE 9.5. Same patient as Fig. 9.4, clinical photograph. Note gangrene of fourth and fifth toes two days after phlebogram.

the pressure in the vein increases so much during injection that the vein bursts. If the deep veins do not fill with a standard technique, such as the one recommended by the author [11], the examination should be abandoned. Intraosseous phlebography or femoral vein injections will often provide sufficient information for clinical management. Even if extravasation occurs from femoral puncture (Fig. 9.7), although local pain results, there is little danger of necrosis (22).

Another cause of bursting of a foot vein during injection is the use of too tight a tourniquet in an effort to obstruct the superficial veins in the search for incompetent perforating veins (Fig. 9.8). Complete occlusion of the superficial system to identify incompetent perforating veins is not essential using a fluoroscopic method [25] but if excessive superficial filling makes interpretation impossible an intraosseous malleolar approach can be used.

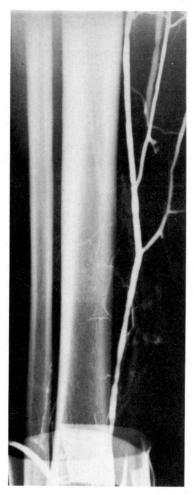

FIGURE 9.6. Phlebogram of patient with greatly swollen leg. She was in coma with a low cardiac output. The deep venous system is totally occluded. The contrast remained trapped in the foot and led to extensive gangrene of foot and leg.

Four patients with skin necrosis following extravasal injection of contrast medium from ascending phlebography have been described by Göthlin & Hallböök [22]. In one of their patients, contrast medium was demonstrated radiographically in the soft tissues a day after the examination. All their patients had diminished arterial or venous circulation to the feet, together with swelling. They suggest that in high-risk patients of this kind, descending phlebography should be carried out by femoral vein puncture. In addition to these cases the writer is aware of four patients who developed severe gangrene of the feet and legs requiring amputation. These patients also had severe circulatory deficiency in the affected limbs.

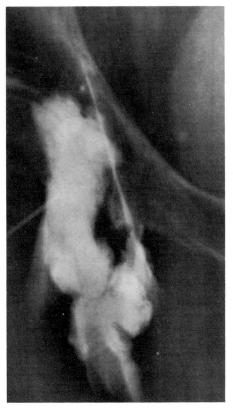

FIGURE 9.7. Femoral puncture with extravenous extravasation. The only complication was local pain.

It is clear that although tissue necrosis from extravasation of contrast medium is rare, its consequences in some patients are so severe that the greatest care must be taken to avoid extravasation. It should be emphasized, however, that in the author's experience of over 3000 examinations of the lower limbs, and the experience of many others, the complication rate relating to phlebography is very small [17, 26–29].

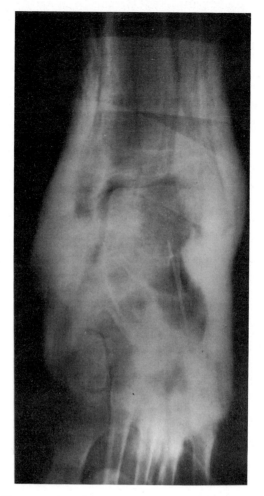

FIGURE 9.8. Very extensive subcutaneous extravasation of contrast medium due to too-tight ankle tourniquet. The needle was correctly positioned but the veins burst under pressure.

(B) INTRAOSSEOUS PHLEBOGRAPHY OF THE LOWER LIMBS AND OTHER SITES

The method requires an injection of contrast medium from a bone marrow or similar needle [30] into the bone cavity. It is a widely used alternative method for the radiographic demonstration of venous circulation when the conventional venous methods are impossible. It has also been used at sites where suitable veins for puncture or cannulation are not readily available.

Such sites include the vertebral plexuses, the azygous veins, the internal mammary veins and the visceral pelvic veins, to name just a few.

It is most commonly used nowadays to demonstrate the veins of the lower limb and pelvis, when ascending phlebography by direct venous injection is impossible due to occlusions or if it is considered undesirable, for instance in the presence of ankle ulceration. As the examination is painful general anaesthesia is usually required.

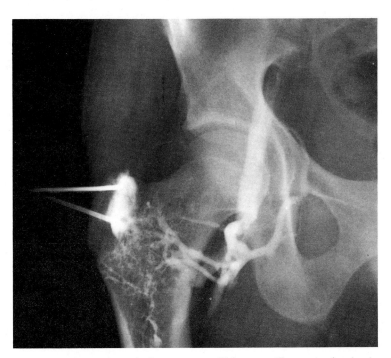

FIGURE 9.9. Pertrochanteric intraosseous phlebogram. Two cannulae *in situ*. The first cannula was placed subperiosteally. The second needle has been inserted leaving the first one in place to avoid leakage of contrast from the initial puncture site at the time of the main injection.

At whatever injection site, care must be taken to ensure that the point of the needle lies within the bone marrow cavity. When the needle is correctly positioned, blood usually oozes from the needle, on withdrawing the stylette, but positive aspiration may be required on occasion. The position of the needle should be checked by a test injection with fluoroscopy before the main one is made. If on the test injection the needle is shown to be incorrectly placed, the first needle should be left in place and another one introduced (Fig. 9.9). Unless this is done contrast extravasates from the first puncture site. Complications due to improper positioning of the cannula include extravasation of contrast into the soft tissues (Figs. 9.10, 9.12), beneath the periosteum (Fig. 9.11) or into a joint. At the end of the examination, as with other forms of phlebography, the contrast medium should be flushed out of the bone marrow cavity with physiological saline to minimize the likelihood of venous thrombosis, and to prevent leakage of contrast into the soft tissues.

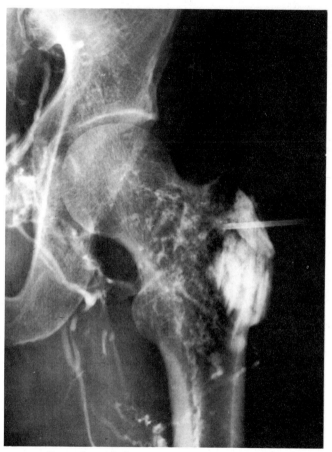

FIGURE 9.10. Pertrochanteric intraosseous phlebogram. Note extravasation of contrast into the muscle planes.

Although the method has been extensively used, serious complications are rare. Schobinger [31] reported only mild idiosyncrasies to the contrast medium in over 1200 examinations. Many patients complain of slight soreness of the site of injection for a day or two after investigation. It has been suggested that aspiration of the contrast at the end of the examination minimizes this [18]. A similar effect is obtained, however, by flushing the contrast from the bone, which is the writer's standard practice [32]. Should contrast be injected beneath the periosteum resulting in a subperiosteal haematoma, the pain at the site may persist for several days or even weeks. If contrast medium is injected into a joint cavity resulting in a haemarthrosis, pain on movement of the joint is likely to occur for several days. Osteomyelitis at the site of

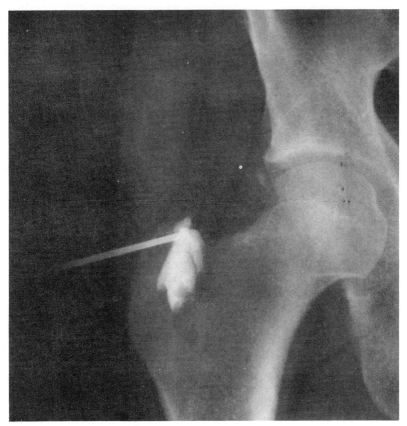

FIGURE 9.11. Subperiosteal injection at pertrochanteric phlebography.

bone puncture is clearly a possibility and full aseptic precautions must therefore be employed throughout the examination. If possible the patient should be investigated in an X-ray room reserved for angiographic procedures. One patient in the writer's series developed septicaemia which fortunately was controlled by antibiotic therapy. Bone infarction has been recorded [33]. This complication has not been encountered by the author, although follow-up films up to a year after the investigations have been carried out in a number of patients. Two patients have been reported who had massive pulmonary embolism during pertrochanteric pelvic examinations but these were considered more likely to be due to general anaesthesia rather than to the procedure itself [14].

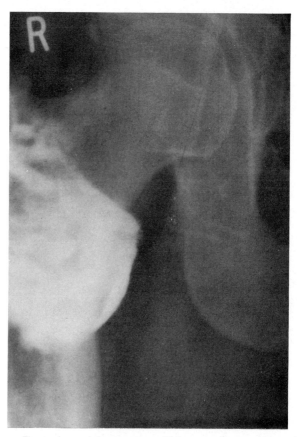

FIGURE 9.12. Pertrochanteric intraosseous phlebogram. Same patient as Fig. 9.11. Extensive extravasation has occurred into the soft tissues despite repositioning of the needle.

The most serious, but rare, complication of intraosseous phlebography is fat embolism. One death from this has occurred in the author's series of 500 intraosseous phlebograms [34]. Another patient with fat embolism which was not fatal has also been described [35]. Although a pressure injection was used in the author's patient, a hand injection was used in the other patient reported. It has been suggested [31] that pressure injections should not be used. However, to produce fully diagnostic radiographs when large

Phlebography

veins such as those in the pelvis are being examined, pressure injections are essential. Any patient developing neurological or pulmonary symptoms or signs twelve hours or later after intraosseous phlebogram should be suspected of having had a fat embolism. Fat globules should be sought in sputum, urine and blood. Treatment of the condition includes continuous oxygen and intravenous injection of low molecular weight dextran. The two cases mentioned here are the only ones reported in the literature. Since the fatal examination, the blood of all patients undergoing intraosseous phlebography has been examined after the examination for the presence of fat globules. None has been found.

When small or thin bones such as ribs, spinous processes, iliac crests, and sternum are punctured, the risk of extravasation of contrast medium is greater and may lead to transient localized pain. A mechanical injector is not necessary at these sites and often local anaesthesia can be used. With hand injections and with the patient awake, it is easier to avoid misplacement of the cannula. Complete penetration of a rib could give rise to a pneumo- or haemothorax. Epidural injection of contrast has been reported during vertebral spinous injections and also anterior mediastinal extravasation from trans-sternal injections [31]. These and similar local complications should be avoided with careful technique and the use of image intensification.

(C) INFERIOR VENA CAVOGRAPHY AND PARIETAL ILIAC PHLEBOGRAPHY

Complications of these procedures are extremely rare. No serious complications have occurred in the author's experience of over 400 examinations. The only minor complication which may occur using the percutaneous femoral technique [32] is a haematoma at the puncture site. This low morbidity has also been the experience of others [36–38]. Thrombosis due to contrast is less likely to occur in these larger veins but femoral thrombosis and pulmonary embolism has been reported [39, 40]. For this reason contrast should be flushed from the veins when the examination is finished. If a guide wire is used to thread a catheter up a femoral vein, perforation of an iliac vein or the inferior vena cava is a possibility. This occurrence is usually without sequelae [38].

(D) SUPERIOR VENA CAVOGRAPHY AND ARM PHLEBOGRAPHY

No complications from these investigations has been encountered by the writer. Minor complications such as contrast extravasation (Fig. 9.13), common to ascending phlebography of the legs and inferior vena cavography may occur but should be avoided with the use of the preventive measures previously described. If a guide wire is used, puncture of a vein is possible. The guide and catheter should be manipulated under screen control. The basilic vein should always be used for a catheterization technique as the cephalic vein joins the axillary vein almost at a right angle and it is often impossible to negotiate this angle with a catheter. If an obstruction is encountered, the catheter should be withdrawn before the main injection is

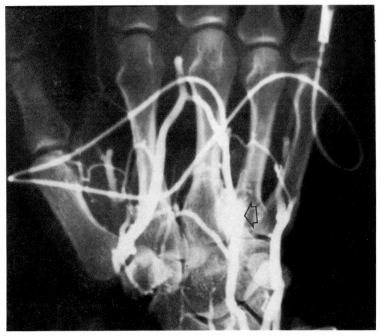

FIGURE 9.13. Hand phlebogram. A little extravasation of contrast medium has occurred around a correctly positioned needle (arrow).

made to avoid rupturing the vein wall. Hand injections should be used whenever possible.

A catheter with side holes is safer than one with a single end opening. The latter may perforate a vein if the jet impinges on the wall. The safest method is to use a blocked-end side-hole catheter but this has the disadvantage that it has to be introduced by a cut down. Venous spasm which grips the catheter and prevents its passage is a rare complication. It is usually due to the use of too large a catheter in too small a vein. Heavy sedation or even general anaesthesia may be required before the catheter can be moved.

(E) SELECTIVE RETROGRADE PHLEBOGRAPHY

General considerations

Selective retrograde catheterization of the tributaries of the inferior and superior vena cavae have been carried out at a number of sites. The usual procedure is to introduce a preshaped catheter from a femoral vein using Seldinger's technique [41] into the inferior or superior vena cava. From there the catheter is manipulated into the tributary under study [48, 49].

Selective phlebography has been applied to the renal, adrenal, hepatic, spermatic, ovarian, ascending lumbar, internal iliac, the azygous veins and veins of the head and neck [42–49]. More recently, subselective studies of the thymic veins [43, 44] and the veins of the mediastinum [50] have been carried out.

The complication common to all these studies is the risk of perforation of the veins being selectively studied, or other tributaries entered inadvertently. Veins are thin walled and easily penetrated, either by an advancing guide wire or a stiff catheter. During manipulation the catheter may stick in a small tributary, such as a lumbar vein, and a large injection into such a vein could lead to rupture. To avoid these complications a guide wire with a soft flexible tip should be used and this advanced through the venous system under screen control. Catheters made of soft materials such as polyethylene are best used rather than stiff materials such as Teflon. Gentle manipulations should be used to minimize the risk of perforation by either catheter or guide wire. Before the main injection is made, a test injection of a small volume of contrast medium under fluoroscopy should be carried out. High-pressure injections through end-hold catheters should be avoided as these may rupture a vein. Thrombus which may develop in the catheter, if dislodged, does not cause complications on the venous side. Should perforation occur this is not usually serious as the vein seals off rapidly. Extravasation of contrast into the tissues only gives rise, as elsewhere, to transient local pain.

Selective catheterization at specific sites

1 Adrenal phlebography

The specific complication is rupture of intraglandular veins giving rise to haematomas. This is particularly likely to occur in patients with primary aldosteronism who have friable veins [51], and is caused by too high an injection pressure [52], or by overfilling the gland with contrast [53].

Unilateral haematomas do not usually give rise to clinical change but are extremely painful [52, 53]. Infarction can, however, occasionally lead to adrenal insufficiency [54].

To avoid venous rupture small amounts of contrast medium, as little as 0.5 ml of meglumine iothalamate 60 per cent, should be injected by hand under screen control. The main injection should always be delivered by hand, the amount of contrast required being 2–3 ml, depending on the side and the position of the catheter. If such precautions are taken the incidence of intra-adrenal extravasation can be considerably reduced. Thus in two recently reported series, there was an incidence of 3.6 per cent in 83 examinations [55], and another series of 40 patients was carried out without any complications [53]. One case of a nonfatal hypertensive crisis following adrenal phlebography in a patient with phaeochromocytoma has been reported [56] but others have examined such patients without incident [53, 55].

As with phlebography elsewhere, venous thrombosis has been reported. Anticoagulants given beforehand are not recommended because of the increased risk of bleeding in the event of venous rupture. Again, as in other forms of phlebography the best method of minimizing this complication is to complete the examination as rapidly as possible and to flush the contrast from the adrenal vein immediately after the examination.

2 Hepatic phlebography

No serious complications have been encountered or reported. Slight discomfort may be noted by the patient after wedge injection, particularly when the catheter is near the liver capsule. Although tissue damage is generally considered minimal, small granulomas have been found at the injection site and a transient rise in serum enzymes noted [57, 58]

3 Renal phlebography

The use of adrenalin by intra-arterial injection to slow the circulation and demonstrate the veins more clearly has been recommended [42]. A possible complication of this is renal infarction. The dose of adrenalin should not exceed 0.2 μg/kg body weight.

4 Pancreatic phlebography

Recently techniques have been described for retrograde phlebography of the pancreatic veins by a transhepatic or transumbilical approach [59].

The only complications which occurred in 33 examinations by the transumbilical technique were small extravasations into the pancreas. These were without sequelae. In two of the seven patients catheterized by the transhepatic route, emergency laparotomy was required for intra-abdominal bleeding. The value and safety of this new method awaits further study.

5 Internal jugular phlebography

Selective catheterization of the internal jugular vein is used to obtain samples for assay from the superior and middle thyroid veins, which drain into it, in patients with suspected parathyroid neoplasms. While small injections of contrast are required for localization of the catheter a phlebogram is not required. No complications of this or similar venous sampling procedures are reported.

Internal jugular vein catheterization is, however, also used to outline the cavernous sinus. Should the cavernous sinus be occluded, which occurs in approximately 50 per cent of the pituitary tumours that have extended into the chiasmatic regions, a high injection pressure into the inferior petrosal sinus, even by hand injection, may lead to rupture of prepontine veins. This may cause partial lateral medullary syndrome or other neurological deficits, depending on the site of the lesion [60].

6 Orbital phlebography

Orbital phlebography by the frontal vein approach is a completely safe procedure. Small haematomas at the puncture site may occur but are easily controlled by digital pressure. Local infection could occur but is unlikely as the scalp is remarkably resistant to local sepsis. Other than haematoma formation and minor extravasation of contrast locally, the author has experienced no adverse consequences of the examination and this has been the experience of others in the large series reported [61, 62]. As the examination is carried out under local anaesthesia a sensation of tension in the forehead or infra-orbitally may be felt by the patient at the time of the injection. This sensation disappears in a few seconds, and is seldom described as painful. Some patients complain of difficulty in breathing through the nose for a short time after each injection of contrast, probably due to transitory swelling of the nasal mucosa.

REFERENCES

1 BERBERICH J. & HIRSCH S. (1923) Die roentgenographische Darstellung der Arterien und Venen am lebenden Menschen. *Klin. Wschr.* 2, 2226–2228.

2 MCPHEETERS H.O. & RICE C.O. (1929) Varicose veins: Circulation and direction of venous flow, experimental problems. *Surg. Gynec. Obstet.* **49**, 29–33.

3 EDWARDS E.A. & BIGURIA F. (1934) Comparison of Skiodan and Diodrast as vasographic media with special reference to their affect on blood pressure. *New Engl. J. Med.* **211**, 589–593.

4 MAY R. (1965) The early X-ray diagnosis of thrombosis. *Nuclear Energy*, **4**, 120–122.

5 WERNER H. & OTTO K. (1962) Hazards and complications in roentgenological venous diagnosis. *Fortschr. Roentgenstr.* **96**, 655–663.

6 KAKKAR V.V. (1972) The ^{125}I-labelled fibronogen test and phlebography in the diagnosis of deep vein thrombosis. *Milbank Mem. Fund. Q.* **50** (Suppl. 2), 206–229.

7 CORRIGAN T.P., FOSSARD J., SPINDLER J. et al. (1964) Phlebography in the management of pulmonary embolism. *Br. J. Surg.* **61**, 484–488.

8 WINZER K., LANGECKER H. & JUNKMANN J. (1954) Zur Frage der Vertraglichkeit von Nieren und Gallekontrastmitteln. *Arztl. Wshnsche.* **9**, 950–952.

9 HÄEGAR K. & NYLANDER G. (1967) Acute phlebography. *Triangle* **8**, 18–26.

10 RABINOV K. & PAULIN S. (1972) Roentgen diagnosis of venous thrombosis in the leg. *Arch. Surg.* **104**, 134–144.

11 LEA THOMAS M., MCALLISTER V. & TONGE K. (1971) Simplified phlebography in deep venous thrombosis. *Clin. Radiol.* **22**, 490–494.

12 EVANS D.S. & COCKETT F.B. (1969) Diagnosis of deep venous thrombosis with an ultrasound Doppler technique. *Br. Med. J.* **2**, 802–804.

13 LEA THOMAS M. (1972) Phlebography. *Arch. Surg.* **104**, 145–151.

14 MAHAFFY R.G., MAVOR G.E. & GALLOWAY J.M.D. (1971) Ilio-femoral phlebography in pulmonary embolism. *Br. J. Radiol.* **44**, 172–183.

15 GÖTHLIN J. (1972) The comparative frequency of extravasal injections at phlebography with steel and plastic cannula. *Clin. Radiol.* **23**, 183–184.

16 BAUER G. (1940) A venographic study of thrombo-embolic problems. *Acta. Chir. Scand.* **84**, (Suppl.) 61, 75 pp.

17 GREITZ T. (1954) The technique of ascending phlebography of the lower extremity. *Acta. Radiol.* **42**, 421–441.

18 DOW J. (1972) *A Text-book of X-ray Diagnosis* (Ed. by SHANKS & KERLEY), 4th edn, pp. 511–564. H. K. Lewis & Co. Ltd., London.

19 GRYSPEEDT L. (1953) Phlebography of the lower limb. *Br. J. Radiol.* **26**, 329–338.

20 GULLMO A. (1956) On the technique of phlebography of the lower limb. *Acta Radiol.* **46**, 603–620.

21 ALMEN T. & NYLANDER G. (1962) Serial phlebography of the normal lower leg during muscular contraction and relaxation. *Acta Radiol.* **57**, 264–272.

22 GÖTHLIN J. & HALLBÖÖK T. (1971) Extravasal injection of contrast medium at phlebography. *Der radiologe* **4**, 161–165.

23 MCALISTER W.H. & PALMER K. (1971) The histologic effects of four commonly used media for excretory urography and an attempt to modify the response. *Radiology* **99**, 511–516.

24 LEA THOMAS M. (1970) Gangrene following peripheral phlebography of the legs. *Br. J. Radiol.* **43**, 529–530.

25 LEA THOMAS M., MCALLISTER V., ROSE D.H. & TONGE K. (1972) A simplified tech-

nique of phlebography for the localisation of incompetent perforating veins of the legs. *Clin. Radiol.* 23, 486–491.
26 BERGVALL U. (1971) Phlebography in acute deep venous thrombosis of the lower extremity. A comparison between centripetal ascending and descending phlebography. *Acta radiol. Diag.* 11, 148–166.
27 DE WEESE J.A. & ROGOFF S.M. (1958) Clinical uses of functional ascending phlebography of the lower extremity. *Angiology* 9, 268–278.
28 WESOLOWSKI S.A., GREENFIELD H., SAWYER P.N. et al. (1965) Diagnostic value of phlebography in venous disorder of the lower extremity. *J. Cardiovasc. Surg.* (Suppl.) pp. 133–151.
29 HUME M., SEVITT S. & THOMAS D. (1970) *Venous thrombosis and pulmonary embolism.* Harvard University Press, Boston.
30 LEA THOMAS M. (1969) An improved intraosseous phlebography cannula. *Br. J. Radiol.* 42, 395.
31 SCHOBINGER R.A. (1960) *Intraosseous venography.* Grune & Stratton, New York and London.
32 LEA THOMAS M. & FLETCHER E.W.L. (1967) The techniques of pelvic phlebography. *Clin. Radiol.* 18, 399–402.
33 ISHERWOOD I. (1972) *Practical procedures in diagnostic radiology* (Ed. by SAXTON & STRICKLAND), 2nd edn, pp. 272–277. H. K. Lewis & Co. Ltd., London.
34 LEA THOMAS M. & TIGHE J.R. (1973) Death from fat embolism as the complication of intraosseous phlebography. *Lancet* 2, 1415–1416.
35 YOUNG A.E., LYNN EDWARDS I., IRVING D. & HANNING C.D. (1973) Fat embolism after pertrochanteric venography. *Br. med. J.* 4, 592.
36 LINDBOM A. (1960) *Modern Trends in Diagnostic Radiology* (Ed. by MCLAREN), 3rd series, pp. 111–123. Butterworth, London.
37 RANNIGER K. & SALDINO R.M. (1968) Abdominal angiography. *Current problems in Surgery*, March, pp. 28–52.
38 HIPONA F.A. (1969) In: *Venography of the inferior vena cava and its branches*, pp. 33–52. The Williams & Wilkins Co., Baltimore.
39 BARTLEY O. (1958) Venography in the diagnosis of pelvic tumours. *Acta Radiol.* 49, 169–186.
40 TERNBERG J.L. & BUTCHER H.R. JR (1965) Evaluation of retrograde pelvic venography. *Arch. Surg.* 91, 607–609.
41 SELDINGER S.I. (1953) Catheter replacement of the needle in percutaneous arteriography. *Acta radiol.* 39, 368–376.
42 KHAN P.C. (1969) In: *Venography of the inferior vena cava and its branches*, pp. 154–224. The Williams & Wilkins Co., Baltimore.
43 KREEL L. (1967) Selective thymic phlebography: a method for visualisation of the thymus. *Br. Med. J.* 1, 406–407.
44 YUNE H.Y. & KLATTE E.C. (1970) Thymic venography. *Radiology* 96, 521–526.
45 ANDERSON R.K. (1951) Diodrast studies of the vertebral and cranial venous systems to show their probable role in cerebral metastases. *J. Neurocurg.* 8, 411–422.
46 FISCHGOLD H., ADAM H., ECOFFIER J. et al. (1952) Opacification des plexus rachidiens et des veines azygos par voie osseuse. *J. Radiol. Electr.* 33, 37–38.
47 JANOWER M.L., DRAYFUSS J.R. & SKINNER D.B. (1966) Azygosography and lung cancer. *New Engl. J. Med.* 275, 803–808.

48 STAUFFER H.M., LaBREE J. & ADAMS F.H. (1951) The normally situated arch of the azygos vein: its roentgenologic identification and catheterisation. *Am. J. Roentg.* 66, 353–360.
49 RINKER C.T., TEMPLETON A.W., MACKENZIE J. *et al.* (1967) Combined superior vena cavography and azygosography in patients with suspected lung carcinoma. *Radiology* 88, 441–445.
50 YUNE H.Y. & KLATTE E.C. (1972) Mediastinal venography: subselective transfemoral catheterisation technique. *Radiology* 105, 285–291.
51 BOOKSTEIN J.J., CONN J. & REUTER S.R. (1968) Intra-adrenal hemorrhage as complication of adrenal venography in primary aldosteronism. *Radiology* 90, 778–779.
52 MIKAELSON C.G. (1967) Retrograde phlebography of both adrenal veins. *Acta Radiol.* 6, 348–354.
53 SUTTON D. (1975) Radiological diagnosis of adrenal tumours. *Br. J. Radiol.* 48, 237–258.
54 JORGENSON H. & STIRIS G. (1974) Hypertensive crisis followed by adrenocortical insufficiency after unilateral adrenal phlebography in a patient with Cushing's Syndrome. *Acta med. scand.* 196, 141–143.
55 MITTY H.A., NICOLIS G.I. & GABRILOVE J.L. (1973) Adrenal venography. *Am. J. Roentg.* 119, 564–575.
56 GOLD R.E., WISINGER B.M., GERACI A.R. & HEINZ L.M. (1972) Hypertensive crisis following adrenal venography in a patient with a phaeochromocytoma. *Radiology* 102, 579–580.
57 WIDMANN W.D., GREENSPAN R.H., HALES M.R. & CAPPS J.H. (1961) A new method for portal venography: retrograde hepatic flushing. *Proc. Soc. exp. Biol. Med.* 106, 540–542.
58 SCHLANT R.C., GALAMBOS J.T., SHUFORD, W.H. *et al.* (1963) The clinical usefulness of wedge hepatic venography. *Am. J. Med.* 35, 343–349.
59 GÖTHLIN J., LUNDERQUIST A. & TYLEN U. (1974) Selective phlebography of the pancreas. *Acta Radiol.* 15, 474–480.
60 HANAFEE W. (1972) Orbital venography. *Radiol. Clin. N. Amer.* 10, 63–81.
61 VIGNAUD J., CLAY C. & BILANIUK L.T. (1974) Venography of the Orbit. *Radiology* 110, 373–382.
62 BRISMAR J. (1974) Orbital phlebography. *Acta Radiol.* 15, 370–382.

CHAPTER 10: PERCUTANEOUS TRANSHEPATIC CHOLANGIOGRAPHY, ENDOSCOPIC RETROGRADE CHOLANGIOPANCREATOGRAPHY AND SPLENO-PORTOGRAPHY

W. B. YOUNG AND E. C. LIM

INTRODUCTION

Percutaneous transhepatic and endoscopic retrograde cholangiography, when successful, can assist the clinician in solving the problem of differentiating hepatocellular from obstructive jaundice; by demonstrating a patent duct system in the one, and mechanical obstruction with distended proximal ducts in the other.

If the duct system is opacified, it will provide detailed information of duct anatomy and may indicate to the surgeon the exact site, extent and likely cause of any obstruction. The investigation also provides information of the amount and position of normal duct, proximal to the obstruction, which may be available for anastomosis or intubation should this be considered necessary. As a general rule more detailed information can be obtained by opacifying the duct system from the proximal side of the obstruction than from below; if the obstruction is complete, this is the only effective approach. However, when the cause of jaundice is uncertain and some form of intrahepatic cholestasis seems likely, retrograde cholangiography is the method of choice, since the success rate of the transhepatic procedure with ducts of normal or decreased calibre is unlikely to be more than 50 per cent, even using a fine needle injection technique. In the presence of an obstruction, especially if the bile is infected, cholangiography, whether from above or below, is a reliably safe procedure only if it is followed within a few hours by surgery to relieve the obstruction. If, therefore, it is thought likely that jaundice is due to duct stones or to post-operative stricture, arrangements should be made to proceed to immediate laparotomy should a mechanical obstruction be shown.

When the cause is likely to be malignant, and especially if a low duct obstruction is suspected, retrograde cholangiography, with its facility for pancreatography and direct biopsy would seem to be the method of choice. If, however, the obstruction turns out to be incomplete, and subsequent drainage of the distended proximal ducts is impaired, the risk of serious infection is not inconsiderable if drainage is delayed. The same applies when a fine needle technique is used for transhepatic cholangiography, though not quite to the same extent, for whereas Gram-ve organisms are probably invariably introduced into the duct from the duodenum, direct liver puncture is invariably sterile. If a needle/polythene catheter method is used, as described by Shaldon, Barber & Young [1] in 1962, the tube is always left to drain the ducts until surgery is undertaken. As a result, we have had no serious infection with this technique in over 600 cases performed since 1960.

When there are strong clinical or radiological indications of malignant low duct obstruction the surgeon will usually want to know not only the exact site and size of the tumour but whether the portal vein is involved and whether the liver is free of metastases. He can then decide either to attempt a radical resection or settle for a more simple drainage procedure; and according to this decision he can plan his procedure, arrange his use of theatre time and order appropriate quantities of blood. The only means of obtaining this additional information preoperatively is by arteriography. If it has already been shown, or there is clinically little doubt, that the lesion arises from the ampulla or head of pancreas, arteriography should be performed *before* cholangiography which can then be undertaken shortly before surgery.

Arteriography is completely unjustified in cases of benign obstruction such as stone, stricture or primary biliary cirrhosis. If one of these conditions is suspected, percutaneous cholangiography is the method of first choice, and if a stone or stricture is shown the case should proceed straight to surgery. Should a malignant obstruction be shown, especially if low and incomplete, it is still safer to proceed to surgery. If, nevertheless, it is thought essential to have arteriographic information of the exact size of the tumour before operation, a flexible polythene or Teflon cannula should always be introduced transhepatically into one of the dilated intrahepatic ducts and left to drain and decompress the ducts until surgery is undertaken.

PERCUTANEOUS CHOLANGIOGRAPHY

This is a simple and safe procedure, easy to perform, wherever there are

X-ray screening facilities, though image intensification is an advantage. Complications are uncommon—less than 2–3 per cent, and are usually avoidable. The mortality rate is low—probably less than 1 in 1000 examinations. As with many diagnostic and operative procedures, the experience and skill of the operator and the correct selection of cases for examination contribute considerably to the safety of the procedure.

CONTRAINDICATIONS AND PRECAUTIONS

In general, unless there are overriding reasons, the examination should not be undertaken if there is active cholangitis, either acute or chronic—or when hydatid disease is a possibility. It should be postponed if possible when liver or kidney function is impaired, especially if there is a likelihood of biliary infection. In particular, it should not be undertaken when the patient's condition does not warrant subsequent surgery sufficient to relieve an extrahepatic duct obstruction if one is shown or, if endoscopic cholangiography is not available, should it fail to demonstrate an unobstructed duct system.

The prothrombin time should be checked and any bleeding tendency should first be corrected. It is also prudent to test if possible for sensitivity to the opaque medium to be used by injecting 0.5 ml intravenously a day or two beforehand. The patient's blood group, ESR, Hb per cent, red, white and platelet count should be ascertained and his BP, temperature and pulse rate checked.

If there is a possibility of biliary infection a preliminary blood culture should be performed and the procedure should be deferred pending the result, and if positive, until organism/antibiotic sensitivity is known. This should be routine when gallstones are suspected, or when there has been previous surgery on the bile ducts, particularly if an anastomosis between duct and small intestine has been carried out. The same applies when a communication between gall gladder and intestine is suspected because of gas in the biliary tree, or if a communication has been shown by barium enema or meal examination.

COMPLICATIONS

Reported complications rarely represent the full picture for it seems probable that a large proportion are never documented. The incidence quoted in different series varies with the selection of cases for examination, the causes found, the experience of the operator and the criteria accepted. The majority of complications described are trivial and rarely lead to serious morbidity;

but some are serious and a number of deaths have been attributed to the procedure.

Puncture of extrahepatic structures

The stomach, duodenum, small intestine, colon, renal pelvis, gall bladder and common bile duct have all at times been unintentionally punctured

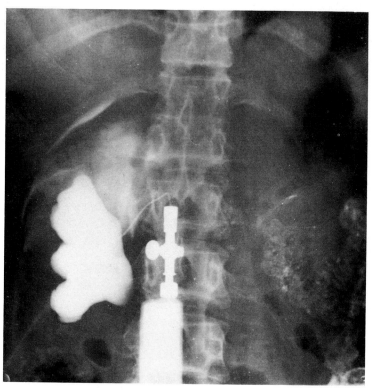

FIGURE 10.1. Percutaneous cholangiography. Puncture of renal pelvis in hydronephrosis. Aspirated fluid was bile stained and thought to be bile. No complications after withdrawal of tube.

(Figs 10.1 and 10.2). With the exception of gall bladder puncture, no serious sequelae have been reported. It is common practice simply to withdraw the needle or tube, discard it and introduce another in a different direction. Though probably unnecessary, it would seem prudent to prescribe prophylactic antibiotics for 24–48 hours if the colon has been punctured. Seldinger [2] described one case in which the inferior vena cava was punctured. This was immediately noticed as the contrast medium was injected and the tube was withdrawn. There were no complications. In another of his cases the injection clearly outlined the portal vein close to the hilum of the liver. This patient developed severe abdominal pain and a shock-like state which lasted

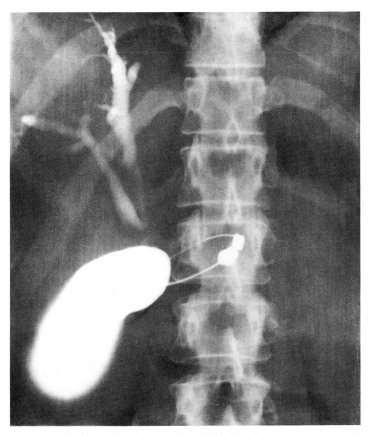

FIGURE 10.2. Accidental intubation of the gall bladder during percutaneous cholangiography in a patient with a carcinoma involving the junction of the right and left hepatic ducts. Immediate surgery was undertaken and bile in the peritoneal cavity was mopped up. The duct system was drained by inserting a catheter through the stricture from below. Uneventful recovery from the operation with no evidence of peritonitis.

for a few minutes, to be followed by rapid enlargement of the spleen, fever, leucocytosis and a fall in haemoglobin level. The patient was transfused and put on antibiotics. The fever and leucocytosis subsided with the passage of a large amount of urine, thought to be a sequel to acute ascites. An eight-year follow-up showed that the spleen was no longer enlarged. It was suggested that the portal vein had been transiently obstructed—possibly by compression from a haematoma—following injury to the hepatic artery. While a traumatic aneurysm of the hepatic artery is a possible complication, as far as we can discover, this has never been reported.

Occasionally the needle or cannula enters the gall bladder and very rarely the common bile duct itself. De Masi et al. [3] reported a series of 30 cases in which the gall bladder was punctured unintentionally in seven but this is an unusually high incidence. Some workers have deliberately tried to inject contrast medium into the gall bladder either by introducing the needle through the liver [4, 5] or under direct vision at peritoneoscopy [6]. If the cystic duct and common duct are freely patent the risk of bile leakage after withdrawing the needle seems to be small (see later under infection).

Intrathoracic injection of opaque medium

The introduction of the tip of the needle through liver and diaphragm in the pleural or pericardial spaces [7], especially where they contain fluid, is not a rare occurrence and the aspiration of bile-stained fluid may easily deceive the operator into thinking that he has entered a bile duct. Seldinger [2] reported four cases in which contrast medium was injected into the pleural cavity and most workers have probably made similar mistakes. This probably occurs more often with an intercostal approach than with an anterior or posterior one. It causes immediate pain and is usually easily recognized on the screen. Occasionally a small pneumothorax is produced but usually requires no special treatment, though Isley & Schauble [4] reported one case in which a tension pneumothorax followed accidental puncture of the lung and required the subsequent insertion of an intercostal tube to relieve it. Intrapericardial injection usually causes severe pain followed by tachycardia. These reactions tend to be transient though pain may be severe enough to require intravenous Fortral or pethidine to relieve it. Occasionally dyspnoea may be sufficiently severe as to require oxygen. No prolonged ill effects have been reported (see page 85).

Shock

A severe fall in blood pressure may occasionally occur during or immediately after the injection. This is usually associated with severe pain due to the mis-siting of the needle tip so that opaque medium is injected outside a duct directly into the region of the coeliac plexus. Reactions to the contrast medium itself are very uncommon with an aspiration technique such as the one we recommend, and, though rare, are more likely to occur with techniques recommended by Wiechel [8] and more recently Okuda [9] whereby contrast is injected as the needle is withdrawn through the liver. As far as is ascertainable, there have been no reports of anaphylactic type shock following injection of water-soluble opaque medium directly into the bile ducts, whether from above or below (see Chapter 16, p. 339).

Bile leakage: peritonitis

In most reports it is stated that bile peritonitis is the greatest single hazard of the procedure. Seldinger [2] quoted an incidence of 2.5 per cent of serious complications, 2 per cent being due to peritonitis. In 1794 cases we have culled from the literature bile peritonitis has been a complication in 1.8 per cent, with an overall mortality of 1.1 per cent. Bile peritonitis is a serious matter, especially in a jaundiced patient who is usually in poor general health with impaired immunological mechanisms; but with the technique we advocate, using a polythene tube which is left in position to drain and decompress the ducts until surgery is undertaken, it should be very uncommon. Bile is normally sterile and can accumulate in large amounts in the peritoneal cavity after blunt trauma, over a period of several weeks, without the development of chemical or bacterial peritonitis. Furthermore, if the duct system is freely patent and not dilated, the risk of bile leakage would seem to be minimal for, if not under abnormal pressure, bile will tend to flow through normal channels rather than along the needle puncture tracts in the liver. If, however, a duct is obstructed the risk of leakage probably increases proportionately to the rise in intraductal pressure. There are also other factors of importance, such as the number of punctures made, the time that elapses before surgery is undertaken to relieve the obstruction, and whether an obstructed duct system is decompressed at the end of the procedure by leaving the cannula in the duct to drain it.

Glenn *et al.* [10] reported a case in which a large amount of bile was found in the peritoneal cavity at laparotomy following an examination performed under general anaesthesia during which breathing had not been adequately controlled: and there have been occasional reports of quite large amounts of bile or blood in the peritoneal cavity due to laceration of the liver. We believe that serious liver damage is unlikely using a Teflon or polythene tube threaded over a fine introducing needle to cannulate a duct, especially if the needle–catheter combination is introduced slowly and steadily into the liver parenchyma while the patient holds his breath on full inspiration and the needle is withdrawn immediately the tube has been fully introduced.

It is commonplace to find some bile in the peritoneal cavity, and occasionally quite large amounts at laparotomy performed immediately after the examination. However, in our experience of more than 600 examinations over a period of fifteen years, peritonitis has never been a serious problem and we have had no deaths from this cause. We attribute this to the fact that, whenever possible, the patient proceeds straight to surgery unless a clearly patent duct system has been demonstrated. Once an obstruction is relieved the risk of leakage is very small and any bile in the peritoneal cavity can be dealt with by normal peritoneal toilet. Moreover, we use a flexible tube which cannot damage the liver should the patient suddenly cough or move during the procedure. Probably most important of all, is that we always leave the tube to drain dilated ducts until surgery is performed.

The risk of bile peritonitis is probably increased if several punctures fail to find distended ducts which are in fact present, especially if the bile is infected; for this may lead to deferment of laparotomy and infected bile may then leak along puncture tracks into the peritoneum. The danger is considerable if a distended duct system has been shown and is not decompressed by immediate surgery, or by percutaneous tube drainage if for any reason it is decided to delay surgery. In such cases, the patient should be watched especially carefully for signs of peritoneal infection or bleeding. If, on the other hand, an unobstructed duct system has been demonstrated there seems to be little risk.

Pain

Pain during the introduction of the needle. Most examinations are undertaken under local anaesthesia such as 1 per cent lignocaine. In spite of this, pain is sometimes experienced during the introduction of the needle or needle-tube combination into the liver. In children and highly nervous adults it may be advisable to give a light general anaesthetic, especially if the patient can proceed immediately to surgery close by should an obstructive lesion be demonstrated.

However, if premedication has been adequate (Na. pentobarbital 50 mg and diazepam 10–20 mg) and providing local anaesthetic is injected through the introducing or exploratory needle, as it is slowly inserted all the way into the liver, the patient rarely complains of any discomfort.

Pain during the introduction of the contrast fluid. This should suggest that the needle is wrongly located and that the injection is subcapsular, or intraperitoneal or pleural—or that it is causing undue distension of the intrahepatic ducts. When subcapsular or perihepatic the pain is usually localized to the right shoulder and when intraperitoneal to the central abdomen. It can usually be relieved by spraying the painful area with ethyl chloride but may occasionally require intravenous analgesics. Occasionally pain is associated with a shock-like reaction and rapid fall in blood pressure. The injection should be stopped at once and the shock treated.

Six patients in Seldinger's series [2] complained of momentary pain as the opaque medium was injected into the duct system and in one it was felt whenever slight pressure was applied to the piston of the syringe. He considered this to be a tension phenomenon for it tended to develop as the biliary tree was filled. This is especially likely to occur if ducts are already distended and the injection pressure should be reduced. Pain of this type is unusual if the ducts are first decompressed by aspirating bile from them for culture and if the injection is not made too forcibly; it is unnecessary, as some recommend, to aspirate as much bile as possible before contrast medium is injected.

Bile peritonitis after puncture of the gall bladder

Intraperitoneal leakage of bile may follow puncture of the gall bladder, whether accidental or deliberate, though if the distal duct system is not obstructed the amount of leak tends to be small and insignificant (Fig. 10.2). If the cystic or common bile duct is obstructed and the bile is infected, the risk of peritonitis is very real and increases with the length of time that elapses between the examination and the surgical relief of the obstruction and peritoneal toilet. This is borne out by De Masi and his colleagues [3] who, as already stated, using a needle technique, inadvertently punctured and injected contrast into the gall bladder in seven out of thirty examinations. In four cases the bile duct was shown to be obstructed, and in three a polythene tube was inserted through the needle before it was withdrawn and left in place to decompress the gall bladder. None of these three cases developed peritonitis and no bile leakage was found at subsequent laparotomy some days later. The fourth case, which was not drained, developed peritonitis and died. In the remaining three cases the ducts were shown to be patent and the needle was withdrawn without any ill effects.

It would therefore seem to be a wise precaution, whenever a distended obstructed gall bladder has been punctured, for a tube to be left *in situ* to decompress the gall bladder until laparotomy is undertaken.

Subphrenic abscess

Glenn & Mujahed [11] reported two cases of subphrenic abscess with one death. There was also one case of chronic external biliary fistula from a puncture site, even though a cholecysto-jejunostomy had been performed to relieve main duct obstruction due to carcinoma. It must be assumed that the drainage procedure had not been very effective.

Decompression procedures

In two cases we have left a tube in an intrahepatic duct to drain a distended duct system when, as a result of the examination, they had been judged inoperable. Intense and distressing pruritus was rapidly relieved and bile continued to drain through the polythene tubes until death ten and fourteen days later respectively. At autopsy there was no evidence of peritonitis in either case.

We therefore recommend that, if for any reason laparotomy is delayed or postponed, once a dilated duct system has been shown the tube should always be left in place as a deliberate drainage procedure until laparotomy can be undertaken. Recently, Tylen [12] has recommended that, when percutaneous cholangiography has demonstrated dilated intrahepatic ducts, a flexible guide should be introduced through the tube as far as the obstruction, and a grey Seldinger arterial catheter should then be introduced over it. After removing the guide, contrast medium can be injected through the catheter to demonstrate the obstruction and duct system proximal to it. If the obstruction is incomplete, an attempt can then be made, by re-introducing the guide, to find a way through the stricture to show the state of the duct below. Another catheter, with appropriate side-holes, can then be introduced to institute temporary or permanent internal drainage of the ducts, depending on whether operative resection is considered feasible or not. In the meanwhile the passage is kept clear by injecting fluid through the externally projecting catheter. The decompression of dilated ducts, by whatever method, will relieve jaundice and pruritus and allow the patient to regain a state in which he can better withstand subsequent surgery. At the same time, it safely provides time for such further investigations considered desirable (e.g. angiography) to be performed, before surgery is finally undertaken. In a series of some one hundred cases, Tylen claimed that there was only one serious complication, a case of bile peritonitis. It is difficult to see why this should have occurred with adequate duct decompression and appropriate antibiotic therapy following culture of aspirated bile.

Liver abscess or portal pyaemia attributable to the procedure would seem to be very rare indeed, and we can find no reports.

Haemorrhage

Patients with prolonged obstructive jaundice are more prone to bleeding than non-icteric patients. This is probably because of low prothrombin levels caused by deficiency of vitamin K which, being fat soluble, depends on the emulsifying action of bile for its absorption. In addition, when large duct obstruction is due to carcinoma of the head of pancreas or ampulla, the tendency to bleed may be exaggerated by increased fibrinogen degradation products (FDP) which have been found in the serum of 61 per cent of all cancer patients and are thought to act by inhibiting the polymerization of fibrin monomers. Certain tumours themselves contain fibrinolytic substances which may increase further the tendency to bleed.

The development of large subcutaneous or subcapsular haematomas at puncture sites has been reported [13, 14] and it is commonplace for small amounts of blood to be found under the liver capsule or in the peritoneal cavity when laparotomy is carried out shortly after the procedure. In our experience, however, active bleeding from puncture sites sufficient to require special surgical action is rare, and is a risk only when there are metastatic deposits in the liver or if the liver has been lacerated as a result of faulty technique. On the other hand, there have been reports of quite large quantities of blood in the peritoneal cavity without the patient showing much in the way of signs to indicate active bleeding. This is another cogent reason for proceeding to immediate laparotomy after the examination, since haemostasis can then usually be easily achieved and any blood in the peritoneal cavity mopped up or sucked out. If for any reason laparotomy is delayed, it is essential to monitor the blood pressure and pulse rate carefully for at least 24 hours after the procedure.

Traumatic communications between the biliary tree and vascular systems

The hepatic arteries and veins run closely related to the bile ducts, the veins lying a little posterior to the ducts. A needle or cannula, especially if introduced from the front or back, is therefore liable to puncture first one and then the other and may possibly produce a communication—temporary or permanent—between the two systems. Wiechel [8] claims that an intercostal approach is less likely to result in a blood/bile fistula than an anterior one, though this is probably largely theoretical. At times, contrast can be seen on the monitor to enter both a duct and vascular channel at the same time. *A communication should be suspected if, following an aspiration technique, a pyelogram is noticed after the injection of contrast medium, when a simple clean intubation of the duct would otherwise seem to have been achieved.*

Normal intraductal pressure is between 8–10 cm H₂O which is lower than normal hepatic or portal venous pressure and considerably lower than hepatic arterial pressure. Therefore, unless the duct system is obstructed and pressure is raised, blood will tend to flow into the duct. Haematobilia and melaena may follow the examination, though severe blood loss is extremely unlikely. Blood may, however, clot in the larger ducts, especially if they are partially obstructed and may even convert a partial obstruction into a complete one. Occasionally, clots may present as filling defects in the contrast filled ducts either on the monitor or on films taken during the examination. These may look like stones. Clots can usually be flushed easily from the ducts at laparotomy or be recognized by change in shape as contrast fluid drains from the ducts. If a clot converts a partial obstruction into a complete one, there may be a reversal of flow from duct to vein with the formation of a permanent fistula. This may not be a serious matter in itself providing that the communication is small but, if the bile is infected, it may lead to septicaemia or portal pyaemia, depending on the various systems involved.

Bacteraemia, septicaemia and endotoxic shock

Elkeles & Mirizzi [15] found that the bile was infected in about one-third of their cases of chronic cholecystitis. The bile is usually infected when there are stones in the ducts [16]. Moreover, in those cases where a stone causes intermittent obstruction of a duct with a diameter exceeding 1 cm and where there is a two to three weeks' history of jaundice, the bile in the common duct is almost invariably infected [17]. It has been claimed that in dogs, ligature of the common bile duct is shortly followed by infection of the bile trapped proximal to the ligature; but this is disputed by Musgrave [18] who found that the bile was usually still sterile two or three weeks after ligature, though it was invariably infected at similar periods after anastomosing the duct to jejunum or duodenum. In man, bile is almost always infected if a fistula develops, or if an anastomosis is performed between duct and intestine, whether small or large. In our experience bile in the common duct is usually also found to be infected after sphincterotomy. On the other hand, bile is almost always sterile above a malignant stricture, whether primary malignant in the duct, or due to carcinoma of the head of pancreas. It may, however, be infected if the growth impairs the sphincter mechanism at the lower end of the common bile duct as, for instance, in some cases of carcinoma of the ampulla.

Hultborn et al. [19] showed that there is direct anatomical communication between biliary canaliculi and the liver sinusoids in obstructive jaundice. Moreover when a main duct is completely obstructed, intraduct pressure may be considerably higher than venous pressure. This may then create a pressure gradient from the biliary to the systemic venous systems if there is any communication between these two systems. It is also interesting to note that as long ago as 1795, William Saunders [20] Senior Physician at Guy's Hospital, showed conclusively for the first time in history that jaundice could result from an obstruction to the flow of bile down the bile duct (by ligaturing the common duct in a dog). He also observed that if water was injected into the dilated duct proximal to the ligature, it required only a little pressure to cause the water to flow back into the liver or general circulation. With more sophisticated and highly expensive X-ray apparatus—or with isotope-labelled fluids and equally costly apparatus—it can be confirmed that this may occur also in man. This must lead one to assume that the pressure under which contrast medium is injected into the ducts may at times be of some importance when causes for septicaemia are sought.

One must assume that there is always a risk of bacteraemia whenever obstruction and jaundice are due to stone or post-operative stricture, even though obstruction may be intermittent. This probably explains the intermittent attacks of fever with jaundice in these conditions. The same applies when jaundice follows the closure of a duct-entero-anastomosis or spincterotomy, whether due to fibrosis, accumulation of biliary mud, or obstruction by retained or re-formed calculi.

The commonest organisms found in infected bile by Scott & Khan [16] were *Escherichia* (68 per cent), *Streptoccus faecalis* (50 per cent) and *Klebs areogenes* (25 per cent)—nearly always in some form of combination. Less often *Enterobacteraerogenes*, *Pseudomonas aeruginosa* or *Proteus* were found.

It is not unusual for there to be a small rise in temperature for 24–48 hours after percutaneous cholangiography. When Gram-negative septicaemia develops it is usually associated with a rapid rise in temperature to somewhere in the 101–104°F (38.3–40°C) range, often accompanied by chills or rigors. Symptoms may develop within one to one and a half hours after examination though more often they become manifest between six and twelve hours. They should always prompt immediate examination and culture of the blood for a causative organism and the administration of intravenous fluids and antibiotics. Occasionally the temperature response remains below 100°F (37.8°C). This type of presentation is particularly dangerous, for its onset may not be recognized until severe endotoxic shock has developed, yet it is likely to be a manifestation of severe septicaemia.

The prognosis in Gram-negative septicaemia worsens with age and the degree of associated liver damage, whether due to cirrhosis or prolonged jaundice. It also varies somewhat with the types of causative bacteria, rising from about 50 per cent, when *E. coli* is the main organism, to about 75 per cent with *Proteus* or *Pseudomonas* [21]. When Gram-negative septicaemia is accompanied by shock sufficient to produce a drop in arterial systolic blood pressure to below 70 mmHg, which happens in 30–40 per cent of cases, the mortality rises to between 80 and 95 per cent for the same organisms. There is then also a serious risk of concomitant acute renal failure and tubular necrosis.

Some writers have suggested that the examination should not be undertaken if there is a history of previous cholangitis due to the likelihood of infection, but each case must be judged on its merits. The anatomical information provided by a successful examination can prove of such value to the surgeon, and lessen the operative risks for the patient to such an extent, that this may outweigh the risk of infection. This is especially the case when the cause of jaundice is in doubt and a traumatic stricture is suspected; or when details of previous surgery on the duct system are not available. Similar considerations apply when liver biopsy and biochemical tests have proved equivocal and the diagnosis rests between non-opaque stones, neoplasm or intrahepatic cholestasis. However, whenever there is a likelihood that the bile is infected, the dangers should always be recognized and fully weighed against the advantages of a successful examination.

The risk of septicaemia seems to be very small if the patient proceeds straight to laparotomy following the examination: once an obstruction is relieved or bypassed, the duct pressure falls and the pressure gradient from vein to duct is re-established. On the other hand, the danger is very real if an obstructed duct system is demonstrated and the ducts are not decompressed, either by immediate surgery or, by the percutaneous introduction into an intrahepatic duct of a cannula of sufficient calibre to allow the free flow of bile.

As an added precaution, whenever there is a possibility that the bile may be infected it is advisable to prescribe preliminary *prophylactic antibiotics* such as tetracycline or gentamycin. Admittedly, in obstructive jaundice or severe liver disease, there is little if any hepatic excretion of these drugs in the bile—though there may be some if the obstruction is intermittent as in some cases due to stones. There is, however, always the risk that if bile is infected, the blood may be suddenly flooded with organisms during or shortly after the procedure, and it would then be reasonable to expect that there may be an advantage if there is already an effective blood level of an antibiotic, or group of antibiotics, to which the likely Gram-negative organisms may be sensitive.

Once a duct is cannulated it should be routine to aspirate bile for immediate bacterial examination and culture so that, if necessary, appropriate antibiotics can be administered without delay.

It is very unusual for septicaemia to develop after laparotomy once duct obstruction has been relieved. If it does occur, consideration should be given to injecting an appropriate antibiotic through a T-tube, if one has been inserted, or to reintroducing a needle or cannula percutaneously into a duct for the same purpose. It has been suggested that antibiotics should be injected through the cannula into the duct at the end of each examination [22]. This might be worth a trial in selected cases in which bile is likely to be infected, though again the choice of selecting the most effective antibiotics is difficult in the absence of information regarding specific organism sensitivity. We have found that, by always leaving the cannula in place to drain the ducts after the examination is concluded and proceeding directly to surgery, there is little risk of serious infection and we have not had a case of Gram-negative septicaemia in over 600 cases performed over a period of fifteen years.

Okuda [9] recommended a technique developed from one originally described by Arner, Hagberg & Seldinger [23] and Wiechel [8] whereby with patient supine, a *fine needle* 15 cm long and of 0.7 mm diameter is introduced horizontally through one of the lower right intercostal spaces into the right lobe of the liver. Contrast medium is then injected under visual control in small quantities as the needle is withdrawn from the liver until filling of a bile duct is noticed. Once this happens, further contrast fluid is injected to demonstrate the duct system as well as possible. With this technique a number of patients experience pain, either local or referred to the right shoulder, as the initial injections of opaque medium are being made, especially it seems when they involve the portal tract. The degree of pain seems to bear some relation to the amount of contrast injected but is rarely severe enough to interrupt the procedure. Complications with this technique are claimed to be uncommon, and it is asserted that the risk of leakage is so small that laparotomy can be safely delayed even when an obstructed duct system has been shown. This is not our experience, and a statement we consider to be unjustified and dangerous. There is no doubt that the technique is an excellent and effective one, for very small ducts may be found and a normal or contracted biliary tree can be demonstrated more often than when a needle/polythene cannula combination is used. If a patent and unobstructed duct system has been shown, the risk of bile leakage after removing the needle is almost non-existent. In the presence of dilated obstructed intrahepatic ducts however, with bile under increased pressure, there is always a risk of leakage along the puncture tract, fine though this may be; and if surgical decompression is delayed there is a far greater risk of bile peritonitis than when a pliable tube is left in to drain the ducts. The risk of Gram-negative septicaemia with this method is also greater than when using the needle/polythene cannula method: contrast medium injected into ducts already distended with bile under increased pressure may lead to infected bile being forced retrogradely into liver sinusoids and into the blood stream as intraductal pressure rises still higher, for it is impossible to aspirate sufficient bile through the needle to lower pressure before the injection of contrast is made. There is also no means of draining and decompressing the ducts after the investigation, unless a polythene cannula is deliberately inserted into a dilated intrahepatic duct for this purpose at the end of the procedure.

It is our practice, and one we strongly recommend, that if duct obstruction is shown and it is decided to defer surgery, to take a delayed film of the upper abdomen one to two hours after the needle has been withdrawn. If this shows contrast medium remaining in dilated ducts it indicates that surgical drainage should be undertaken as soon as possible.

ENDOSCOPIC RETROGRADE CHOLANGIO-PANCREATOGRAPHY (ERCP)

The risks and complications of this procedure are closely related to the experience of the operator and the careful selection of patients, but complications may occur and may even result in death. The complications fall under three main categories:
1 Those connected with the passage of the duodenoscope down the oesophagus and through the stomach and first part of the duodenum.
2 Complications that may follow the injection of the opaque medium through the papilla into the bile or pancreatic ducts.
3 Pulmonary complications which may result from aspiration of infected material from the nasopharynx or upper gastrointestinal tract during or following the examination while the pharynx, larynx or patient is anaesthetized.

The majority of complications are due to too forcible injection into the common bile or pancreatic ducts especially when these are involved in some **pathological process**.

1 *Complications during the passage of the instrument*

(*a*) *Perforation of the oesophagus.* This danger usually only arises when there is unsuspected disease of the oesophagus, such as an ulcer, stricture, diverticulum or neoplasm; or when there is marked kyphosis or scoliosis.

(*b*) *Rupture of oesophageal varices* should be a risk only when unsuspected.

(*c*) *Intubation of the bronchial tree.* This is only likely to occur if the procedure is performed under general anaesthetic or if the patient is given anaesthetic lozenges to suck beforehand. The latter practice should be avoided and if possible the examination should be performed with the patient sedated by sodium amytal 100–200 mg one hour before, and diazepam 10–20 mg intravenously is usually given shortly before the scope is introduced except in chronic cirrhosis or bronchitis. Atropine should be avoided as it renders the throat more prone to trauma with these provisos. Misdirection causes immediate cough reaction and the scope is withdrawn before any damage can be done to a smaller bronchus.

Preliminary PA and lateral films of the chest should be taken routinely to show the state of the lungs, the configuration of the spine and any abnormal dilation or position of the aorta: followed by a barium examination of the upper gastrointestinal tract to show the state of the oesophagus, stomach and pyloric canal. The demonstration of pyloric obstruction sufficient to prevent the passage of the scope into the duodenum, or the demonstration of actual anatomy when details of a previous surgical procedure are not fully available, or when normal anatomy is distorted, may spare the operator much time or frustration, and the patient an uncomfortable, and possibly expensive and fruitless examination.

2 *Complications following the injection of contrast medium into the pancreatic and bile ducts*

(*a*) *Acute pancreatitis.* This is probably the commonest of the more serious complications which may follow the procedure. Almost invariably, the serum amylase level is elevated after an ERCP examination, especially if the pancreatic duct has been opacified. If an injection is made too forcibly into the pancreatic duct, or an unnecessarily large amount is injected, there is always the risk of subsequent acute pancreatitis. The risk is increased if the pancreas or ducts are diseased, and especially high if there is a pancreatic pseudocyst. The risk can be minimized by injecting the smallest volume of contrast necessary to gain the information required, and by keeping the injection pressure as low as possible. The danger is possibly greatest when an attempt is made to force opaque medium through a stricture or past a stone in the common bile duct. In these circumstances, if the bile and pancreatic ducts join close above the tip of the injecting cannula, contrast may inadvertently be forced up the pancreatic duct, in a quantity larger than safe. If, during injection of the pancreatic duct, the appearances raise the possibility of a pseudo-pancreatic cyst, the injection should be stopped at once. Should the contrast fail to drain freely from the ducts after the injection is terminated, the risk of pancreatitis is further increased and appropriate safeguards should be taken.

(*b*) *Rupture of the bile duct.* Occasionally the bile duct may be ruptured if the injection is made under excessive pressure to try to demonstrate the duct above a complete, or nearly complete, obstruction. If this complication is not recognized at the time, and followed by immediate surgery to repair the duct or bypass the obstruction, the risk of bile peritonitis is high; especially if stones have been shown or found previously in the ducts, or if previous duct surgery has been performed. Deaths have occurred although not all have been reported in the literature.

(*c*) *Septicaemia.* The examination may at times be followed by Gram-negative septicaemia. This complication has already been discussed in the section dealing with percutaneous cholangiography, and it seems likely that the danger is even greater following ERCP. The reason for this is probably due to the fact that, whereas bile is rarely infected *above* an obstruction by a carcinoma of the duct or head of pancreas, the duodenum always contains organisms which may be introduced into the duct by the cannula. Bacteraemia following the procedure is not uncommon and septicaemia is therefore always a risk, especially if opaque medium is forced past an obstruction into dilated ducts which then fail to drain freely after the injection is stopped. A delayed film of the right upper abdomen should always be taken after the instrument has been removed and before the patient is returned to bed, to make sure that adequate drainage is taking place. Elias, Summerfield & Dick [24] recommend the routine administration of antibiotics following an examination in which partial duct obstructions has been demonstrated. This would seem a wise precaution, especially if stones have been found in the duct, even though bacterial sensitivity to the antibiotics prescribed cannot be accurately predicted. Liver excretion is seriously impaired in severe obstructive jaundice and there is therefore little chance of sterilizing bile already infected in obstructed ducts. Nevertheless an effective antibiotic might prevent septicaemia should the circulation be flooded with sensitive organisms. It is possible that the injection of an antibiotic, together with contrast, might prevent bile from becoming infected should the organisms be sensitive to it—though this is unlikely ever to be provable. Urgent surgery is advisable whenever a partial obstruction has been shown and the delayed film indicates inadequate drainage of proximal intrahepatic ducts. If for any reason it is decided to delay surgery, serious consideration should be given to the insertion of a catheter percutaneously into a dilated intrahepatic duct to decompress and drain the obstructed ducts till surgery is undertaken.

(*d*) *Damage to the duodenal wall.* An inexperienced operator may inject into a 'false' ampulla or where he assumes the ampulla to be, and a dissection of the duodenal wall results. This should be easily recognized on the screen and serious consequences rarely result.

3 Pulmonary complications

The risk of aspiration pneumonitis following ERCP is small, especially as it is rarely performed under general anaesthesia. In all cases, patients should take no food or fluids by mouth for six hours beforehand. Cough and elevation of temperature following the examination should suggest the possibility of pulmonary infection and the availability of a preliminary chest film for comparison may make a valuable contribution to the diagnosis.

SPLENO-PORTOGRAPHY

Reactions to the intraveneous injection of various escape media have been discussed elsewhere in Chapter 1.

Bleeding

Of the complications related directly to splenic puncture the most serious is bleeding, but with good technique and patient selection this risk can be greatly diminished. The examination should not be performed on patients with bleeding tendencies, or it should be delayed in those in whom the bleeding or prothrombin time is increased, until this has been corrected. It should be avoided, when possible, in the presence of ascites or shortly after abdominal surgery when there is still air in the peritoneal cavity. Bleeding may occasionally be severe in patients suffering from primary biliary cirrhosis and Professor Sheila Sherlock prefers arteriography as an alternative method of demonstrating the portal vein in these cases.

The anatomical position and size of the spleen should be determined as accurately as possible, and the needle should be inserted slowly and steadily into it whilst the patient holds his breath. If a needle only is used, breathing should be restricted to very shallow respiration while the needle is in the spleen. Breathing should be stopped during any manipulations to connect or detach the needle from pressure-recording instruments or syringes.

Damage to the spleen is usually the result of imprecise location of the needle tip. It is therefore advisable to check the position of the needle on the screen by first injecting a small amount of opaque medium—diluted with local anaesthetic—using hand pressure. The full injection should only proceed when one is satisfied that the needle is correctly sited. Otherwise, large amounts of irritating opaque fluid may be injected in undesirable situations. This may then result in severe pain and uncontrollable patient movement, sufficient to damage the spleen, while it is still transfixed by the needle. A conscious patient must be told, and must fully understand, that he must hold his breath while the contrast is being injected even though he may feel some pain or a sudden sensation of warmth. The needle should be removed immediately the injection is completed.

The risk of splenic damage is reduced if the injection is made through a polythene or Teflon catheter, introduced as in percutaneous cholangiography, using the needle solely as an introducer. Objections have been made to this technique on the grounds that a polythene tube is more likely to be extruded by recoil during the injection than a needle. This is largely a theoretical objection, especially if one or two sideholes are made in the catheter close to its tip, and if the position is checked by a test dose before the main injection is made.

A few ml of blood or some clots are commonly found in the peritoneal cavity whenever surgery is performed shortly after the examination, and small subcapsular haematomas are not infrequently noticed. It is also not uncommon for the injection of contrast fluid to cause small haemorrhagic foci or areas of necrosis within the spleen.

Rupture of the spleen is the most serious complication and one to be feared. Fortunately it is uncommon. Figures of between 0.5 and 1.0 per cent are quoted in the literature but the risk should be minimal if proper care is taken. Rupture usually results in severe blood loss. Although bleeding often ceases as the blood pressure falls, it usually recurs once it rises again. Acute rupture of the spleen is an emergency requiring immediate blood transfusion and, not uncommonly, splenectomy. Severe bleeding may occur without frank rupture of the spleen, and the risk is increased if the examination is performed in the presence of marked ascites, shortly after surgery, or after the induction of a pneumoperitoneum. At times, bleeding may be slow and lead to considerable blood loss into the peritoneal cavity before it is detected.

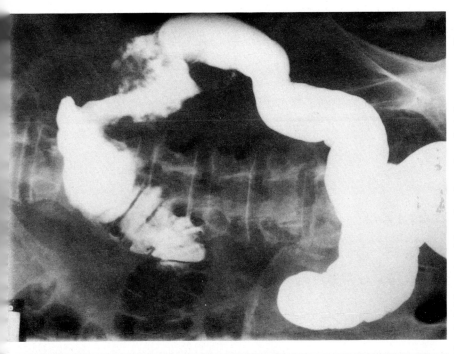
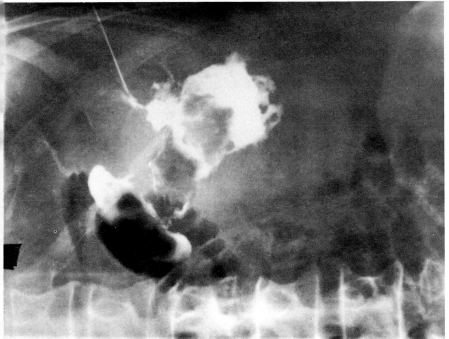

FIGURE 10.3a (left). Spleno-venography. Patient with large liver. Mass in left hypochondrium thought to be splenic but was actually due to carcinoma of the colon. 10.3b (right). Confirmation by barium enema.

On these occasions it may present with gradually increasing anaemia and a steady drop in blood pressure.

When blood collects in the flank it is often associated with pain, tenderness and guarding, but this may be absent in the presence of ascites. It is recommended that, in the presence of ascites, the examination should be performed only when the information likely to be obtained justifies the risk. In these cases as much fluid as possible should be aspirated beforehand if diuretics do not produce the desired result.

After spleno-venography, the patient should always be carefully watched for signs of bleedings, and the pulse recorded every fifteen minutes for the first two or three hours, and then half-hourly up to 48 hours.

If surgery follows shortly after spleno-portography, or if the examination is carried out during surgery, especially where the spleen has been manipulated during the operation, a very careful watch for bleeding must always be kept after the patient is returned to the ward.

Pain

It is not uncommon for patients to complain of pain in the left shoulder or scapular region immediately following the injection and the patient should always be forewarned that this may occur and advised to refrain from sudden movements which may cause him harm. Pain of this kind is rarely severe and usually resolves spontaneously, although spraying the skin over the affected area with ethyl chloride often helps to relieve it. Pain in the shoulder usually indicates leakage of opaque medium under the splenic capsule or into the lienorenal recess.

Pain in the left hypochondrium or flank usually indicates an intraperitoneal injection. This may be so severe as to require the intravenous injection of Fortral or pethidine though it usually passes off gradually without treatment in five to ten minutes. Pain in the left flank which persists—or gradually increases in intensity—or which comes on after an interval following the injection—should alert the operator to the possibility of bleeding. This may be associated with tenderness on palpation, flank dullness to percussion, and muscle guarding.

Pain may at times be more central in location and radiate towards the hilum of the liver. This seems to occur only when the portal vein is obstructed and it may possibly be due to a distension phenomenon.

Injection of opaque medium into organs other than the spleen

Most operators have had experience of this, especially when the spleen is not greatly enlarged and has not been previously demonstrated by isotope scanning. The organs most often involved are the left kidney and left lobe of the liver, though occasionally other organs may be mistaken for the spleen (Fig. 10.3a & b). At times, contrast may be injected into the lung (Fig. 10.4), pleural space or pericardium. A preliminary test injection of a few ml of

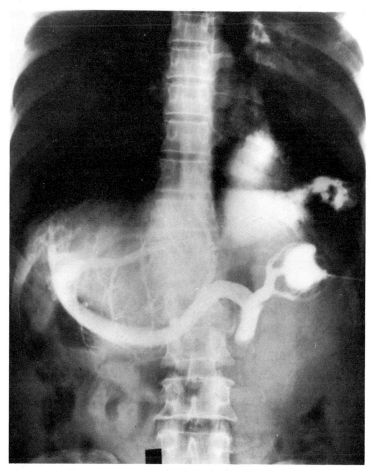

FIGURE 10.4. Spleno-portography. Initial injection of contrast medium into left lung.

contrast should always be made under screen control before the full injection is made and this should reveal any mis-siting of the needle. The results of pulmonary, intrapericardial and hepatic injections have already been discussed in the section relating to percutaneous cholangiography. If a small preliminary injection is mis-sited it rarely produces more than transitory pain and discomfort, but larger quantities may cause intense pain and dyspnoea and result in pulmonary consolidation or pleural effusion. More often, however, the water-soluble medium is rapidly absorbed leaving no trace of the incident.

Although often dramatic, extremely unpleasant and frightening for the patient, and even for the operator, no deaths have as yet been reported from these misadventures.

REFERENCES

1 SHALDON S., BARBER K.M. & YOUNG W.B. (1962) Percutaneous transhepatic cholangiography: a modified technique. *Gastroenterology* **42**, 371–379.
2 SELDINGER S.I. (1966) Percutaneous transhepatic cholangiography. *Acta. Radiol.* Suppl. 253.
3 DE MASI C.J., AKDAMAR K., SPARKS R.D. & HUNTER F.M. (1967) Puncture of the gall-bladder during percutaneous cholangiography. *J. Am. med. Ass.* **201**, 225–228.
4 ISLEY J. & SCHAUBLE J.F. (1962) Interpretation of the percutaneous transhepatic cholangiogram. *Am. J. Roentg.* **88**, 772–777.
5 ZINBERG S.S., BERK J.E. & PLASENCIA H. (1965) Percutaneous transhepatic cholangiography. *Am. J. Digest. Dis.* **10**, 154–169.
6 LEE W.Y. (1942) Evaluation of peritoneoscopy in intra-abdominal diagnosis. *Rev. Gastroenterology* **9**, 133.
7 JAMES M. (1971) Normal or 'negative' percutaneous cholangiogram. *Arch. Surg.* **103**, 31–33.
8 WEICHEL K.L. (1964) Percutaneous transhepatic cholangiography technique and application. *Acta Chir. Scand.* Suppl. 330.
9 OKUDA K. (1974) Non-surgical percutaneous transhepatic cholangiography. Diagnostic significance in medical problems of the liver. *Digestive Diseases* **19**, 21–35.
10 GLENN F., MUJAHED Z. & THORBJARNARSON B. (1967) Percutaneous transhepatic cholangiography. *Ann. Surg.* **165**, 33–40.
11 GLENN F., EVANS J.A., MUJAHED Z. & THORBJARNARSON B. (1962) Percutaneous transhepatic cholangiography. *Ann. Surg.* **156**, 451–462.
12 TYLEN V.G. (1975) Paper read at Symposium on 'Biliary Tract'. Bordeaux.
13 MACHADO A.L. (1971) Percutaneous transhepatic cholangiography. *Br. J. Surg.* **58**, 616–624.
14 TURNER F.W. & COSTOPOULOS L.B. (1968) Percutaneous transhepatic cholangiography. A study of 115 cases. *Canad. Med. Ass. J.* **99**, 513–521.
15 ELKELES G. & MIRIZZI P.L. (1942) Study of bacteriology of common bile duct. *Ann. Surg.* **116**, 360.
16 SCOTT A.J. & KHAN G.A. (1967) Origin of bacteria in the bile duct bile. *Lancet* **2**, 790–792.
17 KEIGHLEY M.R.B. & GRAHAM N.G. (1973) Infective cholecystitis. *J. Roy. Coll Surg. Edin.* **18**, 213–220.
18 MUSGROVE J.E., GRINDLAY J.H. & KARLSON A.G. (1952) Intestinal-biliary reflux after anastomosis of common duct to duodenum or jejunum. An experimental study. *Arch. Surg.* **64**, 579–589.

19 HULTBORN A., JACOBSON B. & ROSENGREN B. (1962) Cholangiovenous reflux during cholangiography. An experimental and clinical study. *Acta Chir. Scand.* **123**, 111.
20 SAUNDERS W. (1795) A treatise on the structure economy and diseases of the liver.
21 ALTEMEIER W.A., TODD J.C. & INGE W.W. (1967) Gram-negative septicaemia: a growing threat. *Ann. Surg.* **166**, 530–542.
22 FELLICI U. (1962) Colangiografia transhepatica percutanea. *Minerva Med. (Torino)* **53**, 858.
23 ARNER O., HAGBERG S. & SELDINGER S.I. (1961) Transhepatick cholgrafi. *Nord. Med.* **65**, 730.
24 ELIAS E., SUMMERFIELD J.A., DICK R. & SHERLOCK S. (1974) Endoscopic retrograde cholangio-pancreatography in the diagnosis of jaundice associated with ulcerative colitis. *Gastroenterology* **67**, 907–911.

CHAPTER 11: ORAL AND INTRAVENOUS CHOLEGRAPHY

G. ANSELL

ORAL CHOLECYSTOGRAPHY

Oral cholecystographic agents are lipid-soluble iodinated organic acids and are absorbed by passive diffusion across the mucosa of the gastrointestinal tract. In the liver, they are conjugated with glucuronic acid to form water-soluble glucuronides. In this form they are excreted mainly in the bile, but back-diffusion of the conjugate also occurs from the liver cell into the blood and approximately 35 per cent of the dose is excreted by the kidneys. There is negligible re-absorption of the contrast medium–glucuronide from the bowel and elimination takes place in the faeces. Factors influencing absorption and excretion of oral cholecystographic media have recently been reviewed by Berk et al. [1]. Iopanoic acid (Telepaque) is probably still the commonest medium in current use. Other commonly used media include ipodate (Biloptin, Oragrafin), tyropanoate (Bilopaque) and iocetamic acid (Cholebrin). Bunamiodyl (Orabilix) has been withdrawn from the North American and U.K. markets, because of its implication in renal failure, but it still appears to be in use on the continent of Europe.

Considering the widespread use of these media, major complications are relatively uncommon but approximately 50 per cent of patients appear to suffer mild or moderate side effects such as nausea, with or without vomiting, diarrhoea, or headache [2]. Other side effects noted included abdominal pain, dysuria, dizziness and urticaria [3, 4]. While there are variations in the reported incidence of side effects in the different studies, there appears to be no significant difference in the individual media, apart from the incidence of diarrhoea. In each of the studies, iopanoic acid is reported to cause a higher incidence of diarrhoea in comparison with the other media, though the actual incidence of diarrhoea quoted for iopanoic acid varied from 15 per cent [2] to 30 per cent [5]. Occasionally, the diarrhoea may be severe and prostrating, especially in elderly patients. Diarrhoea and vomiting may sometimes also be due to the Prosparol fatty meal [6]. Rarely, oral cholecystography may cause severe hypotensive collapse with peripheral vasodilatation [7]. In five men with pre-existing heart disease, an acute episode of coronary infarction or cardiac arrest occurred within a few hours of ingestion of a cholecystopaque [8, 6]. In these cases the question of coincidence cannot be excluded. Nevertheless, it appears possible that there could be an increased risk in such patients. Occasionally, oral cholecystography may be followed by erythematous or urticarial skin rashes or even by angioneurotic oedema. The rash usually occurs within a few hours after ingestion of the medium but may persist for up to two weeks [6, 7]. Experimental work has shown that iopanoic acid causes inhibition of acetylcholinesterase. This enzyme normally metabolizes acetylcholine and it has been suggested that an excess cholinergic effect, resulting from 'enzyme-inhibition', may be a factor in many of the toxic features of oral cholecystographic media, such as diarrhoea, skin or cardiovascular changes [9].

Renal function

Oral cholecystography frequently causes minor elevation of the serum creatinine of 0.1–0.2 mg per cent. This occurred in 46 per cent of patients receiving iopanoic acid and in 38 per cent of patients receiving ipodate: in one patient with renal calculi, the serum creatinine increased from 1.3 mg per cent to 4.2 mg per cent. With a second dose of contrast medium there may be a further increase in the creatinine of 0.1–0.3 mg per cent. Usually, the levels revert to normal within one week [5]. Bunamiodyl appeared to have a more severe effect on renal function and caused significant transitory diminution in the creatinine clearance [10]. There have been at least 80 cases of renal failure following oral cholecystography reported in the literature up to 1972 [11]. In the great majority of these cases the medium concerned was bunamiodyl, but other oral media such as iopanoic acid, ipodate, tyropanoate and pheniodol have also been incriminated. In a number of cases, intravenous iodipamide had been administered in addition to the oral medium. One of the most important of the suspected aetiological factors was *impaired liver function*, such as cirrhosis, with or without jaundice. Decreased biliary excretion of contrast medium in these cases results in an increased excretory load on the kidneys. In the presence of severe liver damage, there may also be some impairment of glucuronide conjugation so that a relatively insoluble form of the contrast medium might be presented to the kidneys. The other factor of major importance was the use of *repeat, double or multiple, doses* in an attempt to force excretion when the gall bladder was not visualized with a single dose of medium. In many cases, these two factors were interacting with nonvisualization of the gall bladder being primarily due to liver disease. *Oral cholecystography is both ineffective and contraindicated in the presence of jaundice.* If liver disease is suspected, even in the absence of jaundice, care should be taken in the use of oral cholecystographic media. Although most cases of renal failure have been associated with an excess dose of the medium, a fatal case has occurred in an elderly patient after only 2 g (four tablets) of iopanoic acid [7]. Transient acute renal failure has also occurred in a patient receiving 3 g of sodium ipodate [12] and in another patient, following two doses of 3 g iopanoic acid [13]. This latter patient had also received tetracycline which may have contributed to the nephrotoxic effect. The histological changes are those of diffuse tubular necrosis. Birefringent green–brown calcium containing crystals may also be present in the kidneys, but the full composition of these crystals remains unknown [14]. In one patient, an intravenous urogram performed on the day after a cholecystogram with 6 g bunamiodyl resulted in bilateral persistent nephrograms with oliguria for 24 hours [15]. It has been suggested that *hypotension*, due to the labile blood pressure in patients with liver disease, may be an aetiological factor [16], but this does not appear to be a major cause in most of the recorded cases [11].

Mudge [17] has shown that cholecystographic media have a powerful *uricosuric action* analogous to that of probenecid, which is used in the treatment of gout. He also pointed out that whereas probenecid was given in divided doses throughout the day, with a liberal fluid intake to prevent uric acid crystalluria, oral cholecystographic media were given at night, when fluid output is normally decreased. Moreover, the manufacturers' literature formerly advocated restricting fluids. Under these conditions a uric acid obstructive nephropathy was a possibility. The uric acid hypothesis still remains unproven as a cause of nephrotoxicity in these cases. Nevertheless, it is important to ensure that patients undergoing cholecystography receive a liberal fluid intake. Aspirin reduces the uricosuric action of cholecystographic media and may possibly be of value in certain clinical circumstances such as hyperuricaemia, if cholecystography is required [18]. On the other hand, salicylate might possibly also interfere with excretion of cholegraphic media by the liver. The uricosuric property of the oral cholecystographic agents also leads to a striking reduction in the serum urate concentration that often persists for five to six days. This is of clinical importance since a diagnostic test of serum urate is of little value for a period of several days after a cholecystogram. In addition, the wide fluctuation in serum urate concentration after cholecystography may tend to precipitate an acute attack of gouty arthritis in patients with a history of gout [19].

In the absence of liver disease the prognosis for renal failure following oral cholecystography is relatively good. However, when there is associated liver disease the renal failure is often complicated by rapidly progressive liver failure [11]. In the majority of the reported cases of renal failure following cholecystography pre-existing renal disease did not appear to be a significant aetiological factor [20]. Nevertheless, the presence of renal disease will diminish the renal reserve and it is reasonable to take this factor into account if oral cholecystography is proposed.

Pseudoalbuminuria may occur in from 12 to 18 per cent of patients following an oral cholecystogram [5]. There is a false positive test for protein when either the sulphosalicylic acid test, nitric acid ring test, or the heat and acetic acid test is used, but the Albutest and Albustix tests are not affected [21]. Pseudoalbuminuria is of no clinical significance but, in a few patients, epithelial casts and RBCs were noted in the urine [5].

Thyroid function

Oral cholecystographic media are excreted more slowly than urographic media and have a longer half-life. This causes a marked increase in the total and protein-bound iodine levels in the blood, and also in the blood level of butanol-extractable iodine. At 30 days the total and protein-bound levels of iodine are still grossly elevated but the butanol-extractable fraction, although elevated, usually lies within physiological limits. Iodine levels may not return to their pre-cholecystogram values for as long as three months [22].

Recent clinical and experimental studies suggest that liberation of iodine from cholecystographic media may actually produce minor alterations of *thyroid function* for up to three months after administration, causing a slight degree of thyroid suppression in euthyroid individuals [23], and increased hormonal synthesis in thyroid adenoma [24]. Two cases of frank thyrotoxicosis have also been reported following oral cholecystography. In both of these cases, symptoms commenced a few days after taking sodium ipodate [25].

Iophenoxic acid (Teridax), an oral cholecystographic agent which is no longer in use, produced a unique effect in that it persisted in the blood for many years, causing markedly elevated blood iodine levels. Passage also occurred through the placenta so that persistently elevated protein-bound iodine levels might occur in an infant, many years after maternal ingestion of the medium [26]. A case of congenital hypothyroidism in an infant born five years after the mother had cholecystography with iophenoxic acid has also been reported [27].

Metabolic factors

When different substances are excreted by the same hepatic mechanism, they may compete if given together. Thus liver tests which depend on this type of excretion, e.g. the BSP test, may give abnormally high values if performed within 24 hours of an oral cholecystogram. For a similar reason there may be a transient slight elevation in the serum bilirubin following cholecystography [21]. The enzyme involved in this excretory mechanism is *glucuronyl-transferase* [1]. Transient slight hyperbilirubinaemia is probably of no significance under normal circumstances but in the first eight weeks of neonatal life there is a deficiency of glucuronyl-transferase in the infant's liver [28]. If cholecystographic media are administered to a lactating mother, excretion in the milk might therefore cause significant hyperbilirubinaemia in the infant. Some drugs, such as phenobarbitone, cause 'enzyme-induction' in the liver so that the amount of glucuronyl-transferase is actually increased. In the experimental animal, pretreatment with phenobarbitone significantly increases the biliary excretion of iopanoic acid. On the other hand, biliary iodipamide excretion is depressed by this form of treatment. This suggests that there is an important difference in the hepatic excretory mechanisms for iodipamide and iopanoic acid [29]. Problems related to competitive inhibition between oral and intravenous cholegraphic media are discussed in the section on intravenous cholangiography. Iopanoic acid is strongly bound to serum albumin [30] and this might in theory result in displacement of other protein-bound drugs and thereby increase their specific therapeutic activity. There are many theoretical possibilities for drug interactions with cholecystographic media in relation to hepatic excretion, protein-binding or enzyme-inhibition, but little attention has so far been given to this subject.

An abnormal appearance of the gall bladder has been reported when cholecystography was performed in a patient receiving cholestyramine. The gall bladder was collapsed and contained unexplained translucencies. Repeat cholecystography one week after discontinuing cholestyramine showed normal appearances [31]. Cholestyramine apparently binds to iopanoic acid preventing its absorption from the gastrointestinal tract. In addition, it causes bile salt depletion and this may also influence the absorption and biliary excretion of cholecystographic media.

Miscellaneous

Accidental overdosage has been reported after 60 tablets (30 g) of Telepaque were dispensed, apparently in error for 6 tablets. The patient consumed all the tablets and on the following morning had extreme nausea with severe diarrhoea and vomiting but no other ill-effects [32]. It appears, however, that no measurements were made in this patient to detect any renal or biochemical abnormalities.

A unique case has been described in which the mucosal folds of the stomach were rendered radio-opaque by deposition of iodine contrast material in the submucosa, apparently as a result of one or more cholecystograms performed in earlier years. The mechanism of this deposition was uncertain but since there were lymphatic cysts in the gastric mucosa it was suggested that there may have been abnormal lymphatic drainage from previous operative procedures on the biliary tract [33].

INTRAVENOUS CHOLANGIOGRAPHY

Idiosyncrasy reactions following the intravenous injection of iodipamide (Biligrafin, Cholografin) or ioglycamate (Biligram, Bilivistan) are, in many respects, similar to those occurring with urographic contrast media and the principles and treatment are similar. These features have been fully discussed in Chapter 1. However, the reactions following intravenous cholangiography are frequently more severe. In the U.K. national survey, the incidence of reactions following intravenous *injection* of iodipamide was: intermediate reactions, 1 in 700; severe reactions, 1 in 1600; and death, 1 in 5000. The incidence of severe reactions and death in this series was therefore approximately eight times that resulting from intravenous urography [34]. Ioglycamate appears to have a marginally lower toxicity than iodipamide [35] but, since the recent introduction of this medium to the U.K., there have been at least two cases of cardiac arrest, one of which was fatal [6]. The major immediate reactions following intravenous cholangiography are hypotensive collapse and bronchospasm, but other reactions include nausea, skin rashes, abdominal pain, diarrhoea and tetany. The hypotensive action of iodipamide is increased by rapid injection [36] and the injection should always be given slowly, taking a minimum of five to ten minutes. If pethidine is administered with iodipamide, a mild 'cardiac infarction-like syndrome' may occur due to spasm of the Sphincter of Oddi and the pain is relieved by intravenous propantheline [6].

In recent years, with the introduction of infusion cholangiography, the contrast medium can be administered over a period of fifteen to thirty minutes or longer. This allows the liver to excrete the cholangiographic medium more efficiently [37]. Although there are, as yet, no large-scale survey figures available for infusion cholangiography, it is probable that the incidence of reactions has been significantly decreased by this technique. However, isolated cases of severe hypotensive collapse, and cardiac arrest following bronchospasm have occurred during infusion techniques, sometimes after only a few ml of contrast medium have been administered [6]. A case of laryngeal stridor has also been reported in a patient receiving an infusion of 40 ml of 52 per cent iodipamide [38]. It is therefore important that patients should remain under constant observation and that full facilities for resuscitation should be immediately available in the event of a reaction.

The dosage of contrast medium used in infusion urography is still the subject of debate. In a large series of cases, Cooperman *et al.* [39] did not find that there was any significant advantage if the dose of 50 per cent iodipamide used in the infusion was increased from 20 ml to 40 ml. Nolan and Gibson [40] claimed that a dose of 1 ml/kg of 50 per cent iodipamide increased the density of the common bile duct and gall gladder shadows, and that the flow of contrast in the bile was prolonged. However, at a very high *total* dosage (above 70 ml of 50 per cent iodipamide) the results tended to deteriorate. There is a transport maximum for the excretion of biliary contrast media by the liver. Experimentally, in the dog, the maximum concentration of iodipamide in the bile showed only a *slight* increase at dose levels above 0.5 ml 50 per cent iodipamide/kg [41]. In the rhesus monkey, using methylglucamine ioglycamate, the maximum concentration of contrast medium in the bile was achieved at infusion rates of 10.5–26 mg per minute (0.012–0.03 ml/kg/min 17 per cent Biligram infusion). At higher dose levels there was actually a slight *decrease* in ioglycamate concentration in the bile [42].

There is therefore little justification for the use of high dose levels. Moreover, it is probable that the enhanced safety margins, achieved by slow infusion techniques, could be at least partly eroded as the dosage of iodipamide is increased. A pyelogram was visible in 105 out of 107 patients receiving an infusion of 40 ml of 52 per cent iodipamide [38]. This would suggest that even at this dosage, the blood level of contrast medium exceeded the excretory ability of the liver so that the excess was eliminated in the urine. Miller *et al.* [43] showed that there is a transport maximum for the excretion of contrast medium through the human liver and they have produced diagnostic cholangiograms using as little as 3–5 ml of 50 per cent methylglucamine ioglycamate in an infusion. The diagnostic value of low-dose infusion cholangiography in patients with normal liver function has been confirmed using doses of 10 ml of 50 per cent iodipamide [44]. However, in patients with a very dilated common bile duct, a dose of 20–40 ml of iodipamide might sometimes be required for adequate visualization of the duct system. A new intravenous cholangiographic agent, meglumine iodoxamate (Cholovue), is claimed to have only half the toxicity and twice the biliary transport maximum of iodipamide. In a preliminary study of eighteen patients with normal liver function, this compound produced impressive and detailed visualization of the biliary tract at doses of 10–20 ml of the 40 per cent solution [45].

Hepatic impairment

In jaundiced patients the results of conventional intravenous cholangiography are, more often than not, disappointing and duct visualization was only achieved in 9.3 per cent of cases when the serum bilirubin exceeded 4 mg per cent [46]. Fuchs *et al.* [47] used a prolonged infusion of 40 ml ioglycamate, administered over a period of twelve hours and achieved diagnostic visualization of the gall bladder or common duct in 7 out of 21 jaundiced patients. The highest serum bilirubin level at which the gall bladder was visualized was 11.1 mg per cent. In these cases, visualization of the biliary tract was more likely to occur when the jaundice was due to liver disease than when it was due to an obstructive lesion [47].

Finby & Blasberg [48] claimed that there was an increased incidence of toxic reactions when intravenous cholangiography was performed immediately after an oral cholecystogram. It was also suggested that the preliminary oral cholecystogram might exert a blocking action on the excretion of iodipamide. In their investigation, however, unusually large doses of oral and intravenous media were used. In the U.K. survey, an oral cholecystogram had been performed within 24 hours in three out of five deaths due to iodipamade. There also appeared to be a possible association between the prior administration of an oral cholecystogram and severe diarrhoea with collapse about four hours after the iodipamide injection [34]. Although there is no conclusive evidence of an increased incidence of reactions following combined oral and intravenous cholegraphy, it is now generally considered wiser for the examinations to be separated by an interval of several days. In animal experiments, when high doses of iopanoic acid or ipodate were administered *concurrently* with iodipamide there was depression of iodipamide excretion. However, the combined excretion of the 'oral' and intravenous media resulted in a *total* iodine concentration which was at least as high as that due to iodipamide alone and was sometimes higher. When iopanoic acid was administered on the preceding day, in a dose equivalent to that used clinically, there was apparently no impairment of iodipamide excretion [49]. This study indicates that combined oral and intravenous cholangiography probably should not adversely affect the degree of visualization of the biliary tract, but it does not exclude the possibility of increased toxicity from the combination of the two media, particularly if there is impairment of hepatic or renal function.

In diabetic patients treated with sulphonyl-ureas such as tolbutamide, there appears to be impairment of biliary excretion of ioglycamate with an increased incidence of side effects [50]. Pretreatment with phenobarbitone also decreases the biliary excretion of iodipamide [29]. It is possible that other drugs which are metabolized by the liver, or which interfere with protein-binding, might sometimes have similar effects. Lindgren et al. [51] noted that oral contraceptives appear to cause impaired excretion of intravenous cholangiographic media with an associated increase in the incidence of side effects. However, since oral contraceptives apparently cause a higher incidence of liver damage in Scandinavia, it is not yet known whether their effect on cholangiographic media is equally applicable in other population groups.

Liver pain may occasionally occur a few hours after intravenous cholangiography [52]. A hepatotoxic reaction has recently been reported following an infusion of 60 ml of iodipamide. Twenty-four hours later, the patient developed acute epigastric pain and nausea with marked elevation of the SGOT and LDH levels which had previously been normal. There was also some peri-orbital oedema which resolved rapidly. The liver enzyme levels gradually subsided over the next four days but they were still above normal when the patient was then discharged asymptomatic [53]. Hepatoxocity appears to be dose-related in that 8.7 per cent patients receiving 20 ml of iodipamide developed abnormal SGOT levels whilst 18.3 per cent developed abnormal SGOT levels after 40 ml of iodipamide [54]. This is another argument for using the lowest dose of iodipamide which will produce diagnostic results.

Renal failure

A few cases of renal failure have been described following iodipamide. In the two cases described by Crafts & Swales [55], one was a dehydrated elderly patient with mild obstructive jaundice and an unsuspected pyelonephritis. The other patient, who had only one kidney, had a laparotomy for a pancreatic cyst 24 hours after the cholangiography. There were mild hypotensive episodes in both of these patients following the injection of iodipamide. In another patient, who developed mild renal failure following an infusion containing 40 ml iodipamide, there was poor biliary excretion of the contrast medium and there was also a urinary infection which was treated by a sulphonamide [56]. Thus, in these three cases, there were one or more possible aggravating factors including dehydration, hypotension, operative trauma, decreased renal reserve and use of sulphonamide. In addition, impaired biliary excretion would have increased the diversion of contrast medium to the kidney. In another patient, who was severely jaundiced, renal failure resulted from a massive overdose of 200 ml iodipamide given in an infusion of 1000 ml 5 per cent dextrose in a fruitless attempt to demonstrate the common bile duct [57]. This case, in particular, illustrates the risk of illogical escalation of iodipamide dosage. Renal failure has also occurred in a patient with chronic hyperbilirubinaemia, believed to be due to the Dubin–Johnson syndrome, following the infusion of 40 ml of 50 per cent iodipamide [58]. Renal biopsy showed acute tubular necrosis. In this patient, there was also increase in the jaundice. Two of these five cases of renal failure required peritoneal dialysis. The other three cases responded to conservative management.

Iodipamide resembles the oral cholecystographic media in that it increases uric acid excretion [20]. Experimental work in animals also suggests that it has a direct nephrotoxic effect if injected into the renal artery, or even following rapid intravenous injection. Moreover, when iodipamide is injected rapidly in patients, there is increase in the size of the kidneys [59]. Ioglycamate appears to be less nephrotoxic than iodipamide experimentally [35]. Nevertheless, renal failure has also been reported following intravenous cholangiography with ioglycamate [60]. Iodipamide causes precipitation of Bence–Jones protein [61] and would therefore be a potential cause of renal failure in patients with myelomatosis.

Urinalysis with Clinitest tablets may give a false positive black copper reduction reaction simulating the appearance in alkaptonuria [62].

Effects on blood

A fatal reaction occurred in a patient with Waldenström's monoclonal IgM paraproteinaemia as a result of infusion cholangiography [63]. The patient collapsed after only 'a few drops' of methylglucamine ioglycamate (Biligram) had been administered intravenously, and at autopsy there was gel precipitation of the plasma. *In vitro* experiments confirmed that IgM paraproteins were precipitated by minute dilutions of ioglycamate. Methylglucamine iodipamide (Biligrafin) also caused some precipitation of IgM paraproteins, but this was less marked than in the case of ioglycamate. IgM paraproteinaemia should therefore be regarded as an absolute contraindication to the use of ioglycamate or iodipamide. *In vitro* testing of the sera of five patients with other types of paraproteinaemia (monoclonal IgG and mixed forms) produced no reaction with these media. Nevertheless, in view of the risk of renal damage discussed in the previous paragraph, biliary contrast media are probably inadvisable in all types of myelomatosis.

Iodipamide may cause decreased coagulability of the blood, probably due to interference with the protein factors responsible for the coagulation mechanisms [64]. In addition to its protein-binding action, iodipamide may cause partial inhibition of various enzyme systems such as acetylcholinesterase, beta-glucuronidase, lysozyme, alcohol-dehydrogenase and glucose-6-phosphate dehydrogenase [65].

As with other contrast media, iodipamide may cause clumping of the red cells. Haemolysis may also occur [65]. Scanning electron-microscopy showed that severe morphological changes may occur in the erythrocytes after only a few minutes' exposure to iodipamide [66].

Miscellaneous

Methylglucamine iodipamide should not be mixed in the same syringe with antihistamines since precipitation may occur [67].

Malarial relapse occurred in two patients following the administration of iodipamide [68]. The authors postulated that the relapses were provoked by the sudden high intrahepatic concentration of iodine. A more plausible explanation is that there was increased fragility of the red cells containing the malarial parasites, as a result of inhibition by the iodipamide of the enzyme glucose-6-phosphate dehydrogenase in the red cell membrane [65].

REFERENCES

1 BERK R.N., LOEB P.M., GOLDBERGER L.E. & SOKOLOFF J. (1974) Oral cholecystography with iopanoic acid. *New. Engl. J. Med.* **290**, 204–210.
2 STANLEY R.J., MELSON G.L., CUBILLO E. & HESKER A.E. (1974) A comparison of three cholecystographic agents. A double-blind study with and without a prior fatty meal. *Radiology* **112**, 513–517.
3 RUSSELL J.G. & FREDERICK P.R. (1974) Clinical comparison of tyopanoate sodium, ipodate sodium and iopanoic acid. *Radiology* **112**, 519–523.

4 PARKS R.E. (1974) Double-blind study of four oral cholecystographic preparations *Radiology* 112, 525–528.
5 TISHLER J.M. & GOLD R. (1969) A clinical trial of oral cholecystographic agents: telepaque, sodium oragrafin and calcium oragrafin. *J. Canad. Assoc. Radiol.* 20, 102–105.
6 Reports to National Radiological Survey.
7 Reports to Committee on Safety of Medicines, London.
8 LITTMAN D. & MARCUS F.I. (1958) Coronary insufficiency associated with oral administration of gall-bladder dye. *New. Engl. J. Med.* 258, 1248–1250.
9 LASSER E.C. & LANG J.H. (1966) Inhibition of acetylcholinesterase by some organic contrast media. A preliminary communication. *Invest. Radiol.* 1, 237–242.
10 WENNBERG J.E., OKUN R., HINMAN E.J. *et al.* (1963) Renal toxicity of oral cholescystographic media. *J. Am. med. Ass.* 186, 461–467.
11 GAUDET M., CORTET P., RIFLE G. & VILLAND J. (1974) L'insuffisance rénale aigue après cholécystographie orale. Revue générale à propos de deux observations. *Rev. Fr. Gastro-Enterol.* 102, 47–58.
12 DAYMOND T.J. (1971) Cholecystography and renal failure. *Lancet* 2, 549–550.
13 SCHIRO J.C. & RICCI J.A. (1971) Transient renal insufficiency secondary to iopanoic acid. *Penn. Med.* 721, 53–54.
14 SETTER J.G., MAHER J.F. & SCHREINER G.E. (1963) Acute renal failure following cholecystography. *J. Am. med. Ass.* 184, 102–110.
15 HARROW B.R. & SLOANE J.A. (1965) Acute renal failure following oral cholecystography. A unique nephrographic effect. *Am. J. Med. Sci.* 249, 26–35.
16 TEPLICK J.G., MYERSON R.M. & SANEN F.J. (1965) Acute renal failure following oral cholecystography. *Acta Radiol. Diagn.* 3, 353–369.
17 MUDGE G. (1971) Uricosuric action of cholecystographic agents. A possible factor in nephrotoxicity. *New. Engl. J. Med.* 284, 929–933.
18 POSTLETHWAITE A.E. & KELLY W.N. (1972) Radiocontrast agents and aspirin. *J. Am. med. Ass.* 219, 1479.
19 KELLY W.N. (1971) Uricosuria and X-ray contrast agents. *New. Engl. J. Med.* 284, 975–976.
20 MUDGE G.H. (1971) Cholescystography and renal failure. *Lancet* 2, 872.
21 SANEN F.J. (1962) Considerations of cholecystographic media. *Am. J. Roentg.* 88, 797–802.
22 JACOBSSON L. & SALTZMAN G.F. (1971) Effect of iodinated roentgenographic contrast media on butanol-extractable, protein-bound, and total iodine in the serum. *Acta Radiol. Diag.* 11, 310–320.
23 CONSTANTINESCU A., NEGOESCU I., DON M. & HELTIANU C. (1973) Effects of the administration of some indigenous radiologic contrast media upon the thyroid function. Clinical and experimental studies. *Rev. Roum. Endocr.* 10, 49–57.
24 MAHLSTEDT J. & JOSEPH K. (1973) Decompensation autonomer adenome der schilddrüse nach prolongierter jodzufuhr. *Dtsch. med. Wschr.* 98, 1748–1751.
25 FAIRHURST B.J. & NAQVI N. (1975) Hyperthyroidism after cholecystography. *Brit. med. J.* 3, 630.
26 SHAPIRO R. (1961) The effect of maternal ingestion of iophenoxic acid on the serum protein-bound iodine of the progeny. *New. Engl. J. Med.* 264, 378–381.
27 DE JONGE G.A. (1965) Vruchbeschadiging door farmaca. *Folia. Med. Neerl.* 8, 65.

28 BOYLAND R. & GOULDING R. (1974) *Modern Trends in Toxicology*, p. 154. Butterworth, London.
29 NELSON J.A., PEPPER W., GOLDBERG H.I. et al. (1973) Effect of phenobarbital on iodipamide and iopanoate bile excretion. *Invest. Radiol.* 8, 126–130.
30 LANG J.H. & LASSER E. (1967) Binding of roentgenographic media to serum albumin. *Invest. Radiol.* 2, 396–400.
31 NELSON J.A. (1974) Effect of cholestyramine on telepaque oral cholecystography. *Am. J. Roentg.* 122, 333–334.
32 HANKINS W.D. (1971) Brief communication: human tolerance of iopanoic acid (Telepaque). *Radiology* 101, 434.
33 BOGSCH A. (1953) Permanent deposition of iodine contrast medium in the wall of the stomach. *Acta Radiol.* 39, 219–224.
34 ANSELL G. (1970) Adverse reactions to contrast agents. Scope of problem. *Invest. Radiol.* 5, 374–384.
35 BRISMAR J., LINDGREN P. & SALTZMAN G.F. (1971) Ioglycamide (Bilivistan) as a contrast medium for intravenous cholecystography. A clinical and experimental investigation. *Acta Radiol. Diagn.* 11, 129–147.
36 SALTZMAN G.F. & SUNDSTRÖM K.-A. (1960) The influence of different contrast media for cholegraphy on blood pressure and pulse rate. *Acta Radiol.* 54, 353–364.
37 WHITNEY B.P. & BELL G.D. (1972) Simple bolus or slow infusion for intravenous cholangiography? Measurement of iodipamide (Biligrafin) excretion using a rhesus monkey model. *Br. J. Radiol.* 45, 891–895.
38 BORNHURST R.A., HEITZMAN E.R. & MCAFEE J.G. (1968) Double-dose drip-infusion cholangiography. An analysis of 107 consecutive cases. *J. Am. med. Ass.* 206, 1489–1494.
39 COOPERMAN L.R., ROSSITER S.B., REIMER G.W. & NG E. (1968) Infusion cholangiography. Thirteen years' experience with 1,600 cases. *Amer. J. Roentg.* 104, 880–883.
40 NOLAN D.J. & GIBSON M.J. (1970) Improvements in intravenous cholangiography. *Brit. J. Radiol.* 43, 652–657.
41 BENNESS G.T. (1971) Improvements in intravenous cholangiography. *Br. J. Radiol.* 44, 559.
42 WHITNEY B.P. & BELL C.D. (1976) Levels of ioglycamate (Biligram) in the bile of the rhesus monkey following intravenous infusion at different dose rates. *Br. J. Radiol.* 49, 118–122.
43 MILLER G., FUCHS W.A. & PREISIG R. (1969) Die infusioncholangiographie in physiologischer sicht. *Schweitz. Med. Wschr.* 99, 557–581.
44 ANSELL G. & FAUX P.A. (1973) Low-dose infusion cholangiography. *Clin. Radiol.* 24, 95–106.
45 SARGENT E.N., SCHULMAN A., MEYERS H.I. et al. (1975) A new contrast medium for cholangio-cholecystography; meglumine iodoxamate. *Am. J. Roentg.* 125, 251–258.
46 WISE R.E. (1966) Current concepts of intravenous cholangiography. *Rad. Clin. N. Amer.* 4, 521–523.
47 FUCHS W.A., PREISIG R. & BURGENER F. (1975) Die langzeit infusionscholangiographie bi patienten mit icterus. *Fortschr. Geb. Röntgenstr.* 122, 148–151.
48 FINBY N. & BLASBERG G. (1964) A note on the blocking of hepatic excretion during cholangiographic studies. *Gastroenterol.* 46, 276–277.
49 GOERGEN T., GOLDBERGER L.E. & BERK R.N. (1974) The combined use of oral chole-

cystopaque media and iodipamide. The effect on iodipamide and iodine excretion in the bile. *Radiology* 111, 543-546.
50 KLUMAIR J. & PFLANZER K. (1976) Der einfluss oraler antidiabetica (sulfonylharnstoffe) auf die ausscheidung intravenoeser gallenkontrastmittel (to be published).
51 LINDGREN P., SALTZMAN G.F. & ZEUCHNER E. (1974) Intravenous cholegraphy after peroral contraceptives. A preliminary report. *Acta Radiol. Diagn.* 15, 217-224.
52 SALTZMAN G.F. (1959) Side effects of biligrafin forte. *Acta Radiol.* 51, 121-127.
53 STILLMAN A.E. (1974) Hepatotoxic reaction to iodipamide meglumine injection. *J. Am. med. Ass.* 228, 1420-1421.
54 SCHOLZT F.J., JOHNSTON D.O. & WISE R.E. (1974) Hepatotoxicity in cholangiography. *J. Am. med. Ass.* 229, 1724.
55 CRAFT I.L. & SWALES J.D. (1967) Renal failure after cholangiography. *Br. Med. J.* 2, 736-738.
56 BROWN R.C. & COHEN W.N. (1973) Acute renal failure following cholangiography. *South. Med. J.* 66, 1142-1144.
57 MCEVOY J., MCGEOWN M.G. & KUMAR R. (1970) Renal failure after radiological contrast media. *Br. Med. J.* 4, 717-718.
58 SPRENT J., SPOONER R. & POWELL L.W. (1969) Acute renal failure complicating intravenous cholangiography in a patient with Dubin-Johnson syndrome. *Med. J. Austral.* 2, 446-448.
59 LINDGREN P., NORDENSTAM H. & SALTZMAN G.F. (1966) Effect of iodipamide on the kidneys. *Acta Radiol. Diag.* 4, 129-138.
60 MEIR W. & HOFFMAN H. (1971) Akutes nierenversagen nach intravenöser cholezystographie. *Munch. Med. Wschr.* 113, 652-655.
61 LASSER E.C., LANG J.H. & ZAWADZKI Z.A. (1966) Myeloma protein precipitation in urography. *J. Am. med. Ass.* 198, 945-947.
62 LEE S. & SCHOEN I. (1966) Black-copper reduction reaction simulating alcaptonuria—occurrence after intravenous urography. *New. Engl. J. Med.* 275, 266-267.
63 BAUER K., TRAGL K.H., BAUER G. et al. (1974) Intravasale denaturierung von plasmaproteinen bei einer IgM-paraproteinämie, ausgelöst durch ein intravenös verabreichtes lebergängiges röntgenkontrastmittel. *Wien. Klin. Wschr.* 86, 766-769.
64 MÜLLER K., HETTLER M. & GERHARD R. (1971) Blutgerinnungsverändenungen durch röntgenkontrastmittel. *Bruns. Bietr. Klin. Chir.* 219, 138-145.
65 LASSER E.C. & LANG J.H. (1970) Contrast-protein interactions. *Invest. Radiol.* 5, 446-451.
66 SCHIANTARELLI P., PERONI F., TIRONE P. & ROSATI G. (1973) Effects of iodinated contrast media on erythrocytes. I. Effects of canine erythrocytes on morphology. *Invest. Radiol.* 8, 199-204.
67 MARSHALL T.R., LING J.T., FOLLIS G. & RUSSELL M. (1965) Pharmacological incompatibility of contrast media with various drugs and agents. *Radiology* 84, 536-539.
68 CROSBY D.J. & STORM A.H. (1966) Malarial relapse induced by intravenous cholangiography. Report of two cases. *Arch. Intern. Med.* 118, 79-80.

CHAPTER 12: BRONCHOGRAPHY

G. ANSELL

Complications due to bronchography may be considered under three main headings:
1 Anaesthesia.
2 Contrast media.
3 Complications of technique.

LOCAL ANAESTHESIA

In a survey covering some 100 000 bronchograms, there were eighteen fatalities and the anaesthetic was considered to be responsible for nearly one half of these cases[1]. Local anaesthesia is probably the commonest individual cause of death during bronchography and, in the great majority of cases in the literature, death has been due to overdosage or excessively rapid absorption of the local anaesthetic. More often than not, these accidents have occurred when bronchography was being undertaken by a casual user with an inadequate appreciation of the hazards of local anaesthetics. As with all drugs, however, a small proportion of patients may be hyper-reactive and develop symptoms with a relatively low dose, so that full precautions for treatment of any reaction must always be available [2].

Toxic effects

At present, the most commonly used local anaesthetic is lignocaine (lidocaine, Xylocaine). In recent years there has been renewed interest in the pharmacology of this drug because of its use for the treatment of cardiac arrhythmias[3]. Since it is administered intravenously for this latter purpose, the doses used are lower than those employed for surface anaesthesia. If the arterial blood levels of lignocaine exceed 6 μg/ml, evidence of central nervous system toxicity is occasionally observed whilst with arterial blood levels above 10 μg/ml such toxicity is frequent [4]. The toxic effects of lignocaine and other local anaesthetics are complex and are mediated mainly through the central nervous system but there may also be secondary respiratory and cardiovascular involvement [5, 8]. When lignocaine is used for surface anaesthesia the toxic effects are dependent on the rate of absorption of the local anaesthetic and on its concentration in the blood (Fig. 12.1).

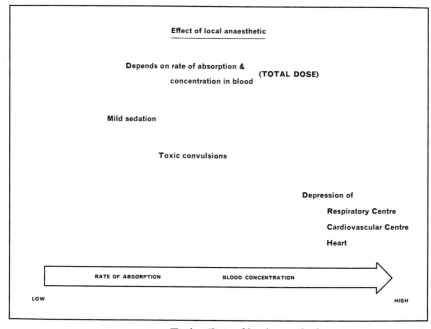

FIGURE 12.1. Toxic effects of local anaesthetics.

With low blood concentrations there is a mild sedative effect on the cerebral cortex but, as the concentration rises, this is reversed by a stimulating action, resulting in muscle twitchings which may progress to toxic *convulsions* with status epilepticus. With still higher blood concentrations of lignocaine, there is *depression of the respiratory and cardiovascular centres* and also direct depression of the heart by a quinidine-like effect. Hypotension, bradycardia, heart block or asystole may occur, or there may rarely be respiratory arrest. When these high blood concentrations are achieved rapidly, due to large doses, or excessively rapid absorption, cardiovascular and respiratory involvement may occur without a preceding convulsive phase.

In the early stages of toxicity there may be various *bizarre symptoms*, such as paraesthesiae, a sensation of cold, dizziness, sleepiness, a feeling of pressure on the chest, auditory disturbances, headache, speech disturbances, nausea, paleness, sweating, or a syndrome resembling alcoholic intoxication. If any of these symptoms occur they should be regarded as a warning of a possible impending, more serious, toxic reaction. The patient should be carefully observed and, depending on the clinical condition, an assessment made as to whether the examination should be temporarily interrupted or even abandoned.

Idiosyncrasy effects

Local anaesthetics may rarely cause hypersensitivity reactions of an allergic type, such as skin rashes, angioneurotic oedema, etc. It is important that these should not be confused with the toxic effects of overdosage, since the treatment is different. Until recently it was believed that lignocaine did not cause allergic reactions but a number of cases have now been recorded [9]. Apart from allergic type reactions, a rare idiosyncrasy effect after injection of local anaesthetic is a transitory rise in blood pressure. This is more likely to occur if sympathomimetic vasoconstrictors are added, e.g. in dental anaesthesia, and particularly in anaesthetics containing 1 in 25 000 noradrenaline. If a patient is also receiving tricyclic antidepressant drugs, there is a risk of a major hypertensive crisis with sympathomimetic agents. Where an added vasoconstrictor is required for local anaesthesia in such patients, felypressin appears to be the drug of choice [10]. Antidepressant drugs may also potentiate other drugs such as diazepam which may be used in the prevention or treatment of local anaesthetic convulsions. While these factors are mainly theoretical, it is nevertheless desirable that before investigations are undertaken in patients receiving antidepressant drugs, the various risks should be carefully evaluated in each individual patient.

Dose factors

The strength and dosage appropriate for different local anaesthetics varies. The user should therefore check the information on these before using an unfamiliar drug. In the case of lignocaine, although larger doses have been used, the manufacturers recommend that a dose of 200 mg (i.e. 10 ml of 2 per cent lignocaine or 5 ml of 4 per cent lignocaine) should not be exceeded in a normal adult. The maximum dose of local anaesthetic is dependent on the body weight of the patient and is considerably less for children (Table

12.1). Similarly, the dose should be reduced in wasted, frail, or elderly patients. Lignocaine is metabolized by the liver so that patients with liver disease are more susceptible to its toxic effects [11, 12]. Patients with congestive cardiac failure and low cardiac output may also develop toxic effects with smaller than normal doses of lignocaine [4]. When higher concentrations of local anaesthetic are used, the rate of absorption and the risk of systemic reactions is increased. For surface anaesthesia of the pharynx and larynx, there is no advantage in using a concentration higher than 4 per cent lignocaine. Concentrations above this level are no more effective as an anaesthetic and are more likely to cause toxic reactions [13]. Some authorities in fact use lower concentrations than this. When the local anaesthetic is administered

TABLE 12.1. Maximum permissible dose of lignocaine in relation to body weight.

Patient's body weight kg		10	20	30	40	50	60	70	80
Maximum dose lignocaine (ml)	1% Soln.	3	6	8.6	11.4	14	17	20	23
	2% Soln.	1.5	3	4.3	5.7	7	8.5	10	11.5
	4% Soln.	0.7	1.5	2.2	2.8	3.5	4.3	5	6.7

(Courtesy Astra Chemicals Ltd.)

directly into the tracheo-bronchial tree, there is a risk of more rapid absorption through the highly vascular pulmonary alveoli. The lignocaine should therefore be diluted for this phase of the examination. With endotracheal administration, the maximum blood concentration of local anaesthetic occurs within 5 to 25 minutes [14]. Excretion is relatively slow with a biological half-life of some two hours, so that at one hour, 73 per cent of the lignocaine may still be present in the body [6]. Administration of a second dose of local anaesthetic may therefore produce a higher peak concentration in the blood. Significant quantities may be absorbed after the oral administration of large amounts of lignocaine, with a peak blood concentration of 1 μg/ml being attained in 15 minutes [7, 12]. Patients are sometimes given a tablet of local anaesthetic to suck before the pharynx is sprayed. This increases the total absorbed dose of local anaesthetic and, even if different drugs are used in the tablet and in the spray, the doses are additive.

In a recent unpublished case of death due to local anaesthesia for bronchography [24], the patient was given two amethocaine lozenges (each containing 60 mg) to suck, after which a medical registrar sprayed the pharynx with a 10 per cent solution of lignocaine from a commercial metered aerosol. It was claimed that only 80 mg[1] of lignocaine had been used before the patient developed convulsions. Unfortunately, these were not treated with either barbiturate or diazepam and the patient died. In this case, there was also a possible history of previous epilepsy but the patient was not under any active treatment for this. The type of aerosol used for this patient was originally introduced for anaesthetization of the tympanic membrane but it has since been recommended for anaesthesia of the larynx. Examination of a similar model showed that the anaesthetic spray, emitted from the tip of the long nozzle, consisted of coarse droplets with an irregular distribution so that it would be difficult to achieve uniform anaesthesia with a small volume of anaesthetic. Moreover, the use of 10 per cent lignocaine significantly increases the risk of toxic absorption. This type of aerosol could therefore be expected to be hazardous in unskilled hands.

If a machine nebulizer is used, it is essential to control the dose of local anaesthetic administered and it is preferable that lower concentrations of the anaesthetic should be used. Wheeler [15] has claimed satisfactory anaesthesia using only 5 ml of 1 per cent lignocaine in a nebulizer.

Prilocaine (Citanest) is claimed to have only 60 per cent of the toxicity of lignocaine and the maximum permissible dose is therefore somewhat larger. However, prilocaine may occasionally cause mild cyanosis due to methaemoglobinaemia. *Diclonine hydrochloride* has been advocated for surface anaesthesia and has been claimed to be virtually devoid of adverse systemic reactions [13, 16] but so far, it does not appear to be widely used.

Prevention of reactions

Undoubtedly, the most important precaution is to *restrict the dose* of local anaesthetic used. The total volume to be used should always be measured out before commencing any examination and the maximum recommended dose should not be exceeded. In this connection the dose of local anaesthetic in any lozenge or lubricant must also be taken into account. If satisfactory anaesthesia has not been obtained within the limits of the maximum recommended dose, the technique has probably been faulty and the examination should be abandoned for another occasion. In surface anaesthesia the addition of adrenaline does not appear to be of any value in potentiating the action of the anaesthetic [13]. When the pharynx and larynx are being sprayed with local anaesthetic, the patient should be instructed to spit out any surplus liquid and not to swallow it, to avoid unnecessary absorption.

[1] There is some doubt about this dose.

The higher concentrations of lignocaine appear to be relatively more toxic due to their more rapid rate of absorption. This is of particular significance when local anaesthetic is injected into the trachea because of the danger of rapid absorption. For this reason, it is preferable to dilute the lignocaine to a 1 per cent concentration for endotracheal use. The larger volume of this dilute solution is also an advantage when an endotracheal catheter is used, since the dead space of the catheter will then be of less significance. For example, if a small volume of 4 per cent lignocaine is used and the patient coughs during the instillation, it may be impossible to decide how much of the anaesthetic has reached the trachea and there may be a temptation to repeat the injection with untoward results.

TABLE 12.2. Prophylaxis of local anaesthetic reactions.

1 Premedication: Diazepam 2–5 mg orally one hour prior to examination.
   Atropine 1 mg I.M. 30 minutes prior to examination.

2 Measure out dose of anaesthetic to be used before commencing and DO NOT EXCEED MAXIMUM DOSE:
   Up to 0.2 g lignocaine (5 ml 4%) for ADULT; less for frail patients and children.

3 Dilute to 1 per cent for intratracheal use.

4 Patient recumbent if possible.

In animal experiments, relatively low doses of diazepam (Valium) of the order of 0.25 mg/kg have been shown to be effective in preventing lignocaine-induced convulsions [17]. It therefore seems that diazepam is the most appropriate drug for premedication. Oral administration of 5 mg, at least one hour before the examination, will have a minimal sedative or depressive effect. If diazepam is being used intravenously, appropriate measures should be available to counteract hypotension or respiratory depression which may occur.

Atropine sulphate in a dose of 1 mg, administered intramuscularly 30 minutes before the examination, decreases the risk of vagal inhibition and, by drying up the secretions, it simplifies anaesthesia. When lignocaine is being administered by injection into the tissues, a concentration of 1 per cent should not be exceeded. As with other operative procedures, it is preferable for the patient to be recumbent to diminish the effect of any drug-induced or psychological hypotension. Recommendations are summarized in Table 12.2.

It is usual for patients to be in the fasting state to avoid the risks of vomiting. Out-patients should be warned that it may be inadvisable for them to drive a car for some hours after the examination. Following bronchography, the patient should be instructed not to eat or drink for at least one hour, and to commence with a small drink of plain water.

Treatment of reactions

This will depend partly on whether the presenting symptoms are neurological, respiratory, or cardiovascular. However, whichever system is involved, immediate treatment should be commenced with 100 per cent oxygen administered by a face mask and bag. At the same time a clear airway should be ensured. Moore & Bridenhaugh [18] claimed that oxygen alone was effective in treating the majority of cases of local-anaesthetic-induced convulsions. They deprecated the use of intravenous barbiturate and they only used muscle relaxants to control convulsions when these were interfering with the oxygenation of the patient. While their results were impressive, this approach represented an extreme point of view; but at the same time it drew attention to the importance of adequate oxygenation, since convulsions increase the oxygen requirements of the brain and muscles. Moreover there may be a vicious circle. Convulsions may cause death partly as a result of anoxia from interference with the respiratory exchange, whilst anoxia itself may give rise to convulsions.

It is important that local anaesthetic convulsions be brought rapidly under control since status epilepticus in these patients may cause foramenal crowding with sudden death. Animal experiments have again shown that intravenous diazepam is highly effective in controlling lignocaine convulsions at a dose as low as 0.1 mg/kg [19]. At this level, diazepam produces only minimal effects on the respiratory and cardiovascular systems. On this basis, diazepam should be considered the mainstay of treatment if convulsions occur. Ampoules of diazepam and a syringe should always be available in the room, to treat an emergency, when local anaesthetics are being used. While the patient is receiving oxygen, the syringe can be loaded with 2 ml (10 mg) diazepam. The injection should be given *slowly* giving only the *minimal dose* required to control the convulsions. In resistant cases, however, it may be necessary to administer larger doses. If respiratory depression occurs, this must be treated by positive pressure ventilation. Hypotension may also require appropriate treatment (Table 12.3). Toxic convulsions may also be treated by the use of muscle relaxants followed by endotracheal intubation and ventilation with 100 per cent oxygen but this usually requires the presence of an anaesthetist. Local anaesthesia is unlikely to be used for bronchography in children but, for the control of convulsions in infants, a dose of up to 0.25 mg/kg diazepam has been recommended.

Acute cardiovascular or respiratory collapse due to lignocaine may rarely occur without preceding convulsions, particularly with high blood levels of the drug. The milder form of hypotension with syncope may respond to posture, oxygen and, if necessary, vasopressors. More severe involvement of the cardiovascular system may produce cardiac arrest which requires

TABLE 12.3. Treatment of local anaesthetic reactions.

Toxic convulsions	Respiratory depression	Cardiovascular depression
Oxygen	Maintain airway	Posture, oxygen
Diazepam I.V. (10 mg in 2 ml). Use smallest effective dose but increase if necessary. (Infants 0.25 mg/kg.)	Oxygen Artificial ventilation	Aramine 0.2–1 ml I.M. Cardiac massage

appropriate treatment. In the treatment of respiratory arrest from lignocaine poisoning, respiratory stimulants such as nikethamide should be used cautiously, if at all, because of their possible convulsive effect.

Bronchography in children will usually be performed under general anaesthesia. This subject is discussed in Chapter 18.

CONTRAST MEDIA

The most widely used bronchographic contrast medium in current use is *Dionosil* (propyliodone), either as an aqueous suspension containing carboxymethylcellulose, or as a suspension in arachis oil. Other contrast media which have been widely used for bronchography at one time or another include *Hytrast* (an aqueous suspension of iopydol and iopydone containing carboxymethylcellulose), *Lipiodol* (iodinated poppy-seed oil), *Viscidodol* (iodinated poppy-seed oil with sulphanilamide powder as a thickening agent), and *barium sulphate*. Instillation of contrast medium into the bronchi may cause general effects which may occur in greater or lesser degree with any of the media used. These effects appear to be mainly due to partial occlusion of the airways and to the presence of contrast residues in the lungs. In addition, the individual contrast media may produce more specific effects and these will be discussed separately.

General effects

It is only to be expected that the introduction of contrast medium into the tracheo-bronchial tree will cause some degree of breathlessness but the extent to which this may occur is not always appreciated. Christoforidis [20] pointed out that up to 40 ml of contrast medium may be used for bilateral bronchography whilst the volume of the conducting system of the tracheo-bronchial tree is only in the region of 150 ml. Measurements of *vital capacity* and *maximum breathing capacity* show an average reduction of 20 per cent immediately following unilateral bronchography and 30 per cent in the case of bilateral bronchography. This is obviously of considerable significance in patients with pre-existing impairment of lung function such as obstructive airways disease. In patients where this is suspected, lung function tests should be undertaken before bronchography and bilateral bronchography should be avoided if the examination shows impairment of more than 50 per cent.

Perfusion scans with radioactive macro-aggregated albumen one hour after bronchography with Dionosil, showed localized perfusion defects in the lungs. These were presumably due to decreased pulmonary blood flow in areas where there was partial bronchial obstruction by residual contrast medium. *Ventilation scans* also showed areas of defective aeration but these did not always coincide with the perfusion defects [21]. After bilateral bronchography with Dionsil oily, there was a significant reduction in *diffusion capacity* and the effects were more marked in patients with extensive pulmonary disease. In some cases diffusion capacity did not return to its original level until 72 hours after the bronchogram, so that operations are best deferred for at least three days after bilateral bronchography. After unilateral bronchography, significant changes in diffusion capacity were less common [22].

In *asthmatic patients*, bronchography may occasionally precipitate bronchospasm with dyspnoea and the narrowed bronchi may be visualized radiologically (Fig. 12.2). In the case reported by Beales & Saxton [23], there was no response to an inhalation of isoprenaline but striking improvement occurred following intravenous injection of 0.5 g aminophylline. This drug should, of course, be injected *slowly* to avoid hypotension but it appears to be particularly useful when bronchospasm occurs during bronchography and it should always be available in the room. Adrenaline, steroids and oxygen should also be available in case they are required. In exceptional cases, bronchospasm and collapse may be followed by transitory loss of consciousness [24].

In young children the small diameter of the bronchi renders them more liable to obstruction by contrast medium. Bronchography in these young patients is usually performed under general anaesthesia and it is important to aspirate as much as possible of the residual contrast medium from the bronchi at the conclusion of the examination. In a retrospective survey of 165 paediatric bronchograms, evidence of complicating segmental *pulmonary collapse* was found in 45 per cent of the cases [25]. Collapse was more common with aqueous media; particularly Hytrast, but aqueous Dionosil was also associated with a higher incidence of collapse. Children with asthma and other allergic states were also more prone to pulmonary collapse. However, the major factor associated with collapse appeared to be the use of halothane anaesthesia with 50 per cent oxygen and 50 per cent nitrous oxide. This combination of anaesthetic gases is highly diffusible and therefore it is rapidly absorbed in the alveoli before it can be replaced by fresh inspired gas. This rapid peripheral absorption of gas predisposes to alveolar collapse, particularly if contrast medium is already producing a partial bronchial block. Experiments in dogs, using large volumes of Hytrast, showed that halothane anaesthesia rapidly produced visible pulmonary collapse when the ventilating gas was 100 per cent oxygen; whereas with room air (20 per cent oxygen, 80 per cent nitrogen) there was no visible collapse. However, even without any visible collapse, bronchial block by contrast medium could produce a drop of 30 per cent in the arterial oxygen saturation [26].

Another factor which may predispose to pulmonary collapse following paediatric bronchography is the structure of the lung in young children. In the adult lung, communications between individual alveoli and between neighbouring bronchioles allow considerable 'air drift' between adjacent areas of lung, so that air-filled lung may even be present distal to an area of bronchial obstruction. These peripheral communications are much fewer in the infantile lung and do not become fully developed until the age of eight years [27]. Bronchography is usually inadvisable before the age of one year and should, if possible, be deferred until after the age of two. Excessive volumes of contrast medium should be avoided. A death has been recorded in a four-year-old child following bronchography with 22 ml of oily Dionosil [28].

Rayl [29] made a careful assessment of the clinical reactions following 746 bronchograms using various contrast media. Approximately one third of all patients had some type of reaction. The incidence of reactions was highest after Hytrast and then in descending order after Dionosil aqueous, Visciodol and Dionosil oily. However, the contrast media which caused the highest incidence of reactions were, in general, those which produced better quality

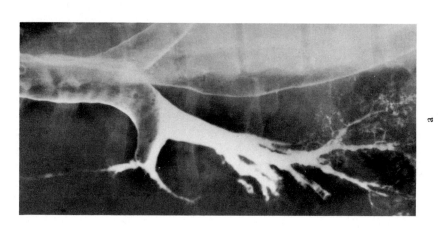

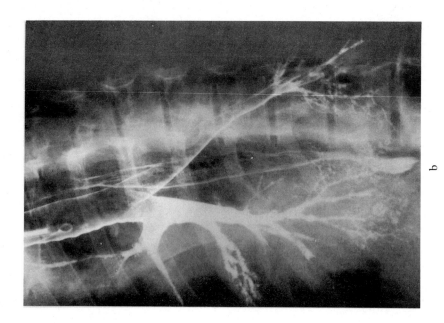

FIGURE 12.2. Female asthmatic, age 24, with contracted right middle lobe. After bronchogram with 15 ml Hytrast, complained of mild breathlessness and an oppressive sensation in the chest. Postero-anterior and right posterior oblique films a and b show fluid level in the right main bronchus. The basal segmental bronchi and their branches are narrowed, irregular and incompletely filled.

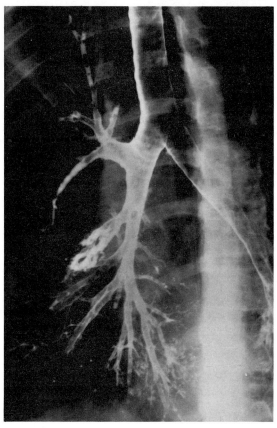

c

c Immediately following intravenous injection of 0.5 g aminophylline, patient no longer dyspnoeic and repeat right posterior oblique view shows that the contrast medium has passed peripherally to outline basal segmental bronchi of normal calibre and smooth outline. Bronchiectasis middle lobe.

(Courtesy Beales and Saxton, *British Journal of Radiology* [23]).

bronchograms. The major factors which appeared to increase the incidence of reactions with all the bronchographic media were (a) the volume used (above 20 ml); and (b) the extent of residual contrast medium visible in the lungs in a chest radiograph taken at 24 hours. The commonest reactions noted were: *headache* (10–21 per cent), *nausea* and/or *vomiting* (10–15 per cent) and *pyrexia* (2–26 per cent). The frequency of pyrexial reactions increased in the second 24-hour period after bronchography but pyrexia was rare after the fifth day. Cold and 'flu-like' symptoms such as a burning sensation in the nose or throat, swelling or redness of the eyes, swelling of the salivary glands, bone pain or skin rashes were attributed to 'iodism'. However, there seems some doubt as to whether all the cases so recorded were genuine cases of iodism since the quoted incidence was 12 per cent for Hytrast and only 3 per cent after Visciodol, whereas one would have expected iodine to be liberated more easily from the iodinated oil in Visciodol rather than from the organically bound media such as Hytrast. Nevertheless, it is possible that with prolonged retention of organically bound contrast media in the tissues some free iodine may be liberated. Other reactions noted in this series were dyspnoea, wheezing, diarrhoea, dizziness, cyanosis, chest pain and pneumonia.

Lipiodol

For many years following the inception of bronchography, iodized oil remained the standard contrast medium. It is now rarely, if ever, used for this purpose but analogous preparations are still used in lymphography, sinography, sialography and occasionally in hysterosalpingography. Opaque residues may remain in the tissues for prolonged periods and may cause a granulomatous reaction [30]. Retained Lipiodol in the lungs presents a characteristic radiological picture of clusters of fine pin-point densities which may subsequently interfere with the interpretation of the chest radiograph.

Because of the relatively loosely bound manner in which iodine is incorporated into the iodized oil, milder cases of iodism were not uncommon, with manifestations such as transient coryza, lachrymation, urticarial or erythematous rashes, excess salivation, swelling of the salivary glands or thyroid gland. Rarely, iodism following the use of Lipidol presented with a more severe syndrome which, in a few cases proved fatal [31, 32]. In these cases, symptoms might develop within four hours of the bronchogram or be delayed for as long as seventeen days. Characteristically, there were widespread severe skin rashes varying from erythematous or urticarial change to pustular acneiform, bullous or petechial rashes with haemorrhage and oedema of the skin and mucous membranes. Other manifestations were purulent conjunctivitis, acute bronchitis, bronchopneumonia, bronchospasm or laryngeal oedema. Histologically the changes were those of a necrotizing angiitis resembling periarteritis nodosa so that steroid therapy would appear to be indicated. Pretesting with potassium iodide was not always reliable in predicting the onset of iodism after Lipidol and in some cases there was a previous history of iodine therapy or Lipiodol bronchography without incident. It is even possible that in these cases the earlier examination might have served to sensitize the patient to iodine.

Visciodol

One of the disadvantages of Lipiodol was its relatively low viscosity which resulted in peripheral or alveolar filling. In an attempt to overcome this, sulphonilamide powder was added as a thickening agent. The commercial preparation Visciodol contained 4.8 g sulphonilamide in 15 ml of iodized oil. This medium resulted in technically satisfactory bronchograms and for a brief period of time it was moderately popular. However, an undue number of patients developed cyanosis which was found to be due to methaemoglobinaemia. In a detailed study [33] it was found that absorption of sulphonilamide from the tracheo-bronchial tree was greater than that occurring after oral administration and that the large amount of sulphonilamide in the bronchographic medium produced methaemoglobinaemia in 29 out of a series of 35 patients. This episode should serve as an important reminder of the hazards which have ensued when new drugs or contrast media have been introduced in the absence of an adequate appraisal of the basic pharmacological data. The reason why sulphonilamide was originally chosen as a thickening agent is not completely clear but doubtless it was assumed that its antibacterial action would be a bonus.

The iodized oil in Visciodol can of course give rise to iodism and transitory pneumonic changes. More recently, an unusual case of delayed granuloma has been described [34]. In 1964 the patient underwent bronchography with Visciodol. Some months later he developed a low-grade pyrexia with an extensive pneumonic infiltration surrounding an area of residual contrast medium in the right upper lobe. The pneumonic process subsided without treatment after a period of 2 months. Some 26 months later, a homogeneous opacity appeared in the right upper lobe surrounding the residual contrast medium. The opacity increased in size over a period of four years, producing a large lobulated mass. On resection in 1971 the mass was found to be a hyalinized lipoid granuloma containing residual contrast medium.

Dionosil

Propyliodone is the diethanolamine ester of diodone. This is hydrolysed in the tissues and the opaque contrast medium is usually absorbed within two days of the bronchogram. However, the disappearances of the opaque medium cannot be regarded as an indication that the nonopaque vehicle in which it was administered has been similarly absorbed. Björk & Lodin [35] showed that oily residues and foreign body granulomata were present in the lungs in seven out of ten rabbits, six months after bronchography with Dionosil oily. These findings have since been disputed and further animal experiments using both aqueous and oily preparations of Dionosil suggest that while histological changes of an acute inflammatory reaction were common in the first few days, these were of brief duration only [36, 37]. Similarly, an investigation of the various vehicles used for bronchographic contrast media showed that poppy-seed oil, arachis oil and sodium carboxymethyl-cellulose, could all individually cause a mild acute foreign body reaction in the rabbit lung but that this was apparently reversible [38]. Holden & Cowdell [39] examined resected specimens of lungs in 49 patients who had bronchograms with Dionosil oily at periods varying from one week to four years prior to the operation. They considered that fatty and granulomatous changes in these lungs were no more frequent than in similar specimens of diseased lungs from patients who had not been subjected to bronchography. However, since the lungs were abnormal, it is impossible to prove that, in fact, none of the histological changes found was due to the contrast medium.

The oily preparation of Dionosil usually causes more peripheral filling, whereas the aqueous preparation is usually more effective in coating the proximal bronchi. It is widely believed that the aqueous suspension is more irritant than the oily preparation, but the findings in a recent comparison of the two preparations suggested that coughing was no more frequent with the aqueous preparation than it was with the oily preparation [40].

Mild pneumonic and pyrexial reactions are not uncommon following Dionosil bronchography and, as with other bronchographic media, these appear to be mainly related to the extent of residual contrast medium in the lungs. Absorbed diodone may be expected to cause any of the reactions known to occur with urographic contrast media (see Chapter 1). A possible example of acquired sensitivity to propyliodone is described by Friedell *et al.* [41]. The patient previously had a bronchogram with Dionosil aqueous without incident. A repeat bronchogram with Dionosil oily performed ten months later, showed bronchiectasis of the left lower lobe. This was resected four days after the bronchograms and showed histological changes of a marked granulomatous reaction with macrophages, epithelial cells and foreign body giant cells. It was suggested that these findings represented an accelerated hypersensitivity reaction following previous sensitization with propyliodone but this interpretation is open to doubt. In another patient, a severe erythematous rash occurred 48 hours after a bronchogram with Dionosil. A previous bronchogram six days earlier, with the same medium, had caused no reaction [24].

Hytrast

This is an aqueous suspension of iopydol and iopydone. It contains sodium carboxymethylcellulose which acts as a wetting and lubricating agent and increases the viscosity. The higher iodine content of this medium (50 per cent w/v) resulted in excellent bronchograms [42–44] but the incidence of post-bronchogram pyrexia was increased, particularly in association with peripheral filling [29]. Occasional severe pneumonias have also been reported after Hytrast [36]. At least two deaths have been reported. One of these [45] is only mentioned briefly. In the other case [46] a 63-year-old woman developed severe respiratory distress immediately after a bilateral bronchogram with 20 ml of Hytrast. Positive pressure ventilation and tracheostomy were required and, despite treatment with steroids and chloramphenicol, death occurred on the sixteenth day. At autopsy, there was massive bilateral consolidation due to a 'crystalline inclusion pneumonia'. Hytrast produces a more severe inflammatory reaction in the lung than Dionosil [37], and the histological changes have been reviewed in detail by Morley [47]. In contradistinction to Dionosil, Hytrast is not hydrolysed in the tissues, so that crystalline residues remain in the alveoli. It is therefore important to minimize alveolar filling with Hytrast. The majority of serious reactions were reported from the U.S. where Hytrast contained only 0.5 per cent carboxymethylcellulose as compared with the European preparation which contains 1.5 per cent. Grainger *et al.* [48] suggested that the lower viscosity of the American preparation might have been responsible for causing increased peripheral filling but, in animal experiments, they were unable to demonstrate any difference between the histological reactions caused by the two preparations of Hytrast.

In the French literature there are a number of cases suggesting the possibility of acquired sensitivity following repeat bronchograms with Hytrast one to three weeks after earlier symptom-free bronchography. In these cases, the second bronchogram was followed by pneumonic reactions, eosinophilia, skin rashes and prolonged pyrexia suggesting a Loeffler syndrome [49, 50]. In one of the patients there was also a positive skin test to Hytrast and a positive lymphoblast transformation test developed. These suggest a true allergic reaction [50]. The effect of Hytrast in causing pulmonary collapse during paediatric bronchography under general anaesthesia has already been referred to earlier. It also appears that in isolated incidents the suspension has thickened after administration, causing occlusion of the smaller bronchi and even of the suction tube. At the time of writing, Hytrast is no longer marketed in the U.S., Canada or in the U.K. but it is still available in some countries in Europe, South America and the Far East [51].

Miscellaneous media

Barium sulphate has been advocated as a bronchographic agent on the grounds that it is free from the risk of allergic reactions, and does not appear to cause chronic histological changes in the lungs [52]. Bismuth subcarbonate, on the other hand, produced small areas of necrosis [53]. The main effect of barium in the bronchi is due to bronchial blockage if large volumes are used [28]. Nevertheless, retained barium in other tissues has been shown to cause granulomatous changes (see Chapter 13), and barium has never become widely popular for bronchography.

Accidental inhalation of hypertonic contrast medium such as Gastrografin may cause fatal pulmonary oedema, particularly in patients with compromised lung function [54, 55]. Pulmonary oedema has also been demonstrated in animals following intratracheal injection of diatrizoate [56]. However, if the animals survive, there does not appear to be any permanent histological damage [28]. Tantalum powder has been used for experimental aerosol bronchography in a number of centres but there is a risk of spark discharge and there have been two recent instances of spontaneous fires, both of which resulted in fatalities [20].

The brief popularity of Visciodol has already been mentioned. Another example of the hazards of introducing new bronchographic media is provided by the episode involving Pulmidol (propyldocetrizoate). This tri-iodinated compound was introduced about 1961 [57] and was found to produce excellent bronchograms. The following year it was withdrawn after the occurrence of fatal convulsions when the medium was used in association with ether or halothane anaesthesia.

COMPLICATIONS OF TECHNIQUE

There are many variations in bronchographic technique. The majority of these involve the use of a soft rubber catheter passed into the trachea. However, injection following puncture of the crico-thyroid membrane is preferred by an increasing number of radiologists. This has the advantage of speed and requires a smaller volume of local anaesthetic. The commonest complication following crico-thyroid puncture is surgical emphysema. To minimize this, the patient should be instructed to press on the puncture site when coughing. Bleeding may occur at the injection site. A severe haematoma might necessitate tracheostomy, whilst late infection of the haematoma can rarely cause abscess formation and even perichondritis of the larynx [58]. An important complication of transcricoid bronchography is the accidental injection of contrast media into the soft tissues of the neck [59, 60]. The majority of these cases have been due to inexperience of the operator but displacement of the trachea by old fibrosis or a short bull neck may make the examination more difficult. It is essential to check that the needle is correctly sited in the lumen of the trachea by aspirating air into the syringe, before injecting contrast medium. This test should be repeated after every 1–2 ml of medium are injected and after every swallowing movement or change in position of the patient, to make sure that the needle has not become displaced. In the majority of cases reported in the literature where a reaction has followed extraneous injection, the contrast medium involved has been Dionosil oily. Usually there is little if any immediate discomfort at the time of the extratracheal injection but after a period varying from a few hours to several days, a chemical inflammatory reaction develops with swelling of the soft tissues of the neck, severe neck pain and tenderness (Fig. 12.3). Depending on the distribution of the injected medium there may be oedema of the vocal cords, hoarseness or even stridor. Characteristically, there is marked stiffness of the neck which may mimic a Kernig's sign and pyrexia may be present. Dysphagia is also common. Contrast medium tracking into the anterior mediastinum may cause chest pain with ECG changes resembling those of pericarditis. If untreated the symptoms at first increase in severity and then gradually resolve over a period of two to three weeks. The contrast medium is usually gradually absorbed from the neck over a period of weeks. Steroid therapy appears to be useful in aborting the clinical syndrome by virtue of its anti-inflammatory action and antibiotics are usually given to prevent secondary infection. Extraneous injection of Hytrast has produced a similar syndrome in at least two cases [24] and, in one of these, the patient also developed an extensive purpuric rash and abdominal pain five days after the bronchogram. However, it is not certain whether the purpura was solely due to the Hytrast since the patient had also received ampicillin for three days. On experimental injections of bronchographic media into the soft tissues, Hytrast and Dionosil aqueous produced a severe granulomatous reaction with fibrosis. Dionosil oily produced a moderate reaction, whilst Lipiodol produced only a mild inflammatory response [37]. The transcricoid route may also be used for the introduction of a Seldinger catheter [61]. This minimizes the risk of spillage in the neck but a recent example of such spillage has occurred with the use of a transcricoid catheter [24].

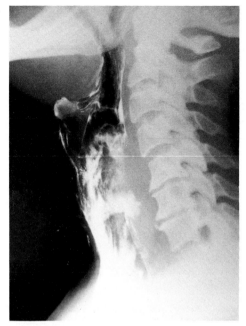

a

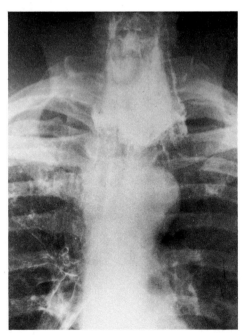

b

Bronchography

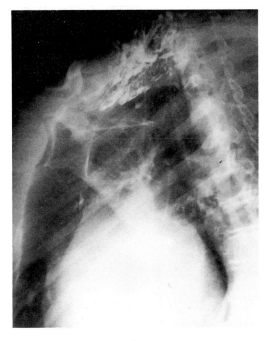

c

FIGURE 12.3. Female, age 53; transcricoid bronchogram with Dionosil oily for investigation of haemoptysis.

(a and b) Lateral and postero-anterior films show extratracheal injection of contrast medium. (c) Oblique view shows contrast medium tracking into anterior mediastinum.

The patient did not experience any significant pain at the time but this developed later and she was unable to sleep at night. Two days later, she was admitted with severe pain in the neck on movement, and headache on bending the head. Her head felt as though it was 'in a block of plaster'. There was also some dysphagia. T = 37.8°C.

Treatment with prednisolone 10 mg 6 hourly and ampicillin produced rapid symptomatic improvement.

REFERENCES

1 OLSEN A.M. & O'NEIL J.J. (1967) Bronchography. A report of the committee on bronchoesophagology, American College of Chest Physicians. *Dis. Chest.* **51**, 663–668.
2 ANSELL G. (1973) *Notes on Radiological Emergencies*, 2nd edn. Blackwell Scientific Publications, Oxford.
3 SCOTT D.B. & JULIAN D.G. (1971) *Lidocaine in the treatment of ventricular arrhythmias.* Livingstone, Edinburgh.

4 HARRISON D.C. & ALDERMAN E.L. (1971) *Lidocaine in the treatment of ventricular arrhythmias*, pp. 178–188. Livingstone, Edinburgh.
5 ÅSTRÖM A. (1971) *Lidocaine in the treatment of ventricular arrhythmias*, pp. 128–139. Livingstone, Edinburgh.
6 HAYES A.H. (1971) *Lidocaine in the treatment of ventricular arrhythmias*, p. 183. Livingstone, Edinburgh.
7 PRESCOTT L.F. (1971) *Lidocaine in the treatment of ventricular arrhythmias*, pp. 205–206. Livingstone, Edinburgh.
8 STEINHAUS J.E. (1962) Toxic reactions to local anesthetics. *J. Am. med. Ass.* **32**, 168–172.
9 LEHNER T. (1971) Lignocaine hypersensitivity. *Lancet* **1**, 1245–1246.
10 BOAKES A.J., LAURENCE D.R., LOVEL K.W., O'NEIL R. & VERILL P.J. (1972) Adverse reactions to local anaesthetic/vasoconstrictor preparations. A study of the cardiovascular responses to Xylestin and Hostacain with noradrenaline. *Br. Dent. J.* **133**, 137–140.
11 SELDEN R. & SASAHARD A.A. (1967) Central nervous system toxicity induced by lidocaine. Report of a case in a patient with liver disease. *J. Am. med. Ass.* **202**, 142–149.
12 ADJEPON-YAMOAH K.K., NIMMO J. & PRESCOTT L.F. (1974) Gross impairment of hepatic drug metabolism in a patient with chronic liver disease. *Br. Med. J.* **4**, 387–388.
13 ADRIANI J. & ZEPERNICK R. (1964) Clinical effectiveness of drugs used for topical anesthesia. *J. Am. med. Ass.* **188**, 711–716.
14 BROMAGE P.R. & ROBSON J.G. (1961) Concentration of lignocaine in the blood after intravenous, intramuscular and endotracheal administration. *Anaesthesia* **16**, 461–478.
15 WHEELER P.C. (1962) A new method of anesthesia for bronchography. *Missouri Medicine* **59**, 419–420.
16 MARTIN B.H., ISRAEL J. & STOVIN J.J. (1972) Anesthesia and intubation for bronchography. *Radiology* **104**, 536.
17 DE JONG R.H. & HEAVNER J.E. (1971) Diazepam prevents local anesthetic seizures. *Anesthesiology* **34**, 523–531.
18 MOORE D.C. & BRIDENHAUGH L.D. (1960) Oxygen: the antidote for systemic toxic reactions from local anesthetic drugs. *J. Am. med. Ass.* **174**, 842–847.
19 MUNSON E.S. & WOGMAN I.H. (1972) Diazepam treatment of local anaesthetic-induced seizures. *Anesthesiology* **37**, 523–528.
20 CHRISTOFORIDIS A.J. (1973) Bronchography. In: *Complications and legal implications of radiologic special procedures.* (Ed. by MEANY T.F., LALLI A.F. & ALFIDI R.J.), pp. 102–116. Mosby, St Louis.
21 SUPRENANT E., WILSON A., BENNETT L., O'REILLY R. & WEBBER M. (1968) Changes in regional pulmonary function following bronchography. *Radiology* **91**, 736–741.
22 BHARGAVA R.K. & WOOLF C.R. (1967) Changes in diffusing capacity after bronchography. *Am. Rev. Resp. Dis.* **96**, 827–829.
23 BEALES J.S.M. & SAXTON H.M. (1968) The radiographic demonstration of bronchospasm and its relief by aminophylline. *Br. J. Radiol.* **41**, 899–901.
24 Reports to National Radiological Survey.
25 ROBINSON A.E., HALL K.D., YOKOYAMA K.N. & CAPP M.P. (1971) Pediatric bronchography: the problem of segmental pulmonary loss of volume. I. A retrospective survey of 165 pediatric bronchograms. *Invest. Radiol.* **6**, 89–94.
26 ROBINSON A.E., HALL K.D., YOKOYAMA K.N. & CAPP M.P. (1971) Pediatric broncho-

graphy: the problem of segmental pulmonary loss of volume. II. An experimental investigation of the mechanism and prevention of pulmonary collapse during bronchography under general anesthesia. *Invest. Radiol.* **6**, 95–100.

27 REID L. (1974) Communication to *The Fleischner Society*, London.
28 FRECH R.S., DAVIE J.M., ADATEPE M., FELDHAUS R. & MCALISTER W.H. (1970) Comparison of barium sulfate and oral 40% diatrizoate injected into the trachea of dogs. *Radiology* **95**, 299–303.
29 RAYL J.E. (1965) Clinical reactions following bronchography. *Ann. Otol. Rhinol. Laryngol.* **74**, 1121–1132.
30 FELTON W.L. (1953) The reaction of pulmonary tissue to Lipiodol. *J. Thorac. Surg.* **25**, 530–542.
31 SCADDING J.G. (1934) Acute iodism following Lipiodol bronchography. *Br. Med. J.* **2**, 1147–1148.
32 SUMNER J., LICHTER A.I. & NASSAU E. (1951) Fatal acute iodism after bronchography. *Thorax* **6**, 193–199.
33 JOHNSON P.M. & IRWIN G.L. (1959) An evaluation of the pharmacological hazards in bronchography. *Radiology* **72**, 816–828.
34 SMITH T.R., FRATER R. & SPATARO J. (1973) Delayed granuloma following bronchography. *Chest*, 122–125.
35 BJÖRK L. & LODIN H. (1957) Pulmonary changes following bronchography with Dionosil oily (animal experiments). *Acta Radiol.* **47**, 177–180.
36 LIGHT J.P. & OSTER W.F. (1966) Clinical and pathological reactions to the bronchographic agent Dionosil aqueous. *Am. J. Roentg.* **98**, 468–473.
37 GREENBERG S.D., SPJUT H.J. & HALLMAN G.L. (1966) Experimental study of bronchographic media on lung. *Arch. Otolaryng.* **83**, 276–282.
38 CHRISTOFORIDIS A.J., NELSON S.W. & PRATT P.C. (1967) An experimental study of the tissue reaction to the vehicles commonly used in bronchography. *Am. Rev. Resp. Dis.* **96**, 249–253.
39 HOLDEN W.S. & COWDELL R.H. (1958) Late results of bronchography using Dionosil oily. *Acta Radiol.* **49**, 105–112.
40 WALKER H.G. & MA H. (1971) Oily and aqueous propyliodone (Dionosil) as bronchographic contrast agents. *J. Canad. Ass. Radiol.* **22**, 148–153.
41 FRIEDELL G.H., KAUFMAN S.A., LAFORET E.G. & STRIEDER J.W. (1962) Granulomatous lung reaction following repeat bronchography with propyliodone. *Am. J. Roentg.* **87**, 847–852.
42 LE ROUX B.T. & DUNCAN J.G. (1964) Bronchography with Hytrast. *Thorax* **19**, 37–43.
43 WRIGHT F.W. (1965) Bronchography with Hytrast. *Br. J. Radiol.* **38**, 791–795.
44 PALMER P.E.S., BARNARD P.J., CUSHMAN R.P.A. & CRAWSHAW G.R. (1967) Bronchography with Hytrast. *Clin. Radiol.* **18**, 94–100.
45 MOUNTS R.J. & MOLNAR W. (1962) The clinical evaluation of a new bronchographic medium. *Radiology* **78**, 231–233.
46 AGAE F. & SHIRES D.L. (1965) Death after bronchography with a water-soluble iodine-containing medium. *J. Am. med. Ass.* **194**, 459–461.
47 MORLEY A.R. (1969) Pulmonary reaction to Hytrast. *Thorax* **24**, 353–358.
48 GRAINGER R.G., CASTELLINO R.A., LEWIN K. & STEINER R.N. (1970) Hytrast: experimental bronchography comparing two different formulations. *Clin. Radiol.* **21**, 390–395.

49 ROCHE G. & GUERBET M. (1968) Bronchographie a l'hytrast compliquée de pneumopathie fébrile. *J. Franc. Med. Chir. Thorac.* **22**, 371-381.
50 MIGUÈRES J., JOVER A. & ABOU P. (1970) Infiltrats pulmonaires labiles de nature allergique après bronchographie a l'hytrast. Le problème de l'allergie a l'hytrast. *Rev. Tuberc. Pneumon. Paris.* **34**, 991-998.
51 DAVISON T.H. (1974) Personal communication.
52 NELSON S.W., CHRISTOFORIDIS A.J. & PRATT P.C. (1964) Further experiments with barium sulfate as a bronchographic contrast medium. *Am. J. Roentg.* **92**, 595-614.
53 NELSON S.W., CHRISTOFORIDIS A. & PRATT P.C. (1959) Barium sulfate and bismuth subcarbonate as bronchographic contrast media. *Radiology* **72**, 829-838.
54 ANSELL G. (1968) A national survey of radiological complications: interim report. *Clin. Radiol.* **19**, 175-191.
55 CHIU C.L. & GAMBACH R.R. (1974) Hypaque pulmonary edema. A case report. *Radiology* **111**, 91-92.
56 REICH S.B. (1969) Production of pulmonary edema by aspiration of water-soluble non-absorbable contrast media. *Radiology* **92**, 367-370.
57 MAXWELL D.R., REID L. & SIMON G. (1961) Properties of a new contrast medium for bronchography: n-propyl 2,4,6-triiodo-3-diacetamidobenzoate (propyl docetrizoate). *Br. J. Radiol.* **34**, 744-747.
58 SAXTON H.M. & STRICKLAND B. (1972) *Practical Procedures in Diagnostic Radiology*, 2nd edn, pp. 132-150. H. K. Lewis, London.
59 ZUCHERMAN S.D. & JACOBSON G. (1962) Transtracheal bronchography. Complications of injection outside the trachea. *Am. J. Roentg.* **87**, 840-843.
60 WRIGHT F.W. (1970) Accidental injection of Dionosil into the neck during bronchography. *Clin. Radiol.* **21**, 384-389.
61 CRAVEN J.D. (1965) A new method of bronchography. *Br. J. Radiol.* **38**, 395-397.

CHAPTER 13: LYMPHOGRAPHY

J. S. MACDONALD

During lymphography several substances are injected into the body, the blue marker dye (Patent Blue Violet), local anaesthetic and the iodized oil (Lipiodol, Ethiodol) which is itself a mixture of substances. There is therefore ample opportunity for allergic reactions. Patent Blue Violet is an intense dye which stains the body fluids blue, hence any urticarial weals will be blue in colour and the appearance can be quite dramatic.

The urine will be blue or green for 48 hours or so after lymphography and fair-skinned patients may turn a dusky blue or green. This is not a complication in the true sense, but it should be noted so that the patient can be warned. The chances of this happening are greater if the higher concentrations of dye are used (10 per cent or 11 per cent), but there is no indication for this except in a few special cases where the primary lymphoedemas are being investigated and difficulty is experienced in finding a lymphatic. The majority of lymphograms are done using 2.5 per cent Patent Blue Violet. General anaesthesia is used in babies and children and as a proportionately larger amount of blue dye is used in these cases, the chances of turning the patient blue are higher. It is important to warn the anaesthetist since this can be mistaken for cyanosis and the mother should also be warned so that she is not worried by the child's appearance.

There will be a dark blue stain at the site of the injection of the Patent Blue Violet. This will gradually clear, but there is considerable individual variation. It may clear in a matter of days in the oedematous limb, in the average patient it will be hardly noticeable in three weeks, but traces may persist for months. Over the years other dyes have been used, but none has proved as satisfactory and some have shown a tendency to persist at the site of the injection for far longer. This may be associated with a greater tendency for protein binding [1].

ALLERGIC REACTIONS

There has been discussion as to whether Patent Blue Violet was capable of producing allergic reactions, particularly since Kinmoth observed none in a series of 2000 lymphograms [1]. Since most people mix local anaesthetic with the dye, it was difficult to be sure to which the reaction was due. Reactions are fewer under general anaesthesia and Kinmonth's series was almost entirely under general anaesthesia. There is however no doubt that Patent Blue Violet can cause an allergic reaction [2] since we have seen this in children under general anaesthesia before any other substance was injected and a member of the staff is also allergic to the dye. A local urticarial reaction can occur at the site of the injection of the dye or it may be generalized. Occasionally a reaction can be seen along the line of the lymphatics. These allergic reactions are usually mild but can be severe with generalized urticaria, mucosal congestion, choking and syncope.

Redman [3] published details of three patients with dermatitis as a reaction to lymphography appearing as a delayed reaction eight, ten and nineteen days after lymphography. This was taken as a local drug reaction of the vascular type to extravasated iodized oil (Ethiodol). All three patients showed tissue extravasation or backflow, but this is a common finding in lymphography and is not usually accompanied by dermatitis.

INFECTION

Since lymphography requires an incision there is the possibility of wound infection. This is rare where an adequate sterile technique is used and happens in fewer than 1 per cent of patients. Even more rarely (less than 1:1000) a lymphangitis may be a complication.

DISCOMFORT AND PAIN

It is not unusual for the patient to feel slight discomfort or tingling at the beginning of the injection as the Lipiodol moves up the leg. Usually, when this reaches the inguinal region, the discomfort passes off and the patient feels nothing further. If the contrast medium is injected at too high a pressure, then the patient may have pain and this is easily remedied by reducing the pressure of the injection. The mechanism producing the pain is probably overdistension of the lymphatics since it was found in this department that pain was produced during the intralymphatic injection of normal saline by merely increasing the pressure, the pain stopping immediately the injection pressure was reduced. Increase of the injection pressure would again produce the pain.

In a series of 4500 lymphograms, we have only come across one patient who experienced severe pain unrelated to pressure and this was so intense that the injection had to be stopped. 0.5 ml of lignocaine 2 per cent was injected directly into the lymphatic. This proved successful and no further trouble was encountered. The injection of the local anaesthetic itself caused severe pain, but this soon passed off. No cause could be found; the lymphogram was normal.

EXTRAVASATION

Extravasation of the Lipiodol from the lymphatics into the surrounding tissues is seen in 5 per cent of patients and is usually asymptomatic and of no significance. Extravasated Lipiodol takes much longer to be absorbed from the tissues than from the nodes. Extravasation can be seen at any level, but it is most usually seen in the upper thighs when the feet are injected. It is much more common in the arm when the lymphatics of the hand are injected and it is also common when the lymphatics of the spermatic cord are injected. Dermal backflow, however, is an abnormal finding. This is the filling of the fine dermal lymphatics and indicates obstruction of the deeper lymphatics. It is seen, for instance, after trauma to the lymphatics and around the edges of a skin graft. Sometimes after trauma and consequent extravasation, or if there has been leakage of Lipidol into the soft tissues at the site of injection, the Lipiodol is seen passing along the perivascular spaces.

OEDEMA

Occasionally there may be some swelling of the limb following lymphography. This is transient, but may be more marked in patients where the lymphatic drainage is already impaired. It is undesirable to have pools of oily contrast medium stagnating in the dilated lymphatics of primary lymphoedema of the hyperplastic type and if this is suspected then a water-soluble contrast medium should be used to show the state of the lymphatics. Dealing with oily contrast medium can impose an added burden on an impaired lymphatic system and the radiologist should assess carefully the indications for lymphography on any oedematous limb. This should only be done in cases where it is thought likely that the lymphogram will contribute directly towards the treatment of the condition. There is, for instance, little indication for lymphography in chronic oedema of the arm following a radical mastectomy. In such cases, if a phlebogram is normal, then the lymphatic system is impaired and there is little point in overburdening this already impaired circulation and possibly also running the risk of infection which will make matters worse.

When the needle is removed from the lymphatic and the wound is stitched there is very little leakage of lymph as is seen from monitoring dressings following the injection of radioactive Lipiodol. In the oedematous limb there may be considerable oozing of oedema fluid until the wound heals. It is therefore better to admit these patients to hospital for lymphography. This is more comfortable for the patient and diminishes the risk of wound infection.

THE LUNGS

The factor which distinguishes lymphography from other vascular contrast investigations is that inevitably one is ending up with an intravenous injection of oil. If a diagnostic volume of Lipiodol is injected then, in the normal patient, the surplus not taken up by the nodes will find its way through the thoracic duct into the subclavian vein and hence through the right side of the heart and into the lungs. We know from isotope studies that an appreciable amount of oil reaches the lungs even when only small volumes (less than 4 ml) are injected into one foot. Gold *et al.* [4] calculated that, if 20 ml of Lipiodol were broken up into micro-emboli 10 μ in diameter it would be possible for every capillary in the average lung to receive two such oil emboli. The tolerances within which we are working are therefore not great. In the early days of lymphography large amounts—totals of 20 ml or more—were frequently used, but it was soon found by the centres which were developing experience of lymphography that the more contrast medium used the higher was the complication rate [5]. The normal lung is an efficient filter for this surplus oil. The globules are trapped in the capillaries and this contrast oil embolism can be shown after lymphography on low kV chest radiographs (Fig. 13.1). While the lungs are dealing with this oil there is evidence of impairment of lung function [4, 6–10]. Gold *et al.* [4] showed a decrease in pulmonary diffusing capacity, pulmonary capillary blood volume and pulmonary compliance. These changes were at a maximum at 37 hours (range 3 to 72 hours). Fraimow *et al.* [10] showed from lung biopsies, taken after lymphography, that initially the oily droplets were widely distributed throughout the pulmonary capillary bed. On the following day less oil was present and it was no longer exclusively in the capillary bed, but scattered in the interstitial tissue. Some oil was also seen free in the air spaces and within macrophages. This latter finding was checked using Ethiodol tagged with I^{131} and, in some cases, quite appreciable amounts were recovered in the sputum.

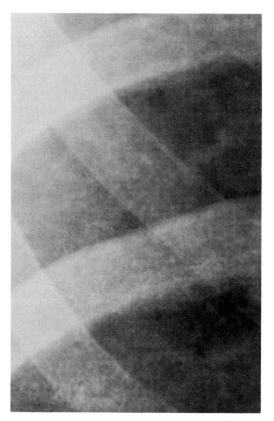

FIGURE 13.1. Close up view of the upper zone of the right lung 24 hours after lymphography, showing the fine, stippled appearance of contrast oil embolism in the lung.

In normal patients there will be no symptoms from pulmonary oil embolism, but in patients with impaired lung function the effects can be severe: most of the reported deaths following lymphography [11] have been in patients with lung disease and impaired lung function. Lymphography should therefore never be done without a preliminary radiograph of the chest and clinical assessment of the lung function. If the chest is clear and there is no history of lung disease or dyspnoea then the lymphogram can be started. If there is any doubt, then lung function studies should be done and the case reconsidered. If the lymphogram is still considered necessary then it is often possible to do this, one side at a time, using a reduced amount of Lipiodol, and leaving an interval of one week between the studies. Often

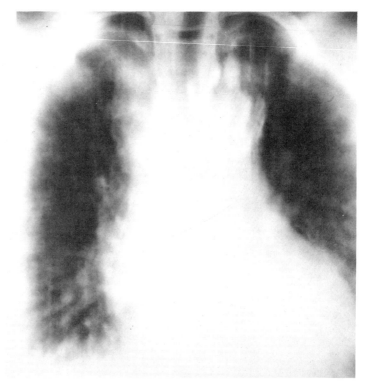

FIGURE 13.2. Tomogram of the mediastinum and lung following lymphography. There are diffuse patchy changes throughout the lungs. These were associated with pyrexia and haemoptysis, but cleared completely. This is probably an example of the chemical pulmonary reaction seen in a few patients.

all the information needed will be obtained from the injection of the one side, but if the second injection is needed then the interval of one week will allow the lungs to recover before being subjected to more oil. If the patient is severely dyspnoeic, and certainly if the patient is orthopnoeic, lymphography should not be done. It would also seem prudent to refuse lymphography in any patient who has active thrombophlebitis, because of the risk of a pulmonary embolism coming at a time when the lung function is already impaired by the oil.

It is difficult to determine the incidence of true pulmonary embolism following lymphography: it is rare and it is difficult to assess the part which lymphography has played. In our series, two patients have died of deep vein thrombosis and pulmonary embolism within three weeks of lymphography. At the time of lymphography there was no clinically suspected deep vein thrombosis in either patient. Another phenomenon which occurs in about 1 per cent of patients is haemoptysis, usually very mild, sometimes no more than a slight staining of the sputum some days after lymphography (three to ten days). It is thought that this is a chemical reaction resulting from the splitting of the oil into fatty acids which have a higher toxicity than neutral fats [12]. The incidence of haemoglobin in the sputum is certainly much higher than this, but it is only a few patients who will notice anything and they will often only mention it in passing. Chest radiographs in these patients will usually show nothing more than the resolving pulmonary oil embolism but, rarely, diffuse ill-defined patchy changes are seen (Fig. 13.2) and there may be an associated pyrexia and definite haemoptysis. Occasionally the picture may be that of pulmonary oedema.

NEUROLOGICAL COMPLICATIONS

However efficient the lung is as a filter, some oil does pass through into the systemic circulation and this is a cause for concern. The evidence is that any droplets which do pass through into the systemic circulation in the normal course of events are so fine that they do not cause embolism. However, larger globules do get through; these have been found for instance in the spleen following splenectomy in the lymphomas. The most serious complication is cerebral oil embolism. This has been reported by several workers [13-16] and in our series of 4500 patients there has been one case, so far unreported. He presented with haemoptysis and was found to have endobronchial malignant lymphoma. The chest radiographs were normal, but the lymphogram showed involved para-aortic and pelvic nodes. The patient had no chest symptoms and only 6 ml of Lipiodol Ultra Fluid was injected into each foot. The patient returned home, but some hours later developed a frontal headache, became confused and uncooperative and seemed to be having difficulty with his right hand. He was admitted to hospital and by the following morning had developed a monoplegia of the right upper limb. He was discharged fourteen days later having completely recovered. No oil was recovered from the urine, no abnormality was seen in the fundi, nor was there any radiological evidence of Lipiodol seen in the brain on skull radiographs as has been reported in other cases [14–16]. There had been no previous radiotherapy to the chest and very extensive subsequent investigation showed no evidence of a cardiac shunt. The effects of cerebral oil embolism can however be much more severe. These can include hypotension, confusion and drowsiness going on to coma which can last for several days. A drop in the serum calcium has also been observed [14]. Oily contrast emboli have been seen in the retinal capillaries at ophthalmoscopy together with retinal oedema and haemorrhages [15].

A significant number of patients, approximately 20 per cent, will complain of mild symptoms on the evening of the lymphogram. These consist of elevation of the temperature (rarely above 100°F) slight headache and possibly nausea. Sometimes they will just feel a little out of sorts and say 'it was just as if I was getting 'flu'. Rarely these symptoms can be much more severe with headache, vomiting and pyrexia. Such symptoms have never satisfactorily been explained, but it would seem likely that they are a CNS reaction to the oil or its breakdown products.

POSSIBLE OTHER EFFECTS

Does lymphography help to spread cancer? Engeset [17] has collected Sternberg–Reed cells from the thoracic duct lymph during lymphography, but there is no convincing evidence so far that lymphography contributes to the spread of cancer, nor is there any evidence to date of immunological impairment following lymphography. In fact, the clinical evidence, for what it is worth, is the other way: it is those tumours where lymphography has been employed most have shown the greatest improvement in prognosis over the last ten years.

Are there any long-term effects on the nodes? Thousands of patients have been subjected to lymphography and have eventually been examined at post-mortem. Here again there is no evidence of long-term effects on the nodes once the oil has been dealt with.

PRECAUTIONS, PREVENTION AND TREATMENT OF COMPLICATIONS

Reactions to the Patent Blue Violet are seldom severe and rarely need more treatment than 4 mg chlorpheniramine (Piriton) orally, repeated after four hours if necessary. If the reaction is severe then 10 mg given intravenously may be more appropriate, but this will depend on clinical judgement. The possibility of producing allergic reactions is common to all injections and the usual precautions must be taken with an emergency tray available where lymphography is being done.

In the very rare case where a patient has a definite history of sensitivity to iodine, it would seem prudent to refuse a lymphogram since the iodine-containing contrast medium remains in the body for long periods, a year on the average in diagnostic amounts, and traces remain for much longer. We do not know exactly how much of a risk this is, but it would seem wiser not to take it (see also Chapter 12).

From the discussion above, it will be clear that the dangers of lymphography, when compared with other contrast investigations, stem from the oil causing complications in the lungs themselves or from the lungs acting as an inefficient filter and allowing through into the systematic circulation oily globules of too large a size which may cause embolization elsewhere, particularly in the brain. The amounts of contrast medium reaching the lungs should therefore be limited as much as possible consistent with the production of diagnostic lymphograms.

Lymphography is contraindicated in patients with cardiac shunts and in those with arterio-venous malformations in the lungs because of the risk of systematic oil emboli. Lymphography is also contraindicated in active thrombophlebitis and in patients with a well-documented history of sensitivity to iodine; these two aspects of the problem have been discussed above.

Too much Lipiodol can reach the lungs through the injection of a small vein instead of a lymphatic. Small veins take up the marker dye and on occasion it may be impossible to differentiate between vein and lymphatic. Control films of the site of injection immediately after the contrast medium begins to flow will decide the matter (Figs 13.3 and 13.4). Lipiodol in the venous system forms globules. If there is still doubt as to whether the injection is into a lymphatic or a vein then a further film taken higher up the leg will decide the matter (Fig. 13.5). Too much Lipiodol may reach the lungs because it is bypassed into the venous system early through lymphatico-venous anastomoses opening up. At the level of the lymph nodes, there is a safety valve mechanism which is not brought into use unless there is an obstruction to the flow of the lymph. If there is an obstruction, and this can be mechanical from previous surgery (Figs 13.6 A and B), or due to tumour replacement of the nodes (Figs. 13.7 A and B), then these channels may open up. If there has been previous surgery or a mass of nodes is suspected, then

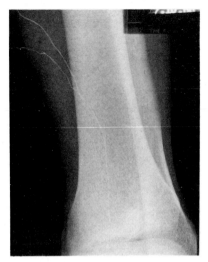 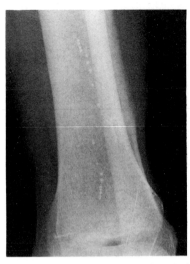

FIGURE 13.3. The typical appearance of Lipiodol in a leg lymphatic. It forms a solid column and the lymphatics tend to branch upwards.

FIGURE 13.4. The appearance of Lipiodol injected by mistake into a small vein. The Lipiodol globulates in the bloodstream.

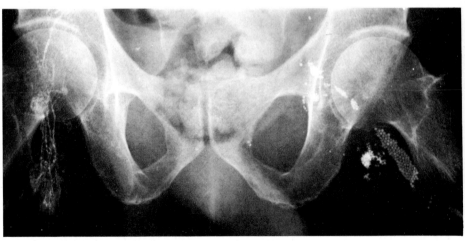

FIGURE 13.5. On the right side the Lipiodol is flowing up the lymphatics, but on the left it is seen globulating in the veins.

further control films should be taken of the affected area, when it is estimated that the Lipiodol has reached that level. If a lymphatico-venous communication is seen then the injection on that side should be stopped.

Too much Lipiodol may reach the lungs because there is less lymphoid

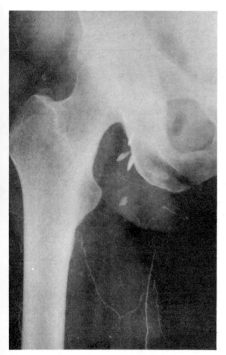
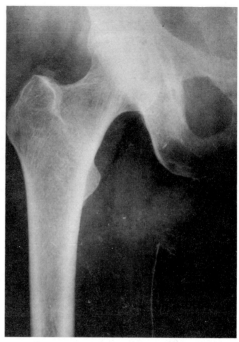

a   b

FIGURE 13.6a. A lymphatico-venous communication following a block dissection of inguinal nodes for malignant melanoma. The lymphatica are seen in the thigh. Numerous collateral channels have opened up, but there is also evidence of a lymphatico-venous communication with the femoral vein. Lipiodol is seen forming globules on the cusps of a valve in the femoral vein. 6b. Same patient ten minutes later, after the injection had been stopped and the leg exercised. The Lipiodol remains in the lymphatics, but has been flushed out of the femoral vein by the flow of blood.

tissue present to absorb it. This may be due to previous treatment which reduced the size of the nodes (normal as well as those involved by tumour), or there may be a congenital anomaly present with congenital absence of groups of nodes. It is necessary to take a further control film of the abdomen when approximately two-thirds of the estimated dose has been injected. If

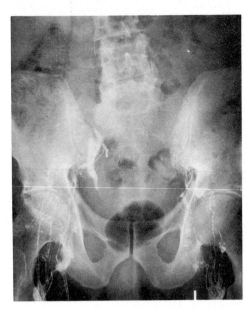

FIGURE 13.7a. A lymphatico-venous communication due to obstruction from massive involvement of the nodes by Hodgkin's disease. The veins are seen filling from the lymphatics on the right side of the pelvis.

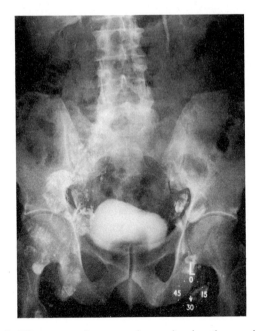

FIGURE 13.7b. The same patient at 24 hours showing the massively involved nodes on the right side of the pelvis (excretion urogram).

Lymphography

at this time the head of the contrast medium has reached the level of L.4, then enough has been injected and the injection should be stopped. The amount of contrast medium to inject is best estimated from the height of the patient. When injecting into the foot we never give more than 8 ml of Ultra Fluid Lipiodol per side even in the largest men. A slightly built patient of 5 ft will need about 5 ml per side and a baby of 18 months will need only 1 ml per side. If the control films suggested above are taken, then the volume given

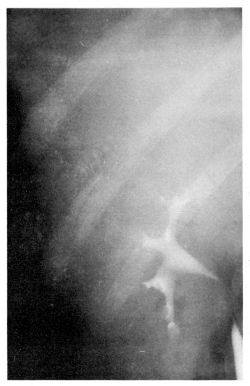

FIGURE 13.8. Hepatic oil embolism. The lipiodol shows as fine stippling in the liver.

can be adjusted for the individual and flooding of the lungs avoided. To do lymphography where there is no facility for taking control films is considered dangerous. The rate of injection should be about 6 ml (per side) in one hour.

In cases of grossly involved para-aortic nodes a lymphatico-portal shunt may develop (approximately 1 in 2000 cases) giving the picture of hepatic oil embolism (Fig. 13.8). This does not cause symptoms.

A preliminary radiograph of the chest is essential before deciding on lymphography and if there is any doubt about the lung function then great care should be taken, with restriction of the volume of the contrast medium used, or the investigation should be cancelled in severe dyspnoea. As pointed out by Davidson [14], therapeutic irradiation of the lungs may contribute to the passage through the lung of larger oil emboli, thus increasing the risk of cerebral oil embolism. No patient who is having irradiation to any part of the lung should have lymphography. It is our practice to wait for six weeks after the irradiation is complete, before attempting a lymphogram.

Patients suffering from pulmonary or cerebral complications should have full supportive therapy. Oxygen is probably the most important single factor, and shock and hypotension should be treated. Heparin has been suggested in fat embolism and seems to be effective clinically. In the case of cerebral oil embolism described above, heparin was used. In this case also the ECG showed right ventricular strain, but there was no evidence of failure. If there is evidence of cardiac failure then it may be necessary to digitalize the patient. In the severe allergic reactions which may occur, and where the lungs or the nervous system are reacting abnormally to the breakdown products of Lipiodol, there is a case for the use of steroids in an attempt to reduce the inflammatory response: they may also help to reduce the intracranial pressure in comatose patients. It should also be remembered that the serum calcium may fall: this should be watched for and treated if necessary.

CONCLUSION

Lymphography has made a considerable contribution to the diagnosis and management of malignant disease and the lymphoedemas. It is, however, potentially dangerous. For the First International Symposium on Lymphology in 1966, statistics on the incidence of serious complications of lymphography were collected from various centres throughout the world. These amounted to 1.3 per cent of lymphograms done on 16 501 patients and this included 18 deaths [11]. These data referred to the early days of lymphography using oily contrast media when too much was given. Often this was given too fast; the importance of patient selection particularly from the point of view of lung function was not realized nor was the importance of radiological control of the injection. The discussion which was initiated by these figures and by the whole section devoted to complications was very valuable and has contributed to the relative safety of the present-day investigation. No experienced centre now has a complication rate approaching this.

All contrast investigations have their complications, but if the precautions outlined above are taken and there is sensible preliminary assessment of the patients, with careful radiological control of the volume of Lipiodol Ultra Fluid injected, the complications, and, in particular, serious ones, should be kept to a minimum.

REFERENCES

1 KINMONTH J.B. (1972) *The Lymphatics*, p. 2. Edward Arnold, London.
2 KOPP W.L. (1966) Anaphylaxis from alphazurine 2G during lymphography. *J. Am. med. Ass.* **198**, 200–201.
3 REDMAN H.C. (1966) Dermatitis as a complication of lymphangiography. *Radiology* **86**, 323–326.
4 GOLD W.M., YOUKER J., ANDERSON S. & NADEL J.A. (1965) Pulmonary function abnormalities after lymphangiography. *New Engl. J. Med.* **273**, 519–524.
5 DOLAN P.A. (1966) Lymphography: complications encountered in 522 examinations. *Radiology* **86**, 876–880.
6 BAERT A.L. (1967) In: *Progress in Lymphology* (Proceedings of the International Symposium on Lymphology, Zurich 1966) (Ed. by RUTTIMANN A.), p. 315. Georg Thieme Verlag, Stuttgart.
7 FABEL H. (1967) In: *Progress in Lymphology* (Proceedings of the International Symposium on Lymphology, Zurich 1966) (Ed. by RUTTIMAN A.), p. 314. Georg Thieme Verlag, Stuttgart.
8 WALLACE S. (1967) In: *Progress in Lymphology* (Proceedings of the International Symposium on Lymphology, Zurich 1966) (Ed. by RUTTIMAN A.), p. 311. Georg Thieme Verlag, Stuttgart.
9 FISCHER H.W. (1969) In: *Lymphography in Cancer* (Ed. by FUCHS W.A., DAVIDSON J.W. and FISCHER H.W.), p. 28. Wm. Heinemann, London.
10 FRAIMOW M.D., WALLACE S., LEWIS P., GREENING R.R. & CATHCART R.T. (1965) Changes in pulmonary function due to lymphangiography. *Radiology* **85**, 231–241.
11 KOEHLER P.R. (1967) In: *Progress in Lymphology* (Proceedings of the International Symposium on Lymphology, Zurich 1966) (Ed. by RUTTIMAN A.), p. 324. Georg Thieme Verlag, Stuttgart.
12 KOEHLER P.R. (1967) In: *Progress in Lymphology* (Proceedings of the International Symposium on Lymphology, Zurich 1966) (Ed. by RUTTIMAN A.), p. 317. Georg Thieme Verlag, Stuttgart.
13 NELSON B., RUSH E.A., TAKASUGI M. & WITTENBERG J. (1965) Lipid embolism to the brain after lymphography. *N. Engl. J. Med.* **273**, 1132–1134.
14 DAVIDSON J.W. (1969) Lipid embolism to the brain following lymphography: Case report and experimental study. *Am. J. Roentg.* **105**, 763–771.
15 RASMUSSEN K.E. (1970) Retinal and cerebral fat emboli following lymphography with oily contrast media. *Acta Radiol. Diagn.* **10**, 199–201.
16 COLLARD M., LEROUX G., NOËL G. & DECLERCQ A. (1969) L'embolie cérébrale

graisseuse diffuse: complication de la lymphographie lipiodolée. *J. Radiol. Electrol.* **50,** 793–802.

17 ENGESET A., HÖEG K., HÖST H., LIVERUD K. & NESHEIM A. (1969) Thoracic duct lymph cytology in Hodgkin's Disease. *Int. J. Cancer.* **4,** 735–742.

CHAPTER 14: HYSTEROSALPINGOGRAPHY

E. BARNETT

Hysterosalpingography involves little risk of complications provided certain criteria are observed and the examination is carried out by an experienced operator, using fluoroscopic control.

The following complications may be recognized:

(1) *Pain*

Although the cervix is relatively insensitive, the patient may experience discomfort when the volsellum forceps are applied to the anterior lip, or when the cannula is inserted. In the presence of tubal occlusion, distension of the Fallopian tubes proximal to the block may induce pain which increases rapidly in severity if the injection of contrast medium continues. The location of the pain will vary according to the site of occlusion. Nonfilling of the tubes due to cornual spasm and distension of the uterus by the contrast medium may also induce discomfort; late onset of pain perhaps several days after hysterosalpinography should raise the suspicion of pelvic infection.

Discomfort, sometimes fairly severe, may be induced by irritation of the peritoneum by the contrast medium. The incidence of pain due to this cause varies with the type of contrast medium used. In the author's experience Endografin (50 per cent or 70 per cent methylglucamine iodipamide) is associated with a fairly high incidence of after pain. Little discomfort is experienced using Salpix (54 per cent sodium acetrizoate with polyvinylpyrrolidone) or Diaginol Viscous (40 per cent sodium acetrizoate with dextran) but the former tends to crystallize readily when cool. Urografin 76 per cent (a mixture of the sodium and methylglucamine salts of diatrizoate) is a very acceptable water-soluble contrast medium for hysterosalpinography. It has a low viscosity and yet produces a relatively dense shadow; moreover peritoneal spread occurs rapidly and induces little discomfort. However, it must be admitted that to some extent the incidence of after pain depends upon the pain threshold of the individual patient [1, 2]. Björk [3] found that the metrizoate dimer (iozamate) caused a lower incidence of abdominal pain, and preliminary studies with Dimer X (methylglucamine iocarmate) suggest that this produces less pain than Diaginol Viscous [4].

(2) *Trauma to uterus and cervix*

Theoretically the uterine wall or cervix may be damaged by the cannula but with an experienced operator this complication is highly unlikely. It is valuable to pass a uterine sound as a preliminary procedure to evaluate the size and direction of the uterine cavity. This will obviate possibility of damage to the uterine wall by the cannula. When the Green Armytage type of cannula is used, placing the rubber cone close to the tip of the cannula will limit the extent to which it can be advanced into the cervical canal. The Leech–Wilkinson type of cannula, which is normally 'screwed' into the external cervical os, is more traumatic than the Green Armytage type.

(3) *Haemorrhage and intravasation*

The occurrence of haemorrhage after hysterosalpinography usually indicates the presence of organic disease, e.g. fibroids or tuberculosis or malignant disease, but it can also occur due to trauma. These conditions predispose to intravasation of the contrast medium into the pelvic venous system. Intravasation may also be induced by the use of excessive pressure when injecting the contrast medium, or by faulty timing of the procedure (Figs 14.1, 14.2 and 14.3). A hysterosalpingogram should not be carried out within the seven days before or after a period, when the endometrium may be very congested or denuded. The procedure should also be avoided after recent curettage or abortion. Using an oily contrast medium, e.g. fluid Neohydriol, the possible danger of pulmonary oil embolism must be considered although this complication is rare. Retinal embolism has been reported after hystersalpingography [5]. Lymphatic intravasation is a rare occurrence and suggests the presence of uterine tuberculosis [6–9]. It is not usually associated with symptoms.

(4) *Chemical and inflammatory reaction*

Foreign body granuloma formation has been described in the peritoneal cavity in association with oily contrast media [10, 11]. But Siegler [12] claims

Hysterosalpingography

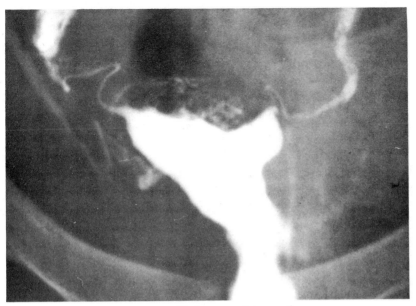

FIGURE 14.1. Early venous intravasation

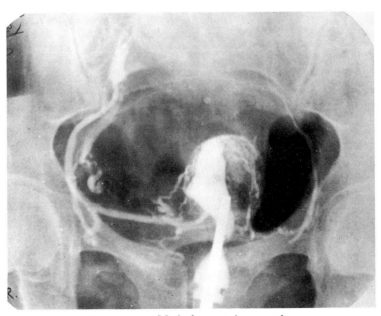

FIGURE 14.2. Marked venous intravasation.

that this complication is rare in the presence of normal tubes with active peristalsis, and that granulomas primarily arise in occluded Fallopian tubes.

Iodine sensitivity is uncommon and usually presents as urticarial or acneiform skin eruptions. It is stated that in nearly every case of allergic reaction, contrast medium is forced into the uterine circulation [13].

The injection of contrast medium can transmit bacteria from the uterus

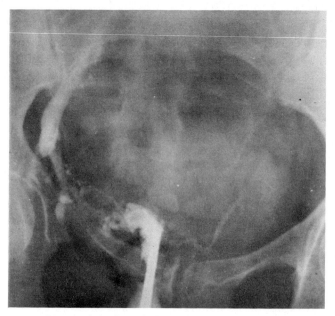

FIGURE 14.3. Utero-tubal tuberculosis. Deformed uterus with venous intravasation.

into the peritoneal cavity, but this complication is uncommon. Nevertheless hysterosalpingography should not be carried out in the presence of a history of recent pelvic infection, unless a suitable course of antibiotic treatment has been completed. As already noted, the onset of symptoms due to bacterial pelvic inflammation following hysterosalpingography may be delayed sometimes for several days.

(5) *Abortion of an early pregnancy*

Occasionally a hysterosalpingogram is inadvertently performed in the presence of an early pregnancy. This occurrence is unlikely if the procedure is carried out at about the seven to tenth day after the beginning of a period, although theoretically there is the danger of damage to the early fetus. The presence of an early pregnancy is usually recognized very early in the examination and, in practice, abortion is unlikely.

(6) *Rupture of an ectopic pregnancy*

The X-ray appearances of tubal pregnancy were described by Neilsen [14] but hysterosalpingography is now considered to be contra-indicated in cases of suspected ectopic pregnancy in view of the danger of tubal rupture, haemorrhage and possible infection. Diagnostic ultrasound is now the investigation of choice. In a series of 21 cases of ectopic pregnancy reported by Korayashi *et al.* [15], the ultrasonic diagnosis was correct in 16. However, the diagnosis of advanced extrauterine pregnancy can be very difficult and hysterography will contribute to the diagnosis by the demonstration of an empty uterus.

It may be concluded that hysterosalpingography is a valuable procedure which involves little risk to the patient when carried out under suitable aseptic conditions by an experienced operator.

REFERENCES

1 BARNETT E. (1955) The clinical value of hysterosalpingography. *J. Foc. Radiol.* 7, 115.
2 BARNETT E. (1972) In: *Practical Procedures in Diagnostic Radiology* (Ed. by SAXTON H.M. & STRICKLAND B.). H. K. Lewis, London.
3 BJORK L., ERIKSON U., INGLEMAN B. & WILBRAND H. (1972) A new contrast medium for hysterosalpingo-pelvigraphy. *Acta Radiol. Diagn.* 12, 891–896.
4 MARKHAM G.C., BOTTOMLEY J.P. & ANSELL G. (1976) Dimer X in hysterosalpingography (to be published).
5 CHARAWANAMUTTU A.M., HUGHES-BURSE J. & HAWLETT J.D. (1973) Retinal embolism after hysterosalpingography. *Br. J. Ophthalmol.* 57, 166.
6 WHITE M.M. (1952) Genital tuberculosis in the female. *J. Obstet. Gynaec. Br. Emp.* 59, 746.
7 KIKA J. (1954) Clinical analysis of 'angiograms' found in the course of hysterosalpingography with special reference to tuberculosis of the female genitals. *Am. J. Obstet. Gynaec.* 67, 56.
8 ROZIN S. (1965) *Uterosalpinography in Gynaecology.* Charles C. Thomas, Springfield, Illinois.
9 ROZIN S. (1952) X-Ray diagnosis of genital tuberculosis. *J. Obstet. Gynaec. Brit. Emp.* 59, 59.
10 RIES E. (1929) Effect of Lipiodol injection on the tubes. *Amer. J. Obstet. Gynaec.* 17, 720.
11 BROWN W.E., JENNINGS A.F. & BRADBURY J.T. (1949) Absorption of radio-opaque substance used in hysterosalpingography. *Amer. J. Obstet. Gynaec.* 58, 1041.

12 SIEGLER A.M. (1967) *Hysterosalpingography*, p. 44. Hoeber Medical Division, Scranton, Pa. Harper & Row, New York.
13 HOLT B.B. & ARMSTRONG J.T. (1970) Dangers and contraindications to hysterosalpingography. *Texas Medicine* 66, 44.
14 NEILSEN B. (1947) Diagnosis of ectopic pregnancy by hysterosalpingography. *Acta Radiol.* 28, 185.
15 KORAYASHI M., HELLMAN L.A. & FILLISTI L.P. (1969) Ultrasound. An aid in the diagnosis of ectopic pregnancy. *Amer. J. Obstet. Gynaec.* 103 1131.

CHAPTER 15: AMNIOGRAPHY AND INTRAPERITONEAL TRANSFUSION

E. BARNETT

Amniography was first introduced by Menees, Miller & Holly in 1930 [1] as a procedure for placental localization, but the water-soluble contrast media available at that time were rather irritant and induction of labour was a common consequence so that the procedure fell into disrepute. However, revived interest in amniography developed primarily as a preliminary procedure to intrauterine transfusion following the report by Liley in 1963 [2]. This was encouraged by the development of more suitable water-soluble contrast media, e.g. Hypaque 45 and Conray 280.

At the present time the major use of amniography in this country is to outline the fetal intestine as a preliminary procedure to intrauterine transfusion of the fetus. The incidental localization of the placenta is also valuable additional information. Other indications are listed by McLain [3], e.g.

(a) The demonstration of uterine deformities, uterine tumours.
(b) Demonstration of major soft tissue abnormalities of the fetus.
(c) The study of the cause of hydramnios and the detection of oesophageal and duodenal atresia of the fetus.
(d) The study of fetal gastro-intestinal motility and therefore possible signs of fetal distress *in utero*.
(e) The definite determination of fetal death *in utero* by the absence of swallowing.
(f) The determination of whether multiple pregnancies are monamniotic or biamniotic.
(g) Study of fetal respiration.
(h) The diagnosis of hydatidiform mole, rupture of amniotic membranes and fetal hydrops *in utero*.

The procedure of intrauterine transfusion of the fetus depends on the fact that erythrocytes are absorbed into the fetal circulation through the peritoneum [4, 5]. Since the introduction of anti-D gamma globulin, there

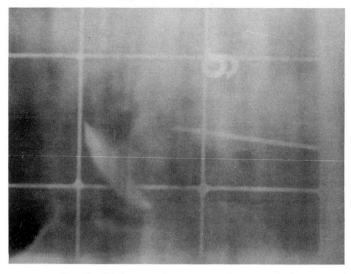

FIGURE 15.1. Localized injection of contrast medium into soft tissues of the fetal back.

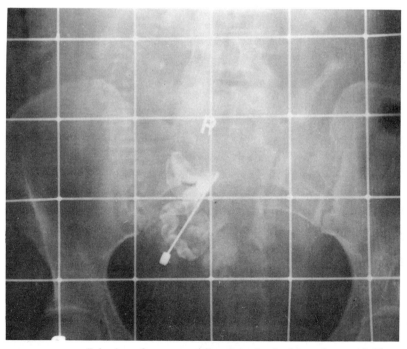

FIGURE 15.2. Injection of contrast medium into the fetal intestine—probably small bowel.

Amniography and Intraperitoneal Transfusion

has been a marked reduction in the number of cases requiring intrauterine transfusion, but this will still be necessary in cases where the mother is already highly sensitized. The complications of this procedure can be divided into fetal and maternal.

FETAL COMPLICATIONS

There have been reports of faulty positioning of the needle in various sites, e.g. pleural cavity, pericardium, fetal soft tissues (Fig. 15.1), liver, small

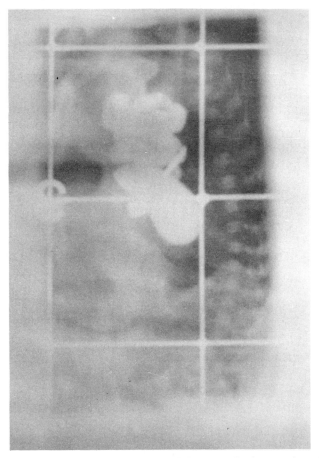

FIGURE 15.3. Contrast medium injected into the fetal stomach and small bowel.

bowel (Fig. 15.2), stomach (Fig. 15.3), bladder, vertebral canal (Fig. 15.4), vascular spaces (Fig. 15.5), colon, kidney, umbilical cord [6–21]. Accidental injection of contrast medium into the subcutaneous tissues has caused sloughing of the skin [22].

The risk to the fetus is greatest when the injection is made into the vertebral canal or into vascular spaces. In the presence of a small baby, the inadvertent insertion of the needle into the soft tissues of the fetus may be

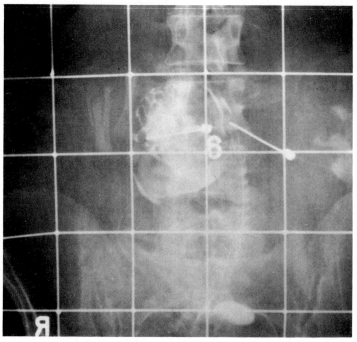

FIGURE 15.4. Injection of contrast medium into the vertebral canal. Second needle correctly placed in the fetal peritoneal cavity.

utilized to steady the baby and permit the correct placing of a second needle in the peritoneal cavity [6, 22]. The risk of traumatic death is at least three times as great when the placenta lies anteriorly [14].

The passage of the needle through the placenta may cause bleeding with or without placental separation. This may allow the admixture of fetal and maternal blood and result in an increase of circulating antibodies with the danger of sudden fetal death [24–26]. Cassady [27] reported four fetal deaths due to blind traversal of an anterior placenta, resulting in laceration of a major feto-placental vascular channel with subsequent fetal exsanguination.

Meconium peritonitis has been reported following intrauterine transfusion [28]. Two cases of thyroid hyperplasia have recently been reported as being possibly attributable to the use of iodized media (Urographin and Ultrafluid Lipiodol), at amniography [29].

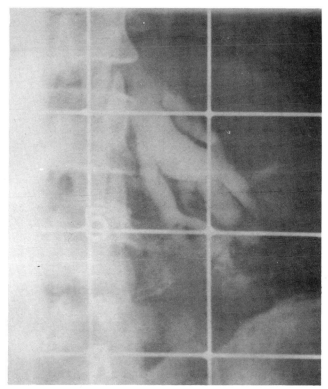

FIGURE 15.5. Contrast medium presumably in chorio-decidual space.

MATERNAL COMPLICATIONS

The risk to the mother is small. It is claimed that infection is probably the most serious maternal risk [14]. In a study of 584 cases reported by Queenan [30] there were 58 cases of infection and 4 subsequently required hysterectomy. The outlining of the fetal soft tissues may also permit the early recognition of developing hydrops by the demonstration of increased thickness of the scalp and enlargement of the fetal abdomen. Increased scalp thickness is also a feature of fetal death. Hepatomegaly and fetal ascites may also be diagnosed by ultrasonic means [31].

Premature rupture of the membranes and onset of labour is a recognized complication following intraperitoneal transfusion, as is haematoma of the abdominal or uterine walls. Theoretically, leakage of amniotic fluid into the peritoneal cavity could cause peritoneal irritation. Amniotic embolism has not been reported following amniocentensis.

PREVENTION OF COMPLICATIONS

The patient's bladder should be empty. Premedication is necessary and the patient should be permitted to relax for some time on the X-ray table before the procedure is commenced in order to reduce the tendency to fetal movement. The day before the intrauterine transfusion, amniocentensis is often carried out for the purposes of spectro-photometric analysis. It is useful to

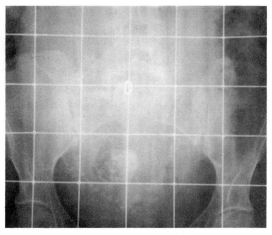

FIGURE 15.6. Normal appearance of ingested contrast medium in the fetal intestine following amniography.

take the opportunity at this time to inject 30 ml of a water-soluble contrast medium, such as Conray 280. If the fetus is alive, it will swallow the opacified liquor and in the absence of oesophageal or duodenal atrasia contrast medium will normally be evident in the fetal bowel when the intrauterine transfusion is carried out next day (Fig. 15.6). This facilitates the correct siting of the Tuohy needle. Severely affected babies may not swallow the liquor. An oily contrast medium may also be used to outline the fetal abdomen; it adheres to the vernix caseosa of the skin of the fetus, thus outlining the soft tissues [10,

32]. Carbon dioxide may be injected as a contrast medium under fluoroscopic control to facilitate the placing of the needle [8, 33].

The placental site should be located prior to amniography [34]. Ultrasound is probably the most widely used method of placental localization at the present time and the diagnostic accuracy is about 95 per cent [35]. However, during the preliminary amniography prior to IUT the placental site may be determined. The risk of traumatic death is at least three times as great when the placenta lies anteriorly [14]. It may not be possible to avoid puncturing the placenta, but if the placenta must be traversed, an attempt should be made to puncture it as near to the edge as possible, avoiding the largest fetal vessels which are likely to be more central.

A wire grid as described by Lees [36] should be applied to the anterior aspect of the abdomen with a marker on the umbilicus, and a preliminary radiograph taken. A 'step ladder' type of grid may be applied to the lateral aspect of the abdomen if desired and a second control film using a horizontal beam is taken. This latter film will indicate the depth to which the needle should be inserted. Prior to the administration of the actual transfusion it is advisable to confirm the correct siting of the needle by passing a fine epidural catheter through the needle and injecting a further 2–5 ml of contrast medium. If the needle is correctly placed the contrast medium is seen to spread over the surface of the fetal intestine, the liver and under the diaphragm (Fig. 15.7). In the presence of ascites the mixture of the ascitic fluid and contrast medium produces a diffuse ground-glass appearance within the confines of the abdomen (Fig. 15.8).

If the back of the fetus is anterior, it may be preferable to delay the procedure for a few days in the hope that the fetus will alter position, or an attempt may be made to change the position of the fetus by external palpation [37]. However, merely rotating the patient into an oblique position may allow the fetus to rotate sufficiently to permit access to the abdomen.

The risks to the fetus arising from intrauterine transfusion are many, and may be serious. Dempster [38] accepts an 8 per cent risk of fetal death from trauma with each transfusion undertaken and a 25 per cent mortality rate in nonhydropic fetuses undergoing the three fetal transfusions usually required from 24 weeks. In the presence of fetal ascites, intrauterine transfusion may cause fetal death due to cardiac failure. Reports from the 53rd Ross Conference on Pediatric Research, 1966, and the International Symposium on the Rh. Problems in 1970, show a stillbirth rate of about one-third, neonatal deaths one-third and survivors one-third [35, 38]. Close co-operation between the radiologist and obstetrician will help to reduce the accepted risk to the fetus.

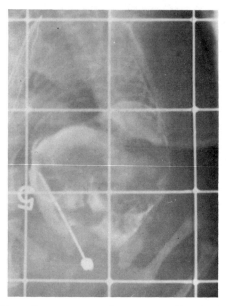

FIGURE 15.7. Normal appearances of contrast medium in the fetal peritoneal cavity. Medium spreading over the surface of bowel loops and the liver and outlining the diaphragm.

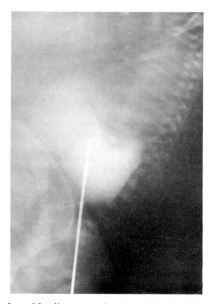

FIGURE 15.8. Ascites. Needle correctly placed in the fetal peritoneal cavity. Admixture of contrast medium with ascitic fluid producing a diffuse ground-glass appearance.

REFERENCES

1 MENEES T.O., MILLER J.D. & HOLLY L.E. (1930) Amniography. A preliminary report. *Am. J. Roentg.* **24**, 363.
2 LILEY A.W. (1963) Intra-uterine transfusion of foetus in haemolytic disease. *Br. med. J.* **2**, 1107.
3 MCLAIN C.R. JR (1964) Amniography: versatile diagnostic procedure in obstetrics. *Obstet. Gynaec.* **23**, 45.
4 MELLISH P. & WOLMAN I.J. (1958) Intraperitoneal blood transfusions. *Am. J. Med. Sc.* **235**, 717.
5 SCOPES J.W. (1963) Intraperitoneal transfusion of blood in newborn babies. *Lancet* **1**, 1027.
6 LIGGINS G.C. (1966) Fetal transfusion by the impaling technique. *Obstet. Gynaec. N.Y.* **27**, 617.
7 CAMPBELL B.L. & STEWART J.H. (1966) Radiology of foetal blood transfusion. *Br. J. Radiol.* **39**, 81.
8 LECKY J.W., BACHORE R.A., TOBIN P.A. & HANAFEE W.N. (1968) Percutaneous intra-uterine fetal transfusion. *Am. J. Roentg.* **103**, 186.
9 TESSARO A.N. & CHASLER C.N. (1968) Amniography as an aid in intra-uterine transfusion. *Am. J. Roentg.* **103**, 195.
10 RAPHAEL M., GORDON H. & SCHIFF D. (1967) Radiologic aspects of intra-uterine blood transfusion. *Br. J. Radiol.* **40**, 520.
11 OGDEN J.A., WADE M.E. & DAVIS C. (1969) Radiological aspects of fetal intra-uterine transfusion. *Radiology* **93**, 1315.
12 GRECH P. & BEVIS D.C.A. (1969) Foetal myelography—an unusual complication in intrauterine transfusion. *Br. J. Radiol.* **42**, 389.
13 GREEN G.H., LILEY A.W. & LIGGINS G.C. (1965) The place of foetal transfusion in haemolytic disease. *Aust. N.Z. J. Obstet. Gynaec.* **5**, 53.
14 FRIESEN R.F. (1971) Complications of intra-uterine transfusion. *Clin. Obstet. Gynaec.* **14**, 572.
15 HOUSTON C.S. & BROWN A.B. (1966) An unusual complication of intra-uterine transfusion. *Canad. Med. Ass. J.* **94**, 1274.
16 HOLMAN C.A. & KARNICKI J. (1964) Intrauterine transfusion for haemolytic disease of the unborn. *Br. Med. J.* **2**, 594.
17 LILEY A.W. (1965) The use of amniocentesis and fetal transfusion in erythroblastosis fetalis. *Pediatrics* **35**, 836.
18 DUHRING J.L., KAFOGLIS K.Z. & GREENE J.W. JR (1966) Intra-uterine fetal transfusion. *J. Am. med. Ass.* **195**, 135.
19 KARNICKI J. (1966) Intrauterine transfusion. *Proc. Roy. Soc. Med.* **59**, 83.
20 MANDELBAUM B. & EVANS T.N. (1969) Life in the amniotic fluid. *Am. J. Obstet. Gynaec.* **104**, 365.
21 CREASMAN W.T., LAWRENCE R.A. & THIEDE H.A. (1968) Fetal complications of amniocentesis. *J. Am. med. Ass.* **204**, 91.
22 SULLIVAN T. & SMITH G.F. (1970) Complication of amniography. *Lancet* **1**, 946.
23 QUEENAN J.T., ANDERSON G.G. & MEAD P.B. (1966) Intra-uterine transfusion by multiple-needle technique. *J. Am. med. Assoc.* **196**, 664.

24 WANG M.Y.W., MCCUTCHEON E. & DESFORGES J.F. (1967) Feto-maternal hemorrhage from diagnostic transabdominal amniocentesis. *Am. J. Obstet. Gynaec.* **97**, 1123.
25 MISENHIMER H.R. (1966) Fetal hemorrhage associated with amniocentesis. *Am. J. Obstet. Gynaec.* **94**, 1133.
26 BOWMAN J.M. & POLLOCK J.M. (1965) Amniotic fluid spectrophotometry and early delivery in management of erythroblastosis fetalis. *Paediatrics* **35**, 815.
27 CASSADY G., BARNETT R. & CEBALLOS R. (1971) Dangers of fetal transfusion—importance of placental localisation. *Am. J. Obstet. Gynaec.* **110**, 672.
28 KARNICKI J. (1968) Results and hazards of prenatal transfusion. *J. Obstet. Gynaec. Brit. Cwlth* **75**, 1209.
29 MUXI M., PÉREZ SOLER J., ALCON S. et al. (1972) Presentacion de dos casos de hyperplasia del tiroides, possiblemente desencadenada par la administracion de products yodados en la amniografia previa a la T.I.U. *Toko–Ginec. Pract.* **31**, 79.
30 QUEENAN J.T. (1969) Intrauterine transfusion. A co-operative study. *Am. J. Obstet. Gynaec.* **104**, 397.
31 GARETT W. & KOSSOFF G. (1973) Ultrasonic diagnosis of fetal abnormalities. *Proceedings of 2nd World Congress on Ultrasonics in Medicine.* Excepta Medica.
32 DAW E. (1970) An assessment of intra-uterine foetal visualisation. *Br. J. Radiol.* **43**, 710.
33 HANAFEE W. & BISHORE R. (1965) Carbon dioxide and horizontal fluoroscopy in intra-uterine fetal transfusion. *Radiology* **85**, 481.
34 PAULS F. & BOUTROS P. (1970) Value of placental localisation prior to amniocentesis. *Obstet. & Gynaec.* **35**, 175.
35 DONALD I. & ABDULLA U. (1968) Placentography by sonar. *J. Obstet. Gynaec. Brit. Cwlth* **75**, 993.
36 LEES L.A. (1966) Grid localisation technique for intra-uterine transfusions. *Br. Med. J.* **2**, 287.
37 POWELL L.C. JR. & SCHREIBER M.H. (1974) Intra-uterine fetal transfusion. *Radiol. Clin. N. Amer.* **12**, 37.
38 DEMPSTER J.W. (1966) Radiological aspects of intra-uterine blood transfusion. *Br. J. Radiol.* **40**, 960.
39 LUCY J.F. & BUTTERFIELD L.J. (eds) (1966) *Report on the 53rd Ross Conference on Pediatric Research. Intra-uterine Transfusion and Erythroblastosis Fetalis.* Ross Laboratories, Columbus, Ohio.

CHAPTER 16: ALIMENTARY TRACT

G. ANSELL

ORAL BARIUM

Barium meal examinations are virtually the bread and butter of radiology and there must be few examinations which have a higher margin of safety in relation to their diagnostic yield. Nevertheless, to many patients, they may still be a source of considerable anxiety. With the use of image intensification, fainting is perhaps less common than in the days of hot, dark, X-ray rooms, but it may still occur and the experienced radiologist often develops a sixth sense which enables him to put the table into the horizontal position before the patient collapses on the floor and possibly sustains an injury. More important, the collapse may occasionally be caused by or progress to cardiac arrest, particularly in elderly debilitated or cardiac patients. In these cases prompt resuscitation may avert a fatal outcome [1]. Patients who have been confined to bed for even a few days may rapidly lose vasomotor tone and be liable to sudden fainting. Loss of vasomotor tone may, of course, also occur in the pyrexial or toxic patient. Where there is any doubt, the patient should be examined in the recumbent position or, if necessary, in the semi-erect position with the table tilted 30° from the vertical. The modern trend to remote control examination might be expected to exacerbate the problem. With the patient supported by retaining straps it is conceivable that a faint in the erect position might not be immediately noticed. The failure to alter the posture might then impair both the blood supply to the brain and the venous return to the heart, with resultant cerebral damage or cardiac arrest. In infants barium meals are sometimes administered after sedation. This can be hazardous if the patient vomits and inhales the barium [1]. Perforation is discussed in the section on barium enemas.

Obstruction

It is well known that a barium meal may impact in the colon and precipitate an acute obstruction at the site of a carcinoma [2], or diverticulitis (Fig. 16.1). It is less often appreciated that there may be prolonged stasis of barium in

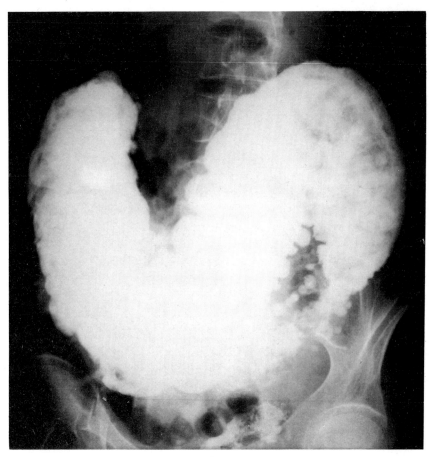

FIGURE 16.1. Female, age 61. Follow-through film four days after barium meal for 'weight loss and anorexia'. Impaction of barium in colon due to extensive diverticular change in descending and pelvic colon. Obstruction relieved by transverse colostomy. Diverticular area resected at later date.

the colon without any obvious obstructing lesion. Prout *et al.* [3] described four elderly patients who were admitted two to six weeks after barium meal examinations with severe constipation and extensive accumulations of barium in the colon shown on straight X-rays. Routine purgation and enemas failed to dislodge the barium but lactulose was effective. In a fifth case in the same series, barium obstructing at a colonic carcinoma was also cleared by lactulose. In small bowel lesions, however, there is probably no significant risk of an obstruction being aggravated by barium, since the large accumulation of fluid should prevent inspissation.

Minor degrees of barium impaction are not uncommon in routine barium meal examinations and patients may suffer considerable discomfort from the passage of rock-like masses of hardened barium. It is a good practice to warn patients of the possible constipating effects of barium and to advise the use of a laxative or liquid paraffin when necessary.

Barium appendicitis

Barium retained in the appendix for a prolonged period may act as an obstructing faecolith and cause acute appendicitis. In three such cases, appendicitis occurred on the 9th, 10th and 105th days after a barium meal [4]. In an extreme example of barium retention, a large opaque *barolith* was present in the caecum two years after a barium meal and required operative removal. The barolith was found to be firmly attached to the mucosa but separation was possible by gentle teasing [5].

Poisoning, etc.

The soluble salts of barium are highly poisonous and in the early days of gastro-intestinal radiology, occasional cases of poisoning occurred due to contamination of barium sulphate by one of the soluble salts. The widespread use of commercial preparations of barium sulphate has largely eliminated this risk but a recent case of fatal barium poisoning has been reported from India, where a pharmacist mis-read a prescription for barium sulphate and dispensed barium sulphide [6]. This caused corrosive changes, vomiting and diarrhoea followed by vascular paralysis and cardiac arrest. In the emergency treatment of these cases, sodium sulphate or magnesium sulphate can be given orally in an attempt to form an insoluble precipitate of barium sulphate. Acute barium poisoning is associated with a rapid and severe decrease of serum potassium, apparently due to a shift from extracellular to intracellular compartments, and recovery has been reported following an intravenous infusion of potassium [7]. Artificial respiration may also be required to counteract temporary paralysis of the skeletal muscles involving respiration.

Suspensions of barium sulphate may become contaminated and growths of *E. coli*, *Klebsiella*, *Strep. faecalis*, *Pseudomonas* and *Clostridia* have been found in barium solutions which have been allowed to stand overnight in reservoirs after mixing. Infection of the barium solution may be associated with a disagreeable odour or taste of the barium. To avoid the growth of contaminants, all barium receptacles should be thoroughly cleansed and dried between use. Alternatively, individual doses of sterilized suspensions of barium sulphate may be obtained in prepackaged form [8].

Mistakes may also arise due to mis-labelling. Felson [9] records an episode where Plaster of Paris was stored in a barium container and administered to four patients in error. The plaster set to form solid gastric casts but these gradually crumbled after a few days and were passed in due course. Accidental administration of washing powders containing alkali in mistake for barium may cause acute corrosive poisoning.

Hypotonic duodenography

Anticholinergic drugs may cause transient side effects such as tachycardia, dryness of the mucous membranes, difficulty in micturition and mydriasis. Because of the latter effect they are contraindicated in glaucoma. Three cases of *acute gastric dilatation* have been reported following the administration of 30 mg Probanthine for hypotonic duodenography [10]. In two of these patients partial obstruction of the third part of the duodenum by a carcinoma of the pancreas was exacerbated by the anticholinergic agent. In the first case, gross gastric dilatation with retained fluid and barium was noted 36 hours after the examination and the patient died from pulmonary embolism. In the second case, massive gastric dilatation occurred at the conclusion of the examination and required treatment by nasogastric suction. No organic obstruction was present in the third patient but acute dilatation of the stomach occurred at twelve hours and was successfully treated by nasogastric suction. It was suggested that partial duodenal obstruction should be regarded as a contraindication and that excessive quantities of air and barium should in any case be avoided. Food should also be withheld for eight to twelve hours after the examination and until straight radiography confirms that the stomach has returned to normal size. Presumably, this complication would be unlikely with hypotonic agents of briefer duration such as Bucopan and glucagon.

WATER-SOLUBLE MEDIA

Gastrografin is a 76 per cent aqueous solution of sodium methylglucamine diatrizoate with 0.1 per cent of the wetting agent Tween 80 and added flavouring. Originally introduced for use in the investigation of a wide range of acute abdominal disorders, it can be of particular value in cases of suspected perforation or acute pancreatitis but is now less frequently used in the investigation of haematemesis or small bowel obstruction.

Hypovolaemia

The major disadvantage of Gastrografin is its hypertonicity, since it has an osmolarity of 1900 milliosmols per litre, approximately six times that of normal serum [11]. This hypertonicity causes fluid to be drawn into the gastrointestinal tract and frequently has a marked cathartic effect. Loss of fluid from the circulation causes hypovolaemia and in experimental animals there may be a decrease in circulating plasma volume of 15 to 30 per cent with an increase in serum osmolarity. Similar changes have also been noted in adult patients following the oral administration of Gastrografin: increases occur in the serum values of protein, calcium, uric acid and increases in the haematocrit of up to 11 per cent [12]. These changes can be particularly serious in cases where the plasma volume is initially low, as for example in dehydrated or malnourished children. Collapse may also occur in debilitated adults. Caution is therefore necessary in the use of Gastrografin: dehydration should be corrected prior to, and following, its administration.

Recently, the hyperosmolar property of Gastrografin has been used with advantage for the nonoperative treatment of uncomplicated meconium ileus. A high Gastrografin enema is administered under fluoroscopic control. Large volumes of fluid are drawn into the bowel, loosening the viscid meconium and allowing this to be passed per rectum. As with oral Gastrografin, this may cause hypovolaemia with an increase in haematocrit and a profound reduction in cardiac output. It is, therefore, essential that the infant should be adequately hydrated before the procedure. Additional intravenous fluids should be administered during the examination and the water balance should be monitored for several hours [13]. Animal experiments suggest that some of the Gastrografin enema may be absorbed and that this may contribute to the increase in the plasma osmolarity [14].

Ileus

Somewhat surprisingly, in view of the cathartic action of Gastrografin, it appears possible that the post-operative administration of this preparation may have been a cause of ileus in 4 per cent of patients [15]. Other authors have also noticed a deterioration in the clinical condition with increase in symptoms and/or abdominal distension in several patients who had received Gastrografin for the assessment of acute abdominal problems [12]. There is some experimental support for this possibility, since it has been found that in 50 per cent of cats who had a laparotomy with handling of the bowel, per-oral administration of Hypaque (diatrizoate) caused prolongation of the small bowel transit time, whereas this effect did not occur with barium sulphate[16].

Inhalation

Accidental inhalation of Gastrografin may cause little in the way of cough or immediate dyspnoea. However, after a brief interval, there may be respiratory collapse and death from pulmonary oedema [17, 18]. The risks are probably higher in patients with pre-existing pulmonary disease and cor pulmonale. Pulmonary oedema has also been produced experimentally in animals by intratracheal injection of diatrizoate [19]. In animals who survive the attack of pulmonary odema, there does not appear to be any permanent lung damage [20]. Gastrografin has been advocated in the past in preference to barium when there was a risk of aspiration. This seems unwise, especially since small amounts of barium are relatively innocuous in the bronchi, providing that it does not cause significant bronchial obstruction. Inhalation of barium after vomiting, presents a more serious risk because of the possibility of the acid gastric contents causing Mendelson's syndrome.

Precipitation

Diatrizoate is precipitated in an acid environment at a strength of 0.1 N hydrochloric acid to form water-insoluble carboxylic acid. In patients with hyperchlorhydria, a solid putty-like precipitate may occur in the stomach after instillation of Gastrografin [21]. Similar precipitation of Gastrografin has occurred inside the balloon of a Sengstaken–Blakemore tube which had remained in the stomach for six days, with resultant difficulty in withdrawal of the tube. In this case it was assumed that hydrogen ions had passed across the rubber membrane of the balloon [22].

Gastrografin can apparently also be precipitated in the achlorhydric stomach after partial gastrectomy and vagotomy, if stomal obstruction is present. In a recently reported case, Gastrografin remaining in the gastric stump for 24 hours formed a solid precipitate, which caused multiple gastric erosions with a massive haematemsesis. At operation, the erosions were restricted to the area of contact of the Gastrografin bolus [23].

Hypersensitivity

It was originally believed that water-soluble iodinated contrast media were not absorbed from the gastrointestinal tract in the absence of a perforation, but this is no longer accepted. There is therefore reason to believe that idiosyncrasy reactions may occur and two case reports are available to support this. In the first case, a five-week-old infant developed widespread urticaria fifteen minutes after a swallow performed with Dionosil oily (propyliodone) [1]. The second case involved a 66-year-old man who received 2 ml of Gastrografin during the course of a jejunal biopsy. He developed acute angioneurotic oedema and this was followed by a cerebrovascular accident which proved fatal [24]. In this case, it is also possible that the medium may have been absorbed through the biopsy site.

BARIUM ENEMA

Perforation

Pyle & Samuel [25] and Seeman & Wells [26] have reviewed many of the important complications of barium enema examinations. The most important hazard is undoubtedly that of perforation with extravasation of barium into the retro-peritoneal tissues, the peritoneum, or both. This complication is probably not excessively rare. In a recent survey of some 250 000 barium enemas [27], there were fifteen cases of intraperitoneal rupture and six cases of extraperitoneal rupture; an incidence of approximately 1 in 12 000 and 1 in 40 000 respectively. It is possible, however, that some minor cases of extraperitoneal leakage may have remained undiagnosed. In a smaller series there were four cases of intraperitoneal rupture in 10 000 enemas, an incidence of 1 in 2250 examinations [39]. In an earlier study among 100 teaching institutes in the United States, Zheutlin et al. [28] collected 53 cases of intraperitoneal rupture of the colon for analysis. In the majority of these cases, perforation occurred *during* the administration of the enema but in seven cases perforation apparently occurred *following* the enema and, in one case, 24 hours later. In four cases, perforation had occurred *prior* to the barium study of the colon. In this series, the mortality rate was 47 per cent in surgically treated patients and 58 per cent in those treated conservatively. Although this difference in mortality was partly due to patients in the conservative group being in a poorer general condition it was nevertheless concluded that early surgical treatment provided the best chance for survival. In the Australian

TABLE 16.1 Barium enema perforations in U.K. survey

Case	Sex, Age	Catheter details	Description of perforation	Follow-up	Additional comments
COLONIC RUPTURE					
1	F 70	20 ml air in balloon catheter.	No complaint but extravasation into sigmoid mesentery seen at fluoroscopy.	No barium in peritoneal cavity. Transverse colostomy. Died 3 months later.	P.M. Ruptured diverticulum. Barium in mesentery caused infarction of colon.
2	F 73	Balloon catheter. Inflation 'painful'.	Collapse and shock when barium enema commenced. Barium in peritoneum but films misinterpreted at first.	Diagnosis not made until 5 hours later. Patient then appeared improved so operation deferred to allow further recovery. Sudden collapse and death 12 hours after perforation.	Previous colostomy for diverticulitis. P.M. Perforation through diverticulum.
3	M 67	Balloon catheter. Normal height of barium.	As barium reached pelvic colon, sudden severe pain. 'Felt something burst.' Severe shock. Flooding of peritoneal cavity seen at fluoroscopy.	Carcinoma pelvic colon. Slit in colon distal to growth and in growth itself. Tumour resected. Death 12 hours later.	P.M. Bowel wall at site of distal perforation was not abnormally thin. No inflammatory change.

4	M 84	Balloon catheter. Discomfort on inflation.	Barium in peri-rectal soft tissues seen at fluoroscopy. No discomfort or shock. Enema discontinued.	Laparotomy 2½ hours later. Transverse colostomy, pelvic drainage, excision mucocoele of gall bladder. Collapsed and died 13 hours after perforation.	Free spread of barium in mesentery after flow discontinued. Small quantity of barium entered peritoneal cavity through sigmoid diverticulum. Patient 'wasted'—atrophic changes in rectal wall and mesentery.
5	F 78	Balloon catheter.	Slight discomfort during enema but initial films normal. Lower abdominal discomfort and faintness after evacuation but soon recovered and went home. Collapsed later and re-admitted. Post-evacuation film showed barium in retroperitoneal tissues.	Colostomy. Alive.	Sigmodiscopy 3 weeks earlier was normal.
6	M 67	Bulbous polythene cannula. Difficult insertion.	Para-rectal extravasation of barium.	Temporary colostomy for 4 months. Well at follow-up 9 months after perforation.	Small tear in rectal wall due to difficult cannulation.

342   Chapter 16

TABLE 16.1 (*continued*)

Case	Sex, Age	Catheter details	Description of perforation	Follow-up	Additional comments
7	F 71	Balloon catheter.	Became shocked in ward 3 hours after barium enema. Radiography showed barium in peritoneal cavity.	Colostomy. Death	Previous diverticular disease. P.M. showed perforation through stercoral ulcer at recto-sigmoid
8	M $\frac{16}{12}$	Type not stated. Normal pressure.	Barium enema for reduction of intussusception. Barium entered peritoneal cavity at site of intussusception.	Laparotomy. Death.	Prolonged history of symptoms.
9	F 79	Balloon catheter.	Extravasation of barium into peri-rectal tissues and peritoneum seen at fluoroscopy.	Colostomy. Developed septicaemia. Died following day.	Previous operation for rectal prolapse. Tight anal canal.
10	F 81	Not self-retaining.	Barium held up in upper rectum. No pain, Enema abandoned. Collapsed 4 days later. Straight radiograph showed retro-peritoneal and ? intra-peritoneal barium.	Died on 12th day.	P.M. Diverticulitis pelvic colon and peri-colic abscess. Death attributed to coronary infarction and L.V.F.

11	F 54	Balloon catheter. Inflated by Higginson syringe. Discomfort.	Enema showed diverticulitis. Discomfort and slight bleeding after examination but no evidence of extravasation in A.E. film. Returned following day with pelvic cellulitis. Repeat straight radiograph showed extravasation of barium.	Died 3 weeks later.	Chronic myxoedema. Incompletely treated.
12	F 53	Rubber catheter. Balloon not inflated.	Extraperitoneal extravasation of barium.	Defunctioning colostomy. Died approx. 14 days later.	P.M. Peritonitis. History 7 years' steroid therapy for asthma and sarcoidosis.

VAGINAL RUPTURE

13	F 78	Balloon catheter in vagina. Height of barium 3 ft.	Fluoroscopy without image intensification suggested ballooning of rectum. Complaint of 'lower abdominal discomfort'. Bleeding noted and examination abandoned. Films interpreted as extraperitoneal leak from rectal perforation.	Patient became shocked after return to ward. Thought to have bleeding 'P.R.'. Laparotomy. No intraperitoneal barium. Transverse colostomy. Died 3 weeks later.	Diagnosis of vaginal perforation not made until P.M. which showed 2 in. tear in posterior vaginal wall. Vagina atrophic. History of hypothyroidism.

TABLE 16.1. (continued)

Case	Sex, Age	Catheter details	Description of perforation	Follow-up	Additional comments
14	F 72	Balloon catheter in vagina. Height of barium 3 ft.	Fluoroscopy without image intensification. Abnormal spread of barium noted. Examination discontinued. Films showed extraperitoneal spread of barium, barium also in bladder. Venous intravasation.	Profuse vaginal bleeding. Surgical repair. Died 24 hours after perforation.	P.M. Barium in peritoneum, veins and lungs.
15	F 62	Balloon catheter in vagina.	Venous intravasation recognized on image intensifier. Examination immediately abandoned. No complaint of pain by patient. Vaginal bleeding. Speculum examination showed tear left side vaginal fault.	Chest radiographs did not reveal any barium in lungs. Treated by antibiotics only. Patient left hospital on 4th day. No sequelae. Well at follow-up.	Previous colostomy for diverticulitis pelvic colon. Catheter initially in rectum but slipped out. Accidentally replaced in vagina.

series [27] a surprisingly high survival rate of 87 per cent was achieved in cases of intraperitoneal perforation.

Twelve major cases of intraperitoneal or extraperitoneal rupture of the colon and three cases of vaginal rupture were reported to the U.K. national radiological survey. These are summarized in Table 16.1. The combined mortality rate for intraperitoneal and extraperitoneal perforations of the colon in these cases was as high as 80 per cent. These patients were nearly all elderly and this suggests that atrophic changes in the bowel may have been a factor predisposing to injury and also to the high mortality. In this respect, it is also of interest that in the two patients who were aged only 54 and 53, one patient had myxoedema and the other patient had been treated with steroids for several years. Both of these factors might have affected the tensile strength of the tissues. Prolonged steroid therapy, in particular, may cause atrophic changes in the tissues. Once the perforation had occurred, there could also be an increased risk from spread of infection and an impaired response to stress due to adrenal suppression. One of the patients with vaginal rupture was also being treated for hypothyroidism.

Another important feature which emerged from this series was the difficulty of diagnosis in some cases. Whereas intraperitoneal perforation usually caused sudden severe pain and shock, extensive extraperitoneal spread could occur without any appreciable discomfort in the early stages, since the rectal mucosa above the pectinate line is insensitive to pain. These cases may, however, be followed by sudden collapse after a period of several hours. In some cases also, the unfamiliar radiological appearance may lead to delay in diagnosis. Typical radiographs illustrating the appearances of intraperitoneal and extraperitoneal rupture are shown in Figs 16.2 and 16.3. An obstructing carcinoma in the rectum may increase the risk of rupture, particularly if a balloon catheter is being used. In the early stage of perforation, a 'translucent membrane' of bowel wall may be visible separating the subserosal collection of barium from that in the rectum [29].

In his classic paper, Burt [30] measured the bursting pressures of different areas of the colon removed at autopsy in a series of patients. When a section of colon is inflated by pneumatic pressure, perforation is preceded by a longitudinal split in the serosa and muscularis. Then, as the distension increases, the mucosa herniates through the split and eventually ruptures. Animal experiments have subsequently shown that bursting pressures on segments of colon excised after death are comparable with those measured *in vivo* [31]. The bursting pressures in Burt's series are shown in Table 16.2 and it will be seen that the rectum has the highest bursting pressure whereas the caecum is the weakest area. These are average pressures. The lowest bursting pressure in the series was in a 30-year-old patient who died following an appendicectomy. In this patient, the caecum ruptured at a pressure of only 50 mmHg. The highest pressure was sustained in the rectum of a 5½-year-old infant which required a pressure of 646 mmHg before rupture

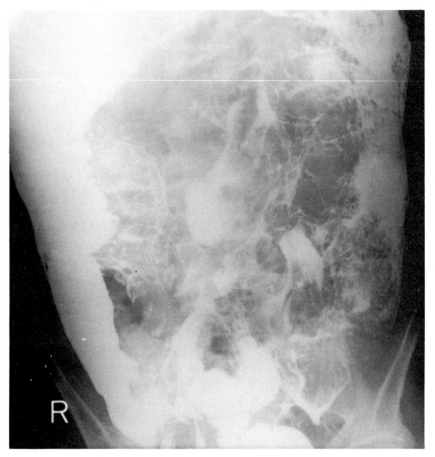

FIGURE 16.2. (Case 3, Table 16.1) Male, age 67. Perforation of pelvic colon by barium enema with severe shock and flooding of peritoneal cavity. Barium shown in peritoneum and pooled in right para-colic gutter. (Courtesy of *Clinical Radiology*.)

occurred. Thus, the narrower the diameter of the bowel, other things being equal, the higher the pressure it can sustain. However, the condition of the bowel wall is also important and if it is thinned, ulcerated, or even unduly rigid, it will be more prone to perforation.

It has been suggested that when colonic haustra are present, they indicate

Alimentary Tract

that tone is still present in the colonic wall so that there is a capacity to respond to increased volume: correspondingly, when the haustra are no longer present, there is probably a reduced safety factor. In a recent case, perforation of the ascending colon occurred when the enema reservoir was

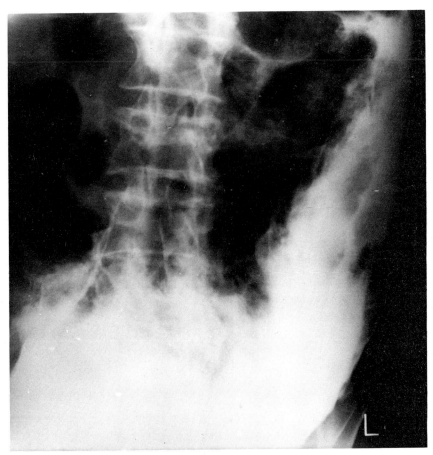

FIGURE 16.3. (Case 4, Table 16.1) Male, age 84. Perforation of sigmoid with extra-peritoneal leakage of barium into pelvi-rectal soft tissues and pelvic mesocolon. No immediate symptoms. (Courtesy of *Clinical Radiology*.)

raised to 4 ft (120 cm) in an attempt to reflux barium into the terminal ileum. The perforation in this case was preceded by disappearance of the colonic haustra and atropine had also been administered before the examination [32].

The theoretical hydrostatic pressure resulting from a 20 per cent low viscosity solution of barium is 71 mmHg when the reservoir is 3 ft (90 cm) above the rectum; 97 mmHg at 4 ft (120 cm); and 123 mmHg at a height of 5 ft (150 cm). However, the effects on intracolonic pressure are more complex and depend partly on the rate of flow and on the distensibility of the colon. The resting colonic pressure is usually below the hydrostatic pressure but transient colonic spasms or straining may increase the pressure by an additional 170 mmHg. Moreover, raising the level of the reservoir to 4 or 5 ft increases the incidence of such spasms. However, when high viscosity barium containing methylcellulose is used, the rate of flow is more sluggish and a height of 5 ft does not appear to cause high-pressure colonic spasms [31]. In general, when low-viscosity barium solutions are used, a height of 2 ft is claimed to be safest and least painful but, providing that the colonic haustra are still present, there are exceptions when a higher pressure may be required [32].

TABLE 16.2. Average bursting pressure in the colon (mmHg) [30].

	Serosa	Mucosa
Rectum	246	278
Sigmoid	201	260
Transverse colon	115	160
Caecum	80	136

Average pressures in autopsy specimens from 17 patients ranging in age from $5\frac{1}{2}$ months to 78 years. (Pressures have been converted from pounds per sq.in. to mmHg.)

In the majority of barium enema perforations recorded in the literature and in at least thirteen of those in Table 16.1, a balloon catheter had been used. Overdistension of the balloon with air may damage the rectal wall: in one case, rupture of the balloon preceded colonic rupture [28]. In two other cases of perforation, the balloon was unusually high in position and was actually sited at the recto-sigmoid junction. This area is much less distensible than the rectal ampulla so that inflation of the balloon is more likely to damage the bowel wall [33]. The experimental work on colonic and bursting pressures suggests additional mechanisms by which balloon catheters may increase the risk of perforation in susceptible patients. If the patient develops a high intracolonic pressure from spasm or straining, or if there is a stricture which limits the proximal flow of barium, the balloon may prevent the evacuation of barium which might otherwise act as a safety valve. While there are occasions when a balloon catheter is necessary, the routine use of such a catheter in each and every patient is to be deprecated. Even in patients with poor sphincter control, it is sometimes possible to perform an enema without a self-retaining catheter by the simple expedient of turning the patient into the prone position. In those cases where a self-retaining catheter cannot be avoided, a carefully measured volume of air should be injected into the balloon preferably under the control of the radiologist.

Perforation may also occur without the use of a self-retaining catheter and such a catheter was used in only six out of thirteen cases in one series [26]. An abrasion of the anterior rectal wall may be caused by the enema tip or by the preceding cleansing enema. In some cases the cleansing enema may itself result in perforation and, if the patient complains of undue symptoms, straight X-rays of the abdomen should be performed *before* commencing the barium examination. Instrumentation such as sigmoidoscopy may also traumatize the rectal mucosa, particularly if a biopsy has been performed. In these cases, it is preferable to defer the barium enema for a week or two. If, nevertheless, it is decided that a barium enema is required in a patient who has recently had a sigmoidoscopy, with or without a biopsy, it is important for the radiologist to have the relevant information. It has also been suggested that a preliminary straight radiograph should be performed to exclude possible evidence of retroperitoneal gas before commencing the enema [34]. Air insufflation should also be avoided in these cases. It is probable that a number of minor extraperitoneal extravasations of barium cause no symptoms and remain undiagnosed. Since perforation and extravasation of barium may sometimes only be visible on the post-evacuation film, the routine inspection of this film as soon as it has been taken, can contribute to early diagnosis, particularly if the patient is initially symptomless.

Barium enemas performed through a colostomy are a particular hazard and accounted for 12 out of 53 perforations in Zheutlin's series [28]. This is understandable when one considers the fragile nature of the attachment of the colostomy to the abdominal wall and peritoneum. Balloon catheters are a particular hazard in these patients [26]. It has been suggested that if a balloon catheter is used, the inflated balloon should be compressed against the *outside* surface of the colostomy. A disposable plastic colostomy bag may be used as an alternative to a balloon catheter. A soft rubber catheter is threaded through the colostomy bag before this is placed in position, and the catheter is inserted gently for only a short distance into the colostomy. The position of the appliance is arranged so that any overflowing barium which is collected in the bag does not obscure the site of interest (Fig. 16.4).

Infants form a group who are more susceptible to perforation by barium enemas [26]. Balloon catheters should not be used in these patients and an excessive head of barium should be avoided. Perforation of the colon may occur during hydrostatic reduction of an intussusception and abdominal palpation should be avoided during this procedure [35]. In infants with ilial atresia, the ileum distal to the block may communicate with the peritoneal cavity so that extravasation into the peritoneum may occur with a retrograde small bowel enema [36].

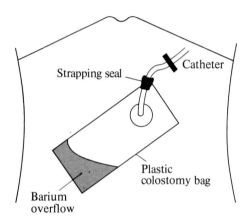

FIGURE 16.4. Method of administering barium enema through a colostomy using a disposable colostomy bag. The catheter is threaded through the bag before application.

BARIUM PERITONITIS

Cochran et al. [37] investigated the effects of barium in the peritoneal cavity of dogs. The most serious changes and highest mortality were produced by a combined mixture of barium and unsterile faeces, an analogous situation to that in patients who have a barium enema perforation. However, even sterile barium sulphate by itself has an adverse effect in the peritoneal cavity and, in those animals which survived, granulomatous changes occurred. A commercial preparation (Starbarium) had a higher mortality and produced more serious changes than the USP barium sulphate, suggesting that the suspending agent in the commercial preparation was particularly harmful. In a more recent investigation by Nahrwold et al. [38] large volumes of sterile USP barium were injected into the peritoneal cavities of dogs. In untreated animals, there was 100 per cent mortality. Removal of the barium by laparotomy produced only a marginal decrease in the mortality rate to 88 per cent. However, in dogs which received large volumes of intravenous fluid at one and six hours after the injection of barium, together with laparotomy, the mortality rate was reduced to 9 per cent. The major factor responsible for this improvement was undoubtedly *hyperhydration*. At laparotomy, considerable exudation of fluid was apparent in the peritoneal cavity suggesting that decrease in the plasma volume was a major factor in the high mortality rate of untreated animals. In the surviving animals, severe peritoneal adhesions developed. Following on this work, a poor-risk patient with a barium enema perforation was treated by hyperhydration with large volumes of intravenous fluid (4 litres per day) and recovered without laparotomy. Recovery in three other cases of barium enema perforations was also attributed to aggressive intravenous fluid therapy in addition to surgery and antibiotics [39]. In experimental peritonitis, Gram-negative endotoxin passes through the damaged intestinal wall into the peritoneal cavity and endotoxic shock from this source appears to be a major cause of death [40]. It seems possible that the high mortality of barium peritonitis may also be due to liberation of endotoxin as a result of damage to the intestinal wall by the barium and faeces in the peritoneum. Moreover, prevention of hypovolaemia by intravenous fluids has also been shown to be of importance in eliminating endotoxin [41]. The sudden collapse and death of cases 2, 3 and 4 (Table 16.1) approximately 12 hours after perforation would be consistent with endotoxic shock.

On the basis of this recent experimental work, a rational approach to the treatment of barium enema perforation may be expected to significantly improve the hitherto unfavourable outcome. Vigorous intravenous fluids therapy should be commenced as soon as the diagnosis is made and should be continued for several days. Where possible, laparotomy should be undertaken as soon as the patient's condition permits. A diverting caecostomy should be performed with peritoneal toilet but major surgery should be avoided. In addition to systemic antibiotic therapy there may be a case for injecting a nonabsorbable antibiotic such as Kanamycin into the bowel to destroy the Gram-negative bacteria which are the source of the endotoxin [40]. Systemic steroids also appear useful in the prevention and treatment of endotoxic shock [41]. Major extraperitoneal extravasations of barium should be treated along the same general lines.

BARIUM GRANULOMA

In the survey by Zheutlin *et al.* [28], 30 per cent of those patients who survived an intraperitoneal extravasation of barium eventually developed peritoneal adhesions causing narrowing of the bowel, and sometimes required repeated surgery. Ureteral obstruction due to periureteric barium granuloma may also occur as a late complication [42]. Small extravasations of barium may pass into the perirectal tissues and remain undiagnosed at the time. Symptoms may occur one to three weeks later [43, 44] or be delayed for several months. There is usually an indurated tender mass on the anterior wall of the rectum and, in the early stages, proctoscopy may show an irregular ulcer with a dirty grey base. In a later case, presenting at six months [45], sigmoidoscopy showed a polypoid haemorrhagic lesion with granular haemorrhagic mucosa between 5 and 12 cm from the anus. The endoscopic findings in these cases may easily be mistaken for malignancy, but barium sulphate crystals with a macrophage reaction can be detected in the biopsy specimen and a straight X-ray reveals the presence of barium in the interstitial tissues. There may be a low-grade chronic infection and even abscess formation but symptoms usually subside with conservative treatment even though barium persists in the soft tissues. When rectal barium granulomas were induced experimentally in dogs, epithelialization occurred despite retained barium. On the other hand, debridement of the edges of the ulcer actually delayed healing. Squamous metaplasia may occur at the ulcer site and absorbed barium may pass to the regional lymph nodes [46].

VENOUS INTRAVASATION

Venous intravasation of barium may rarely occur during barium enema examination and cause massive pulmonary embolism of barium. Cove & Snyder [47] report a recent case and tabulate the findings in eight other cases reported in the literature. An additional case of barium embolism is also mentioned in the Australasian survey [27]. Only two out of the ten patients in the literature survived and one of these [48] developed a liver abscess following the intravasation of a small amount of barium into the portal circulation. In nearly all these cases, balloon catheters had been used. Venous intravasation may be detected at fluoroscopy causing a medusa-like appearance [49] and barium may be seen to pass into the heart and pulmonary circulation [50]. There may be associated intravasation into the extraperitoneal soft tissues but this is not invariable. In most fatal cases in the literature, death usually occurred within a matter of minutes and there was apparently little that could be done apart from the administration of oxygen. By contrast, Cases 14 and 15 (Table 16.1) showed no immediate pulmonary symptoms and Case 15 recovered completely. If venous intravasation can be detected early during fluoroscopy, it would be rational to reverse the procedure used for air embolism by tilting the table 'feet down' and turning the patient on to his *right* side in an attempt to prevent the passage of barium into the pulmonary circulation. It appears that minimal degrees of barium embolization may occur without symptoms and in a recent case report [51], small pulmonary emboli of barium were noted following a barium meal examination for haematemesis in a patient with Hodgkin's disease who had

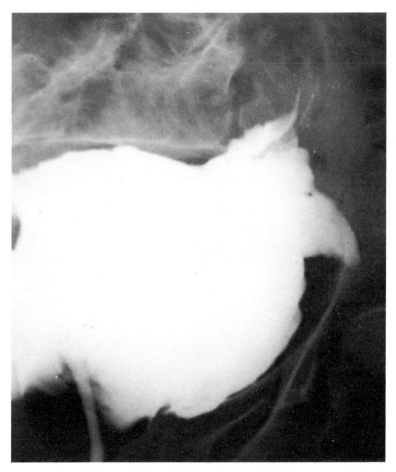

FIGURE 16.5. (Case 13, Table 16.1) Female, age 78. Misplacement of balloon catheter into vagina. Vaginal rupture with extra-peritoneal leakage of barium. This appearance may be confused with a rectal perforation.

previously had radiotherapy. The patient later died from recurrent haematemesis and the presence of barium in the pulmonary vessels was confirmed histologically. The duodenum was involved by Hodgkin's tissue and it was assumed that barium had been absorbed through mucosal ulceration which was present. Another possibility in this case which was not considered, is

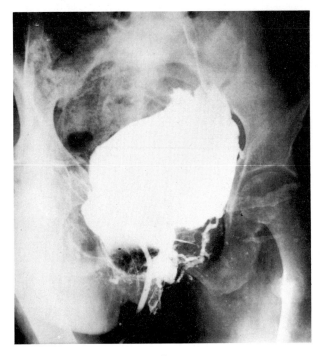

a

FIGURE 16.6

that there may have been a lymphatico-venous shunt associated with glandular obstruction, similar to that which may occasionally be demonstrated during lymphography.

VAGINAL RUPTURE

Vaginal rupture with venous intravasation of barium occurred in three cases where a barium enema had been inadvertently administered into the vagina using a balloon catheter [27, 49, 52]. Vaginal rupture appears to be relatively rare, and the high incidence of venous intravasation may possibly be due to the build-up of pressure in a closed cavity with escape of barium prevented by the inflated balloon. Three new cases of vaginal rupture are recorded in Table 16.1. In each of these, a balloon catheter had been used and in two cases the fluoroscopes were not equipped with image intensifiers. In Case 13, the examination was discontinued when the patient complained of lower abdominal discomfort and blood was found on the X-ray couch. The radiograph (Fig. 16.5) was interpreted as showing an extraperitoneal leak of barium and, perhaps not unreasonably, this was assumed to be due to bowel

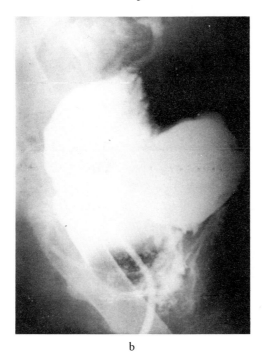

b

FIGURE 16.6. (Case 14, Table 16.1). Female, age 72. Misplacement of balloon catheter into vagina. The balloon is at the introitus. a Rupture of vagina with barium in pelvic soft tissues. There is venous intravasation of barium. b Lateral view shows barium in the iliac vein. The bladder is also shown filled with barium. Died 24 hours later.

damage. Laparotomy and transverse colostomy were performed but no lesion was detected in the sigmoid colon. The patient died three weeks later and at autopsy a 2-inch tear was found on the posterior vaginal wall. In Case 14, venous intravasation of barium occurred (Fig. 16.6) and the examination was discontinued. The condition was at first relatively symptomless and there was no immediate collapse. Profuse vaginal bleeding then occurred and, despite surgery, the patient died approximately eight hours later. In Case 15 (Fig. 16.7), there was obvious venous intravasation but this did not appear to cause any symptoms and repeated chest radiographs did not reveal any barium in the lungs. The patient made an uncomplicated recovery.

RETROPERITONEAL EMPHYSEMA

This appears to be a very rare but not necessarily grave complication. In two detailed case reports and in a review of five other cases from the literature, all the patients made an uneventful recovery [25, 53]. Retroperitoneal emphysema was usually a complication of air insufflation for double contrast examinations, but in one patient extensive interstitial emphysema was noted in a post-evacuation film in the absence of air insufflation. During the insufflation, gas may be seen to lie outside the bowel. Subsequently, this passes up into the retroperitoneal tissues and on delayed films, twelve to twenty-four hours later, there may be extensive mediastinal emphysema

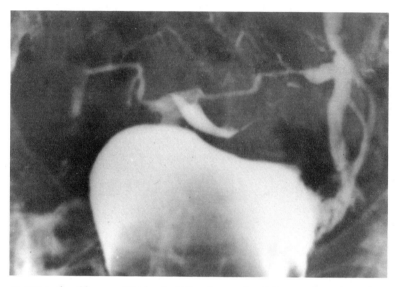

FIGURE 16.7. (Case 15, Table 16.1) Female, age 62. Balloon catheter in vagina. Venous intravasation. Symptomless apart from vaginal bleeding. Uncomplicated recovery.

followed by surgical emphysema in the neck. In one patient there was a 2 cm tear in the rectum but, in the majority of the cases, sigmoidoscopy was either normal or revealed only mild superficial lesions. Where a tear is present in the rectum, a temporary colostomy may be indicated but in the other cases, conservative treatment with antibiotic cover appeared adequate. A case of *pneumoperitoneum* has also been reported following a double contrast enema but symptoms were minimal and the condition resolved without treatment [54]. Gas embolism is a theoretical complication of double contrast enemas but in their review of the literature, Pyle & Samuel [25] were unable to find any recorded cases.

PORTAL VEIN GAS

The presence of gas in the portal venous system of the liver is usually an indication of invasion by gas-forming organisms and is then associated with a grave prognosis. However, gas may very rarely occur in the portal venous system following air insufflation for double contrast enemas in patients with chronic ulcerative colitis [55]. In the three reported cases in which this had occurred, there did not appear to be any serious adverse effect on the clinical condition. In one of these patients, there was also a coincidental bowel perforation with free intraperitoneal air and this responded to antibiotic therapy. The mechanism by which gas enters the portal vein in these cases is obscure but it is possible that it may be a form of air embolism with gas entering a hypervascular ulcerated area in the colon. This might then be analogous to the case of barium intravasation into the portal vein which also occurred in a patient with ulcerative colitis [48]. Portal vein gas embolism may also occur in infants undergoing umbilical catheterization and these cases also appear to have a benign prognosis [56].

WATER INTOXICATION

The hazard of water intoxication in megacolon is now well established [25, 57–59] and should be preventable, but unless this condition is borne in mind, occasional cases may still occur [27]. The typical history is that a tap water or barium enema has been administered with incomplete evacuation and a further cleansing enema may be given with similar effect. Water is absorbed by the large surface area of the colonic mucosa with resulting hydraemia, hyponatraemia and cerebral oedema. Clinically, the patient becomes drowsy and apathetic. Vomiting may occur and abundant dilute urine may be passed. The skin may be flushed but nevertheless appears cold and bathed in sweat. Despite the appearance of shock, the blood pressure is usually not lowered. Convulsions, coma and death may supervene. In some cases, the use of prostigmine has been thought to be an aggravating factor since it increases the colonic contractions with a resultant rise in the hydrostatic pressure which, in turn, leads to increased absorption of water. This drug should therefore be avoided in such cases [60]. Water intoxication has also been reported in a child with a normal colon following repeated enemas. In this patient, however, there was some impairment of renal function which impeded the excretion of the absorbed water [61].

The most important preventive measure is to limit the volume of fluid administered by enema in patients with megacolon or chronic constipation.

It has been suggested that sodium chloride might be added to the enema solution but there is a theoretical possibility that this may also be absorbed and cause pulmonary oedema, particularly if there is any cardiac or renal insufficiency. In established cases of water intoxication with cerebral oedema, urgent treatment is required by intravenous infusion of hypertonic (10 per cent) sodium chloride. If this leads to pulmonary oedema, intravenous frusemide or venesection may be required. At the opposite extreme, *hypernatraemic dehydration* has been recorded in a case of megacolon following the administration of four 'Fleet' hypertonic phosphate enemas [62].

TANNIC ACID

For a number of years after 1946, suspensions of tannic acid were widely used in barium enemas to improve the quality of the post-evacuation film. This mainly resulted from the stimulation of colonic contraction, but other properties attributed to tannic acid were inhibition of mucin secretion, precipitation of proteins in the superficial mucosa, and increased adherence of barium sulphate to the mucosa. In 1963 McAlister *et al.* [63] described a death from hepatic necrosis in a child following a barium enema containing 0.75 per cent tannic acid. In addition, this patient had also had two preparatory enemas, both containing 0.75 per cent tannic acid. At autopsy, tannic acid was demonstrated in the liver. A retrospective review then revealed two other deaths from liver failure in children, apparently due to barium enemas containing tannic acid. In two of these three cases there was reflux of the barium tannic acid mixture into the small bowel and delayed evacuation. Partly as a result of these findings, Lucke *et al.* [64] reviewed their records and found that in three adults and two children who died from liver failure, there was strong presumptive evidence that this was due to barium enemas containing 2 per cent tannic acid. In all of these eight patients, symptoms of abdominal pain, vomiting, drowsiness and irritability occurred at intervals varying from a few hours to two days after the barium enemas, and death occurred between the second and seventh days.

Experiments in animals have confirmed that tannic acid can be absorbed from the colon causing both liver and renal damage, but this appears to be dependent on both the concentration of tannic acid used and the length of time that it is retained. With enemas retained over a period of two hours, relatively low doses of 0.2 g/kg tannic acid caused prolongation of thiopentone anaesthesia, possibly due to impaired metabolism of the anaesthetic following liver damage [65]. In a further series of investigations [66–68], enemas containing 1 per cent tannic acid administered to rats did not produce detectable blood levels of tannic acid even when they were retained for one hour, whereas a 2.5 per cent concentration retained for this period caused inflammatory changes in the colon and liver damage. At higher concentrations, colonic ulceration and liver damage occurred after shorter periods of retention. It was concluded that 1 per cent concentrations appeared safe but that multiple cleansing enemas, even with saline, may damage the colonic mucosa and predispose to tannic acid absorption. In two large case reviews covering 75 000 barium enemas, no definite association between tannic acid and liver damage was established [69].

More recently, Kemp Harper *et al.* [70] performed serial liver function studies in 90 patients who had barium enemas containing 1 per cent tannic acid. They were unable to demonstrate any significant effect on liver function in these patients. They concluded that enemas containing tannic acid in a concentration of 1 per cent were both useful and safe, provided that the enema solution was freshly prepared and that tannic acid was not included in the bowel preparation before the enema.

It would appear that in those cases where there is a good diagnostic indication for the use of tannic acid the risk is small, but time alone will tell. Many radiologists consider that tannic acid produces an improvement of mucosal coating which contributes considerably to the diagnostic accuracy of the enema. Assessment of mucosal coating is, however, subjective and, using the double-blind technique in a large survey, no statistical evidence of improved mucosal coating was demonstrated with 0.25 per cent tannic acid [71]. In yet another investigation, improvement of mucosal coating was claimed with enemas containing 0.25 per cent tannic acid but a similar degree of improvement was also noted when 1 per cent carboxy-methylcellulose was used in the enema [72].

MISCELLANEOUS

Studies of electrocardiograms recorded by radiotelemetry in patients undergoing routine barium enema examinations showed that transient ECG changes commonly occurred, particularly in elderly patients, or in patients with heart disease [73, 74]. These changes appeared to be most common during the evacuation phase, but they were also noted during the filling phase, particularly if the patient had to strain to retain the enema; during gas insufflation; and during inflation of the balloon catheter. The abnormalities noted included arrhythmias, conduction defects and RS-T depression. Although there were frequently no accompanying symptoms, some of the ECG abnormalities were potentially hazardous. ECG abnormalities and sudden death can, of course, also occur as a result of alternative types of investigations such as sigmoidoscopy [75].

During the routine administration of barium enema, faeces and mucus may be seen to float up the plastic tube and even to enter the enema reservoir. The whole of the enema equipment is, therefore, liable to contamination and it may become a potential source for transmission of pathogenic organisms which are excreted in the bowel [76]. Confirmation that infection may indeed occur by this route is provided by an experiment in which poliomyelitis vaccine was administered in individual barium enemas to seven patients. In six of these patients there was a subsequent rise in serum antibody to the vaccine [77].

Exacerbation of colonic obstruction by oral barium is well known, but occasionally acute large bowel obstruction may be precipitated by a barium enema [2]. Fatal perforation of a carcinoma of the pelvic colon has occurred as a result of using double strength Senokot for out-patient preparation, prior to a barium enema [17]. Perforation has also resulted from a silicone foam enema in a patient with a radiation stricture of the sigmoid [78].

In two infants undergoing barium enema examinations at different hospitals, faulty clamps on the tubing resulted in massive reflux of barium into the small bowel and thence into the stomach. Vomiting occurred with inhalation of barium and one of the infants died three days later from bronchopneumonia. In both cases the accidents occurred after the conclusion of the fluoroscopic examination while additional films were being taken. A similar incident in another infant resulted in a fatal rupture of the caecum [79].

ENDOSCOPY

The introduction of fibreoptic instrumentation has considerably extended the scope and frequency of endoscopic procedures in the gastrointestinal tract and, in addition to oesophagoscopy and gastroscopy, these now also include duodenoscopy, retrograde cholangiopancreatography and colonoscopy. The undoubted increase in diagnostic accuracy resulting from these procedures is nevertheless associated with some measure of discomfort to the patient and a somewhat higher complication rate than routine gastrointestinal radiology, so that the selection of patients for endoscopy requires appropriate clinical judgement. Complications such as aspiration or perforation may not be immediately apparent at the time of endoscopy and may only be detected on subsequent radiological examination.

PERORAL ENDOSCOPY

Aspiration pneumonia

In upper gastrointestinal examinations, where there is a combination of pharyngeal anaesthesia in heavily sedated subjects, there is an inherent risk of aspiration of vomit or gastric contents, particularly during the withdrawal of the instrument; several cases of aspiration pneumonia have been described [80]. In a few patients, cinefluorography has shown delayed clearing of contrast medium from the hypopharynx as late as two days after endoscopy, and this abnormal functional state may also contribute to the development of aspiration pneumonia [81].

Perforation [81]

Contrary to early expectation, it now appears that the risk of perforation has *not* been reduced as a result of the introduction of the completely flexible fibrescope and may, if anything, have been increased somewhat. Fibregastroscopy is quoted as having a perforation rate of 0.074 per cent with a mortality of 0.019 per cent; whereas in earlier studies using the semiflexible lens-gastroscope, the perforation rate varied from 0.05 to 0.061 per cent and the mortality from 0.004 to 0.014 per cent. This increase in incidence may partly be a function of the more widespread current use of endoscopy.

The risks of perforation are highest in the oesophagus, particularly after biopsy, or where there is inflammatory or neoplastic involvement of the oesophageal wall. However, perforations may also occur in a normal oesophagus. The usual sites for perforation are at the anatomical areas of narrowing or at organic strictures. The commonest site is probably in the lower third of the oesophagus but perforation of the cervical oesophagus is also frequent and may be complicated by a retropharyngeal abscess and mediastinitis. The other important area of narrowing is at the level of the aortic arch and tracheal bifurcation. Frequently, even an experienced endoscopist may be unaware that a perforation has occurred. Depending on the site of the perforation, air may be seen in the prevertebral space of the neck, or in the mediastinum. There may also be widening of the mediastinum, pneumothorax, pleural effusion or even a pneumoperitoneum. Water-soluble contrast medium may be valuable in demonstrating the actual site of the perforation.

In the stomach, the commonest site for endoscopic trauma lies just distal to the gastro-oesophageal junction on the posterior gastric wall. Injury may range from intramural haematoma formation, to a true perforation into the lesser sac with an associated pneumoperitoneum. It appears that many of these posterior perforations may be self-sealing so that the site of perforation may sometimes not be demonstrable by water-soluble contrast medium or even by laparotomy. Anterior perforations on the other hand present a more dramatic picture with peritonitis or localized abscess formation. With fibreduodenoscopy, perforations of the duodenal cap may occur [82].

Pancreatic and biliary complications

In the experimental animal retrograde pancreatography with large volumes of sodium diatrizoate causes extensive pancreatic damage. With moderate volumes of contrast medium, there is only a minimal inflammatory change in the pancreas although there may be elevation of the serum amylase. The elevation of the serum amylase is less marked when the medium is injected at low pressure and Trasylol may have a very slight protective effect [83, 84]. In clinical use, retrograde pancreatography may give rise to abdominal discomfort and occasionally to severe abdominal pain [80]. There is often a rise in serum amylase between three and eighteen hours after the procedure but this usually returns to normal within two to five days. Elevation of serum amylase is more likely to occur when the fine pancreatic ducts or acini have been opacified [85].

In a series of 853 retrograde cholangio-pancreatographies, Heully *et al.* [82] quoted an overall incidence for complications of approximately 2 per cent. This included two cases of acute pancreatitis, four cases of cholangitis, one case of infective cholecystitis and two cases of gangrenous cholecystitis, one of which was fatal. They suggest that the concentration of contrast medium used for this procedure should be reduced to 30 per cent (sodium methylglucamine diatrizoate). A fatal case of pancreatitis has also been recently reported [86]. For a more detailed consideration of pancreatic and biliary complications see chapter 10.

Miscellaneous

There is as yet no entirely satisfactory method of sterilizing fibreoptic endoscopes. In particular there is a risk of transmitting serum hepatitis and it is therefore usual to avoid endoscopy in patients with evidence of Australia antigen. Other infections such as bacteraemia and cholangitis have also occurred following retrograde cholangiography [80]. In three cases of acute leukaemia, *pseudomonas* infection was transmitted by the fibrescope used for oesophagoscopy. Following biopsy in two of these cases, there was a fatal *pseudomonas aeruginosa* septicaemia. As a result of this experience, gas sterilization of the fibrescope is advocated, particularly in patients with decreased resistance to infection, and a full aseptic technique should be used in these patients [89].

If a large volume of air is introduced during endoscopy, a condition of *pseudo-acute abdomen* may occur with severe abdominal pain, distension and diffuse tenderness. Unless this is correctly diagnosed radiologically, laparotomy may be undertaken for suspected perforation [81]. Other complications include impaction of the fibrescope which may be curved back on itself, and swelling of the salivary glands due to accidental injection of air [81]. As with all forms of instrumentation, there is a possible risk of reflex myocardial involvement.

COLONOSCOPY [87]

Perforation

This is the commonest complication of colonoscopy. The incidence rate reported in a number of large series varies from 0.4 per cent to 1.9 per cent. The majority of perforations occur during mechanical manipulations, particularly in the sigmoid loop, or following electrosnare polypectomy if the colonic wall is accidentally involved in the area of coagulation. Perforation may also occur following air insufflation when there is increased friability of the colonic wall, or following unrecognized impaction of the tip of the colonoscope in a large diverticulum. The commonest sites of perforation are in the recto-sigmoid and at the junction of the sigmoid and descending colon. *Intraperitoneal perforations* with pneumoperitoneum require prompt operative repair. Fortunately, they are usually recognized immediately, but in the aged or infirm patient the diagnosis may initially be missed.

Extraperitoneal perforation is often less readily recognized clinically and occasionally there may be a delay of several days in reaching the diagnosis. Extraperitoneal perforation usually responds to conservative treatment with antibiotics. It has been suggested that routine abdominal radiographs should be taken in all cases following colonoscopy to detect occult perforation. In retroperitoneal perforations the distribution of gas may also give an indication of the site of the perforation [88]. Mural perforation by the colonoscope at the site of a stricture may result in a localized abscess.

Miscellaneous

Other reported complications of colonoscopy include colonic haemorrhage, haematoma formation in the mesentery, and incarceration of the colonoscope in an inguinal hernia. Avulsion and rupture of the spleen has also occurred, apparently due to intussusception of the splenic flexure on withdrawal of the instrument [87].

REFERENCES

1 Reports to National Radiological Survey.
2 KILLINBACK M. (1964) Acute large bowel obstruction precipitated by barium X-ray examination. *Med. J. Austral.* **2**, 503–508.
3 PROUT B.J., DATTA S.B. & WILSON T.S. (1972) Colonic retention of barium in the elderly after barium-meal examination and its treatment with lactulose. *Br. Med. J.* **4**, 530–535.
4 YOUNG M.O. (1958) Acute appendicitis following retention of barium in the appendix. *Arch. Surg.* **77**, 1011–1014.
5 DIXON G.D., FERRIS D.O. & HODGSON J.R. (1967) Unusual complication of barium studies: Report of a case of adherent cecal barolith. *Am. J. Roentg.* **99**, 106–111.
6 GOVINDIAH D. & BHASKAR G.R. (1972) An unusual case of barium poisoning. *Antiseptic* **69**, 675–677.
7 BERNING J. (1975) Hypokalaemia of barium poisoning. *Lancet* **1**, 110.
8 AMBERG J.R. & UNGER J.D. (1970) Contamination of barium sulfate suspensions. *Radiology* **97**, 182–183.
9 FELSON B. (1973) Radiologist on the rocks. *Semin. Roentg.* **8**, 361–363.
10 GELFAND D.W. & MOSKOWITZ M. (1970) Massive gastric dilatation complicating hypotonic duodenography. A report of three cases. *Radiology* **97**, 637–639.
11 HARRIS P.D., NEUHAUSER E.B.D. & GERTH R. (1964) The osmotic effect of water soluble media on circulating plasma volume. *Am. J. Roentg.* **91**, 694–698.
12 ELMAN S. & PALAYEW M.J. (1975) Assessment of biochemical and hematologic changes related to oral administration of an iodinated contrast medium. *Invest. Radiol.* **8**, 322–325.
13 ROWE M.I., FURST A.J., ALTMAN D.H. & POOLE C.A. (1971) The neonatal response to gastrografin enema. *Pediatrics* **48**, 29–35.
14 ROWE M.I., SEAGRAM G. & WEINBERGER M. (1973) Gastrografin-induced hypertonicity. The pathogenesis of a neonatal hazard. *Am. J. Surg.* **125**, 185–188.
15 DAVIES N.P. & WILLIAMS J.A. (1971) Tubeless vagotomy and pyloroplasty and the 'gastrografin test'. *Am. J. Surg.* **122**, 368–370.
16 NOONAN C.D. & MARGULIS A.R. (1970) Small bowel transit time of water soluble iodinated contrast media and barium sulfate in cats with simulated surgical abdomen. *Am. J. Roentg.* **110**, 334–342.
17 ANSELL G. (1968) A national survey of radiological complications: interim report. *Clin. Radiol.* **19**, 175–191.
18 CHIU C.L. & GAMBACH R.R. (1974) Hypaque pulmonary edema. A case report. *Radiology* **111**, 91–92.

19 REICH S.B. (1969) Production of pulmonary edema by aspiration of water-soluble nonabsorbable contrast media. *Radiology* 92, 367-370.
20 FRECH R.S., DAVIE J.M., ADATEPE M., FELDHAUS R. & MCALISTER W.H. (1970) Comparison of barium sulfate and oral 40% diatrizoate injected into the trachea of dogs. *Radiology* 95, 299-303.
21 ROSS L.S. (1972) Precipitation of meglumine diatrizoate 76% (Gastrografin) in the stomach. Observations on the insolubility of diatrizoate in the normal range of gastric acidity. *Radiology* 105, 19-22.
22 HUGH T.B., HENNESY W.B., GUNNER W. et al. (1970) Precipitation of contrast medium causing impaction of Sengstaken–Blakemore oesophageal tube. *Med. J. Austral.* 1, 60-61.
23 GALLIJANO A.L., KONDI E.S., PHILLIPS E. & FERRIS E. (1976) Near-fetal hemorrhage following gastrografin studies. *Radiology* 118, 35-36.
24 Reports to Committee on Safety of Drugs, London.
25 PYLE R. & SAMUEL E. (1960) An evaluation of the hazards of barium enema examinations. *Clin. Radiol.* 11, 192-196.
26 SEAMAN W.B. & WELLS J. (1965) Complications of the barium enema. *Gastroenterology* 48, 728-737.
27 MASEL H., MASEL J.P. & CASEY K.V. (1971) A survey of colon examination techniques in Australia and New Zealand, with a review of complications. *Australas. Radiol.* 15, 140-147.
28 ZHEUTLIN N., LASSER E.C. & RIGLER L.G. (1952) Clinical study on effect of barium in the peritoneal cavity following rupture of the colon. *Surgery* 32, 967-979.
29 SPECTOR G.W. & SUSMAN N. (1963) The roentgen recognition of intramural perforation following barium enema examination in obstructing lesions of sigmoid. *Am. J. Roentg.* 89, 876-879.
30 BURT C.A.V. (1931) Pneumatic rupture of intestinal canal with experimental data showing mechanism of perforation and pressure required. *Arch. Surg.* 22, 875-902.
31 NOVEROSKE R.J. (1964) Intracolonic pressures during barium enema examination. *Am. J. Roentg.* 91, 852-863.
32 NOVEROSKE R.J. (1972) Perforation of a normal colon by too much pressure. *J. Ind. St. M.A.* 65, 23-25.
33 NOVEROSKE R.J. (1966) Perforation of the rectosigmoid by a bardex balloon catheter. Report of 3 cases. *Am. J. Roentg.* 96, 326-331.
34 FIELDING J.F., LUMSDEN K. & SHUKART W.A. (1973) Large-bowel perforations in patients undergoing sigmoidoscopy and barium enema. *Br. Med. J.* 1, 471-473.
35 DIBBELL D. & COHN R. (1966) Perforation of the colon during hydrostatic reduction. *Am. J. Surg.* 111, 715-717.
36 WOLFSON J.J. & WILLIAMS H. (1970) A hazard of barium enema studies in infants with small bowel atresia. *Radiology* 95, 341-343.
37 COCHRAN, D.Q., ALMOND C.H. & SHUCART W.A. (1963) An experimental study of the effects of barium and intestinal contents on the peritoneal cavity. *Am. J. Roentg.* 89, 883-887.
38 NAHRWOLD D.L., ISCH J.H., BENNER D.A. & MILLER R.E. (1971) Effect of fluid administration and operation on the mortality rate in barium peritonitis. *Surgery* 70, 778-781.

39 GARDINER H. & MILLER R.E. (1973) Barium peritonitis. A new therapeutic approach. *Am. J. Surg.* **125**, 350–352.
40 CUEVAS P. & FINE J. (1972) Role of intraintestinal endotoxin in death from peritonitis. *Surg. Gynaec. Obstet.* **134**, 953–957.
41 CUEVAS P., ISHIYAMA M., KOIZUMI S. *et al.* (1974) Role of endotoxaemia of intestinal origin in early death from large burns. *Surg. Gynaec. Obstet.* **138**, 725–730.
42 HERRINGTON J.L. (1966) Barium granuloma within the peritoneal cavity: ureteral obstruction 7 years after barium enema and colonic perforation. *Ann. Surg.* **164**, 162–166.
43 CARTER R.W. (1963) Barium granuloma of the rectum. *Am. J. Roentg.* **89**, 880–882.
44 LULL G.F., BRYNE P. & SANOWSKI R. (1971) Barium sulfate granuloma of the rectum. A rare entity. *J. Am. med. Ass.* **217**, 1102–1103.
45 CAMERON H.C. (1964) Barium proctitis. *Proc. Roy. Soc. Med.* **57**, 399–400.
46 RAND A.A. (1966) Barium granuloma of the rectum. *Dis. Colon Rect.* **9**, 20–32.
47 COVE J.K. & SNYDER R.N. (1974) Fatal barium intravasation during barium enema. *Radiology* **112**, 9–10.
48 ISAACS I., NISSER R. & EPSTEIN B.S. (1950) Liver abscess resulting from barium enema in a case of chronic ulcerative colitis. *N.Y. St. J. Med.* **50**, 332–334.
49 ZATKIN H.R. & IRWIN G.A.L. (1964) Non-fatal intravasation of barium. *Am. J. Roentg.* **92**, 1169–1172.
50 ROSENBERG L.S. & FINE A. (1959) Fatal venous intravasation of barium during a barium enema. *Radiology* **73**, 771–773.
51 MAHBOUBI S., GOBEL V.K., DALINKA M.K. & CHO S.Y. (1974) Barium embolisation following upper gastrointestinal examination. *Radiology* **111**, 301–302.
52 GEIPEL A. Quoted by Cove & Snyder. (Ref. 47.)
53 BRUNTON F.J. (1960) Retroperitoneal emphysema as a complication of barium enema. *Clin. Radiol.* **11**, 197–199.
54 MOWAT P.D. (1967) Pneumoperitoneum following double contrast enema. *Br. J. Radiol.* **40**, 230–231.
55 KEES C.J. & HESTER C.L. (1972) Portal vein gas following barium enema examination. *Radiology* **102**, 525–526.
56 SWAIM T.J. & GERALD B. (1970) Hepatic portal venous gas in infants without subsequent death. *Radiology* **94**, 343–345.
57 JOLLEYS A. (1952) Death following a barium enema in a child with Hirschprung's disease. *Br. Med. J.* **1**, 692–693.
58 Editorial (1959) The hazard of water enemas. *Lancet* **1**, 559–560.
59 Ibid. (1963) **1**, 983.
60 STEINBACH H.L., ROSENBERG R.H., GROSSMAN M. & NELSON T.L. (1955) The potential hazards of enemas in patients with Hirschprung's disease. *Radiology* **64**, 45–50.
61 PETERSON C.A. & CAYLER G.G. (1957) Water intoxication. Report of a case following a barium enema. *Am. J. Roentg.* **77**, 69–70.
62 FONKALSRUD E.W. & KEEN J. (1967) Hypernatremic dehydration from hypertonic enemas in congenital megacolon. *J. Am. med. Ass.* **199**, 584–586.
63 McALISTER W.H., ANDERSON M.S., BLOOMBERG G.R. & MARGULIS A.R. (1963) Lethal effects of tannic acid in the barium enema. Report of three fatalities and experimental studies. *Radiology* **80**, 765–773.

64 LUCKE H.H., HODGE K.E. & PATT N.L. (1963) Fatal liver damage after barium enemas containing tannic acid. *Canad. Med. Ass. J.* **89**, 1111-1114.
65 SINGH J. & BOYD E.M. (1966) Thiopental anesthesia and tannic acid diagnostic enemas. *Canad. Med. Ass. J.* **95**, 558-562.
66 RAMBO O.N., ZBORALSKE F.F., HARRIS P.A., RIEGELMAN S. & MARGULIS A.R. (1966) Toxicity studies of tannic acid administered by enemas. I. Effects of enema-administered tannic acid on colon and liver of rats. *Am. J. Roentg.* **96**, 488-497.
67 HARRIS P.A., ZBORALSKE F.F., RAMBO O.N., MARGULIS A.R. & RIEGELMAN S. (1966) Toxicity studies on tannic acid administered by enemas. II. The colonic absorption and intraperitoneal toxicity of tannic acid and its hydrolytic products in rats. *Am. J. Roentg.* **96**, 498-504.
68 ZBORALSKE F.F., HARRIS P.A., RIEGELMAN S., RAMBO O.N. & MARGULIS A.R. (1966) Toxicity studies on tannic acid administered by enema. III. Studies on the retention of enemas in humans. IV. Review and conclusions. *Am. J. Roentg.* **96**, 505-509.
69 Editorial (1966) Tannic acid in colon examinations. *Br. J. Radiol.* **39**, 401-402.
70 KEMP HARPER R.A., PEMBERTON J. & TOBIAS J.S. (1973) Serial liver function studies following barium enemas containing 1% tannic acid. *Clin. Radiol.* **24**, 315-317.
71 LASSER E.C., GERENDE L.J. *et al.* (1966) An efficiency study of Clysodrast. *Radiology* **87**, 649-654.
72 PEREZ C.A. & FRIEDENBERG M.J. (1967) Comparison of carboxymethylcellulose, tannic acid and no additive in barium examinations of the colon. *Am. J. Roentg.* **99**, 98-105.
73 BERMAN C.Z., JACOBS M.G. & BERNSTEIN A. (1965) Hazards of the barium enema examination as studied by electrocardiographic telemetry: preliminary report. *J. Am. Geriat. Soc.* **13**, 672-686.
74 EASTWOOD G.L. (1972) ECG abnormalities associated with the barium enema. *J. Am. med. Ass.* **219**, 719-721.
75 FLETCHER G.F., ERNEST D.L., SHUFORD W.F. & WENGER N.K. (1968) Electrocardiographic changes during routine sigmoidoscopy. *Arch. Intern. Med.* **122**, 483-486.
76 STEINBACH H.L., ROUSSEAU R., MCCORMACK K.R. & JAWETZ E. (1960) Transmission of enteric pathogens by barium enemas. *J. Am. med. Ass.* **174**, 1207-1208.
77 MEYERS P.H. & RICHARDS M. (1964) Transmission of polio virus vaccine by contaminated barium enema with resultant antibody rise. *Am. J. Roentg.* **91**, 864-865.
78 AMBERG J.R. (1967) A hazard of silicone foam diagnostic enema. Report of a case of perforation of the colon. *Am. J. Roentg.* **99**, 96-97.
79 DE CARLO J. (1960) Complications associated with diagnostic barium enemas. *Surgery* **47**, 965-969.
80 BLUMGART L.H. & SALMON P.R. (1973) Fibreduodenoscopy and transpapillary cholangiopancreatography. In: *Recent Advances in Surgery* (Ed. by TAYLOR S.), No. 8, pp. 36-38. Churchill Livingstone, London.
81 MEYERS M.A. & GHAHREMANI G.G. (1975) Complications of fiberoptic endoscopy. (1) Esophagoscopy and gastroscopy. *Radiology* **115**, 293-300.
82 HEULLY F., GAUCHER P., JEANPIERRE R. & LAURENT J. (1974) Nécrose vésiculaire survenue après cholangiographie rétrograde par duodénoscopie, à propos d'un cas. *Ann. Gastroentérol. Hépatol.* **10**, 471-474.
83 WALDRON R.L. (1968) Reflux pancreatography. An evaluation of contrast agents for studying the pancreas. *Am. J. Roentg.* **104**, 632-640.

84 WOLLOWICK H.E., WALDRON R.L. & WOO S.A. (1971) The protective effect of trasylol versus contrast-media-induced pancreatitis. *J. Canad. Ass. Radiol.* **22**, 40–43.
85 KASUGAI T., KUNG N., KOBAYASHI S. & HATTORI K. (1972) Endoscopic pancreato-cholangiography. *Gastroenterology* **63**, 217–234.
86 CLASSEN M. Quoted by Blumgart & Salmon. (Ref. 80.)
87 MEYERS M.A. & GHAHREMANI G.G. (1975) Complications of fiberoptic endoscopy. II. Colonoscopy. *Radiology* **115**, 301–307.
88 MEYERS M.A. (1974) Radiological features of the spread and localisation of extra-peritoneal gas and their relationship to its source. An anatomical approach. *Radiology* **111**, 17–26.
89 GREEN W.H., MOODY M., HARTLEY R. *et al.* (1974) Esophagoscopy as a source of pseudomonas aeruginosa sepsis in patients with acute leukaemia: the need for sterilization of endoscopes. *Gastroenterology* **67**, 912–919.

CHAPTER 17:
MISCELLANEOUS PROCEDURES

G. ANSELL

PNEUMOGRAPHY

Retroperitoneal pneumography is now less frequently employed than formerly. In 1956 Ransom *et al.* [1] carried out a nationwide survey of urologists in the U.S. who performed retroperitoneal pneumography. They were able to collect details of 58 deaths due to gas embolism and 64 severe non-fatal cases of gas embolism. The risk of embolism following peri-renal injection by the lumbar route appeared to be approximately four times greater than when the pre-sacral route was used. Gas embolism might result from accidental introduction of the needle into a vein or, alternatively, a vein distal from the puncture site might be torn as the gas separated the layers of tissue in the retroperitoneal space. Oxygen did not appear to be any safer than air whilst helium, which was the least soluble gas, was probably the most dangerous. Carbon dioxide is twenty times more soluble than air or oxygen in the body, and nitrous oxide is also highly soluble [2]. Carbon dioxide or nitrous oxide are the safest gases for use in pneumography: nevertheless, two cases of cardiac arrest have occurred during peritoneal insufflation of carbon dioxide for laparoscopy and sterilization. These were believed to be due to gas embolism as a result of needle puncture of the pregnant uteri. Tubal insufflation apparatus was used in both cases and this may result in gas pressures exceeding 300 mmHg if there is partial obstruction of the needle. In these two cases, cardiac arrest occurred within seconds of the commencement of gas flow [3]. The main disadvantage of using these highly soluble gases for pneumography is their rapid rate of absorption which makes the radiological examination less satisfactory. As a compromise, 400 to 500 ml of carbon monoxide or nitrous oxide may be used to initiate the separation of the tissues, which is the most crucial phase, after which the injection is continued with oxygen [4].

The clinical manifestations of gas embolism will be determined mainly by its location in the cardiovascular system. Minute amounts of air in the *left ventricle* may be fatal, since the air may then pass via the aorta to impact in the coronary or cerebral arteries with resulting ischaemia [5]. Left-sided air embolism may result following right venous embolism if there is a patent foramen ovale, or if air accidentally enters a pulmonary vein during the needling of a pneumothorax.

A gas embolus following retroperitoneal pneumography usually travels via the venous system to the *right side of the heart* and, in this location, a somewhat larger volume of gas may be tolerated. The effects of right-sided air embolism have been extensively studied in dogs [6, 7]. Gas in the right ventricle enters the main pulmonary artery to produce an air lock with acute cor pulmonale, hypertension and respiratory distress which often manifests as polypnoea. Mortality is highest when the experimental animal is in the right lateral recumbent position and conversely, the prospects of survival are appreciably improved if the animal is turned so that its left side is lowermost. In this latter position, gas in the right ventricle is displaced away from the outflow tract and the circulation of blood can be resumed. The air pocket in the right ventricle is then gradually broken up so that only small amounts of air enter the pulmonary arteries with each contraction.

Before commencing pneumography, a stethoscope should be strapped over the praecordium and this should remain in position throughout the examination so that if gas embolism occurs, the first trace of air in the heart will be detected immediately, by its characteristic churning sound. Frequently, however, the clinical signs of embolism may only manifest themselves after the patient has assumed an upright position and the air has travelled up through the venous system to the heart or brain. There may then be signs of circulatory collapse, loss of pulse, pallor or cyanosis, and hyperpnoea followed by respiratory arrest.

Any change in the patient's condition during pneumography should be regarded as indicating possible gas embolism and the patient should immediately be placed in the *left lateral position with the head of the table lowered* to prevent gas from travelling upwards to the brain. This procedure may result in rapid recovery, but several hours should elapse before the patient is allowed to sit up again, otherwise the embolism may recur. This is probably the single most valuable immediate measure of treatment for gas embolism but a fatal case of gas embolism has even occurred during pneumography with the patient in the left lateral position [1].

In experimental animals with gas embolism, the heart continues to beat long after respiration has ceased [7]. Therefore, positive pressure ventilation with pure oxygen can play an important part in treatment. The inhalation of pure oxygen may also promote the re-absorption of air and thereby help to shrink the air bubbles [8]. If a pressure chamber is available, hyperbaric oxygen could be even more valuable in this respect.

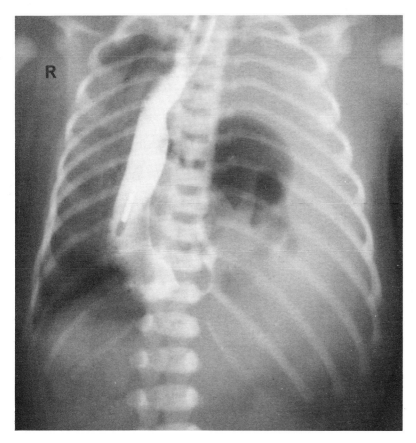

FIGURE 17.1. Oesophagogram in a neonate with a large left-sided diaphragmatic hernia. A diagnostic pneumoperitoneum was attempted through a needle puncture in the left lower quadrant of the abdomen and resulted in death from immediate air embolism. Autopsy showed that the needle had punctured the liver which had been rotated downward into the left side of the abdomen as a result of the herniation.

Hypotension, if severe, should be treated with vasopressors. The differential diagnosis between cardiac arrest and low cardiac output may present a particular problem in gas embolism and, in these cases, cardiac auscultation will probably be more valuable than palpation of the pulse. If the circulatory failure does not respond to treatment by posture in the left lateral position and to vasopressors, providing that the heart sounds indicate the presence of air in a contracting heart, it would be reasonable to try to assist cardiac output by external cardiac massage in the head-down position. If, however, there is true cardiac arrest, it should be assumed that there is probably air in the coronary arteries. This tends to produce resistant ventricular fibrillation and requires immediate thoracotomy with aspiration of air out of the ventricle. Direct cardiac massage is then performed in an attempt to displace the air from the coronary arteries.

The vertebral plexus of veins provides an alternative route by which air may occasionally pass from the pelvis to the venous sinuses of the brain where it may produce neurological sequelae.

Occasional cases of retroperitoneal infection may occur following presacral pneumography and these have included a fatal case of retroperitoneal phlegmon and a severe ischiorectal abscess [1]. Gas embolism is also a recognized hazard in pneumoperitoneum [9, 10]. The risks are particularly high if penetration of a solid viscus has not been recognized and gas is instilled through the needle (Fig. 17.1).

If large volumes of carbon dioxide are used in a pneumo-peritoneum, absorption of the carbon dioxide may increase the blood Pa,CO_2 and may give rise to cardiac arrhythmias. Such changes are less marked when nitrous oxide is used for the pneumoperitoneum [11]. It has been suggested however that, if nitrous oxide is used for laparoscopy, there might be a remote risk of an explosive mixture forming in the peritoneal cavity and that this could constitute a hazard if electrocautery is used [12].

ARTHROGRAPHY

In experienced hands, pneumo-arthrography of the knee achieves an overall diagnostic accuracy of 90 per cent and demonstrates meniscal tears correctly in 97.5 per cent of cases [13]. There are few recorded complications. Occasionally, a joint effusion may occur and may persist for several days, usually disappearing spontaneously [14]. In a series of children, the incidence of hydrops following arthrography with Urografin was 2 per cent. In one of these cases the hydrops persisted for one month but cleared after intra-articular hydrocortisone [15]. A gurgling sensation may persist in the knee for up to 36 hours after the examination [16]. Excessive injection of air may cause painful muscle emphysema which clears in two days [14]. In one patient, a pneumomediastinum was noted several hours after the injection [17]. A rare case of air embolism had been reported following the injection of oxygen into a knee joint but the technique employed was faulty [18]. As with all examinations involving the use of iodinated contrast media, systemic reactions may occur including generalized urticaria and bronchospasm [14]. Lindblom encountered two cases of infective arthritis in 4000 cases [19]. Two cases of 'abacterial' arthritis have also been reported. These healed after immobilization and antibiotic therapy [20, 21].

PERCUTANEOUS NEEDLE BIOPSY

In a series of 326 patients at the Cleveland Clinic, aspiration biopsy in the thorax yielded a diagnostic accuracy of 87 per cent [22]. A pneumothorax occurred in one-third of the patients but only 4 per cent of the patients required drainage. Brief small haemoptyses occurred in 10 per cent of patients. The risks appeared to be higher when a cutting type of needle (Vim–Silverman or Franklin type) was used and a number of deaths from bleeding have been reported. In one elderly patient who coughed during the insertion of a Franklin needle, there was an immediate massive haemoptysis. This occurred within 20 minutes and, at autopsy, one litre of blood was present in the pleural cavity. Another patient coughed up 500 ml of blood following a drill biopsy but recovered [23]. Norenberg et al. [24] described two deaths from endobronchial haemorrhage due to lung biopsy with Franklin and Vim–Silverman needles. They also reviewed eleven other deaths due to percutaneous needle biopsy of the lung. Most of these were due to endobronchial haemorrhage but five were due to gas embolism and one to pneumothorax. In the majority of these cases a cutting type needle had been used but in two of the cases of gas embolism, and in the death due to pneumothorax, an 18 gauge needle was used for aspiration biopsy.

Meyer et al. [25] draw attention to the risks of excessive sedation during lung biopsy and report one such case where a patient died after only a small haemoptysis. At autopsy the trachea and bronchi contained 80 ml of clotted blood. It is therefore important for the patient to be alert enough to clear blood and other secretions from the tracheo-bronchial tree. Patients on anticoagulant therapy or with other haemorrhagic tendencies constitute a particular risk.

There have been very few instances where tumour has spread along the track of the biopsy needle when a cutting type of needle has been used [26] but it is believed that spread is unlikely when a fine biopsy needle is used [27].

Mesothelioma has a particular predilection to local invasion and there is a greater risk of spread along the needle track [28].

Fragments of skin may be implanted by the tip of the biopsy needle and two cases of epidermal inclusion cyst have been reported in the breast, twelve and fifteen years respectively after needle biopsy [29].

Needle biopsy of the kidney may be followed by an arterio-venous fistula [30] and an arterio-biliary fistula with massive haemobilia has been reported following needle biopsy of the liver [31].

TRANSCATHETER BIOPSY

Bronchial brush biopsy appears to have a relatively low incidence of complications. In a series of 693 patients, Fennessey *et al.* [32] had two cases of massive haemoptysis leading to death. There were six cases of pneumothorax but only two required intubation. Staphylococcal septicaemia occurred in one patient and mild pneumonic infection in seven. The pneumonic episodes were usually related to bronchography when this was performed after the biopsy. Thirty to forty per cent of patients having bronchography and biopsy developed febrile episodes but these were rare in the absence of bronchography. Two patients had local anaesthetic reactions, and in one patient the brush was lost but was retrieved with forceps. Recent severe haemoptysis and suspected vascular lesions were regarded as absolute contraindications to bronchial brush biopsy. Thrombocytopenia and other bleeding diatheses, or severe dyspnoea at rest, were considered to be relative contraindications.

An isolated case of seeding of a squamous cell carcinoma of the lung has been reported following bronchial brush biopsy [33]. The examination was performed via the cricothyroid route and the secondary deposit appeared approximately twelve weeks later in the cricothyroid region at the puncture site into the trachea. It was suggested that seeding would be less likely in examinations performed via the nasal or oral routes.

SIALOGRAPHY

Injection of contrast media for sialography should be discontinued as soon as the patient experiences pain. Over-injection may lead to rupture of the alveoli and experimental work suggests that oily contrast media may produce a foreign body reaction if extravasation occurs [34]. In Sjogren's syndrome, where the gland may show marked atrophy with fatty infiltration, there is probably a higher risk of rupture of the ductal system or even of the gland capsule, but this complication was not encountered using hydrostatic sialography [35]. There is commonly slight swelling of the gland following sialography but, in one patient, painful swelling of the parotid gland persisted for three days after injection of 1 ml of Urografin 76 per cent [21]. In another patient, surgical dilatation for a submandibular sialogram caused an unsuspected false passage. Endografin (70% methylglucamine iodipamide)

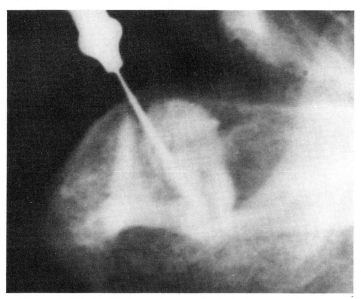

FIGURE 17.2. Submandibular sialogram. Endografin extravasated through a false passage causing sterile induration of the floor of the mouth which lasted for two weeks.

was injected and extravasated into the soft tissues (Fig. 17.2). This caused sterile induration of the floor of the mouth which lasted for two weeks before resolving [21]. As with all radiological contrast media, sensitivity reactions may occur and a severe case of iodism with oedema of the glottis has occurred after the injection of only a drop or two of iodized oil [21].

DISCOGRAPHY

Although discography was described as far back as 1948 [36], its use has been somewhat restricted. The most serious complication is that of 'discitis' [37].

The patient develops pain within a few days of the examination and may have a mild pyrexia. The pain may be extremely severe and relieved only by immobilization. After a period of several weeks, the disc space becomes narrowed and there may be sclerosis of the vertebral margins. It appears probable that the majority of these cases are due to infection of the disc with organisms of low pathogenicity which are able to flourish in the avascular tissues of the disc. Cloward [38] reported one case of discitis in 1500 lumbar discograms but the risk of infection in cervical discograms was approximately ten times greater, particularly with the antero-lateral approach, when the needle might be contaminated by accidental puncture of the oesophagus. In the series by Fenström [39] the incidence of complications following lumbar discography was 14.5 per cent and included disc herniation; headache, varying from a few days to more than a year; and arachnitis.

More recently, interest has developed in the treatment of pathological vertebral discs by the combination of discography and intra-discal injection of chymopapain which hydrolyses the disc protein causing dissolution of the disc [40]. In approximately 1 per cent of cases, hypersensitivity to the chymopapain may cause anaphylactic reactions with severe shock. The treatment of these reactions is similar to that of contrast media reactions. Chymopapain is toxic if injected into the subarachnoid space. Animal experiments suggest that large doses of hydrocortisone and immediate decompression of the cord may have some value in treatment [40].

MURPHY'S LAW

If the design of equipment, or the method of performing a procedure, provides scope for a serious error to be made then, sooner or later, someone will inevitably make that error. Conditions in an X-ray department provide ample scope for the operation of Murphy's law, particularly during the performance of special procedures which may require the undivided attention of the radiologist so that he may have to rely to a considerable extent on the efficiency of his supporting staff. If the procedure has to be performed under conditions of poor illumination, an additional hazard is thereby introduced. Some of these complications have already been described in other sections of this book but several instructive cases were reported during the course of the National Radiological Survey. These illustrate some of the problems involved and the need for constant vigilance.

At left carotid arteriography, 10 ml of Conray 420 was injected by error instead of Conray 280. This resulted in an immediate convulsion with loss of consciousness for two minutes. Following this, the patient appeared well for about five hours and then developed a right hemiparesis with dysphasia. There was some deterioration over a period of six hours, but full recovery occurred within twenty-four hours. The delayed onset of symptoms in this case suggests that cerebral oedema may have played a part and, in similar circumstances, a large dose of corticosteroid might usefully be administered systemically at an early stage. It is impossible to overemphasize the importance of a double check to ensure that the correct material is injected. Some ampoules, e.g. those of Biligrafin and Urografin, may be easily confused if the labels are not checked, so that the more toxic iodipamide may be injected in place of diatrizoate.

On another occasion a radiologist had satisfactorily performed the first injection of a carotid angiogram but, after the second injection, no contrast material was visible in the cerebral arteries. A film of the neck was immediately taken and confirmed that there had not been an extravascular injection. It then emerged that a trainee anaesthetist had noticed that the ampoule of contrast medium had fallen over in the bowl of water in which it was being warmed and, in an attempt to be helpful, she had restored the ampoule to its upright position. As a result, the second injection had consisted of non-sterile warm water which had entered the open ampoule. The patient was treated with antibiotics and kept under observation and, fortunately, no sequelae developed.

Two patients in a neurosurgical ward woke up in the morning with petechial rashes on their faces (Fig. 17.3). The startled ward staff at first suspected an infectious fever until it was realized that the rashes were probably due to cerebral angiography performed on the previous day. At the same time, a new nurse on the cerebral angiography suite was observed to be powdering her hands over the open bowl of saline used for flushing the syringes, prior to putting on her sterile gloves. The hazards of cotton fibre and glove powder embolism during angiography [41, 42] are now well known and the use of closed systems for flushing syringes should eliminate this hazard. This is discussed in greater detail in Chapter 2. Recently, it has been shown that small glass particles from an opened ampoule may cause microemboli and that microscopic particles of plastic may also be derived from disposable syringes, but no clinical complications have yet been recorded [43].

Another example of the risk of using open bowls for injection material has recently occurred [44]. The contrast medium for cerebral angiography had been emptied into a gallipot, apparently to save time during the injection. An identical gallipot contained cetrimide for skin sterilization. The syringe was accidentally filled from the gallipot of cetrimide and the solution was

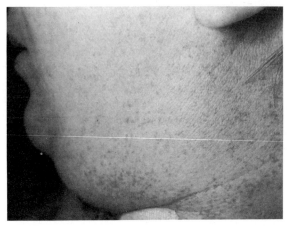

FIGURE 17.3. Petechial rash following carotid arteriography due to glove powder embolism. A new nurse in the angiography suite had powdered her hands over the open bowl of saline used to flush the syringes.

injected into the cerebral circulation. Respiratory failure developed within five minutes and the patient died three days later without recovering consciousness. It was subsequently learned from the manufacturers that cetrimide was highly toxic when administered by injection.

Problems may also arise with syringes and other equipment. Using resterilizable nylon syringes, it may be difficult to see air within the syringe. This resulted in the injection of approximately 5 ml air causing air embolism during cerebral angiography. There was transitory respiratory arrest which responded rapidly to intubation and oxygen administration. The patient was at first unconscious and hemiplegic but this cleared over a period of one week. Air may apparently be trapped on the inside walls of metal syringes after repeated use, and has caused air embolism. This problem was eliminated after the internal walls of the syringes were repolished.

Occasional attempts have been made to resterilize 'disposable syringes' as an economy measure. This is unwise and possibly dangerous. Plastics are complex materials and inexpert resterilization may liberate toxic substances [45].

Even new equipment may not be free from risk. A reputable manufacturer distributed a batch of guide wires without prior sterilization. The guide wires were sealed in plastic packages similar to previously marketed pre-sterilized wires. Although the faulty packages were not actually labelled as being sterile, they were accepted as such and had been used in routine angiography, apparently without recognizable complications.

REFERENCES

1 RANSOM L.R., LANDES R.R. & McLELLAND R. (1956) Air embolism following retroperitoneal pneumography; a nation-wide survey. *J. Urol.* **76**, 664–670.
2 STAUFFER H.M., DURANT T.M. & OPPENHEIMER M.J. (1956) Gas embolism. Roentgenologic considerations including the experimental use of carbon-dioxide as an intracardiac contrast material. *Radiology* **66**, 686–692.
3 SCOTT D.B. (1970) Some effects of peritoneal insufflation of carbon dioxide at laparoscopy. *Anaesthesia* **25**, 590.
4 SAXTON H.M. & STRICKLAND B. (1972) *Practical Procedures in Diagnostic Radiology* (2nd edn), pp. 160–168. H. K. Lewis, London.
5 GEOGHEGAN T. & LAM C.R. (1953) The mechanism of death from intracardiac air and its reversibility. *Ann. Surg.* **138**, 351–359.
6 DURANT T.M., LONG J. & OPPENHEIMER M.J. (1947) Pulmonary (venous) air embolism. *Amer. Heart J.* **33**, 269–281.
7 OPPENHEIMER M.J., DURANT T.M. & LYNCH P. (1953) Body position in relation to venous air embolism and the associated cardiovascular respiratory changes. *Amer. J. Med. Sci.* **225**, 362–373.
8 FINE J. (1963) Air embolism. *Lancet* **1**, 605.
9 HOLLANDER A.C. (1951) Air embolism in pneumoperitoneum. *J. Am. med. Ass.* **147**, 568–569.
10 BURMAN D. (1956) Air embolism during pneumoperitoneum treatment. *Thorax* **11**, 49–56.
11 SCOTT D.B. & JULIAN D.G. (1972) Observations on cardiac arrhythmias during laparoscopy. *Brit. med. J.* **1**, 411–413.
12 ROBINSON J.S., THOMPSON J.M. & WOOD A.W. (1975) Laparoscopy explosion hazards with nitrous oxide. *Brit. med. J.* **3**, 764–765.
13 NICHOLAS J.A., FREIBERGER R.H. & KILLORAN P.J. (1970) Double-contrast arthrography of the knee. Its value in the management of two hundred and twenty-five knee derangements. *J. Bone Jt. Surg.* **52A**, 203–220.
14 RICKLIN P., RUTTIMAN A. & DEL BUONO M.S. (1971) *Meniscus Lesions. Practical Problems of Clinical Diagnosis, Arthrography and Therapy.* Grune & Stratton, New York.
15 STENSTRUM R. (1968) Arthrography of the knee joint in children. Roentgenologic anatomy, diagnosis and the use of multiple discriminant analysis. *Acta Radiol.* Suppl. 281.
16 FREIBERGER R.H., KILLORAN P.J. & CARDONA G. (1966) Arthrography of the knee by double contrast method. *Am. J. Roentg.* **97**, 736–747.
17 WEINER S.N. (1967) Contrast arthrography of the knee joint: a comparison of positive and negative methods. *Radiology* **89**, 1083–1086.
18 KLEINBERG S. (1927) Pulmonary embolism following oxygen injection of the knee. *J. Am. med. Ass.* **89**, 172–173.
19 LINDBOLM K. (1948) Arthrography of the knee, a roentgenographic and anatomical study. *Acta Radiol.* Suppl. 74.
20 FICAT P. (1957) *L'arthrographie opaque du genou.* Masson et Cie, Paris. Quoted by Stenstrum. (Ref. 15.)
21 Reports to National Radiological Survey.

22 LALLI A.F. (1973) Roentgen-guided aspiration biopsies of thoracic, renal and skeletal lesions. In: *Complications and Legal Implications of Radiologic Special Procedures* (Ed. by MEANEY T.F., LALLI A.F. & ALFIDI R.J.). C. V. Mosby, St Louis.
23 MCCARTNEY R.L. (1974) Hemorrhage following percutaneous lung biopsy. *Radiology* 112, 305–307.
24 NORENBERG R., CLAXTON C.P. & TAKANO T. (1974) Percutaneous needle biopsy of the lung: report of two fatal complications. *Chest* 66, 216–218.
25 MEYER J.E., FERRUCCI J.T. & JANOWER M. (1970) Fatal complications of lung biopsy and report of a case. *Radiology* 96, 47–48.
26 WOLINSKY H. & LISCHNER M.W. (1969) Needle track implantation of tumour after percutaneous lung biopsy. *Ann. Int. Med.* 71, 359–362.
27 ENGZELL U., ESPOSTI P.L., RUBIO C. et al. (1971) Investigation of tumour spread in connection with aspiration biopsy. *Acta Radiol. (Ther.)* 10, 385–398.
28 GODWIN M.C. (1957) Diffuse mesotheliomas with comment on their relation to localised fibrous mesotheliomas. *Cancer* 10, 298–319.
29 GERLOCK A.J. (1974) Epidermal inclusion cyst of the breast associated with needle aspiration biopsy. *Radiology* 112, 69–70.
30 ELKIN M., MENG C-H. & DEPAREDES R.G. (1966) Roentgenological evaluation of renal trauma with emphasis on renal angiography. *Am. J. Roentg.* 98, 1–26.
31 ATTIYEH F.F., MCSWEENEY J. & FORTNER J.G. (1976) Haemobilia complicating needle liver biopsy. A case with arteriographic demonstration. *Radiology* 118, 559–560.
32 FENNESSY J.G., LU C-T., VARIAKOJIS D. et al. (1973) Transcatheter biopsy in the diagnosis of diseases of the respiratory tract. An evaluation of seven years' experience with 693 patients. *Radiology* 110, 555–561.
33 CURRY J.L., REGALADO R.D., HEIMBACH J.G. et al. (1974) Implantation transfer of pulmonary carcinoma to the trachea: a complication of percutaneous catheter biopsy. *Chest* 64, 163–165.
34 RUBIN P. & HOLT J.F. (1957) Secretory sialography in diseases of the major salivary glands. *Am. J. Roentg.* 77, 575–598.
35 WHALEY K., BLAIR S., LOW D.M. et al. (1972) Sialographic abnormalities in Sjogren's syndrome, rheumatoid arthritis, and other arthritides and connective tissue diseases. A clinical and radiological investigation using hydrostatic sialography. *Clin. Radiol.* 23, 474–482.
36 LINDBLOM K. (1948) Diagnostic puncture of intervertebral discs in sciatica. *Acta orthop. Scandinav.* 17, 231–239.
37 GARDNER W.J., WISE R.E., HUGHES C.R. et al. (1952) X-ray visualisation of the intervertebral disc with a consideration of the morbidity of disc puncture. *A.M.A. Arch. Surg.* 64, 355–364.
38 CLOWARD R.B. (1963) Cervical discography. *Acta Radiol. Diagn.* 1, 678–688.
39 FERNSTRÖM U. (1960) A discographical study of ruptured lumbar intervertebral discs. *Acta chirurg. Scandinav.* Suppl. 258.
40 WILTSE L.L., WIDELL E.H. & YUAN H.A. (1975) Chymopapain chemonucleolysis in lumbar disk disease. *J. Am. med. Ass.* 231, 474–479.
41 ADAMS D.F., OLIN T.B. & KOSEK J. (1965) Cotton fibre embolism during angiography. A clinical and experimental study. *Radiology* 84, 678–681.
42 YUNIS E.J. & LANDES R.R. (1965) Hazards of glove powder in renal angiography. *J. Am. med. Ass.* 193, 304–305.

43 BREKKAN A., LEXON P.E. & WOXMOLT G. (1975) Glass fragments and other particles contaminating contrast media. *Acta Radiol. Diagn.* 16, 600–608.
44 Woman patient died after angiogram injection error. *Rad* p. 2 (March, 1976).
45 ANSELL G. & PITCHFORD A.G. (1974) Dangers of re-sterilized disposable syringes in urography. *Brit. J. Radiol.* 47, 300.

CHAPTER 18: ANAESTHETIC PROBLEMS

S. LIPTON

Many chapters in this book mention anaesthesia. The scope of this one will be limited to any hazard occurring in the X-ray department, or developing after radiology involving the anaesthetist. Some knowledge of the underlying processes will be given but further and more detailed information is provided in the relevant chapters. Hazards of local anaesthesia are discussed in Chapter 12.

The problems facing the anaesthetist in the X-ray department are fourfold and are due to:
1 Condition of the patient.
2 Anaesthetic technique.
3 Radiographic procedure.
4 Contrast media.

CONDITION OF THE PATIENT

Simple procedures

Most procedures in the X-ray department are simple ones not involving the use of contrast media. It must not be forgotten that any of the usual emergencies may arise, such as an epileptiform attack, syncopal faint, asthmatic attack, coronary thrombosis or cardiac arrest. There is nothing unusual in the treatment of such patients. Most of the routine treatments required will be mentioned in the course of this chapter, and each X-ray department will normally have a 'drill' for dealing with these relatively standard emergencies [1]. It is axiomatic that each X-ray department will have a supply of the drugs, syringes, needles and equipment in the special packs for the individual emergencies that have been foreseen; that there are written instructions on

Trauma

Further problems arise with patients who are in the X-ray department following trauma. Open and closed fractures must be dealt with in an appropriate way. There may be large haematomas or loss of blood with corresponding shock and intravenous fluids or blood transfusion will be necessary. Fat embolism is a possibility under these circumstances. Head injuries form a group which has difficult problems. This is because there may be a period of unconsciousness after the initial trauma and during this time aspiration of stomach contents can occur. This produces the well-known *Mendelson syndrome* in which there is an asthmatic response with bronchospasm due to the inhalation of acid gastric contents [2]. Immediately following the aspiration, or sometimes after a few hours, cyanosis, bronchospasm, dyspnoea and tachycardia occur. There are moist sounds in the chest and there is a characteristic radiographic appearance. Severe cases progress to pulmonary oedema. In addition to these problems which are due to the acidity of the gastric contents, there may be solid particles in the lungs as well.

Treatment. Once the airways are contaminated, bronchoscopy and lavage with suction must be carried out, 10 ml saline being injected via the bronchoscope (or endotracheal tube) and aspirated until the washings are clear. Large doses of hydrocortisone are indicated and aminophylline and other bronchodilators are given as required.

Unconscious or semiconscious patients must be transported in a safe fashion. This means that the possibility of vomiting and inhalation, and airway obstruction is kept in mind. A lateral position, preferably on the left side, with the head dependent is essential and if there is deepening unconsciousness a cuffed endotracheal tube should be passed and inflated. Often if a patient is being transported from an outlying district to a special centre it is reasonable to request that the airway be protected in this way during the journey. Only too frequently does a patient appear at the X-ray department with contaminated lungs, having started the journey in a satisfactory condition.

Often injuries are multiple, for instance, the combination of a head injury with a chest injury is one which puts the patient in great hazard. The important feature here is to realize that unless there is an immediate threat to life from a rapidly increasing intracranial pressure, treatment to the chest should be performed first. Cyanosis and increased venous pressure from the chest lesion will increase intracranial pressure, which will not reduce until the chest condition is relieved. Relief of a pneumothorax or a flail chest may be necessary as a matter of urgency.

Complicated procedures

Patients undergoing these are not usually in first-class physical condition. In all cases there is a morbidity and mortality associated with each procedure. In no technique is this more obvious than in angiocardiography, where the patients can rarely be classified as healthy. This applies to all the various X-ray techniques. However, angiocardiography can be taken as an example.

In *angiographic procedures*, cardiovascular pathology is frequently present. Generally, the more severe the disease the greater is the need for angiography, and it is in this situation that angiography presents its greatest risk. Braunwald & Swan [3] gave figures for major complications in angiocardiography of 3.6 per cent and, if coronary arteriography is included, the complication rate is 4.5 per cent. They give an overall mortality of 0.45 per cent and there was a 6 per cent mortality in infants under two months. These figures emphasize the problem of heart disease in early life, though of course there are many complications of cardiac catheterization which occur in patients of all ages [4–6]. Infants seriously ill with suspected cardiac anomalies are now regarded as medical emergencies and potential surgical emergencies [7, 8]. More of these babies in heart failure are being presented for cardiac catheterization. Approximately one-third of the infants born with congenital cardiac defects die in the first month of life [9, 10]. This is the most difficult period of life in which to distinguish the congenital cardiac defects from the respiratory distress syndrome and other respiratory disease. Lambert [7] estimated in 1386 autopsies that congenital cardiovascular defects produced death or severe distress in 165 newborn infants. Twenty-two of the deaths occurred following cardiac surgery and another seven following cardiac catheterization with angiocardiography. Eighty-five of the 165 patients (52 per cent) died in the first week. The two most frequent causes of distress and death were the hypoplastic left heart syndrome and complete transposition of the great vessels. Most of the patients with these two defects died in the first week of life. In hypoplastic left heart cases 28 out of 37 died during the first week and in 25 complete transpositions of the great vessels, two-thirds died during the first week. Multiple major cardiac defects ranked third in Lambert's list of cardiac abnormalities causing early death or severe distress.

Pneumoencephalography and ventriculography are procedures where there is an added risk. In these patients there is often an increase in intracranial pressure, for instance due to hydrocephalus or to a space occupying lesion, and this may be increased during the investigation.

Lymphangiography in patients with no pulmonary disability does not produce any increased respiratory symptoms. This does not apply to patients with pre-existing pulmonary disease. In these patients the pulmonary embolization of the oil-based contrast medium which occurs during the procedure may produce marked respiratory distress [11].

Bronchography. Local anaesthetic drugs are used for this procedure and reactions to these local drugs occur due to their rapid absorption from the bronchial tree. Usually any difficulties arise from the use of excessive quantities with resultant toxic effects. In children, bronchography is usually performed under general anaesthesia and the small diameter of the bronchi renders them more liable to obstruction by the contrast medium, with resultant segmental collapse. This appears to be more likely to occur when halothane is used with 50 per cent oxygen and 50 per cent nitrous oxide. [12]. As much of the contrast medium as possible should be aspirated at the conclusion of the examination.

THE ANAESTHETIC TECHNIQUE

General considerations

In the past 30 years a situation has been reached where the number of patients regarded as being unfit for general anaesthesia has become extremely small [13]. However, in general anaesthesia for operative procedures there is a small but definite mortality to which the anaesthetic makes some contribution. Harrison [14] considered 178 000 anaesthetics. There was an immediate operative mortality of 414. In 58 of these the anaesthetic was believed to have had a significant contribution. This represents an incidence of 0.33 per 1000 anaesthetics. There is considerable controversy in radiological procedures as to whether local, general, or in some cases spinal anaesthesia provides the minimal morbidity and mortality. For instance McAfee [15] made a reasonably complete review from the English literature, the available foreign literature, and the Johns Hopkins' series, in aortography. Although realizing that there was a trend towards the use of local anaesthesia and mentioning one authority who preferred to use spinal anaesthesia, he reported very few complications as a result of general anaesthesia. A similar picture seems to hold true for all the other radiodiagnostic procedures except angiocardiography. It would appear that there is an inherent danger in this variety of investigation that does not exist in the others. It is logical to expect this. To visualize the cavities of an abnormal heart, to catheterize the coronary arteries, to alter the haemodynamics of the pulmonary circulation is a dangerous procedure. It is therefore most necessary to review anaesthetic techniques in order to reduce complications to a minimum, and to ensure that those that do occur are efficiently treated.

With a few noticeable exceptions it does not matter what particular sedative drugs or anaesthetic techniques are used, as long as they are safe methods. It must be admitted, however, that what may be a safe method in the hands of one physician may be dangerous with another of different experience. Manners [8] discussed and reviewed this whole subject in relation to anaesthesia for diagnostic procedures in cardiac disease. He covers methods involving no sedation, mild sedation, heavy sedation; general anaesthesia with or without intubation, and with or without intermittent positive pressure ventilation. These techniques can therefore be applied to any of the radiodiagnostic procedures.

The aim of the anaesthetist is to present the radiologist with an immobile patient during the radiological examination. In general, adults and older children do not require general anaesthesia, while young children and anxious adults do. Some problems are obvious at the pre-operative visit. Lack of injection sites; a large tongue; the large head of hydrocephalus; cyanosis and cardiac failure or pulmonary dysfunction.

Strict attention must be paid to tissue hypoxia and correction of acid-base balance. In most units preoperative digitalization is required as essential in heart failure. No procedure will be undertaken until these measures are taken. Acidosis is corrected pre-operatively, either with 8.4 per cent sodium bicarbonate or 0.6 tris buffer (Tham). The dose of base is calculated as weight in kilograms × base deficit × 0.3 = mEq base. Blood gas analysis is repeated before operation in angiocardiography so that further correction can be made before induction of anaesthesia. Premedication for infants can follow one of two methods. That of Mustard [16] who premedicates, for example, with 0.02 mg per kilogram atropine. Cyanotic infants receiving morphine sulphate 1 mg per 5 kg and are transported to the operating room in an oxygen enriched isolette with the head tilted upwards. The operating table is provided with a warming blanket and the infant is placed 3 ft below an infra-red lamp. After pre-oxygenation the infant is rapidly intubated, oxygenated and paralysed (with succinyl choline or curare). Anaesthesia is maintained by a Jackson–Rees' modification of Ayres' T-piece with moderate hyperventilation and intermittent doses of succinyl choline. Halo-

Anaesthetic Problems

thane is avoided in the presence of heart strain or failure and nitrous oxide is avoided in the presence of arterial desaturation. Isoproterenol or epinephrine, 4 mg per ml, are available. Rarely a slow infusion of epinephrine is given to alleviate severe heart failure.

In an alternative technique which is described later, no premedication is given, intravenous drugs being given in the anaesthetic room [17].

Drug interactions and contrast media

It is reasonable to mention briefly some problems now which are mentioned again later or expanded in other chapters. They all concern the combination of contrast media with the anaesthetic or effects due to the toxicity of the contrast media itself. As the results of contrast toxicity—bronchospasm or pulmonary oedema, for instance—are complications which can be produced or precipitated by the anaesthetic agents themselves, it is impossible to decide which agent is responsible. All that can be done is to treat the condition as due to contrast medium toxicity if one of these substances has been used. Thus large doses of steroids will be given if this combination happens. Convulsions due to the contrast medium are rare under general anaesthesia but may occur under local anaesthesia if excessive doses of contrast media have been used.

There may be interactions between anaesthetic drugs and contrast media which produce prolongation of anaesthesia. This was originally seen with the older varieties of contrast media such as Urokon and the intravenous barbiturates. It is probably due to displacement of the barbiturate by protein binding effects of the Urokon [43, 49]. Nembutal anaesthesia appeared to confer some degree of protection on rats in the period immediately after injection of a lethal dose of Urokon [50].

Finally there will be effects due to interaction of the anaesthetic drugs themselves with other drugs which the patient is taking routinely or given specially for the radiographic technique. For instance, a patient may be on anti-hypertensive drugs, which will be potentiated by adrenergic blockers. The combination will also be potentiated by a general anaesthetic.

Some problems develop during the radiographic procedure. Apart from those complications due to the contrast medium itself, these anaesthetic problems are no different from any other anaesthetic complication. They are now briefly mentioned.

During the induction

Possible complications at this stage of the procedure cover the whole field of anaesthesia and would require a treatise to describe in detail. The important point to bear in mind is that an anaesthetic induced in the X-ray department should be carried out as safely as in the operative anaesthetic room. This means that the room used for induction in the X-ray department must be well equipped, and have enough personnel to ensure that, if anything does go wrong on the anaesthetic side, resuscitation can go forward immediately.

There are two particular areas where special danger lies. Firstly, it is imperative that any safety warning devices used either on the patient (pulse, respiration), or on the anaesthetic equipment (oxygen) should be maintained in good order and tested before each case. The other risk is well known and involves intubating the right main bronchus. This is particularly hazardous in children and babies, where not only may the left main bronchus be occluded but the right upper lobe bronchus also.

During the sedation or anaesthetic

The whole process of the care and protection of the patient is a continuing one. It is wise to ensure that the vulnerable parts of the patient such as eyes, ears and nose are safe; that vulnerable pressure points are protected; that, for instance, the arm cannot drop over the edge of the table, or be trapped by moving X-ray equipment during the procedure.

It is most important the proper 'drill' is carried out to ensure that the correct patient is being investigated, and that the planned investigation is the one that is to be performed. It is, of course, recognized that a decision as to the procedure is often made at the time of the investigation.

The problems that arise can be divided roughly into those due to technical factors, those due to mainly respiratory, cerebral or cardiovascular factors, and there is a miscellaneous group.

Technical factors are of the greatest importance. Maintenance of homeostasis at as normal a level as possible is the key. Oxygenation, electrolyte and fluid balance, normocapnia and acid-base balance are monitored and corrected if necessary. Blood pressure, pulse and respiration rate, ECG, and EEG, are often required and in angiocardiography are regarded as mandatory by some authorities.

Simple anaesthetic faults must not be forgotten. Such developments as displacement or blockage of intravenous or arterial drips and catheters occur with some frequency. Similarly disconnection, kinking, displacement and blockage of the endotracheal tube or the associated connecting tubes must be speedily corrected. Finally there may be an anaesthetic gas failure.

Respiratory factors

Alterations in rate may originate in various ways but if cyanosis results, the essential treatment is to maintain oxygenation. Intubation and controlled respiration may be necessary for this. Bronchospasm can be exceedingly difficult to treat, and if contrast media have been used it will be impossible to distinguish between an anaesthetic, idiopathic or contrast media reaction producing the bronchospasm. However, if a contrast medium has been used, hydrocortisone in large doses must be given. Bronchospasm arises in patients with asthma even when they have had preoperative treatment (bronchodilators, oral aminophylline and/or ephedrine; oral prednisolone). During anaesthesia bronchospasm can be due to inadequate suppression of reflexes, that is, too light an anaesthetic, and in this case deepening of the anaesthetic will help. Treatment for other cases is by aminophylline 0.25 g in 10 ml intravenously and slowly as vasodilation will cause hypotension. Intravenous steroids (hydrocortisone 100 mg) can be life-saving and may be repeated in high dosage if necessary. Adrenaline given slowly, 0.5 ml 1/1000 solution subcutaneously or intramuscularly, can be used.

Malignant hyperpyrexia. The combination of bronchospasm, rigidity of muscles (usually beginning in the jaw) and hyperthermia is due to the rather rare condition of malignant hyperpyrexia. It is important to recognize this as soon as possible as the mortality is high and increases the longer the condition is unrecognized. These patients have an inborn fault of muscle metabolism and any family history of unexplained deaths under anaesthesia should be regarded with suspicion.

The treatment is withdrawal of any halogenated anaesthetic agent, no further use of succinyl choline as a relaxant and the intravenous injection of procaine amide. Procaine can be used as an alternative to procaine amide which produces a larger fall in blood pressure. A loading dose of 30–40 mg/kg procaine amide intravenously is given, followed by 0.2 mg/kg/minute intravenously until muscle rigor relaxes.

From the beginning of resuscitative measures, hypoxia is treated by hyperventilation with 100 per cent oxygen. Sodium bicarbonate is given to combat acidosis. Serum potassium levels are monitored. They rise before treatment and may fall dangerously with treatment. ECG monitoring is standard procedure, frusemide and or mannitol may be used to maintain urinary output, and with this regime isoprenaline may be necessary to maintain blood pressure [18].

Pulmonary oedema. Pulmonary oedema arises from allergic reactions: bronchospasm, asthma, cerebral disease, hypervolaemia and cardiovascular alteration in the pulmonary blood supply. Treatment is by intravenous aminophylline, oxygenation with intermittent positive pressure respiration, 20 mg Lasix (frusemide) intravenously, atropine and treatment of acute cardiac failure. It must be emphasized in the treatment of pulmonary oedema that as in the case of bronchospasm, if contrast media have been used steroid therapy is important.

Cerebral factors

In encephalography, as mentioned in Chapter 7, there is a rise in intracranial pressure when air is injected, and this is discussed in that chapter. However, some remarks are necessary and it may be as well to mention here that there is an inevitable morbidity associated with pneumoencephalography and further information and references can be obtained from an article by White [19]. Briefly this states that in 50 patients examined seven days after pneumoencephalography headache was present in 78 per cent, neck stiffness in 34 per cent, pyrexia in 38 per cent, vomiting in 34 per cent, tachycardia in 74 per cent, a change of conscious level in 18 per cent, and abnormal neurological signs in 30 per cent. Thus it will be seen that there are many complications which may occur after anaesthesia but which are not related to the anaesthetic.

There is a basic decision to be made in pneumoencephalography as to whether sedation or a general anaesthetic is to be used. Adults are usually a much simpler problem for sedation than children. In paediatric encephalography an anaesthetist should be present to look after the child and take control if necessary. This subject is adequately dealt with by Nicholson [20].

General anaesthesia for babies. This is considered a speciality in itself. At the Liverpool Children's Hospital, the general anaesthetic technique is as follows [17]. Babies are given no premedication at all, since it has been found that complications are higher if morphine or Vallergan is given. If these drugs are given the babies tend to become hypotensive, probably due to heavy sedation in the sitting position. Intravenous atropine is given in dosage of 0.15 mg in babies up to two years; 0.30 mg up to six years and 0.60 mg over six years. Thiopentone in dosage of 25 mg per stone or 6 kilos, and tubocurarine chloride 5 mg per stone or 6 kilos are also given. Thirty per cent oxygen and seventy per cent nitrous oxide are the inhalation gases. No halothane is given as this raises the intracranial tension, as mentioned later. A Jackson-Rees modification of Ayres' T-piece circuit is used. Flow of respiratory gases is adjusted to $1\frac{1}{2}$ litres per stone body weight with a minimum of 3 litres per minute. All cases are ventilated, usually manually.

Reversal of the relaxant drugs is by neostigmine 0.5 mg per stone mixed with atropine 0.12 mg per stone and given as one injection intravenously. The average length of encephalography in babies is 45 minutes.

Changes in intracranial pressure. Some complications produced during encephalography are due to the anaesthetic and some due to hypotension produced by the sitting position, but most complications occur after the injection of air. This tends to cause a pressure rise, especially in children and babies. When sitting up, this pressure rise is reasonably well contained but on lying down there may be a marked rise in blood pressure. For instance, in babies the systolic pressure may rise to 180 mm of mercury and diastolic to 120 mm. Presumably this is due to systemic vasoconstriction resulting from increased hydrostatic pressure in the cranium when lying down. In fact blood pressure changes in small children, blood pressure can fluctuate widely in unpredictable fashion every few minutes. This may be related to air around the vasomotor centres in the third and fourth ventricles, or to the increased pressure in the skull. Encephalography and angiography are common procedures in neurosurgical and neurological practice. When these are performed under general anaesthesia it should be remembered that there is a more marked difference in the intracranial pressure in the presence of a space occupying lesion than when there is no such lesion. It has been shown [21] that the volatile anaesthetic agents, halothane, trichloroethylene and methoxyflurane produced a greater rise in intracranial pressure in patients with intracranial space occupying lesions than in patients with normal cerebrospinal fluid pathways. Blood pressure was reduced and cerebral perfusion fell markedly. The authors comment that the increased intracranial pressure would be expected to accentuate internal brain hernias, and if the skull were opened, to result in difficult operating conditions and external herniation of the brain. The greatest rises in pressure were in patients with frontal lesions who also showed clinical evidence of raised pressure before operation. The effect of both halothane and methoxyflurane on the CSF pressure and on perfusion pressure is dose dependent. In normal anaesthetic concentrations methoxyflurane has a somewhat greater effect on intracranial pressure than halothane or trichloroethylene, though this difference is not significant. Trichloroethylene has much less effect on perfusion pressure because blood pressure is much less affected with this agent.

Fitch [22] pointed out that halothane, in the presence of an experimental cerebral tumour, produced considerable transtentorial pressure gradients before unilateral pupillary dilatation occurred. Since pupillary dilatation is frequently the sign which indicates the presence of an acute expanding intracranial mass, the clinical conclusion to be drawn is that the injudicious use of halothane may on occasion lead to brain shift and tentorial impaction in patients with undiagnosed space occupying lesions. Thus, a patient with multiple injuries and an undiagnosed intracranial haematoma may be brought to theatre to have, for example, surgical treatment of fractured limbs. The administration of a halothane anaesthetic may then lead to sudden brain impaction. To some extent (but not entirely) these dangers can be mitigated by controlled respiration and hypocapnia.

The effect of other anaesthetic agents on cerebral spinal fluid pressure has been investigated. For instance, nitrous oxide has a similar effect to halothane in the presence of intracranial space occupying lesions [23]. However, it was found that moderate hyperventilation could efficiently counteract the intracranial hypertension produced by nitrous oxide. This may account for this effect of nitrous oxide on intracranial pressure not being reported more often in the past. Henriksen [23] speculates that little is known about intracranial pressure in patients with recent head injuries and varying degrees of brain oedema, but it must be suspected that they develop pressure increases similar to the patients described with space occupying lesions when exposed to nitrous oxide inhalation. As such patients are quite often given a general anaesthetic for neuroradiological examination this is of importance. Fluroxene produces increased cerebrospinal fluid pressure which tends to respond to hyperventilation and hypocarbia.

Most of the intravenous anaesthetics also affect intracranial tension in man. Althesin [24] and the barbiturates [25] reduce intracranial pressure while ketamine [26] increases it.

Neuroleptanalgesic drugs which reduce intracranial pressure may be the drugs of choice in patients with space occupying lesions [27].

Cardiovascular factors

Blood loss is not usually a problem in adults but with small children and infants relatively small amounts of blood loss can form a disproportionate percentage of the total blood volume. In these cases blood loss must be estimated and replaced. A useful formula for estimating blood volume in infants is 84 ml per kg of body weight, thus a newborn baby has a blood volume of about 300 ml and a one-year-old about double this [28]. The same applies to adults at roughly 80 ml per kg of body weight, or 5.6 litres in an adult weighing 70 kg [29]. Although in adults a slow intermittent loss of blood does not produce a fall in blood pressure until about three pints have been lost, rapid loss of one pint of blood can produce hypotension. Thus speed of loss as well as volume is important; this is why a relatively small blood loss in an infant or neonate can be exceedingly dangerous.

Cardiac arrhythmias. Cardiac irregularities and conduction disturbances are ever-present possibilities during cardiac catheterization. Wennervold [30] found that in 4413 catheterizations, supraventricular tachycardias occurred during 62 catheterizations in 57 patients. Atrial fibrillation occurred in 20 cases and atrial flutter 13 times in 12 patients. These supraventricular rhythms ceased in less than 24 hours. In some cases the attacks were treated with digitalis or quinidine. Short periods of conduction disturbance occurred in the form of complete atrioventricular block and right bundle branch block. Ventricular fibrillation occurred in two patients.

Cardiac arrhythmias arising under these circumstances should be regarded as potentially dangerous. There are modern drugs which may be very useful in controlling and treating such arrhythmias. They act selectively by blocking beta adrenergic receptors, and this prevents these receptors being stimulated by catecholamines or by sympathetic nervous discharge. It will be remembered that hypercarbia is thought to cause release of catecholamines and this may explain its relationship to arrhythmias. Light anaesthesia and surgical stimulation may cause sympathetic overactivity [31]. Thus adequate ventilation and a suitable depth of anaesthesia or sedation is a necessity for any procedure such as angiography where cardiac arrhythmias are an inherent hazard of the procedure. *Beta-adrenergic blocking drugs* isolate the heart from the effect of sympathetic stimulation. The heart rate is slowed, the velocity of contraction is reduced and atrioventricular conduction is delayed. These drugs also block beta-adrenergic receptors in the bronchial tree and peripheral circulation and increase bronchoconstrictor and vasoconstrictor mechanisms. The parasympathetic system is relatively stimulated as sympathetic transmitters are inhibited.

Caution is required in the use of beta blockers in congestive heart failure, particularly after acute myocardial infarction. Increased vagal action often present after myocardial infarction is enhanced in the presence of beta blockers and may lead to marked fall in cardiac rate and output. These drugs should not be prescribed for patients with defective atrioventricular conduction, or in asthmatics in whom they may precipitate bronchospasm. Patients with advanced myocardial failure should be fully digitalized and it must be remembered that the beta blockers have a synergistic effect with digitalis in slowing the heart rate in fibrillation.

There are three beta blocking drugs available: propanolol and oxyprenalol which are nonselective and short acting, and practolol which has a more prolonged action. Practolol[1] is more cardio-selective as it mainly blocks excitatory receptors and has a lack of peripheral action compared to the others. Thus propanolol and oxyprenalol are better for treating angina and hypertension as they produce a larger fall in cardiac output, while practolol is better for treating arrhythmias.

Arrhythmias arising during cardiac catheterization or cardioangiography tend to occur during passage of the catheter through the stenosed valves. They do not usually occur through normal valves. Halogenated anaesthetics should be avoided as they will increase the irritability of the heart, and predispose to arrhythmias. These are usually ventricular extrasystoles but as long as they are not persistent or do not affect the cardiac output the procedure can continue. If the catheter is removed they usually settle anyway. Arrhythmias continuing after withdrawal of the catheter can be treated in children by giving sodium bicarbonate 8.4 per cent intravenously on the basis that there is a -5 base deficit, half of this being given in the initial dose. Adults can be dealt with on a similar basis.

If this fails, then lignocaine 1 mg/kg injected intravenously as a bolus can be tried. Procaine amide can be used instead but it not advocated as it depresses conductivity. If lignocaine proves of no value then a beta blocker should be used, practolol being the drug of choice.

If a severe supraventricular tachycardia is present the patient can be fully digitalized. If still present, a direct current shock can be used, preferably synchronized. If this is not possible then a direct current shock can be used without synchronization. The power of shock to use initially is 40–60 joules in a child; 200 joules in an adult. If ventricular fibrillation occurs then continue with the direct current shock until it reverts [17].

Cardiac arrest. Any patient with arrhythmias which do not respond, may go on to cardiac failure and eventually cardiac arrest. This is not by any means the sole cause of cardiac arrest. It can, for instance, arise in respiratory failure, in blood loss, and for no apparent reason at all. It is shown by alteration in the appearance, such as pallor, cyanosis, unconsciousness or convulsions. There is an absent pulse and dilated pupils.

Treatment consists of ventilating the patient with 100 per cent oxygen. The legs are raised to increase the venous return and three severe thumps or bangs are made with the base of the hand over the praecordium. Some authorities advocate as many as fifteen thumps at one-second intervals. If there is no response within fifteen seconds of the onset of arrest then external cardiac massage is performed.

This is done by placing the patient supine on a hard surface, if he is not

[1] Practolol has now been withdrawn.

already on one; an X-ray table is, of course sufficient. If necessary the floor can be used. The heart is then rhythmically compressed between the sternum and the vertebral column. Initially a rate of 60 per minute can be used for a few minutes moving the sternum about 3–4 cm each time. If there is a succession of helpers this rate can be continued indefinitely but, if not, then a more realistic rate of external cardiac massage must be adopted (say 30/min) or the operator will become exhausted. Pulmonary ventilation is continued at the rate of about one chest inflation in six external compressions. These measures are carried out to the exclusion of all others. If the immediate number of personnel available is limited, then help cannot be sent for until these measures have been carried out. When help is forthcoming an ECG machine and defibrillator is obtained and connected. An intravenous drip of 8.4 per cent sodium bicarbonate is set up and 50–150 ml given. This is to combat the acidosis which develops when the circulation fails, due to metabolites still being formed in the tissues.

In older males with rigid costal joints, chronic bronchitis and perhaps weak ribs, it is not unusual for some (and on occasion all) ribs to be broken. Conversely in an infant with a very flexible chest wall it is more likely that the midsternal compression (done with the thumb) will damage the internal organs if not very carefully controlled.

If performed satisfactorily it should be possible to palpate the pressure wave produced by each cardiac compression. If this is absent and there are no other corroborating signs such as return of colour and contraction of pupils, then the technique should be quickly reviewed. If any alterations do not have a rapid effect (2 minutes at most) then direct cardiac massage should be done.

MISCELLANEOUS

Narcotic overdose. Overdose of drugs used in premedication or sedation has to be kept in mind. In general terms, depression of respiration is counteracted by oxygenation, controlled respiration, and support of the circulation. Often, children having encephalograms or other neuroradiological investigations are severe epileptics and are receiving all sorts of medication to control this. Therefore, in these patients it is reasonable to avoid heavy predmedication and halothane, and in this way problems due to combinations of drugs and summation of drug effects do not arise.

Respiratory depression caused by narcotic drugs can be counteracted by levalorphan tartrate. The dosage to reverse the respiratory depression caused by say, morphine, would be one-twentieth of the morphine dosage. That for pethidine is one-hundredth. This means that levalorphan would be given

0.5–1 mg intravenously and this would be repeated at three to five minute intervals. If pentazocine is used for sedation and produces respiratory depression the antidote is naloxone or a nonspecific stimulant such as nikethamide should be used. The usual narcotic antagonists such as levalorphan are not effective with pentazocine.

Hypercarbia and hypocarbia. A responsibility of the anaesthetist during general anaesthesia, which can scarcely be said to be a complication, is that of keeping the P_{CO_2} at normal levels (as far as possible) during angiography. CO_2 has a marked effect on cerebral blood vessels and hypercarbia produces a very different type of angiographic X-ray picture from hypocarbia. If the radiologist wishes to show particular filling of one or other artery or branch he may wish a higher concentration of carbon dioxide to be used. Conversely, on occasion, cerebral tumours can be more easily distinguished if a low carbon dioxide level is used. The anaesthetist under these special circumstances will be guided by the requirements of the radiologist.

Renal function. Finally in this section there must be mention of the fact that in renal disease there is the possibility that the use of methoxyflurane in the anaesthetic mixture may increase the renal dysfunction. In patients with no previous renal disability the use of methoxyflurane may produce a high output renal insufficiency [32]. In the twelve patients in Mazze's series who received methoxyflurane there was a significant decrease in uric acid clearance. These results were not confirmed by those of Robertson [33] who found that apart from the possibility of a rise in serum urea on the second day after surgery, there was no other change which might have clinical significance. However, Black [34] in paediatric practice found significant reduction in body weight and increases in plasma uric acid and urea concentrations in the postoperative period in normal children who had inhaled 0.5 per cent methoxyflurane for 60 minutes. Cousins [35] showed that when the concentration of methoxyflurane exceeds the level of 2.0 MAC hours, i.e. the minimum anaesthetic concentration for surgical anaesthesia multiplied by the duration, then renal damage occurs. Other studies have shown that the damage is additive in the presence of other nephrotoxic drugs [36] such as gentamycin. There is therefore a theoretical possibility that methoxyflurane anaesthesia might potentiate the nephrotoxicity of contrast media used in angiography (see Chapter 4). The position of methoxyflurane in clinical practice needs re-evaluating. Meanwhile, other anaesthetic agents can be used. Other causes of renal damage are hypovolaemia, hypotension, dehydration with oliguria which may even result from diuretic therapy, diabetes, etc. If contrast media are to be administered, these factors should be considered pre-operatively.

Oral and cholecystographic media may rarely cause renal failure, particularly in the presence of liver disease (Chapter 11). Since these media are excreted relatively slowly, operative procedures which may predispose to shock or oliguria are best deferred for several days after their use, where possible.

RADIOGRAPHIC PROCEDURE

All methods which involve the *use of catheters* have potential complications. These include perforation of a vessel wall and extravasation of blood or contrast medium outside the vessel. Perforation in this way is not necessarily painful but if the patient is under general anaesthesia there is no possibility of them showing that there is pain. Sometimes there is the problem of removing a catheter when spasm of the vessel makes this difficult and painful.

Associated with catheterization is the development of a thrombus either during catheterization with the thrombus 'wiped off' the catheter on withdrawal and producing an embolus, or to thrombus developing at a later postoperative stage due to trauma of the vessel wall. In this group will be included cases where catheterization has produced further damage to an already damaged intimal surface. It may happen that an atheromatous plaque breaks off during catheterization. This type of damage could produce, say, a hemiplegia after a carotid angiogram.

Perforation of the myocardium into the pericardial cavity is a rare complication [30, 37, 38]. There may be praecordial discomfort from stretching of the pericardium. Large effusions may exert pressure on surrounding structures such as the bronchi and lungs producing reflex cough and dyspnoea. Finally as the pressure in the pericardial sac rises, cardiac filling diminishes, venous pressure rises, the ventricular stroke output decreases and hypotension occurs. Tachycardia and reflex vasoconstriction maintain cardiac output and blood pressure until cardiac tamponade occurs. Coronary blood flow is reduced and heart failure results. Some cases fail to recover after decompression and tamponade should be regarded as a medical emergency.

Air embolus. Any system involving catheterization of a blood vessel with injection of contrast medium and flushing agent down the catheter can produce the very dangerous hazard of air embolus [8]. Air injected intra-arterially is of particular danger in the carotid circulation and small amounts in the cerebral circulation can produce widespread permanent effect. However arterial air embolism is rare because of the high pressure necessary to introduce the air into an artery. On the other hand, little if any pressure is necessary for air to enter the venous system. All that is required is that a negative pressure develops in the veins. In general this complication involves one of the large veins, often the patient is sitting or in a reverse Trendelenburg position, thus increasing the negative pressure in the raised portion of the body. The danger is increased during light anaesthesia when deep respirations occur or following coughing and straining, when again a large inspiration tends to increase the negative pressure. Controlled respiration increases the negative pressure and the possibility of air embolism particularly in the sitting position. This is emphasized by Hunter [29].

Small amounts of air are quite capable of being dealt with by the body but large amounts are immediately life threatening. Hunter considered this complication in any unexpected fall of blood pressure of 10 mmHg, or in any rise in pulse rate of 10 beats per minute, or in any alteration of spontaneous respiration.

Auscultation will show the presence of air in the heart. The use of a Sonicaid enables diagnosis of air embolism to be made long before auscultation is useful.

If air embolism is suspected the first measure is to place the patient in the head down position on the left side if possible (to displace air from the outflow tract of the right ventricle). Later measures will be artificial respiration with or without direct cardiac massage and aspiration of air as proves necessary [39]. Hypotension will be treated after these preliminary measures have been taken.

It should also be borne in mind that if a Spitz Holter valve is being reviewed, air must not be put into the ventricles as there is a direct pathway from the ventricle to the right atrium. Carbon dioxide should be used instead of air and the inhaled gases must contain 5–10 per cent CO_2. In this way the CO_2 in the ventricle stays there as a gas. When the procedure is completed the CO_2 in the inhaled gases is discontinued, and the patient is hyperventilated. In this way the CO_2 in the ventricles will be absorbed and rapidly disappears.

There is another hazard which tends to occur in *children and babies*, though it can occur in adults. They may develop airway obstruction after bilateral carotid angiograms. Swelling may develop markedly and occurs not only externally on the skin surface, but internally among the deeper structures. Intubation may be necessary, and rarely laryngostomy or tracheostomy.

There is always a *radiation hazard*, not only to the patient and X-ray personnel but also to the anaesthetist, anaesthetic nurse and technicians. It occurs in particular in those patients not intubated, where the airway is maintained by the anaesthetist's hand on the chin. This hand may receive a significant dose of radiation during angiocardiography. This also applies in children. Norris [40] estimated the total dose of radiation delivered to the patients chin as 10 15 mR/min. He mentions this as being a greater exposure than that noted by Norton [41].

From time to time the *lights in the X-ray room* may be dimmed or blacked out to provide extra definition during screening or in using an image intensifier. Extra care must be taken at this time by the anaesthetist. There are alarm systems both for monitoring the anaesthetic gases and the patient. They are helpful but reliance on them cannot be absolute: there is nothing that can replace direct observation of the patient (such as a finger on the pulse).

CONTRAST MEDIA

Reactions to contrast media and their treatment are considered in detail in Chapter 1. Acute reactions pose an immediate hazard of greater or lesser degree depending on their severity. Their clinical characteristics and incidence have been described by many authorities [42–48]. Patients who react to contrast media do so in one of five basic ways, though there is a certain amount of overlap. These are: *angioneurotic oedema*; *a rash of allergic type*; *an asthmatic response*; *cardiovascular collapse* and *cerebral effects*. These are meant to be broad categories, for instance the asthmatic response would include bronchospasm, pulmonary oedema and might progress to respiratory failure and unconsciousness. The cardiovascular group includes the hypertensive crises of phaeochromocytoma and cardiac arrest. The cerebral group includes cerebral oedema and convulsions. Some of these reactions have already been mentioned briefly in the earlier sections of this chapter as have the problems of drug interactions and renal damage. Some other aspects which may influence the care of the patient by the anaesthetist also require consideration.

Hypertonicity

All the water-soluble contrast media at present in use are hypertonic. Thus there are changes due to the osmolality of these substances. Hypervolaemia develops, serum electrolytes are altered, there is an acidosis. A rise in left ventricular end diastolic and left atrial pressures can lead on to pulmonary oedema. In infants this hypertonicity produces dehydration. Neonates have relative impairment of glomerular filtration and large volumes of contrast medium injected rapidly will cause a relative increase of toxicity. This is shown by haemodynamic changes. There is a depression of cardiac contractility, there are alterations in blood pressure, cardiac output and pulse rate, and changes in the ECG. The electrocardiographic changes tend to be transient. However, ventricular fibrillation can occur but is more likely in right coronary injections due to the distribution of the contrast to the conducting system.

Cerebral reactions

These are mainly concerned with convulsions and cerebral oedema. Convulsions are controlled by intravenous thiopentone, hydrocortisone, oxygen and controlled respiration if it is necessary. Cerebral oedema is treated by immediate intravenous steroids and if there is a raised intracranial pressure mannitol 20 per cent is given intravenously. The dose is 1.5–2 g/kg given over a period of 30 to 60 minutes and it is necessary to catheterize the patient immediately as mannitol produces a vigorous diuresis. There will be a hypervolaemia for a time with a raised blood pressure, and this may temporarily exacerbate the symptoms. Lasix 20 mg intravenously can also be used instead in conditions where there is an element of congestive cardiac failure.

Spinal cord damage

Paraplegia may rarely occur following aortography and is due to an undiluted bolus of contrast medium being diverted to the spinal cord circulation [50]. This occurs when there is an organic blockage preventing contrast flowing into the peripheral circulation. It also occurs when using vasopressor drugs. The mechanism is the same, in both cases it is easier for the contrast to flow into a low-pressure area than into a high one. The vessels of the spinal cord circulation do not take part in the general vasoconstrictor response, and the contrast passes into the low-pressure spinal circulation. Even physiological stimuli which raise blood pressure may have this effect, such as breath holding.

Accidental injection of water-soluble contrast media into the subarachnoid space can produce severe central nervous system irritation with extensor spasms. The patient is postured to gravitate the contrast to the lumbar region and spinal needles are inserted and the subarachnoid space is irrigated with normal saline. Narcotics are given to control the pain and diazepam is given to control the muscle spasms. If this is ineffective, muscle relaxants and controlled respiration will be required.

REFERENCES

1 ANSELL G. (1973) *Notes on Radiological Emergencies*, 2nd edn. Blackwell Scientific Publications, Oxford.
2 MENDELSON C.L. (1946) The aspiration of stomach contents into the lungs during paediatric anaesthesia. *Am. J. Obstet. Gynae.* 52, 191.
3 BRAUNWALD E. & SWAN H.J.C. (1968) Co-operative study on cardiac catheterisation. *Circulation* 37, Suppl. 3.
4 HOHN A.R., GILETTE P. & WEBB H.M. (1970) Heart catheterisation in infants and children. *J.S.C. Med. Ass.* 66, 6.
5 FIELDMAN E.J., LUNDY J.S., DUSHANE J.W. & WOOD E.H. (1955) Anaesthesia for children undergoing diagnostic cardiac catheterisation. *Anaesthesiology* 16, 868.
6 MUNROE J.P., DODDS W.A. & GRAVES H.B. (1965) Anaesthesia for cardiac catheterisation and angiocardiography in children. *Canad. Anaesth. Soc. J.* 12, 67.
7 LAMBERT E.C., CANENT R.V. & HOHN A.R. (1966) Congenital cardiac anomalies in the newborn. *Paediatrics* 37, 343.
8 MANNERS J.M. (1971) Anaesthesia for diagnostic procedures in cardiac disease. *Br. J. Anaes.* 43, 276.
9 MACMAHON B., MCKEOWN T. & RECORD R.G. (1953) The incidence and life expectations of children with congenital heart disease. *Br. Heart J.* 15, 121.
10 RICHARDS M.R., MERRIT K.K., SAMUELS M.H. & LANGMANN A.G. (1955) Congenital malformations of the cardiovascular system in a series of 6,053 infants. *Paediatrics* 15, 12.
11 FRAIMOW W., WALLACE S., LEWIS P., GREENING R.R. & CATHCART R.T. (1965) Changes in pulmonary function due to lymphangiography. *Radiology* 57, 427.
12 ROBINSON A.E., HALL K.D., YOKOYAMA K.N. & CAPP M.P. (1971) Pediatric bronchography: the problem of segmental pulmonary loss of volume. I. A retrospective survey of 165 pediatric bronchograms. *Invest. Radiol.* 6, 89.
13 Editorial (1968) Hazards of cardiac catheterisation. *Lancet* 2, 547.
14 HARRISON G.G. (1968) Anaesthetic contributory death—its incidence and causes. Part 1, incidence. Part 2, causes. *S. Afr. Med. J.* 42, 514, 544.
15 MCAFEE J.G. (1957) A survey of complications of abdominal aortography. *Radiology* 68, 825.
16 MUSTARD W.T., BEDARD P. & TRUSLER G.A. (1970) Cardiovascular surgery in the first year of life. *J. Thor. Cardiov. Surg.* 59, 6.
17 ABBOTT T.R. (1974) Personal communication.
18 ISAACS H. & BARLOW M.B. (1973) Malignant hyperpyrexia. *J. Neurol., Neuros. Psychiat.* 36, 228–243.
19 WHITE Y.S., BELL D.S. & MELLICK R. (1973) Sequelae to pneumoencephalography. *J. Neurol., Neuros. Psychiat.* 36, 146.
20 NICHOLSON J.R. & GRAHAM G.R. (1969) Management of infants under six months of age undergoing cardiac investigations. *Br. J. Anaesth.* 41, 417.
21 JENNETT W.B., BARKER J. & MCDOWELL D.G. (1969) Effect of anaesthesia on intracranial pressure in patients with space occupying lesions. *Lancet* 1, 62.
22 FITCH W. & MCDOWELL D.G. (1971) Effect of halothane on intracranial pressure gradients in the presence of intracranial space-occupying lesions. *Br. J. Anaesth.* 43, 904.

23 HENRIKSEN H.T. & HALSLEV JORGENSEN P. (1973) The effect of nitrous oxide on intracranial pressure in patients with intracranial disorders. *Br. J. Anaesth.* **45**, 486.
24 TURNER J.M., CORONEOS N.J., GIBSON R.M., POWELL D., NESS M.A. & MCDOWELL D.G. (1973) The effect of althesin on intracranial pressure in man. *Br. J. Anaesth.* **45**, 168.
25 WOLLMAN H., ALEXANDER S.C. & COHEN P.M. (1967) Cerebral circulation and metabolism in anaesthetised man. *Clin. Anaes.* **3**, 1.
26 GARDNER A.E., OLSEN B.E. & LICHTIGER M. (1971) Cerebrospinal fluid pressure during dissociative anaesthesia with ketamine. *Anaesthesiology* **35**, 226.
27 FITCH W., BARKER J., JENNETT W.B. & MCDOWELL D.G. (1969) The influence of neuroleptanalgesic drugs on cerebrospinal fluid pressure. *Br. J. Anaesth.* **41**, 800.
28 LEE J.A. & ATKINSON R.S. (1965) In: *Synopsis of Anaesthesia*, p. 563. John Wright & Son, Bristol.
29 HUNTER A.R. (1964) *Neurosurgical Anaesthesia*, pp. 79, 97. Blackwell Scientific Publications, Oxford.
30 WENNEVOLD A., CHRISTIANSON I. & LINDENEG O. (1965) Complications in 4,413 catheterizations of the right side of the heart. *Am. Heart J.* **69**, 2.
31 Editorial. (1971) *Br. J. Anaesth.* **43**, 1.
32 MAZZE R.I., SHUE G.L. & JACKSON S.H. (1971) Renal dysfunction associated with methoxyflurane anaesthesia. *J. Am. med. Ass.* **216**, 2, 278.
33 ROBERTSON G.S. & HAMILTON W.F.D. (1973) Methoxyflurane and renal function. *Br. J. Anaesth.* **45**, 55.
34 BLACK G.W., COPPEL D.L., HUGHES N.C. & LOVE S.H.S. (1973) Anaesthesia for cleft palate surgery. *Br. J. Plast. Susg.* **22**, 343.
35 COUSINS M.J. & MAZZE R.I. (1973) Methoxyflurane nephrotoxicity, a study of dose response in man. *J. Am. med. Ass.* **225**, 1611.
36 MAZZE K.I. & COUSINS M.J. (1973) Combined nephrotoxicity of gentamicin and methoxyflurane anaesthesia in man. *Br. J. Anaesth.* **45**, 394.
37 ESCHER D.J.W., SHAPIOR J.H., RUBINSTEIN B.M., HORWITH E.S. & SCHWARTZ S.P. (1958) Perforation of the heart during cardiac catheterisation and selective angiocardiography. *Circulation* **18**, 418.
38 LURIE P.R. & GRAJO M.Z. (1962) Accidental cardiac puncture during right heart catheterisation. *Paediatrics* **29**, 283.
39 HUNTER A.R. (1964) *Neurosurgical Anaesthesia*, p. 80. Blackwell Scientific Publications, Oxford.
40 NORRIS W. (1963) Basal narcosis or sedation for cardiac catheterisation. *Br. J. Anaesth.* **35**, 358.
41 NORTON K.L. & KUBOTA T. (1960) Experiences with cardiac catheterisation using halothane compressed air anaesthesia. *Anaesthesiology* **21**, 374.
42 ANSELL G. (1970) Adverse reactions to contrast agents. *Invest. Radiol.* **5**, 374–384.
43 ANSELL G. (1968) A national survey of radiological complications: interim report. *Clin. Radiol.* **19**, 175–191.
44 COLEMAN W.P., OCHSNER S.F. & WATSON B.E. (1964) Allergic reactions in 10,000 consecutive intravenous urographies. *South M. J.* **57**, 1401–1404.
45 PENDERGRASS H.P., HILDRETH E.H., TONDREAU R.L. & RITCHIE D.J. (1960) Reactions associated with intravenous urography: discussion of mechanisms and therapy. *Radiology* **74**, 246–254.

46 WITTEN D.M., HIRSCH F.D. & HARTMAN G.W. (1973) Acute reactions to urographic contrast media. *Am. J. Roentg.* **119**, No. 4, 830–840.
47 WEIGEN J.F. & THOMAS S.F. (1958) Reactions to intravenous organic iodide compounds and their immediate treatment. *Radiology* **71**, 21–27.
48 PENDERGRASS H.P., TONDREAU R.L., PENDERGRASS E.P., RITCHIE D.J., HILDRETH E.A. & ASKOWITZ S.I. (1958) Reactions associated with intravenous urography—historical and statistical review. *Radiology* **71**, 1.
49 LASSER E.C., ELIZONDO-MARTEL G. & GRANKE R.C. (1964) The roentgen contrast media potentiation of nembutal anaesthesia in rats. *Am. J. Roentg.* **91**, 453.
50 Editorial (1973) Spinal cord damage after angiography. *Lancet* **2**, 1067–1068.

CHAPTER 19: RADIATION PROBLEMS

G. M. ARDRAN AND H. E. CROOKS

Injuries due to X-rays were reported the year following their discovery. In 1903 Dr Albers-Schonberg [1] discovered that X-rays could sterilize animals, and he appreciated the fact that if sterilization could take place it was important to know what changes would occur in germ cells not actually killed but affected by the radiation. Landmarks in diagnostic X-ray protection have recently been reviewed [2]. It must be stressed that without proper precautions X-rays are today just as dangerous as they ever were, indeed, there is a greater potential danger because of their vast increase in use, the greater intensity of radiation possible with modern equipment and the complicated and lengthy examinations which are now possible and valuable.

In diagnostic X-ray departments radiation problems may arise for personnel, patients and others. These will be discussed briefly with respect to maximum permissible dose limits, potential hazards, the radiation doses which may be received and the principal methods of dose reduction. Special cases and the more potentially hazardous procedures will be considered in greater detail. The problems of radioisotopes and the nonionizing radiations used will not be discussed but some of the problems relating to the radioactive contrast medium colloidal thorium dioxide (Thorotrast) will be briefly mentioned.

RADIATION UNITS

The biological effectiveness of a given absorbed dose of one type of ionizing radiation is not necessarily the same as that of an equally absorbed dose of another type of radiation (e.g. neutrons). For protection purposes, *maximum permissible doses* refer to all types of ionizing radiation and the units are given in terms of *dose equivalent*, taking into account the differences in the relative biological effectiveness of the radiation. The unit of dose equivalent is the *rem*. In the diagnostic X-ray range, the quality factor is unity and, therefore, the value in rems is the same as the value in rads. The *rad (r)* is the unit of absorbed dose and 1 rad is equivalent to 100 ergs/gm or 10^{-2} J kg^{-3}. The *Röntgen (R)* is the unit for measuring exposure; 1 R is approximately equal to 1 r in air which for practical purposes is equivalent in this energy range to soft tissue; thus in the diagnostic X-ray energy range the terms rem, Röntgen and rad are virtually interchangeable.

It must be remembered that an exposure in R takes no account of the area or volume of tissue irradiated. For this reason, when comparing radiation exposure during examinations which involve variable areas or sites of irradiation, the unit $R \times cm^2$ is used to give an indication of the total quantity of radiation incident upon the patient; this can be fairly readily measured in practice [3]. The rad again takes no account of the volume of tissue irradiated. It must either be assumed that the whole body is uniformly irradiated or that one is referring to a specified organ or part; thus, since the gonads are relatively small and, when irradiated, may be assumed to be uniformly irradiated, it is correct to refer to a gonad dose in rads. Doses in rads can be calculated for other specific organs but one needs to know that they were completely or uniformly irradiated; alternatively allowances should be made for non-uniformity. The dose to the organ can then be given as an average dose in rads or the dose to a part of the body in *g/rads*, this being the product of the average dose and the mass of tissue irradiated.

When fluoroscopic procedures are carried out, e.g. gastrointestinal examinations, the incident radiation is falling on different areas of the body and the beam size may be continuously variable and unknown. The output from the tube incident upon the patient in the centre of the beam may be measured in R for the duration of the examination but, since that exposure does not always fall on the same area of the patient, one cannot take the total exposure in R and then say that the patient has received that dose in rads. This simple fact is often forgotten and it must be pointed out that if one irradiates a number of separate small areas, say 10, each with 1 rad, one cannot add these together and say that the patient has received 10 rads. The dose absorbed by the patient will in fact be infinitely less than if the whole body had received 1 rad.

Again, it must be pointed out that diagnostic X-radiation quality is such that it is rare for more than 10 per cent of the incident beam to pass through the patient. It is not uncommon for only 1 per cent to reach the recording medium and with low energy radiation as used for mammography only 0.1 per cent may reach the film. Under these circumstances the incident dose in rads does not give the radiation absorbed by the part in rads; the average absorbed dose must be calculated by taking into account the rapid absorption with depth.

MAXIMUM PERMISSIBLE DOSE

1 Radiation workers

The maximum permissible radiation exposures to personnel are laid down by I.C.R.P. [4, 5] and by local regulations, e.g. the Code of Practice in the

TABLE 19.1. Maximum permissible doses for occupationally exposed persons.

Part of body	Annual dose	Quarterly dose
Gonads, red bone marrow and whole body	5 rems[1]	3 rems (1.3 rems to the abdomens of women of reproductive capacity)
Bone, thyroid and skin of whole body	30 rems	15 rems
Hands, forearms, feet and ankles	75 rems	40 rems
Any single organ (excluding gonads, red bone marrow, bone, thyroid and skin of whole body)	15 rems	8 rems

[1] If the person was occupationally exposed before 18 years of age, subsequent exposure of the gonads, red bone marrow and whole body should be controlled so that the dose accumulated up to 30 years of age does not exceed 60 rems.

U.K. [6]. These are given in Table 19.1. In practice they are best considered as 100 mrem per week for all organs other than the hands and feet. The I.C.R.P. limit of 75 rem per year for the hands should not be regarded as a dose which can be received every year during the working life of the individual since this dose could eventually cause dermatitis. It is therefore only applicable for limited periods.

The vast majority of diagnostic radiation workers in the U.K. receive far less than the maximum permissible dose, in fact less than 1.5 rem/yr. Any individuals who are likely to receive more than 1.5 rem/yr must be designated as 'radiation workers'; they should be monitored and have annual medical examinations and blood counts. If the workers receive less than 1.5 rem/yr, it is only necessary for an initial medical examination and blood count but they should be subjected to surveillance in the form of radiation monitoring of the area in which they work; in practice this is best done by personal monitoring, e.g. film badge, or thermoluminescent or other dosemeter, to ensure that the limits are not exceeded.

Pregnancy

Problems may arise with female radiation workers should they become pregnant. I.C.R.P. states that female radiation workers of reproductive capacity should *not* be allowed to receive the limit of 3 rem in a period of 13 weeks but should be restricted to 1.3 rem in 13 weeks so that in the event of their becoming pregnant, the fetus will not have received more than 1 rem before diagnosis. Thereafter, their work should be regulated so that the fetus does not receive more than 1 further rem during the remainder of the pregnancy. This should be achieved if the worker (in the diagnostic energy range) does not receive more than the maximum permissible dose (100 mR per week on the skin of the abdomen); because of the rapid absorption the fetus will receive much less. However, whenever possible, pregnant workers should not engage in those procedures which are likely to give the maximum permissible exposure, on moral grounds rather than on strictly scientific ones. In diagnostic radiology this should not be difficult to achieve. It is the workers' responsibility to inform the radiation safety officer or head of the department if they suspect that they are pregnant, and female staff should be instructed in this responsibility.

2 *Ancillary workers*

This class of individual is one who is not directly working with ionizing radiation but who may be in the vicinity when irradiation is carried out. These individuals should not receive more than 1.5 rem/yr and the area in which they work should be under strict surveillance. Often the best way of controlling their dose levels is to have them carry personal monitoring devices: they should be instructed how to follow the correct protection procedures.

3 The public

Members of the general public should not be allowed to receive more than 0.5 rem/yr.

4 Patients

No levels are laid down for maximum permissible exposure to patients. All radiation exposures should be kept as low as practicable, consistent with an adequate diagnostic result. X-rays should not be used when there is an adequate less-hazardous alternative; for example diagnostic ultrasonics for the diagnosis of problems relating to pregnancy or the use of isotopes such as Indium for the diagnosis of placental site which may give a lower radiation dose than radiography. About 85 per cent of the population dose from man-made radiation results from diagnostic X-ray procedures. If the population dose is not to rise with the increase in other radiation activities such as the use of atomic energy, etc. the dose to the population from X-ray diagnosis should be progressively reduced if at all possible. With care, we know that it can be.

When patients are not being irradiated deliberately in the course of their diagnosis or treatment, they are regarded as members of the public and as such should not receive more than 0.5 rem/yr.

POTENTIAL HAZARDS

PERSONNEL

Local effects

Skin injuries to the fingers may arise from the practice of holding dental films in the patient's mouth during exposure. This practice is forbidden to operators, but it still occurs. Likewise, fingers should be kept out of the useful beam during injection procedures. If there is any doubt about the doses received in any particular examination, the individuals should be monitored with thermo-luminescent dose meters of the lithium fluoride or borate type.

Some procedures such as fluoroscopy with over-table X-ray tubes may cause significant doses to the eye, head, neck and thyroid. It may also be necessary, from time to time, to monitor the dose to the feet and lower legs of operators carrying out fluoroscopy with under-table X-ray tubes; the results will indicate whether any additional precautions are necessary.

Genetic effects

Provided that the worker does not receive more than 5 rem/yr to the body in the region of the gonads, this is not considered to be a significant hazard to the individual but will only be an addition to the total population genetic dose. The use of the protection precautions laid down by I.C.R.P. [7–9] should be adequate in this respect.

General effects

These are the production of malignancies in any organ, the production of leukaemia and possibly premature ageing. Provided that the individual does not exceed I.C.R.P. limits of exposure, these risks may be regarded as negligible.

PATIENTS

Local effects

Injury to the skin, e.g. erythema or desquamation, should not occur in any examination other than as the result of an accident. Erythema has been reported from mammographic examinations when a tungsten target beryllium window tube was used without additional filtration [10]. The commonest accident which still occurs from time to time, arises from the fluoroscopy switch being left in the 'on' position when it was thought that it was switched off. Fluoroscopy switches should be of the 'dead man' type which cannot be operated without being pressed by the hand or foot and so arranged that this cannot accidentally occur. Equipment which is not of this type should be modified. Examinations such as extensive fluoroscopy and radiography for coronary artery disease and cardiac pacemaker insertions may approach the levels of erythema dose but adequate monitoring of procedures should ensure that this does not occur. Fluoroscopy for the diagnosis and setting of fractures is deprecated and, if carried out, requires competent monitoring. High radiation exposures, e.g. 10–15 R may be given to the cornea of the eye during tomography of the middle ear. The doses may be reduced to about one-tenth of these figures if lead eye-shields are used (*vide infra*). Exposures to the thyroid in extensive examinations, e.g. cineradiography of the neck, may also require consideration.

Pregnancy

The irradiation of an early pregnancy during the last ten days of the menstrual cycle and for the next few weeks, that is during the period of major organogenesis, may result in fetal abnormalities. Following the work of Dr Alice Stewart and others, it is now almost universally accepted that radiation to a fetus at any time from conception to term can cause leukaemia or other neoplastic disease during the lifetime of the child. For this reason the irradiation of a known pregnancy should only be carried out if there are good medical reasons for doing this and it should be carried out with the minimum radiation exposure consistent with an adequate diagnostic result. Problems arise when the abdomen or pelvic region of a female of reproductive capacity are irradiated for other purposes when it is not known that the patient is pregnant. Every precaution must be taken to see that the patient is not pregnant. The natural incidence of childhood neoplasia and leukaemia is about 1 in every 1200 live births. The best estimates indicate that a dose of 1–4 rem will double the natural incidence. Diagnostic examinations of the pelvic regions can result in a dose to the fetus ranging from about 0.25 rem up to perhaps 5 rem or more in an extensive examination. This means that an average examination might double the natural incidence of leukaemia or other malignancies in childhood. This could also be expressed by saying that should leukaemia or neoplasia arise in a child who was radiographed as a fetus there is a fifty–fifty chance that the radiation caused it. All the evidence indicates that an early pregnancy is just as susceptible as a later one.

In order to explain the problem and to suggest a means of tackling it we quote a letter which was sent to all doctors in the Oxford area together with a copy of the form which we suggested might be used, to deal with these difficulties [11, 12]. (See Appendix 1.) It is the duty of the X-ray department and whoever carries out the examination, e.g. the radiographer or radiologist, to act as the final check and be satisfied that every necessary precaution has been taken to ascertain that the patient is not pregnant.

Despite every reasonable precaution, from time to time it will be found that a patient's lower abdomen has been irradiated and that subsequently she is found to have been a few days' or weeks' pregnant. A decision will then have to be taken as to what, if anything, should be done. Estimates of the dose received by the fetus may be made from whatever data may be available; often this can only be made approximately, with errors of 100 per cent or more. Usually, however, the dose is found to be of the order of 1 to 3 rem. If for other reasons the woman does not desire the pregnancy to continue, the irradiation may provide additional grounds for an abortion. It has been suggested that a known dose of 10 rem or above should be a very strong indication for an abortion. However, if the dose has only been 1 or 2 rads and the woman particularly wishes the pregnancy to continue, the risk of abnormality due to the radiation is small. Unfortunately, it is likely that any abnormality in the child will subsequently be attributed rightly or wrongly to the irradiation, and it must be remembered that 1 per cent or more of all fetuses do have some abnormality. It is therefore advisable that every step should be taken to see that a pregnancy is not unwittingly irradiated, so that this difficult problem does not arise. We have known patients who have suffered severe emotional upset for years, because they have consented to have their child aborted; others will have years of worry wondering if anything is going to happen.

General effects

Irradiation to the body can cause leukaemia or other malignancy. Long-term follow-up of patients given therapeutic radiation for ankylosing spondylitis and of the Japanese survivors of the atomic bombs has shown that more cases of neoplasia may be eventually produced than there were cases of leukaemia in the earlier years. Goss [13] has reviewed the latest published data on the Japanese atomic bomb survivors and suggests that the risk of leukaemia is 30 per million persons per rad whereas the risk is 100 per million persons per rad for all types of cancer. There is now little doubt that neoplasia can be produced in almost any organ, usually after 20 years or more, in a small number of people following irradiation. For this reason it is important to restrict any unnecessary radiation as far as practical. However, if the criteria for carrying out the examination have been carefully considered and there is no reasonable alternative available, the hazard may be regarded as small in relation to the immediate benefits to the patient from a diagnostic examination. It is considered likely that more deaths may occur from neoplasia in the individuals irradiated than might occur from the same exposure to the gonads resulting in deaths in future generations.

Genetic effects

Irradiation of the patient's gonads may result in mutations or chromosome aberrations resulting in dominant or recessive abnormalities in future generations or in fetal death. It is considered desirable that the increase in mutation rate due to man-made radiations should not exceed that occurring naturally and, for this reason, the patient's gonads should not be irradiated unnecessarily. It is suggested [14] that if it is not practical to protect the male gonads and they have been in the direct beam of radiation, the subject should be recommended to defer the deliberate conception of a child for at least 6 months. Usually in diagnostic procedures the irradiation of the gonads is not of primary concern to the individual patient but merely adds to the total population genetic burden. It has been suggested [15, 16] that irradiation of the ovaries of a female prior to conception may increase the incidence of Down's Syndrome. To date [17] this has not been confirmed.

HAZARDS TO OTHERS

Potential hazards may arise to persons other than radiographic staff or patients, e.g. waiting patients, ancillary staff in the X-ray department, nurses and parents or others who may on occasion have to hold or restrain seriously ill patients or children. Precautions should be taken to see that they are not in the useful beam and that they are protected against scatter. Surgeons, anaesthetists and cardiologists or others likely to receive significant doses should be monitored and, if they could receive more than 1.5 r per year they should be designated as radiation workers. Designation, however, should not be used as a means of being lax about taking precautions. Potential hazards are primarily genetic, which need not significantly affect the individual but only add to the population genetic burden; such personnel are unlikely to receive high local doses or receive sufficient radiation for this to be significant from the point of view of the production of malignancy. Precautions should also be taken to see that potentially pregnant females are not irradiated, particularly when holding children or other patients, and to see that the same individuals do not repeatedly hold children. Various devices have been designed to hold infants and babies in position during radiography so that no other individual need be irradiated. These should be used whenever possible.

DOSES RECEIVED FROM DIAGNOSTIC PROCEDURES

Personnel

All personnel in a diagnostic X-ray department who are likely to receive radiation in the course of their work must be monitored. Most diagnostic workers receive less than 1.5 r per year and it is possible for many to receive less than 0.25 r per year. In those situations where they receive more than 1.5 r per year, particularly if the volume or nature of the work would have been expected to result in only a small dose, investigations should be carried out to determine how and when they receive the bulk of this exposure. If possible, steps should be taken to reduce their exposure, especially in the case of female workers. Film badges should be worn underneath the protective apron at waist level to cover most statutory requirements. However, when the work involves potentially hazardous procedures such as the use of over-table X-ray tubes for cardiac catheterization and angiography, it may also be desirable to wear badges at a head, neck or upper arm level outside any protective garment in order to check on the doses which may be received by the upper part of the individual's body. Likewise when the fingers for any reason are close to the field of operation, T.L.D. dosemeters should be worn on the fingers from time to time, to check on finger dose. For those carrying out extensive fluoroscopy or unusual examinations, it may also be desirable to monitor the feet or lower legs with film badges or T.L.D. dosemeters. The findings will indicate whether further action is required.

The doses reported from cardiographic procedures vary, but in many instances they are relatively high and potentially likely to exceed maximum permissible doses. Under these circumstances, surveys should be carried out to ascertain what the doses are and where the radiation comes from, so that appropriate action can be taken.

Patients

Skin exposures (R), gonad doses (r) and absorbed doses in kg rads or in Rcm^2 in various investigations have been published. Selected figures are given in Appendix 19.2. Fluoroscopic times should be recorded though it is better to record the exposure in terms of $R \times cm^2$ using an integrated Röntgen area product meter (Diamentor). This should be done for those examinations which are likely to give high doses and for those examinations carried out by nonradiological personnel in order that safety officers may have some control over what is being done. Cinefluorography frequently gives relatively high exposure to patients and every effort should be made to reduce the duration of filming, the area irradiated and the frame rate, consistent with adequate examination. Measurements should be made of the exposures so that appropriate action may be taken if necessary. Where videotape recordings are adequate for the purpose, these should be used in preference to cinefluorography since the doses can be significantly less.

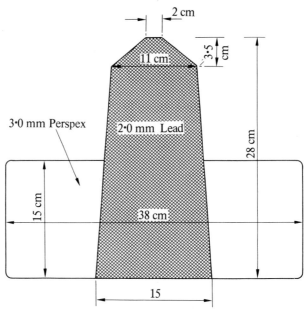

FIGURE 19.1A. Diagram of the Oxford Fig Leaf to show dimensions of the Perspex and lead covering.

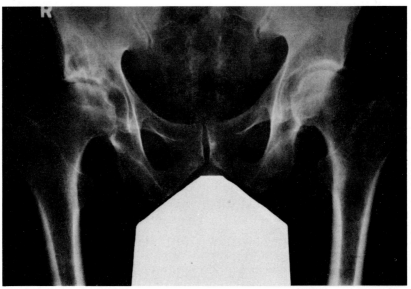

FIGURE 19.1B. The Fig Leaf in use when the bony parts must be visualized. For pyelography or other circumstances when the pubic bones need not be visualized, the Fig Leaf is turned round so that the lower straight edge covers the symphysis; the useful beam should also be restricted to this level.

At all times the patient's gonads should be protected, either by seeing that they are kept out of the useful beam or as far from the useful beam as possible, or shielded by lead. A simple male gonad protector for use with patients lying horizontally on the table has been described [18]. This device, known as 'the Oxford Fig Leaf' (Fig. 19.1A and 19.1B) is simple, inexpensive and about 90 per cent effective in 90 per cent of individuals: it, or a lead sheet, should be placed on the table during horizontal fluoroscopic procedures, such as barium enemas, to prevent accidental movement of the beam over the male gonads.

Others

Apart from the workers and the patients, every effort should be made to see that others, and this includes patients when they are not being deliberately irradiated in the course of their investigation, are protected as far as practicable. Measurements should be made to see that patients in waiting rooms, cubicles, etc., do not receive any significant radiation exposure, likewise, if there is more than one examination site in an X-ray room, steps should be taken to ensure that radiation from the examination of one patient does not fall on adjacent patients.

METHODS OF DOSE REDUCTION

Since many diagnostic examinations have to be carried out with radiation workers or their assistants in the X-ray room and not in a separately protected room, one can consider the methods of dose reduction both for staff and patients simultaneously. Radiographic quality and radiation exposure are determined by several interdependent parameters and these are summarized below:

1 Education and discipline

It cannot be too strongly stressed that one of the principal methods of ensuring a low dose in diagnostic examinations is to ensure that those who request the examinations make certain that what they request is necessary and that the area to be examined is as accurately localized as possible. It is the duty of the X-ray department to see that the examinations requested are carried out with the minimum exposure practicable. Radiologists have the responsibility of maintaining a high standard of work and of consulting with their clinical colleagues to see that unnecessary examinations are not requested. They are frequently responsible for deciding upon the type of equipment to be used, the techniques employed, the type and varieties of films and intensifying screens and the nature of the processing to be used. These factors can very significantly affect the radiation doses which will result. It cannot be too strongly stressed that the radiographer will be responsible for ensuring the use of small radiation beams and of accurate work necessitating as few repeat examinations as possible. A low standard of radiography inevitably means unnecessary radiation to the patients.

2 *The reduction of radiation leaks from X-ray tubes and beam-limiting devices*

This is especially important whenever operators have to work close to the X-ray tube, particularly with fluoroscopic or injection procedures with over-table tubes. Leaks may often be reduced to lower limits than laid down by I.C.R.P. with advantage.

3 *The reduction of off-focus radiation*

This is particularly important in improving the quality of the recorded image but also in reducing leaks from beam limiting devices [19, 20].

4 *'Coning' or collimation of the beam*

The X-ray beam should be restricted to the *minimum area* necessary for adequate diagnosis. This is one of the most important measures and it is desirable that the edge of the beam should be seen on all films as an indication of the total area irradiated. The edge of the beam should always be kept as far from the gonads as possible. Light beam diaphragms used with care can be a very great help in achieving these objectives. They must be checked from time to time to see that the beam of radiation coincides with the light beam. We prefer gonad shields to be applied to the patient and shielding devices are best not incorporated in the light beam diaphragm. In order to reduce extra-focal radiation and to reduce radiation leaks from the diaphragm, the tube port should be reduced so that it only covers the 'largest field at the shortest distance', using the device of Kemp & Nicholls [19, 21]. However, rectangular beam light beam diaphragms are not always the most desirable means of beam limitation. Sometimes those with circular beams may be better and on other occasions fixed rectangular or circular cones may be used independently or fitted to light beam diaphragms. For example, circular tubular metallic cones are best for conventional intra-oral dental radiography. Equipment solely for chest radiography may have the beam permanently limited to the largest sized film and simple diaphragms inserted for reduction of the sides and the bottom of the beam to fit smaller cassettes for smaller patients.

5 Residual useful beam and secondary radiation

These should be absorbed as close to the source as possible.

6 The use of added filtration

I.C.R.P. [7] states that the total tube filtration, inherent plus added, for normal diagnostic work including dental radiography must be equivalent to not less than the following:
 (i) 1.5 mm of aluminium at voltages up to and including 70 kV.
 (ii) 2.0 mm of aluminium at voltages above 70 kV and up to and including 100 kV.
 (iii) 2.5 mm of aluminium at voltages above 100 kV.

For examinations such as mammography, the total filtration must not be less than 0.5 mm aluminium equivalent. Where X-ray generators can be used between 50 and 150 kV, problems may arise as to which filter should be employed. It is very inconvenient to have to change the filter as one goes above 70 kV or above 100 kV; this could lead to the incorrect filter being used and hence to under- or over-exposed films which might have to be repeated. It might likewise be inconvenient to permanently use a filter designed for the highest kV of the generator if in fact much of the work required a lower kV. In our opinion the maximum filter for the maximum kV should be permanently *in situ* if the quality of films is considered adequate. However, it must not be forgotten that provided a 1.5 mm filter is used, increasing the kV will inevitably reduce the dose for the same recording system and object. Although additional filtration will reduce the dose further, primarily the skin exposure, we personally would not regard it as serious if this additional reduction in skin exposure was not obtained. In the past it has been difficult to readily check the total filtration of the beam or to measure inherent filtration of the intact tube: this is essential to obtain good radiographic quality and optimum reduction in dose. These measurements are now fairly simply carried out [22]. The discrepancies and difficulties which previous investigators found have been largely due to neglect of the target angle. At 80 kV, the target filtration varies from about 1 mm Al equivalent with a 6° target to 0.1 mm Al with a 35° target. Neglect to state the target angle renders many of the published output curves for different X-ray qualities inaccurate.

7 Speed of recording medium

The fastest 'film screen combination' suitable for a particular examination is an important method of dose reduction: whether this is measured as skin dose, absorbed dose, gonad dose or dose to the operators. In fluoroscopy, efficient fluorescent screens with good dark adaptation or preferably the use of efficient image intensification are essential. It must not be forgotten that however efficient a system, satisfactory dose reduction will not be obtained unless the operator uses the smallest beam area practical and fluoroscopes for the minimum time consistent with a good standard of work. The fluoroscopy switch should not be in the 'on' position continuously but should be released intermittently, the observations should be made and thought processes continued with the beam switched off.

8 Anti-scatter devices

The use of a stationary or moving grid necessitates an increase in X-ray exposure of two to four times, depending on the efficiency of the grid. It must be pointed out that when using a focused grid it is essential that the central ray be accurately centred perpendicular to the grid, since 3° off centre may require an increase of 25 per cent in exposure to maintain the same film density. Large-area high-ratio parallel grids will inevitably cut off primary radiation at the sides unless the focus grid distance is very large. With moving grids, it is essential that they shall be so adjusted that the exposure takes place symmetrically about the centre point. An air gap can be used as an alternative to a grid for such examinations as the cervical spine, shoulder, knee, chest and sinuses. Though this requires a longer focus film distance to reduce unnecessary enlargement it will reduce the patient exposure to the level obtainable without a grid, and in some cases will produce a sharper image due to effectively reducing the focal spot size [23].

9 Optimum focus film distance

This is important primarily in fluoroscopy to reduce the incident dose by the inverse square law.

10 Kilovoltage

For a given recording system, the higher the kV the more favourable the 'input exit dose ratio'. Thus for example, for an examination of the lumbar spine at 72 kVp the input exit dose ratio might be 200 to 1, while at 90 kVp the ratio would be approximately 150 to 1 with proportional reduction in skin exposure but with less reduction in volume dose. With calcium tungstate intensifying screens the peak response is reached at about 130 kVp; above this level, the dose tends to increase because the screens are less effective in absorbing radiation [24]. Increasing the voltage to about 140 kV reduces the skin dose, the volume dose and the gonad dose whether the gonads are in or out of the beam *provided* other factors such as size of beam, grids, films and screens remain the same for the same optimal film blackening. If, however, in order to use the higher kV one has to resort to higher ratio grids or slower films and intensifying screens, these may increase the volume dose or gonad dose even if the skin dose is still reduced. In the past it has been difficult to check the kVp one uses in practice and this has made it difficult to compare the performance of one X-ray generator with another or to discover changes in a generator with use; this problem is now fairly simple [22, 25, 26].

11 Processing

Maximum energy developers should be used under optimum conditions.

The routine application of the above principles to each particular examination should ensure the lowest dose to both workers and patients, whether skin exposure, integral body or gonad dose.

12 Individual protection

If due attention has been paid to the above eleven points and they have been applied as far as practicable in any given case, the reduction of dose will affect both patient and operator.[1] Further protection of the workers may be obtained by the use of protective shields or garments or by increasing their distance from the scene of operations. Further protection of the patients' gonads may be obtained by the use of gonad shields.

SPECIAL SITUATIONS

General radiography

This gives relatively low doses to the patient provided that the points given

[1] Apart from high kV techniques which may increase scatter radiation to the operator.

above are considered and applied especially with reference to the use of fast systems and small X-ray beams whenever possible and provided that appropriate methods are taken to protect the gonads.

Fluoroscopy

It is generally considered that fluoroscopy should not be used when films can be taken instead. However, measurements have shown that for examinations of the gastrointestinal tract, it depends on the radiologist and the nature of the case as to whether the largest proportion of the dose is given by fluoroscopy or by the associated radiography [27]. The tendency to use unnecessarily large beams during fluoroscopy, particularly the use of rectangular fields, completely or over-covering a circular image intensifier, is prevalent. In the past it was essential to use small beams during fluoroscopy, in order that the scatter might be reduced and the contrast enhanced, so as to have an acceptable image. With modern television systems the contrast can be enhanced electronically and the necessity for small beams from this point of view is now not so great. This means that frequently higher radiation exposures in terms of $R \times cm^2$ are given for the fluoroscopy with modern equipment than used to be given without image intensification. Likewise it must be remembered that fluoroscopic exposures (measured in terms of $R \times cm^2$) with a small beam can often result in less radiation than that caused by one large radiograph. The use of $R\ cm^2$ dose meters is recommended not only to monitor nonradiologists when using the equipment but also to train radiologists, since this gives them an immediate indication of the beneficial effects of speed and of keeping the beam size small. Without such equipment radiologists may be unaware of the magnitude of the dose they are delivering. Fluoroscopic timers are useful to an individual operator but are of little value in comparing the doses given by different operators particularly since beam size can be considerably more important than total fluoroscopic time. Modern fluoroscopic equipment is often supplied with automatic brightness control to ensure that the monitor image has the same brightness regardless of the thickness of the patient. It is desirable to check irradiation outputs with such equipment in order to determine whether the radiation levels at which they operate are not too high. Likewise, some equipments have been found to increase the exposure when the beam size is reduced; these devices, though very convenient, may result in unnecessary patient irradiation. Image intensifier systems can deteriorate both in quality and in intensification factor with use. It is advisable to have these checked periodically. The operator should check whether he has to use higher kVs or mAs on average to obtain the desired result or, if he is employing a R × cm² meter, whether the readings are increasing for similar examinations. If there is any evidence that the intensification factor is decreasing, a full examination should be carried out to check whether the television system requires attention, whether the intensifier is deteriorating or whether the X-ray generator is no longer performing satisfactorily with regard to output, kV or mA. New installations should be checked to give a base line to determine deterioration in the future. Image intensifier systems may deteriorate until one is using the same dose factors that one would without intensification. Before this stage is reached it is desirable that the intensifier should be replaced.

Angiography

Serial or cine angiography, particularly with enlargement, may result in moderately high doses to the patient. Cine angio-cardiography may give skin doses up to 100 R [28, 29]. Rowley [30] has reported the skin exposure and R cm² values in cardiographic investigations with cine speeds up to 200 frames per second. The more extensive examinations have resulted in skin exposures of up to 76 R for the fluoroscopy alone, and up to 100 R for both fluoroscopy and cine. The maximum R cm² exposure for the entire study approached 6000 R cm². The dose to the principal operators, particularly with over-table tubes, may give exposures under the protective apron of 50 mR [31] though with care and an under-table tube average exposures under the apron may be as low as 2 mR [32]. It is essential that a particular procedure shall be adequately investigated so that one knows precisely what the situation is, in order that the necessary steps can be taken to see that the procedure is as safe as practicable.

When over-table tubes are in use it is particularly essential to see that radiation leaks from the X-ray tube and beam limiting devices are reduced to the minimum, that beam sizes are kept as small as possible and never larger than the input screen of the intensifier. It is essential to use shielding as close to the source as possible without unduly restricting the operator, in order to reduce the spread of scatter. Most equipment alters the size of the beam with rectangular movable diaphragms but the X-ray tube port should be limited to give a circular beam not larger than the intensifier at average working distance. The movable diaphragms should then be used to restrict the beam *within* this field whenever possible. In most instances the dose to the operator is largely due to scatter: reduction of beam size reduces this approximately proportionally.

Thyroid

Extensive radiography, fluorography or cinefluorography in the region of the thyroid should be monitored, particularly in children since the thyroid is a relatively radio-sensitive organ.

The eye

Diagnostic procedures do not normally give any significant dose to the eyes provided they are excluded from the useful beam whenever this is practicable; however, tomography of the middle ear with multiple sections of both ears

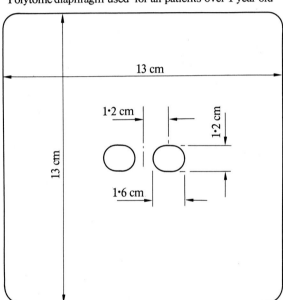

FIGURE 19.2. Dimensions of the beam-limiting apertures placed at 33 cm from the tube focus of the polytome when tomographing both middle ears simultaneously.

has been reported as giving doses to the cornea of 10 to 20 r [33]. We have confirmed these figures and also the statement that the dose may be reduced to about one-tenth by radiographing the patient in the posterior-anterior rather than in the antero-posterior position; this is technically inconvenient and similar results can be obtained by the use of 0.25 to 0.5 mm lead eye-shields [34]. The image of these eye shields does not interfere with radiographic quality and there is no need to use thicker lead.

Normally films of both middle ears are taken simultaneously with each tomographic section. If a diaphragm is inserted in the beam localizing device, to give two small apertures just sufficient to cover each middle ear, then both eyes are not in the useful beam all the time [35]. The two small apertures are made in a lead-covered plate fixed to the beam limiting device of the Polytome 33 cm from the tube focus; the oval apertures are 1.5 cm long and 1 cm high with their centres 1.2 cm from the midline (Fig. 19.2). These apertures appear in practice to be suitable for all individuals over the age of one year; smaller apertures closer together are used for babies. Using this device, together with suitable choice of kV (e.g. up to 90 kV) and other technical factors, the dose to the eye for eight to twelve tomographic sections may be reduced by nearly 50 per cent [36]. These remarks apply particularly to the Polytome in which there is a degree of magnification but in principle they are also relevant to all other tomographic equipments.

Relatively high doses to the eye may also be given in serial cerebral angiography with enlargement techniques and in some cine procedures. Whenever there is doubt, measurements should be made and appropriate steps taken. The eye should be excluded from the useful beam in dental radiography, preferably by the use of open-ended tubular metallic cones with a 5–6 cm diameter [37].

Irradiation of the female pelvis

In children and women of reproductive capacity, irradiation of the female pelvis should be kept to the minimum and precautions taken as described previously to see that the patient is not pregnant. Apart from keeping fluoroscopy to the minimum and taking the minimum number of films with the fastest system consistent with good results, there are only a few other precautions which can be taken.

Firstly, if the ovaries do not need to be in the useful beam, the edge of the useful beam should be kept as far away from the gonads as possible. Haybittle [38] showed that if one assumed that the ovary was halfway between the back and the front of the patient, then if the ovaries were 5 in. away from the edge of the beam they receive approximately the same dose as the film, if they were 10 in. away they receive one-tenth of the film dose, and if 15 in. away, one-hundredth, but every inch less than 5 in. approximately doubled the dose. In young individuals when multiple exposures have been made, e.g. follow-up from congenital dislocation of the hip, it has been suggested that following the initial film a piece of 1 mm lead should be cut to fit the inside of the brim of the pelvis in an attempt to shield the ovaries. This is not easy to do accurately in practice and malposition of the lead may result in repeat exposures which completely obviate the intention; likewise because of Haybittle's observations it will be seen that even when correctly positioned the ovary dose is not very significantly reduced. Of course if only one side of the pelvis needs to be radiographed, the use of small beams may well reduce the dose to the opposite ovary.

When carrying out pelvimetry it may be possible to use two small rectangular beams directly centred over the edge of the pelvis with a stationary film and a tube shift; this enables direct measurements to be made upon the film without having to compensate for enlargement factors. Recently, Virtama *et al.* [39] have described a technique using an image intensifier and 70 mm camera for this purpose with significant reductions of fetal dose to about one-tenth. This type of precaution may be a little tedious but can significantly reduce fetal and gonad dose and is worthwhile.

Particular care should be taken in such examinations as barium enemas, retrograde pyelographies, cystographies and hystero-salpingography. Most salpingographies are carried out for the investigation of infertility. It is fairly well known that women who fail to conceive and as a result adopt a child not infrequently become pregnant themselves after a short period. We have also reason to believe that some apparently infertile women, having finally decided to have their infertility investigated, likewise soon become pregnant. For this reason every precaution should be taken to see that the patients for salpingography are not pregnant when examined. We think that the times of ovulation should be checked by recording the patient's temperature and that they should be advised to see that they do not become pregnant between the time of ovulation and the salpingography being carried out. This is particularly important because it is inadvisable to carry out the examination immediately after the cessation of the menstrual period to avoid the possibility of contrast medium or air embolus.

Mammography

The breast is an organ which is prone to neoplastic change and mammography as frequently carried out may give relatively high doses to the breast tissue. Various techniques have been reported with doses in the range of 0.4 R to 11.5 R. It is difficult to determine the actual doses from the writings of various authors since it is not always clear whether the doses are skin doses, absorbed doses or doses per film or for a complete examination. The doses will vary with the type of nonscreen film used or if one or two intensifying screens are employed, with the target material, the kV, the thickness and type of filter and the thickness of the breast. The doses are particularly high with the use of molybdenum target radiation and beryllium window tubes and with the use of nonscreen films. I.C.R.P. lays down that at least $\frac{1}{2}$ mm Al equivalent filtration should be used for mammography. Beryllium window tubes should not be used without additional filtration. Though molybdenum target radiation [40–42] gives excellent contrast, it should be remembered that with a breast 5 cm thick only about 1 per cent of the radiation is transmitted to the film and, with thicker breasts, this transmission may be reduced to 0.1 per cent. With molybdenum target tubes it is often advocated that molybdenum filters or tube windows should be used. This reduces the radiation below the molybdenum characteristic (17.5 Kev) in a similar manner to a thin aluminium filter. However, the molybdenum filter reduces the proportion of radiation above the characteristic energy to a greater degree than an aluminium filter. This higher energy radiation is primarily reducing contrast of the radiograph rather than adding significantly to patient dose [40].

Male gonads

The male gonads should be kept out of or as far away as possible from the useful beam. Approximately one-third of the significant population genetic dose from diagnostic radiology comes from irradiation of the male gonads. Of this, 80 per cent comes from examinations involving the pelvic region [43]. Gonad shields may be used in many examinations, for instance, a simple lead shield (the Oxford Fig Leaf) (Fig. 19.1) has been used for many years to shield the gonads of patients lying upon the X-ray table for examinations of the pelvis and lower abdomen. No method is completely effective but this simple procedure reduces the dose to the male gonads to about 10 per cent of what would be obtained if the gonads were in the beam. We believe this technique should be employed whenever practicable in all males over the age of about 8 years. In children under this age, it is probably more simple to wrap the scrotum in thin lead foil. Various workers have described lead boxes to cover most of the gonads. These are superior to the use of a simple fig leaf but they are difficult to apply in practice, particularly when the staff are of the opposite sex. It would be a wise precaution to advise a man whose gonads have been in the direct beam, to avoid deliberately conceiving a child for a period of at least six months. The effects will then be reduced, because of the rapidity of cell division in the male germ cells. In examinations such as cystography, every precaution should be taken to shield the gonads where possible but this is often impractical.

Xerography

When xerography was initially introduced in 1947 it suffered the disadvantage that radiation exposures were two or three times higher than those of conventional nonscreen film techniques. Recently, with the advent of the Rank Xerox equipment, radiation exposures have been reduced and they are now more comparable to conventional exposures. Most workers are agreed that Xerography gives very good results in examinations of the breast, equal or superior to any other technique. The method has the advantage of giving good results with Tungsten anode tubes at 45 to 50 kV. This results in lower doses than techniques using 25 to 35 kV with slow nonscreen film.

Xerography can be valuable in the radiography of the larynx, pharynx and associated organs with or without xerographic tomography and is establishing a useful role in other areas of the body. It must not be forgotten that despite its advantages, the radiation exposures required are those for a medium speed nonscreen film and may result in exposures 15 to 50 times greater than would be used for these examinations using film screen combinations.

Ion radiography

The Xonixs corporation of America has recently demonstrated a new system of radiography whereby the X-ray beam, on emerging from a patient, falls on a thin chamber of high pressure gas as large or larger than the area of the recording medium. Ionization occurs in the gas and the free electrons are directed by a potential difference so that they fall on a sheet of plastic material the same size as a conventional X-ray film. This material is then covered with black powder and the image is formed in a similar manner to the xerographic image. After exposure the film is 'processed' so that the black powder image is permanently fixed. The image is similar in appearance to a conventional X-ray film and can give edge enhancement. It is claimed that the definition is superior to that of a medium speed film taken with medium speed intensifying screens and the radiation exposure is the same or about half that used for conventional radiography. It is too early to say whether this method will prove completely satisfactory in practice, but preliminary work shows considerable promise.

Tomography

A single tomographic film requires approximately the same radiation exposure as a conventional film of the same area, though the gonad dose may be increased if the useful beam is pointed towards the gonads for at least half the exposure. Since multiple films may be taken, however, the dose from the complete procedure may be significantly high and every effort should be taken to keep the number of films and the area irradiated to the minimum. Tomography of the middle ear and problems of the cornea and lens of the eye have been discussed separately. In order to reduce exposure dose the approximate site of deep-seated lesions should be determined by direct radiography. Multi-section tomography is often claimed to give a lower total dose than the same number of films taken individually. However, in order to get the same exposure of each film a combination of different speed intensifying screens has to be used. The patient exposure will, therefore, be governed by the slowest intensifying screen in the system. If more than two or three films are used, the total dose may well be reduced by a factor of two or three but if only two or three films are being taken, there may be no significant dose reduction.

Small format radiography

The use of photofluorography for examination of the chest involves a higher radiation exposure than full size radiography because only a proportion of the light from the fluorescent screen can be collected to expose the film, even with large, wide-aperture mirror systems. The comparison of the exposures by this method with full-size films depends on the speed of each system but, in general, it can be said that modern 70 or 100 mm systems using wide-aperture mirror optics (Odelca system) result in exposures of the order of twenty times those obtainable with medium speed film–screen combinations. This increase is partly due to light loss, partly due to the shorter focus screen distance and partly due to the routine use of anti-scatter grids. The gonad dose for similar areas irradiated are likewise increased but can still be less than 1 mr per exposure. When these systems are used it is essential to see that the beam size is not fixed at the maximum for all subjects. Provision should be made to reduce the size of the beam at the sides and bottom so that only the lung field area is irradiated and the beam edge kept as far from the gonads as possible. Because of the radiation problems involved, many would not use this system for children or for pregnant women.

The use of an image-intensifier, preferably of the caesium-iodide type, and 70 or 100 mm films now produces films of sufficient quality for many barium examinations and for some other purposes. The patient dose per film can be less than that for conventional radiography but, when making comparisons, it is important to compare the photofluorographic doses with full-size films of a comparable quality and not with slow systems giving superior quality radiographic detail.

Videotape

Videotape recordings can be made with fluoroscopic exposures and will give considerably less dose than cineradiography. This technique should, therefore, be used providing that the results are adequate for diagnostic purposes. Various workers [44] including ourselves [45] have investigated the technique of recording one fluoroscopic television scan and immediately playing this back with the radiation switched off so that it could be studied without irradiating the patient; subsequently, after a second or more, the patient is again scanned and a fresh image stored and viewed. Theoretically, this could result in fluoroscopy of the more slowly moving parts being carried out with dose reductions of one-twenty-fifth or less of conventional techniques. We found this unsatisfactory because quantum mottle is excessive. When viewing a continuously irradiated object the persistence effect of the eye summates the mottle and reduces the effects. In order to overcome this the individual scans had to be done with radiation intensities increased by a factor of about ten to maintain image quality. This meant that the complicated equipment was giving only small reductions in patient dose. Videodisc recorders can now be adopted to provide instant replay of a 'still-frame'. This may be useful in reducing radiation doses in certain circumstances. For example, in orthopaedic operations using a mobile image-intensifier, a brief fluoroscopic exposure can be recorded and the position of a reduced fracture can then be studied for a more prolonged period using the replay facility [49].

Radiography of the chests of pregnant women

In the U.K. routine radiography of the chests of pregnant women is no longer carried out because of the poor yield of significant pathology, and this obviates even the small dose of radiation which the patient or fetus might receive. Chest radiography should, however, still be performed whenever there is any medical reason and in those individuals who, by virtue of their origin or family history, might make it advisable.

Dental radiography

Intra-oral film dental radiography can give relatively high local doses [37], though with modern high-speed dental films and the use of small beams the dose is well within acceptable limits. When examining the whole mouth pan-oral dental tomography is to be preferred both for convenience and because of the lower total dose. Dental cones should preferably be of the open-ended tubular metal type, the beam at the end of the cone should not exceed 6.5 cm and can, with care, be as small as 5 cm. The cone and patient should be so positioned that residual radiation does not fall on the gonad area of the patient. When the upper teeth are radiographed the patient's head should be tilted so that the residual beam is not directed towards the patient's pelvis. If this is not possible, or as an additional precaution in children and pregnant women, lead protective aprons should be used to cover the relevant parts of the patient.

THORIUM DIOXIDE (THOROTRAST)

The late effects of Thorotrast administration serve as a timely reminder of the unfortunate results which may ensue when there has been inadequate consideration of already known fundamental knowledge. Thorotrast is a 25 per cent colloidal solution of thorium dioxide stabilized to be miscible with blood and other body fluids. It was introduced in 1928 and as far as the 'immediate' effects were concerned it was one of the best radiographic contrast media ever developed, giving good contrast and having few, if any, pharmacological actions. Unfortunately thorium dioxide is naturally radioactive and gives off alpha particles and the daughter disintegration products give off gamma radiation. It has a half-life of 1.4×10^{10} years; it is not excreted significantly from the body and it is taken up by the reticuloendothelial system where it remains radioactive thereafter [46–48]. Even angiographic doses of 20 ml have resulted in the development of neoplasms twenty or more years later. For this reason, it has now been abandoned for routine work and is only used on rare occasions in people with short life-expectancy. Because of its retention in the reticulo-endothelial system Thorotrast was used in relatively large doses (60 ml or more) to opacify the liver to ascertain the presence of secondary deposits, tumours, etc., since these did not take up the material as does the normal liver. This procedure was particularly hazardous. If Thorotrast was accidently injected into the tissues instead of into the blood vessel, it remained in the adjacent tissue planes causing severe local fibrosis, nerve palsies and local neoplasia: these can arise relatively quickly within three years of the injection. Thorotrast has also been used to opacify the nasal sinuses, the breast by injection into the lactiferous ducts, sinus tracts, gall bladder and the pelvis of the kidney by retrograde pyelography; other uses included ventriculography, salpingography and examinations of the lachrymal ducts, salivary ducts and lymphatics. Though an excellent contrast medium for these purposes, the material tended to be retained, and cases of neoplasia have been reported from this cause, usually about 30 years later. Ill effects were reported from the middle 1930s onwards. Malignancies in other organs usually arose after a number of years and were reported from 1940 onwards. The unfortunate effects of Thorotrast have been extensively studied and are still being followed up, and this is giving valuable information on the effects of retained radioactive material in man.

Twenty years after the injection of 20 ml of Thorotrast the alpha dose rate in the liver is of the order of 500 m/rads per week, the spleen about 1000 m/rads per week and the total skeleton about 250 m/rads per week. Over a period of 28 years 20 ml of Thorotrast has been estimated to give doses of 600 rads to the liver, 1500 rads to the spleen and 300 rads to the bone marrow. The gamma doses are only a few per cent of the alpha dose.

Colloidal thorium dioxide was also used (Umbrathor) as a contrast medium for barium enema examinations. This form of the material was very useful since it tended to adhere to the mucous membrane and gave good double contrast films. Since it was not absorbed it did not give rise to late effects but was abandoned when other techniques were developed which were adequate and less expensive. It has also been used for pyography, that is for the injection or cerebral abscesses when it outlined the walls of the cavity so that their eventual reduction in size could be followed radiographically; it has been abandoned for this purpose since the cheaper non-toxic sterile barium sulphate is just as satisfactory. Thorotrast remains a useful contrast medium for experimental angiography in animals because of its lack of pharmacological effects though it has the disadvantage of not being excreted by the kidneys and consequently repeated injections make the experimental animal increasingly radio-opaque.

THE INTRODUCTION OF S.I. UNITS INTO THE RADIOLOGICAL SCIENCES

The units used for measuring radiation dose are now in the process of being changed to the International System of Units (S.I.).

The new unit of absorbed dose is called the gray (Gy) and is equal to 1 joule/kilogram that is to say 1 rad = 1 centi-gray (1/100 Gy). The unit of exposure, at present the Röntgen, has not been given a special name in the S.I. system but its magnitude is 1 coulomb/kilogram equivalent to approximately 3.876×10^3 R. These changes are compatible with the continued use of air-filled ionization chambers as the basic instrument in radiation dosimetry.

REFERENCES

1 ALBERS-SCHONBERG H. (1903) Uber eine bisher unbekannte Wirkung der Rontgenstrahlen auf den Organismus der Tiere. *Munchener Med. Wochenschrift Munchen.* **50**, 1859–1960.
2 ARDRAN G.M. (1971) The society and X-ray protection. *Radiography* **37**, No. 439, 157–165.
3 ARDRAN G.M. & CROOKS H.E. (1965) The measurement of patient dose. *Br. J. Radiol.* **38**, 766–770.
4 I.C.R.P. (1964) *Recommendations of the International Commission on Radiological Protection.* Publication 6. Pergamon Press.
5 I.C.R.P. (1966) *Recommendations of the International Commission on Radiological Protection.* Publication 9. Pergamon Press.
6 D.H.S.S. (1972) *Code of Practice for the Protection of Persons Against Ionising Radiations arising from Medical and Dental Use.*
7 I.C.R.P. (1970) *Recommendations of the International Commission on Radiological Protection.* Publication 15. Pergamon Press.
8 I.C.R.P. (1970) *Recommendations of the International Commission on Radiological Protection.* Publication 16. Pergamon Press.
9 I.C.R.P. (1973) *Recommendations of the International Commission on Radiological Protection.* Publication 22. Pergamon Press.
10 WRIGHT D.J., NICHINI F.M. & STAUFFER H.M. (1971) Beryllium window tubes for mammographic examinations. *Br. J. Radiol.* **44**, 480.
11 ADRAN G.M. & KEMP F.H. (1972) Radiography of potentially pregnant females. Letter to *Br. Med. J.* 18th Nov., p. 422.
12 ARDRAN G.M. & KEMP F.H. (1973) Radiography of potentially pregnant females. Letter to *Br. Med. J.* 14th April, pp. 119–120.
13 GOSS S.G. (1974) The risk of death from radiation-induced cancer as estimated from the published data on the Japanese atomic bomb survivors. *Nat. Radiological Prot. Board. Report R.20.* H.M.S.O.
14 MOLE R.H. & ARDRAN G.M. (1971) Environmental hazards. *Mod. Medicine.* **16**, No. 5, p. 334.
15 UCHIDA I.A. & CURTIS E.J. (1961) A possible association between maternal radiations and mongolism. *Lancet* **2**, 848–850.
16 UCHIDA I.A., HOLINGE R. & LAWLER C. (1968) Maternal radiation and chromosome aberrations. *Lancet* **2**, 1045–1049.
17 STEVENSON A.C., MACON R. & EDWARDS K.D. (1970) Maternal diagnostic X-irradiation before conception and the frequency of mongolism in children subsequently born. *Lancet* **2**, 1335–1337.
18 ARDRAN G.M. & KEMP F.H. (1957) Protection of the male gonads in diagnostic procedures. *Br. J. Radiol.* **30**, 280.

19 ARDRAN G.M. (1957) The implications of the White Paper. *Radiography* 23, 206.
20 ARDRAN G.M., CROOKS H.E. & O'LOUGHLIN B.J. (1964) The speed of systems and image quality. *The reduction of patient dose*, ch. 15, pp. 222–233. Charles C. Thomas, Springfield, Illinois.
21 ARDRAN G.M. & KEMP F.H. (1957) Reduction of radiation dose administered during chest radiography. *Tubercle* (London) 38, 403–410.
22 ARDRAN G.M. & CROOKS H.E. (1972) The measurement of inherent filtration in diagnostic X-ray tubes and the effect of target angle on X-ray quality. *Br. J. Radiol.* 45, 599–602.
23 ARDRAN G.M. & CROOKS H.E. (1964) The reduction of scatter fog in chest radiography. *Br. J. Radiol.* 37, 477–9.
24 ARDRAN G.M. & CROOKS H.E. (1962) Dose in diagnostic radiology: The effect of changes in kilovoltage and filtration. *Br. J. Radiol.* 35, 172–181.
25 ARDRAN G.M., CROOKS H.E., BURT A.K. & PEAPLE L.H.J. (1969) Comparison of a radiographic measurement of X-ray beam quality with spectral measurements. *Br. J. Radiol.* 42, 757–761.
26 ARDRAN G.M. & CROOKS H.E. (1968) Checking diagnostic X-ray beam quality. *Br. J. Radiol.* 41, 193–198.
27 ARDRAN G.M. (1969) Patient dose in fluoroscopy. In: *T.V. & diagnostic radiology* (Ed. by MOSELEY R.D., JR. & RUST J.H.), ch. II. Aesculapius Pub. Co., Birmingham, Alabama.
28 GOUGH J.H., DAVIS R. & STACEY A.J. (1968) Radiation doses delivered to the skin, bone marrow and gonads of patients during cardiac catheterisation and angiocardiography. *Br. J. Radiol.* 41, 508–518.
29 ARDRAN G.M., HAMILL J., EMRYS-ROBERTS E. & OLIVER R. (1970) Radiation dose to the patient in cardiac radiology. *Br. J. Radiol.* 43, 391–394.
30 ROWLEY K.A. (1974) Patient exposure in cardiac catheterisation and cinefluorography using the Eclair 16 mm camera. *Br. J. Radiol.* 47, No. 555, 169–178.
31 MALSKY S.J., ROSWITT B., REID C.B. & HAFT J. (1971) Radiation exposure to personnel during cardiac catheterisation. A prelim. study. *Radiology* 100, 671–674.
32 ARDRAN G.M. & FURSDON P.S. (1973) Radiation exposure to personnel during cardiac catheterisation. *Radiology* 106, No. 3, 517–8.
33 CHIN F.K., ANDERSON W.M. & GILBERTSON J.D. (1970) Radiation dose to critical organs during petrous tomography. *Radiology* 94, 623–627.
34 DOBRIN R., BECKER M.H. & GENIESER N.B. (1973) Radiation protection of the cornea and lens during petrous-bone tomography. *Radiology* 109, 201–204.
35 SANGSTER J.M. (1966) Tomography today—Part 3. The polytome in practice. *X-ray Focus (Ilford)* 7, No. 3, 21–25.
36 DIXON-BROWN A. (1974) Personal communication.
37 ARDRAN G.M. & CROOKS H.E. (1959) Observations on the dose from dental X-ray procedures with a note on radiography of the nasal bones. *Br. J. Radiol.* 32, 572–583.
38 HAYBITTLE J.L. (1957) The effect of field size on the dose to the patient in diagnostic radiology. *Br. J. Radiol.* 30, 663–665.
39 VIRTAMA P., LAUTEALA I. & HALKOLA L. (1973) Reduction of radiation dosage in pelvimetry with the 70 mm image amplifier camera. *Internat. Congress of Radiol.* Madrid, October.
40 ARDRAN G.M., MARSHALL M., CROOKS H.E. & PEAPLE H.J. (1973) A comparison

of X-ray quality and output from molybdenum and tungsten targets. *Excerpta Medica. Int. Congress Series* No. 301, p. 424.
41 MARSHALL M., PEAPLE H.J., ARDRAN G.M., & CROOKS H.E. (1974) A comparison of X-ray quality and output from molybdenum and tungsten targets. *Br. J. Radiol.* 48, 31–39.
42 ARDRAN G.M. & CROOKS H.E. (1971) Some technical problems with diagnostic radiography at 20 to 40 kVp. *Br. J. Radiol.* 44, 625–630.
43 H.M.S.O. (1960) *Radiological hazards to patients*, p. 85.
44 DOCKER M. & ASTLEY R. (1969) A video tape recorder used to reduce patient dose during X-ray screening. *Br. J. Radiol.* 42, 358–363.
45 ARDRAN G.M. & CROOKS H.E. (1969) Dose reduction in fluoroscopy. *Br. J. Radiol.* 42, 554.
46 SWARM R.L. Colloidal thorium dioxide. In: *Internat. Encyclopaedia of Pharmacol. & Therapeutics II* (Ed. by KNOEFEL P.K.), Ch. 12, pp. 431–441. Pergamon Press, Oxford.
47 MUTH H., EDELMANN L., GRILLMAIER R., HERZFELD U., HEYDER J., KAUL A., KOEPPE P., KUNKEL R., OBERHAUSEN E., ROEDLER H.D. & WERNER E. (1972) Radiation dose in different organs of thorotrast patients. In: *Assessment of Radioactive Contamination in Man*. International Atomic Energy Agency, Vienna (Austria), pp. 457–467.
48 Thorotrast: A Bibliography of its Diagnostic Use and Biological Effects. (1964) 151 pp. International Atomic Energy Agency, Vienna (Austria).
49 ZATZ L.M., FINSTON R.A. & JONES H.H. (1974) Reduced radiation exposure in the operating room with video disc radiography. *Radiography* 110, 475–477.

APPENDIX 19.1

Dear Doctor,

Radiography of Potentially Pregnant Females

In June 1972 the Department of Health and Social Security published a revised Code of Practice for the Protection of Persons against Ionising Radiations arising from Medical and Dental Use (London, H.M.S.O.). In the preface (para. 3) it is stated that 'although the arrangements recommended relate primarily to institutions they should be applied, as far as practicable, by all medical and dental practitioners'.

The accompanying circular HM(72)39 from the Department imposes upon Boards of Governors and Regional Hospital Boards the duty of implementing the code. The bulk of code applies to those who administer the radiation, whether for diagnosis or treatment, but Section 7, dealing with the protection of the patient, applies also to those who request that such procedures be carried out. In particular we are concerned with paragraphs 7.3.1. which read as follows: 'In all women of reproductive capacity the clinician requesting the examinations should consider the possibility of an early stage of pregnancy. The date of the last menstrual period should be entered on the request form and it is the responsibility of the clinician requesting the examination to ascertain this. To reduce the likelihood of irradiation of a pregnancy, examinations involving the lower abdomen should, if practicable, be carried out within 10 days following the first day of the menstrual period.'

Radiological Protection Committees are considering how these problems can best be dealt with to ensure that no embryo or foetus receives more radiation than is absolutely essential. The code and the circular do not give specific details as to how the problem might be adequately tackled.

Following the work of Dr. Alice Stewart and others it is now almost universally accepted that radiation to the fetus can cause leukaemia or other neoplastic disease. The hazards may be estimated in simple terms with sufficient accuracy for the present purpose in the following way:

The natural incidence of childhood neoplasia and leukaemia is about 1 in every 1200 live births. The best estimates indicate that a dose of 1–4 r will double the natural incidence. Diagnostic examinations of the pelvic regions can result in a dose to the fetus ranging from about 0.25 r up to perhaps 5 r or more in an extensive examination. This means that an average examination might double the natural incidence of leukaemia etc. This could also be expressed by saying that should leukaemia or neoplasia arise in a child who was radiographed as a fetus there is a fifty–fifty chance that the radiation caused it. All the evidence indicates that an early pregnancy is just as susceptible as a later one.

The code of practice clearly places the responsibility for finding out whether the patient might be pregnant or not on the referring clinician and requests that the date of commencement of the last menstrual period should be entered on the request form. Experience has shown that it is almost impossible for the X-ray department to cope with this problem without further assistance. Many individuals do not know when their next period will occur and booking clerks cannot be expected to make these inquiries and arrange to book the patient within the first ten days of the beginning of the next period (when the patient is unlikely to be pregnant) without considerable delays and difficulties. It is therefore suggested that the referring clinician, when requesting an X-ray or isotope investigation involving the area from the diaphragm to the knees, should ascertain if there is any possibility of pregnancy. He should also state whether the examination is of such urgency that it should be carried out regardless of whether the patient is pregnant or not. If the examination is not of immediate urgency or if it is of a type for which an appointment must be booked, it is suggested that the referring physician should explain to the patient that because of a small hazard to the fetus it is best if the examination is carried out when the patient is not pregnant and the patient therefore should take the necessary steps to see that she does not become pregnant until after the examination.

To avoid difficulties when the patient arrives at hospital either for an outpatient appointment or with a view to admission it is further suggested that in any case in which there is a possibility of a diagnostic radiological investigation being required, the patient likewise should be advised not to become pregnant until after the possibility of a radiological examination has passed, and it should be explained to her that if she fails to do this it will be her responsibility if any unfortunate consequences should occur.

Doctors requesting diagnostic X-ray examinations involving the abdomen or pelvis of women of reproductive capacity are requested to ensure that one of the enclosed forms [see below] is duly filled in. This should be sent to the X-ray Department when the appointment is made or the examination is to be carried out if this is done without an appointment. It will then be filed with the patient's notes. Children between the ages of puberty and 16 years and patients who have language difficulties or whose mental capacity is too low to comprehend the problem, should best be examined, if possible, during the menstrual period.

Some may only consider it necessary to use Part A of the form; however, in order to be sure that the patient understands and to warn the patient not to become pregnant before attending the hospital, it is thought best that the matter should be discussed by the referring doctor with the patient and it is desirable that he should indicate that he has done this by signing Part B of the form.

We appreciate that this problem will cause a little additional work but we think it is the simplest solution and will in the end be time saving to all concerned and we hope for your full collaboration.

Yours etc.
Radiological Hazards Committee

OXFORDSHIRE AREA HEALTH AUTHORITY (TEACHING) RECORD FORM FOR FEMALE PATIENTS BETWEEN PUBERTY AND THE MENOPAUSE	Hospital No. .... Surname .... First Names .... Date of Birth ....Tel. No. ....

This form should be completed when the clinician requests an X-ray examination of any part between the diaphragm and the knees or a radio isotope investigation.

Will you please answer the following questions:—

PART A

To be completed by the patient

(1) Is there any possibility you have recently incurred a risk of pregnancy (State Yes or No)   ....

(2) Please enter date of the 1st day of the last menstrual period   ....

I understand that it is important that I should avoid possible pregnancy until my X-ray examination has been carried out. I undertake to inform the staff of the X-ray Department if there is any chance I might have become pregnant.

If the answer to question (1) in Part A above is Yes, or if you have difficulty in understanding or completing this form, please take it back to the doctor who requested the X-ray examination, for completion of Part B.
If the answer is no, bring the form with you when you attend for your X-ray examination.

....
Signature of patient

....
Date

PART B

To be completed by the Doctor:—

Please indicate which of the alternative procedures you wish to be adopted.

(1) My patient M/s .... has been asked to undergo an immediate X-ray examination. The questions relating to pregnancy have been discussed.

OR

(2) My patient M/s .... is referred for a booked appointment. I have explained that she should not run a risk of pregnancy until the examination has been performed.

....
Date   Signature of Doctor

The Code of Practice for the Protection of Persons against Ionising Radiation (1972) requires that the clinician requesting the examinations should consider the possibility of an early pregnancy.

When you have completed Part B,
Please send this form with the Request Card to the Dept. of Radiology.

Date of patient appointment in X-ray   ....

Name of Radiographer   ....

This form is to be retained in the patients record.

RC 161

APPENDIX 19.2

Region	Position	kVp	mAs	Total filter al. equiv. (mm)	Focal film distance (cm)	Focal skin distance (cm)	Anti-scatter method	Area of skin[1] irradiated (cm^2)	Skin dose with back scatter (mr)	Integral dose (g/r)	Gonad Dose[2] Male (mr)	Gonad Dose[2] Female (mr)
Single finger[3]	P.A.	45	10	1.5	120	118	—	26	33	1.6	0.001	0.001
Metacarpals or wrist[3]	P.A.	45	16	1.5	120	115	—	75	58	13.0	0.001	0.001
Elbow[3]	P.A. or lat.	50	25	1.5	120	113	—	144	100	52.0	0.001	0.001
Knee	A.P.	75	16	5.0	120	103	P.B.	250	38	42.0	0.06	0.008
Skull	Occipito frontal	85	64	5.0	120	95	P.B.	240	180	290.0	0.003	0.003
Sinuses	Occipito mental	120	12	2.0	182	158	A.G.	128	80[4]	52.0	0.001	0.001
		(85)	(30)	(3.0)	(87)	(60)	(P.B.)	(128)	(300)[4]	(168.0)	(0.003)	(0.003)
Chest	P.A.	96	9	4.0	365	327	A.G.	800	5	20.0	0.005	0.008
Lumbar spine and sacroiliac joints	P.A.	85	80	4.0	120	95	P.B.	380	280	460.0	3.2	15.0
	Lat.	85	125	4.0	120	88	G.	340	510	740.0	7.2	72.0
Pelvis	A.P.	85	100	4.0	144	113	P.B.	550	220	520	15.0[5]	52.0
Intravenous pyelography	A.P.	76	50	2.5	120	93	O.G.	300	200	246.0	0.03	0.04
	Bladder	76	50	2.5	120	97	O.G.	300	190	232.0	3.8[5]	20.0
	Oblique	76	60	2.5	120	80	O.G.	240	270	270.0	0.04	0.06
Pregnant abdomen	A.P.	80	60	3.0	—	100	P.B.	800[6]	310	1020.0	—	—
	Lat.	75	100	3.0	—	90	O.G.	800[6]	240	780.0	—	—
Barium meal	P.A. stomach	140	4.5	5.0	80	50	G.	146	110	85.0	0.04	0.23
	Serial film	146	6.0	5.0	80	50	G.	40	130	28.0	0.025	0.16
	Fluoroscopy	96	1.0 mA	5.0	80	50	G.	70	820/min[7]	270.0	0.3/min[7]	4.0/min[7]
	Follow thro.	120	6	5.0	120	98	G.	800	40	160.0	0.6	4.8

Barium enema											
Fluoroscopy	90	0.5 mA	5.0	80	50	O.G.	70	320/min[7]	100.0	1.0/min	29.0/min[7,8]
A.P.	146	4.6	5.0	120	91	G.	580	56	182.0	3.0	20.0
Oblique	146	5.8	5.0	120	82	G.	470	90	230.0	5.0	35.0

(1) Measured directly on skin suface.
(2) Measurements using Mix D Manikin and directly.
(3) With the exception of these regions all radiation assumed to have been absorbed by the patient.
(4) This airgap technique has superseded the Potter Bucky. It will be noted that the filtration has been reduced below that recommended by I.C.R.P. Using the air gap, in view of the large reduction in dose, we might be permitted to disregard the ruling regarding filtration.
(5) Using the Oxford Fig Leaf.
(6) Estimated on the assumption that 15 × 12 film was fully covered.
(7) Using beam reduced to cover on average half the maximum field area.
(8) Using full field with ovaries in the direct beam for half the exposure.

P.B. = Potter Bucky.
A.G. = 6 in. air gap.
G. = Smit Golden Stationary grid 15 : 1 ratio, 110 lines per in.
O.G. = Oscillating grid.

COMMENTS

The above table gives the dose to the patient in selected radiographic projections per film and the dose per minute during fluoroscopy for relatively low dose techniques as advocated in the text for the average sized patient. Doses, however expressed (e.g. skin dose, integral dose or gonad dose) may differ in practice by factors of 2, 4, 8 or even 20 or more if different films, intensifying screens, anti-scatter methods or processing are employed.

Different kVs and filters may alter the skin dose and to a lesser extent the integral dose. Different areas irradiated will alter the Rcm^2 exposure and the integral dose in g/rads.

The gonad doses will be very dependent on whether the gonads are included in the useful beam or not or precisely how close they are to the edge of the useful beam and whether or not any gonad protector is used.

Doses in fluoroscopy depend on the fluoroscopic time, the factors used, the efficiency of the fluoroscopic system and the method of automatic brightness control, if fitted, and the beam area used.

The above figures are given as a guide. Doses in actual practice must be measured since frequently preset kVs are inaccurate, filtration unknown or inaccurate, mA incorrect and timers erratic.

For practical purposes doses in Rcm^2 may be converted to g/rads and vice versa assuming all the radiation is absorbed in the patient, if one remembers that using 2 mm Al equivalent filter at 50 kV, 1000 Rcm^2 = 3350 g/rads, at 80 kV 1000 Rcm^2 = 4200 g/rads and at 150 kV 1000 Rcm^2 = 5470 g/rads.

The reported values of Rcm^2 and g/rads per examination vary widely: for example Ardran [27] showed that typical barium meal examinations with both fluoroscopy and films could vary on average from about 250 Rcm^2 to 1300 Rcm^2, barium enema examinations giving about 900 Rcm^2. It was also shown that doses from the fluoroscopy could be less than the dose from the films taken or vice versa according to the operator and the type of examination. Other workers have given figures of several thousands, even tens or hundreds of thousands of Rcm^2 for similar examinations.

CHAPTER 20: ISOTOPE TECHNIQUES

E. S. WILLIAMS

The clinical use of radioactive materials (radiopharmaceuticals) leads to hazards which can be divided broadly into two groups:
1 Hazards related to exposure to ionizing radiation;
2 Hazards related to the chemical or physical nature of the vehicle carrying the nuclide emitting the radiation.

COMPLICATIONS CAUSED BY IRRADIATION

Hazards of the first kind are essentially similar to those when X-rays are used and are dealt with in Chapter 19. The code of practice there referred to deals also with the safe use of radiopharmaceuticals and everyone using the agents must be familiar with the relevant parts of the code. As compared with the use of X-rays, the use of radioactive tracers as diagnostic agents, while presenting a similar type of hazard, has certain contrasts.
(a) Once the radiopharmaceutical has been administered to a patient the exposure to ionizing radiation continues for a time determined by the physical and biological half-life of the particular material used. This is contrasted with the situation when an X-ray machine is used, the exposure then ending as soon as the machine is switched off.
(b) Although there is normally an organ which concentrates the radiopharmaceutical to a greater extent than do other parts of the body, such concentration is far from complete and the whole body is irradiated. In addition, depending on metabolic handling and route of excretion, organs other than those under investigation may receive a significant dose of radiation. In contrast, an X-ray beam can be confined to the area of interest and, apart from the effect of 'scatter', only tissue in the beam will be irradiated.
(c) Preparations of radiopharmaceuticals have to be manipulated both

chemically and physically. They are normally in solution and therefore can be spilled. Thus, in strong contrast to the use of X-rays, there is a danger of ingestion, of external contamination of the body, of the clothing, and of the surroundings, thus leading to a hazard of continuous irradiation unless stringent precautions are observed. Because of this hazard, the term 'unsealed sources' has been coined to designate such 'spillable' sources of ionizing radiation. Those 'occupationally exposed' are, of course, at greater risk than are patients. Rules for the use of unsealed sources appear in the code of practice.

Radiation hazards arising as a result of the medical use of radionuclides are, in the United Kingdom, the concern of the Department of Health & Social Security Isotope Advisory Panel. Whether the material is to be used for clinical research or for an established routine purpose, it is essential to apply to this panel for approval of the proposals. Although the panel was set up as a consequence of the implementation of the first Atomic Energy Act, its role is advisory and its recommendations are not mandatory. However, should a hospital employee or patient take legal action concerning a believed overdose of ionizing radiation or of alleged incompetent use of radioactive isotopes, then the case would probably not be defensible if the advice of the D.H.S.S. Panel had not been sought before instituting the work in question. A similar situation would be probable if the panel's advice had been obtained but had not been acted upon.

Concern about radiation hazard to the CNS was expressed a few years ago [1]. The use of a ^{131}I-labelled agent should be restricted to those investigations extending over a period of a few days and a short-lived nuclide used where shorter term studies are likely to provide the essential information needed for the optimum management of the patient. In view of an infant's small CSF volume, an agent using a short-lived nuclide is here highly desirable but in hydrocephalus, not only is the CSF volume larger, but the essential data sought can often only be obtained using ^{131}I as the label. The hazard here referred to has, as in all cases, to be weighed against the comparative hazard of alternative investigations as well as against the clinical value for the particular patient of the information likely to be derived from the investigation.

These potential hazards in the use of radioactive materials are now so well understood that, *provided all the rules of best practice are observed*, the actual hazard is very small indeed. In about a quarter of a century of routine clinical experience in the United Kingdom, no medical complication has arisen as a result of radiation injury consequent upon the diagnostic use of these materials.

COMPLICATIONS CAUSED BY PHARMACEUTICALS

Until the middle 1960s virtually all reports dealing with the safety of unsealed sources referred only to the radiation hazards. Accounts of specific adverse effects also dealt only with the results of overexposure to ionizing radiation, where the unsealed source had been used as a therapeutic agent. A scattered literature then began to appear recording adverse reactions to the vehicle carrying the source of radiation and similar reports were from time to time informally mentioned in discussion with colleagues.

During approximately the last decade there has been an explosive increase in the clinical usefulness of imaging using radioactive tracers. In part, this has been a result of improved instrumentation, but a significant factor has been the introduction of a variety of vehicles for the radioactive tracer chosen. During roughly the same period, methods have been developed to make short-lived nuclides available at the site of use by elution from a column containing the longer-lived parent. This gave new impetus to the search for chemicals having desirable organ-specific qualities which could be quickly and easily labelled with such nuclides. Thus there arose the current, still expanding, situation of a wide range of materials of various chemical and physical forms being in routine clinical use.

Inevitably the incidence of medical complications increased and 'adverse reactions' to the administration of radiopharmaceuticals were reported more frequently. In this context the syndrome designated as 'adverse reaction' is similar to that occasionally seen in the use of X-ray contrast media administered intravenously. The reaction may be immediate or of late onset (a few hours), it may be mild and transient, or range in severity to rapid collapse and death.

The patient may first complain of feeling hot and flushed and on examination hyperaemia of the face and upper part of the body may be seen. On the other hand, the patient may complain of feeling faint, of nausea, and show the usual signs of a vaso-vagal attack. Headache may be the first symptom. The reaction may not progress beyond the complaint of one or more of these features with recovery in a few minutes, but the patient's condition may become rapidly worse with loss of consciousness and even cardiac arrest.

A survey carried out in the United Kingdom [2] confirmed this syndrome to be not only very variable in its symptomatology and in its severity, but remarkably variable in its incidence between centres using radiopharmaceuticals. Interpersonal variability is likely to be due to undefined patient sensitivity because aliquots of the same preparation which had given rise to

TABLE 20.1. Incidence of adverse reactions.

Agent*	Approximate incidence: one reaction in:	Approximate mean incidence†	Number of adverse reactions reported
Iron precipitate	2 500 107 4 2 2 150 100 2 (2 in 1500–2000 plus a few 'mild' reactions)	1 in 97	45
Sulphur colloid	128 300 250 1000 25 183 44 136	1 in 258	22

an adverse reaction have often been reported to have been used for the investigation of other patients on the same day without ill effect. The great variation of the apparent incidence of adverse reactions could be related to differing definitions of what constitutes such a reaction but could also be due to minor local variants of the recipe used for the on-site production of radiopharmaceuticals.

Table 20.1 gives some idea of the incidence of adverse reactions. It is taken from the paper reporting the survey referred to above [2].

Since this survey was carried out evidence has accrued which suggests that preparations of iron-precipitate for lung scanning are much less likely to give rise to an adverse reaction if the amount of iron used is reduced to a minimum. The most recent recipes take into account this finding. At the time of the survey, death had occurred in six patients in whom a lung scan had been undertaken using an iron hydroxide precipitate. In no case was a

TABLE 20.1 (contd)

MAA	6, 8, 55, 11, 108, 27 (one in eight years' experience)	1 in 36	8
DTPA	3000		1
Gold colloid	very rare		1 (some years ago due probably to gelatin then used but now discontinued)

* The radioactive tracer was either $^{99}Tc^m$ or $^{113}In^m$ in all preparations except the gold colloid, which was labelled with ^{198}Au.

† This should be considered as a 'worst figure' because it excludes the effect of the 'nil returns' where the work load was not quoted and those where the incidence was given in qualitative terms only.

cause and effect relationship established and all of the patients who died were very ill at the time the lungs were scanned.

It is likely that a lung scan using macro-aggregated albumin (MAA) has from time to time been carried out on equally ill patients but although there were adverse reactions (see table), no deaths were reported during the period of the survey. In spite of this, the user of MAA for lung scanning should be warned of the risks of using this agent and, indeed, of the possible risks of using any form of lung scanning agent, if pulmonary hypertension is present. J. O. Williams [4] reported a case of a serious reaction immediately after the intravenous injection of MAA, progressing to death six hours later. The patient had severe chronic pulmonary hypertension due to disease of the pulmonary vascular bed. Dr. Williams drew attention in his paper to the absence of similar reactions in patients where pulmonary hypertension was due to back pressure, as in mitral stenosis and in acute pulmonary hypertension caused by massive embolism.

The importance of a careful clinical evaluation of each patient referred for investigation using radiopharmaceuticals is emphasized both by this case report and by the literature quoted in references [2] and [4]. Such diagnostic agents are potentially dangerous and the fact that, in a particular user's experience, serious reactions are very rare indeed should not lull him into a false sense of security in the belief that a sudden emergency will not happen among the patients he is investigating.

The reader should, however, be aware that no data are available on the number of patients who die within (say) half an hour of presenting a clinical picture equivalent to that of the patients who died as reported in the U.K. survey and in other reports, but on whom no investigation using a radiopharmaceutical had been carried out.

The American Society for Nuclear Medicine has set up an Adverse Reaction Registry and in 1973 twenty-two adverse reactions associated with the administration of radiopharmaceuticals were reported to it. Only one of these was related to the use of an iron hydroxide preparation. In common with the report of the British survey the American report makes it clear that not all reactions were necessarily the result of the administration of the radiopharmaceutical but the time relationship was such that they *might* have been the result of such administration.

That the incidence (and possibly the severity) of adverse reactions is probably related to the minute details of the recipe used for the production of the radiopharmaceutical is supported by Australian experience. Up to about the middle of 1973 only one reaction (not fatal) had been noted in over 27 000 injections of labelled iron hydroxide produced by the Australian Atomic Energy Commission, although several reactions had been recorded where the material had been locally prepared in the hospital using it.

Attention has been drawn to the possible high radiation dose (high relative to that incidental to most investigations using radioactive isotopes) when certain nuclides are administered intrathecally but it should be stressed that this route presents hazards unrelated to radiation damage and which have resulted in complications unfortunate for the patient. Although uncommon, aseptic meningitis has been reported (sources are quoted in reference [2]). It is imperative that the highest standards are adhered to when preparing materials for intrathecal use. The radiopharmaceutical must be of the highest available specific activity to minimize the quantity of foreign material introduced into the CSF and it must be sterile and pyrogen-free. Since an agent labelled with a short-lived nuclide cannot be tested for these qualities before use, the highest standards of competence and integrity are especially required in those preparing a radiopharmaceutical for intrathecal use.

PRECAUTIONS

The most important precaution is to ensure that all staff are familiar with emergency procedures and the role that they themselves are expected to play should a dramatic complication arise. For example, it should be clearly understood by, say, a junior member of the staff that even if the patient complains of only mild discomfort, the attention of a more experienced colleague should be immediately directed to the patient. This more experienced colleague should not delay calling medical aid if there is the slightest possibility that the symptoms suggest the onset of an adverse reaction to the use of the radiopharmaceutical employed. This is one reason why it is important that Nuclear Medicine investigations should be under the close supervision of a medical man and not carried out by scientific staff with a medical man attending only to report on the results obtained.

Since the adverse reactions are similar to those of diagnostic radiological practice, the general directions laid down in such a handbook as *Notes on Radiological Emergencies* [3] should be familiar to all those using radiopharmaceuticals as diagnostic agents. It may be helpful to tabulate the main precautions:

1   Radiopharmaceuticals used should be of reputable manufacture and all local manipulations should be carried out by well-trained staff so that the material administered will be sterile and pyrogen free.

2   Every prepared dose should be clearly labelled with the name of the nuclide, the chemical and physical form of the pharmaceutical, the amount of radioactivity in milli- or micro-curies, the time and date of preparation, the name of the member of staff preparing it, and the name of the patient for whom the dose is intended.

3   A full record of the administration should be entered in the patient's notes. This should include the date on the dose container label and also the time and date of administration, the route of administration and the name of the person administering. If a reaction occurred, full details should also be included.

4   A medically qualified member of the staff should always be at hand when a radiopharmaceutical is administered.

5   An emergency tray should be available at the place where the radiopharmaceutical is administered and every person concerned with this work should know exactly where it is kept. Such a tray should contain items such as are detailed in the booklet *Notes on Radiological Emergencies* [3].

6   All staff concerned should know how to call the cardiac arrest team and what immediate action to take with the patient should cardiac arrest occur.

REFERENCES

1 Correspondence on 'Radiation Dose in Isotope Encephalography' (1968) SEAR R. & COHEN M., *Lancet* 1, 249; BULL J., *Lancet* 1, 357–358; BROCKLEHURST G., *Lancet* 1, 358; DI CHIRO G., *Lancet* 1, 526; McALISTER J.M., *Lancet* 1, 526–527.
2 WILLIAMS E.S. (1974) Adverse reactions to radiopharmaceuticals: A preliminary survey in the United Kingdom. *Br. J. Radiol.* 47, 54–59.
3 ANSELL G. (1973) *Notes on Radiological Emergencies.* 2nd edn. Blackwell Scientific Publications, Oxford.
4 WILLIAMS J.O. (1974) Death following injection of lung scanning agent in a case of pulmonary hypertension. *Br. J. Radiol.* 47, 61–63.

CHAPTER 21: DIAGNOSTIC ULTRASOUND

E. BARNETT

Ultrasound is now well established as a valuable diagnostic tool, particularly in obstetrics and gynaecology. It is the investigative method of choice in the assessment of pelvic masses [1]; a high degree of accuracy is claimed in the localization of placenta [2]; and in monitoring fetal growth by serial biparietal measurement [3-6]. In more recent years major developments have taken place in the application of ultrasonic diagnostic techniques to many other diagnostic problems.

Ultrasound has now been generally accepted as part of the diagnostic armamentarium of the radiologist and obstetrician and, in the foreseeable future, every major X-ray and obstetric department will include ultrasonic equipment. Therefore, it is important at this stage to evaluate as far as possible the safety of ultrasound, while realizing that it may be many years before conclusive evidence can be obtained in humans by observation of successive generations.

An extensive amount of experimental work has been carried out in animals and more recently in humans. Some controversial results have been obtained but, nevertheless, on the whole the conclusions are favourable.

Andrews [7] insonated frog and perch spawn using continuous ultrasound for 24 hours and failed to show any damage as far as the rate of development or macroscopic appearances were concerned. Woodward [8] exposed pregnant mice to a wide spectrum of ultrasonic intensities, pulse lengths, and pulse frequencies, and failed to show evidence of a significant hazard. Also Smyth [9] was unable to demonstrate visceral, brain or teratogenic damage in rats and mice exposed to diagnostic intensities of ultrasound. While Bleaney et al. [10] exposed CHLF hamster cells to 1.5 mHz continuous and pulsed ultrasound, and found no reduction in survival and no direct effect on cell reproductive integrity with intensities up to 8.8 W/cm^2 of continuous exposure for one hour, or with peak intensities of 15.0 W/cm^2 of pulsed ultrasound.

On the other hand, Connolly [11], using high peak intensities and longer pulse periods (200 W/cm² for 10 μs) found significant changes in litter size and an increased incidence of fetal anomalies in mice insonated during pregnancy. Similarly, Taylor & Dyson [12] insonated chick embryos using a frequency of 1 mHz. They found an increase in fetal abnormalities using a peak intensity of 40 W/cm², but this change was absent when the intensity was reduced to 10 W/cm². These authors had previously shown in 1971 that insonation of the spinal cord of rats using peak intensities of 25 or 50 W/cm² at frequencies of 0.5 mHz to 6 mHz resulted in paraplegia and/or gross haemorrhage in the cord. The ultrasound was pulsed to avoid thermal effects and 10 ms pulses were used at an interval of 100 ms. It was noted that damage was maximal at the lowest frequency used, i.e. 0.5 mHz, and decreased with increasing frequency. Using a frequency of 5 mHz, no damage was evident.

The controversial reports of experiments with ultrasound in animals had little impact upon the general medical user of diagnostic ultrasound. It was generally considered by medical workers that the intensity of ultrasound which would be required to produce tissue and cell damage in humans was far above that which would ever be used in medical diagnosis. However, the report by Mackintosh & Davey in 1970 [13] of the finding of a considerable increase in chromosome aberrations in human blood cultures exposed to continuous ultrasound from a fetal heart detector for one to two hours, caused widespread concern. This report stimulated other experimental work in humans [14–20]. Watts, Hall & Fleming [19] attempted to reproduce the findings of Mackintosh & Davey. They insonated human peripheral blood with continuous wave ultrasound and also pulsed ultrasound, within a frequency range of 1–2.5 mHz and up to 8 W/cm² total power output. They found no increase in chromosome aberrations even when the insonation time was increased to twenty hours. Similarly Abdulla *et al.* [21] found no increase in chromosome aberrations as compared with controls, by insonating human lymphocyte cultures for two to eight hours using a frequency of 2 mHz and gradually increasing the intensity from 23 mW/cm² to 3.5 W/cm². Experiments by Buckton & Baker [18] produced similar results.

Investigations into the effects of ultrasound on the fetus are of particular significance. A study of the effect of continuous and pulsed ultrasound on maternal and fetal chromosomes in patients admitted for hysterotomy for termination of pregnancy, showed no increase in the number of chromosome aberrations in blood cultures from mothers and fetuses when compared with controls [14]. Other workers found no increase in chromosome aberrations in cultures of blood from babies insonated *in utero* using a Sonicaid FM2 monitor for periods ranging from 1 hr 15 min to 9 hr 25 min [22]. The analysis of a series of 1114 patients considered to have normal pregnancies at the time of ultrasonic examination showed no increased evidence of abnormalities in exposed fetuses [23].

It cannot be denied that tissue and cell damage can be produced by high intensity ultrasound. Ultrasonic vibration can lead to appreciable temperature rise and also to the class of phenomena described by the general term 'cavitation', but the extent and specific circumstances in which each of these effects may occur, particularly in living tissue, are still very far from being established [24].

In medical diagnosis using pulsed ultrasound the range of frequencies used is 1-10 mHz. The duration of each pulse is about 1.5 μs. The intensity peak is approximately 1 W/cm^2 but this will vary with the type of apparatus. The NE Diasonograph at maximum power produces an average intensity of 0.8 mW/cm^2 but a peak intensity of 14 W/cm^2 can be achieved [14]. The Doppler apparatus used in medical practice emits a beam of 22 mW/cm^2 intensity at 10 cm.

The experimental evidence considered collectively suggests that tissue and cell damage is more likely to occur with continuous ultrasound at high intensity. It is possible that a threshold to damage may exist and such levels may not be the same for different tissues and under different conditions [25]. For example, further studies by Mackintosh & Davey [26] again using human blood cultures, showed that there was a threshold intensity of 8.2 W/cm^2 below which no chromosomal damage was detected. Taylor & Dyson [12] in their work with chick embryos also found that the reduction of the intensity to 10 W/cm^2 eliminated the occurrence of fetal abnormalities, but they had previously shown that continuous insonation with relatively low intensity ultrasound produced complete but reversible red cell arrest. Theoretically, using pulsed ultrasound, the use of high intensity, long pulse waves and short pulse intervals should be avoided. Frequency is probably not as important as intensity, although there is experimental evidence of spinal cord damage in rats maximal at 0.5 mHz and absent at 5 mHz [27].

From a practical point of view it may be accepted that there is little likelihood of fetal or maternal damage using ultrasound at the frequencies, intensity and pulse repetition frequencies normally used for medical diagnosis. However, it is probably preferable, in the first few weeks of pregnancy, to restrict ultrasonic examination to cases where there is a specific clinical indication for the examination, but under such circumstances the operator should have no hesitation in carrying out serial scans.

If any risk to the fetus does exist it is more likely to occur with continuous ultrasound at high intensity. Although the intensity of ultrasound with Doppler apparatus is relatively low it is considered logical that caution should be exhibited in the use of prolonged continuous fetal heart monitoring with ultrasound.

REFERENCES

1 MORLEY P. & BARNETT E. (1970) The use of ultrasound in the diagnosis of pelvic masses. *Br. J. Radiol.* 43, 602.
2 DONALD I. & ABDULLA U. (1968) Placentography by sonar. *J. Obstet. Gynaec. Brit. Cwlth.* 75, 993.
3 CAMPBELL S. (1968) An improved method of fetal cephalometry. *J. Obstet. Gynaec. Brit. Cwlth.* 75, 568.
4 CAMPBELL S. (1969) The prediction of fetal maturity by ultrasonic measurement of the biparietal diameter. *J. Obstet Gynaec. Brit. Cwlth.* 76, 603.
5 CAMPBELL S. (1972) Ultrasound in obstetrics. *Br. J. Hosp. Med.* 8, 541.
6 CAMPBELL S. & NEWMAN G.B. (1971) Growth of the fetal biparictal diameter during normal pregnancy. *J. Obstet. Gynaec. Brit. Cwlth.* 98, 513.
7 ANDREWS D.S. (1964) Ultrasonography in pregnancy. An enquiry into its safety. *Br. J. Radiol.* 37, 185.
8 WOODWARD B., POND J.B. & WARWICK R. (1970) How safe is diagnostic sonar. *Br. J. Radiol.* 43, 719.
9 SMYTH M.G. (1966) In *Diagnostic Ultrasound* (Ed. by GROSSMAN C.C., HOLMEW J.H., JOYNET C. & PURNELL E.E.), p. 296. Plenum Press, New York.
10 BLEANEY B.I., BLACKBURN P. & KIRKLEY J. (1972) Resistance of C.H.L.F. hamster cells to ultrasonic radiation of 1.5 mHz frequency. *Br. J. Radiol.* 45, 354–357.
11 CONNOLLY C.C. (1968) Action of ultrasound from a schock excited transducer on the foetus of the mouse. British Accoustical Society Meeting, June 1967. Quoted by HILL C.R. *Br. J. Radiol.* 41, 561–569.
12 TAYLOR K.J.W. & DYSON M. (1972) Possible hazards of diagnostic ultraound. *Brit. J. Hosp. Med.* 8, 571–577.
13 MACKINTOSH I.J.C. & DAVEY D.A. (1970) Chromosome aberrations induced by an ultrasonic foetal pulse detector. *Br. Med. J.* 4, 92–93.
14 ABDULLA U., DEWHURST C.J., CAMPBELL S., TALBERT D., LUCAS M. & MULLARKEY M. (1971) Effect of diagnostic ultrasound on maternal and foetal chromosomes. *Lancet* 2, 829–831.
15 BOYD E., ABDULLA U., DONALD L., FLEMING J.E., HALL A.J. & FERGUSON-SMITH M.A. (1971) Chromosome breakage and ultrasound. *Br. Med. J.* 2, 501–502.
16 BOBROW M., BLACKWELL N., UNRAU A.E. & DEANEY B.I. (1971) Absence of any observed effect of ultrasonic irradiation on human chromosomes. *J. Obstet. Gynaec. Br. Cwlth.* 78, 730–736.
17 COAKLEY W.T., HUGHES D.E.C., SLADE J.S. & LAWRENCE K.M. (1972) Chromosomes observations after exposure to ultrasound. *Br. J. Radiol.* 45, 328–332.
18 BUCKTON K.E. & BAKER N.V. (1972) An investigation into possible chromosome damaging effects of ultrasound on human blood cells. *Brit. J. Radiol.* 45, 340.

19 WATTS P.L., HALL A.J. & FLEMING J.E.E. (1972) Ultrasound and chromosome damage. *Br. J. Radiol.* 45, 335-339.
20 LOCH E.G., FISCHER A.B. & KUWERT, E. (1971) Effect of diagnostic and therapeutic intensities of ultrasonics on normal and malignant human cells in vitro. *Am. J. Obstet. Gynec.* 110, 457.
21 ABDULLA U., TALBERT D., LUCAS M. & MULLARKEY M. (1972) Effect of ultrasound on chromosomes of lymphocyte cultures. *Br. Med. J.* 3, 797.
22 LUCAS M., MULLARKEY M. & ABDULLA U. (1972) Study of chromosomes in the newborn after ultrasonic foetal heart monitoring in labour. *Br. Med. J.* 3, 795-796.
23 HELLMAN L.M., DUFFUS G.M., DONALD I. & SUNDEN B. (1970) Safety of diagnostic ultrasound in obstetrics. *Lancet* 1, 1133-1135.
24 HILL C.R. (1968) The possibility of hazard in medical and industrial applications of ultrasound. *Br. J. Radiol.* 41, 561-569.
25 DONALD I. (1974) The safety of using sonar. *Dev. Med. Child Nenrol.* 16, 90-92.
26 MACKINTOSH I.J.C. & DAVEY D.A. (1972) Relationship between intensity of ultrasound and induction of chromosome aberrations. *Brit. J. Radiol.* 45, 320-327.
27 TAYLOR K.J.W. & POND J.B. (1972) A study of the production of haemorrhagic injury and paraplegia in rat spinal cord by pulsed ultrasound of low mega hertz frequencies in the context of the safety for clinical usage. *Br. J. Radiol.* 45, 343-353.

CHAPTER 22: ELECTRICAL AND MECHANICAL HAZARDS IN THE X-RAY DEPARTMENT

G. R. HIGSON

ELECTRICAL HAZARDS

The effects of electric shock

The effects of electricity on the human body are dependent on the current, the frequency, the area of contact and the duration of contact. Electric shock resulting from contact with the unbroken skin of the body is known as *macroshock* and, for this condition, the effects are well understood and have been described in detail [1]. In most circumstances we are concerned with electric currents at the mains supply frequency of 50 Hz or 60 Hz and the effects perceived, on average, by a normal healthy person suffering a hand-to-hand shock may be approximately related to the value of the current according to Table 22.1.

The organ of the body which is most sensitive to electric current is the heart whose delicate control system may be disturbed by very small electric currents. When the current is applied hand-to-hand it is diffused through the body tissues and only a small fraction may find its way through the heart. However, when a patient is undergoing cardiac catheterization or monitoring via internal electrodes, the electric current may be applied directly to the heart and in this condition, known as the *microshock* condition, ventricular fibrillation may result from very small currents indeed.

There is considerable controversy concerning the value of the largest current which can be passed through the heart without causing ventricular fibrillation. The establishment of this value, which is of essential importance in the development of electrical safety standards, is impeded by the difficulty of carrying out appropriate experiments on man and because the introduction of a catheter or electrode into the heart vessels may cause fibrillation without any current being applied at all.

Most experimental work has been carried out on animals, particularly dogs, but recent work on humans has been reported by Starmer & Whalen [2] and by Raftery et al. [3]. For many years 0.1 mA has been accepted as the maximum safe level for electric currents flowing through the heart but a swing to values as low as 0.01 mA was prompted by reports in the United States of very large numbers of patient deaths from microshock. Recent opinion appears to be hardening around a value of 0.05 mA [4, 5].

TABLE 22.1. Macroshock effects of electric currents.

The level of current which is generally accepted as the standard for electrical safety of medical equipment under macroshock conditions is 0.5 mA (half the threshold of perception).

50 Hz Current	Perceived Effect
1 mA	Threshold of perception.
10–20 mA	Muscle stimulation resulting in the inability to let go.
50 mA	Pain, possibly fainting, exhaustion, mechanical injury. Heart and respiratory functions continue.
100–300 mA	Ventricular fibrillation starts. Respiratory function continues.
6 amps	Burns if current density is high. Sustained myocardial contraction followed by normal heart rhythm. Temporary respiratory paralysis.

Methods of reducing electric shock hazards

Rules governing the safe construction of electro-medical and radiological equipment have been developed in recent years and most countries have national standards containing such rules. (In the United Kingdom, detailed recommendations for electro-medical equipment were first issued in 1963 [6].) There is a large measure of agreement between the various national standards but they differ in some essential points of detail. However, a working group set up under the auspices of the International Electro-technical Commission has been working on this topic and is expected to publish within the next two years electrical safety standards for electro-medical equipment which will be accepted throughout most of the world.

The methods of securing mains operated electro-medical equipment from macroshock hazards are essentially the same as those used for other electrical equipment—primarily a combination of insulation and earthing. Double insulation, i.e. the provision of two separate layers of insulation each of which is separately capable of withstanding the imposed electric stresses, is occasionally used as an alternative to earthing. Double insulation has become popular for household appliances and portable power tools but may be ineffective in wet environments unless an isolating transformer (see below) is incorporated in the power supply. In the hospital, where the presence of blood and urine, or the need to wash down, are commonplace, earthing of metal enclosures is most commonly employed—invariably for X-ray equipment.

However, even the provision of insulation and earthing cannot be regarded as sufficient protection unless the reasons for their provision are fully understood and questions of degree are taken into consideration. In the United Kingdom, and most other countries, one wire of the electricity supply (the neutral) is kept very close to earth potential and the other wire ('live', 'line' or 'phase') is at a high potential relative to earth. In the event of a fault, this potential could appear on the case of the equipment and drive unsafe currents through anyone making a connection between the equipment and earth. One way of avoiding this danger is to use an isolating transformer in the power supply. Such a transformer removes the reference to earth of the power supply but, to ensure safety, it is necessary to monitor the degree of isolation and this complexity makes this system too expensive for normal hospital use and it will not be discussed further here. In the conventional three-wire power supply system the purpose of the earth connection is to conduct fault currents and cause a fuse or circuit breaker in the live wire to disconnect the electricity supply without allowing the equipment case to reach a dangerously high potential. A second function of an earth connection, which is less generally understood, is to conduct leakage currents.

Leakage current is the name given to undesired electric currents flowing under normal conditions in the equipment casing or other parts not intended to have any electrical connections. Part of this leakage current arises from conduction across the insulation which can never be perfect, but the majority arises from capacitive coupling between the power wiring and components and the metal casing of the equipment. Such leakage currents are generally quite small and are conducted safely away by the earth connection to the equipment casing but they can never be completely eliminated and their presence leads to serious constraints on equipment designers and on the installation and use of electro-medical equipment. In the first place, should the earth connection of the equipment break, the easiest path to earth for the leakage current may be through the body of the person handling the equipment. It is therefore necessary for equipment designers to reduce capacitive coupling and increase insulation resistance so that leakage currents do not exceed the safe levels for micro- or macroshock according to the intended use of the equipment. In the second place, even when all earth connections are secure, hazards may arise from deficiencies in the earthing system of the installation [7]. It is well known that the earth terminals of different socket outlets, even in the same room, may not be at the same potential. This is particularly so in the case of older hospitals where additional outlets may have been added at various times and connected to the original system only at some remote point, or where corrosion may have added resistance into the earth lines. In such situations it is quite possible for resistance of a few tenths of an ohm to develop between the equipment casing and 'true earth'. A fault current of only 1 amp flowing from the equipment through this earth line may raise the case of the equipment to a potential of a few tenths of a volt above that of another piece of equipment or some other metal fitting in the room. As the resistance of the human body is approximately 1000 ohms, a current of several hundred micro-amps could therefore flow through a patient fitted with low-resistance electrical connections such as internal leads if he formed part of the current path. For these reasons, it is important to ensure that rooms in which electro-medical equipment is to be used are provided with earthing systems which ensure that all earthed surfaces are held very close to the same potential.

These considerations become particularly important when a patient with internal electrodes or catheters is to be protected against the risk of microshock. In this situation the most vital technique for protection of the patient is the isolation of any equipment connected to the heart, from earth or from possible dangerous currents, by a resistance of some ten million ohms. The reason for the isolation is to ensure that neither fault currents nor leakage currents can find a path to earth through the patient's heart. If such currents should arise from faults in equipment, or if a failure of an equipment earth wire should occur, the potential of the patient may rise well above earth but he remains safe (see Figure 22.1).

Equipment for general radiography

The electrical supply for the X-ray apparatus will generally be brought into the X-ray room in the form of a 3-phase supply with, in the United Kingdom, voltages of 415 volts between the phases and 240 volts between each phase and neutral. The supply will be brought to an isolator which enables the apparatus to be completely disconnected from the supply and an earth terminal will also be provided at this point. Before connecting the X-ray apparatus to the electrical supply the installer should ensure that there is a satisfactory local earth—this is a requirement of X-ray installation contracts in the UK. The condition of an earth point can be checked by carrying out a neutral–earth loop impedance test or a line–earth loop impedance test. The neutral–earth loop test consists of injecting a current pulse along the neutral wire to the substation where the neutral is earthed, through the earth path back to the earth terminal in the room. The line–earth loop test is carried out in a similar way, the path in this case consisting of the phase line, a transformer winding, the earthed neutral point and the path to the earth terminal. These tests, which are described in detail elsewhere [8] should be repeated at intervals of not more than three years.

From this point all wiring at mains potential within the X-ray room should be contained within protective conduit or trunking. This conduit or trunking may be of metal, in which case it should be connected to earth. It may also be connected to the X-ray apparatus and used as a protective earth conductor but, because of the difficulty of checking the earth resistance resulting from such connections and because of the possibility of corrosion, particularly at joints, during the relatively long life of an X-ray installation, it is a requirement for British hospitals that each item of equipment must be earthed by a system of insulated earth continuity wires contained within these conduits or trunking and arranged to connect the apparatus to the main earth terminal in such a way that the resistance between each item of the equipment and the main earth terminal does not exceed 0.1 ohm. The X-ray equipment itself should be so constructed that all exposed metal is electrically continuous so that there is a low resistance path between every piece of the apparatus and the main earth terminal. The earth continuity of the apparatus should be checked at every routine service.

An exception to the rule that all the electrical wiring should be contained within protective conduit or trunking is provided by the high-tension cables between the X-ray generator and the X-ray tube. Because of the need to change the position of the X-ray tube easily, these cables must be flexible and a special form of cable construction has been developed. The high potential for the X-ray tube is usually supplied from a transformer/rectifier unit with its centre point earthed and two high-tension cables are used carrying voltages equally disposed above and below earth potential. These cables consist of a central core of stranded copper surrounded by a thick layer of insulating rubber. As a protection against failure of this rubber insulation, an outer metallic braiding is incorporated which is itself protected with

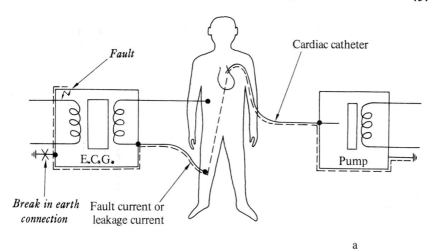

FIGURE 22.1. Prevention of microshock by electrical isolation of equipment.

a. With nonisolated equipment, a fault or leakage current may pass through the patient's heart; the fluid-filled cardiac catheter serving as a low resistance path to earth.

b. Isolated equipment offers no path to patient for currents arising from equipment faults, and no path to earth for currents arising from accidental electrical contact to patient.

a sheath of tough insulating plastic. The cables are terminated in special connectors which are designed to enable this outer braiding to be efficiently connected to the body of the tube head or transformer casing. Both the tube head and the transformer casing should, of course, be efficiently earthed quite separately.

This use of a secure and protected earth connection for all the X-ray apparatus can be applied to all fixed equipment and provides an effective protection for both patients and staff against the risk of an insulation breakdown within the apparatus—the main source of macroshock hazard. This protective earth system would, of course, normally be linked with an overcurrent protective device such as a circuit breaker in the main supply line. Although in most X-ray installations most of the equipment is fixed, it is not uncommon to find hand or foot operated switches connected to the main body of the equipment by flexible cables so that they can be used by the operator from any convenient position. Such switches are usually given very hard duty and it is difficult to ensure a reliable earth connection to them. Furthermore, the risk of breakage due to accidental dropping is so great that even all-insulated construction is not a satisfactory answer and it is advisable to employ only low voltages (e.g. 12 or 24 V) in these switches, using them to operate relays or solid-state control elements in the main circuit.

Besides the earthing arrangement that has been described, the fixing of the equipment to the structure of the room may provide one or more subsidiary earth paths for the X-ray equipment. Because of this multiplicity of earth paths it is generally impossible to determine the leakage current of the equipment, i.e. the total current flowing to earth when the equipment is operated. However, because of this very multiplicity of earth paths, this value is not important but it is important to check what might be called the effective leakage current, i.e. the current which might flow through any person—patient or operator—forming a connection between any two pieces of equipment in the room. It is therefore necessary to make current measurements between the exposed metal parts of all units in the installation and any other earthed equipment or fixtures in the room. In the normal X-ray room it is necessary that these currents should be less than 0.1 mA and this is not generally difficult to achieve. Indeed it is unusual for such measurements to exceed about 0.01 mA. Because these currents are flowing between parts which have low-resistance connections to earth, the value measured is dependent on the resistance of the measuring instrument. A resistance of 1000 ohms (approximately that of the human body) is generally regarded as acceptable for measuring instruments used for this purpose.

Equipment for cardiac catheterization

The X-ray equipment in the cardiac catheterization room should be installed in accordance with the same rules as discussed in the previous section, but in this case great attention must be paid to the measurements of leakage current flowing between different items of the installation. In the catheterization room these currents must not exceed 0.1 mA. Other items of equipment, such as an ECG, a contrast medium injector and a blood-pressure transducer, which may be connected to the patient will be supplied from socket outlets which may be on a different electrical supply circuit from that of the X-ray equipment. It is essential that line–earth or neutral–earth loop impedance measurements should be made on every socket outlet in the room and that the values for the various outlets should not differ by more than 0.1 ohm and that the earth of the socket outlets system should be linked with the earth of the X-ray equipment system. A useful guide to the testing of sockets is given in a recent issue of *Health Devices* [9].

The patient fitted with a fluid-filled cardiac catheter has a relatively low impedance connection direct to the electrically most sensitive part of his body and is particularly vulnerable to faults in any of the equipment which may be used on him or even to dangers arising from handling of the equipment or the patient. In this situation the concept of patient isolation should be rigorously applied. Until recent times it has been necessary for the high-gain amplifiers used in physiological measurement to have a low-impedance connection to earth in order to reduce the effects of interference. In equipment with such an earth connection, the leakage current could flow through the patient in the event of any fault in the earthing of the equipment. This leakage current could flow through the patient and to earth through contact between any part of the patient's skin and earthed metal, through the earth connection on the output terminal of another piece of monitoring equipment, or through contact with any other person who may himself be touching the earthed case of a piece of equipment. Modern electronic techniques have permitted improvements in the design of sensitive amplifiers, and patient monitoring equipment can now be obtained in which the circuit connected to the patient is itself isolated from the earthed metalwork of the casing and from the electrical power supply by impedances of several million ohms.

Equipment of this kind (i.e. complying with Hospital Technical Memorandum No. 8) should always be used with catheterized patients but the use of the correct type of monitoring equipment is not the only requirement for patient isolation. The metal frame of the catheterization table should have a high-resistance connection to earth suitable only for leaking away static charges but, if it is fitted with a protective earth connection it should be covered with insulating drapes as should any metal parts of the X-ray equipment which may come into contact with the patient. Because of the risk from static electrification in anaesthetizing areas there is an upper limit to the resistance of the materials to be used. A material such as cotton, which has a sheet resistivity of the order of 10^{10} ohms/square at 55 per cent relative humidity is generally recommended for drapes. It should be remembered that if these drapes become soaked with blood or other fluids they may become ineffective.

The injector pump itself must constitute neither an earth path nor the source of a hazardous electric current. The majority of those in use today use an all-plastics cylinder and piston and the resistance between the body of the contrast fluid and earth is normally about 100 million ohms.

The staff should appreciate that they themselves may unwittingly provide a current path to the patient and should avoid touching him and any electrical equipment at the same time. Preferably, they should wear rubber gloves to avoid such connections. They must always be on the alert and must investigate any untoward incidents—for instance, a disturbance of the trace on an ECG monitor may be a sign of danger [10].

Endoscopy equipment

Endoscopy equipment must have metal parts isolated from earth and, if constructed with fibre optics, should comply with British Standard 4284 [11].

Mobile X-ray equipment

The important feature of a mobile X-ray unit, from the point of view of electrical safety, is that its earth protection is dependent upon only one wire in a flexible cable. This is in marked contrast to a fixed installation with its protected strain-free earth connection and its multiplicity of subsidiary earth paths. The leakage current of a 300 mA mobile unit is about 1 mA and that of mobile image intensifiers—becoming increasingly popular for theatre and cardiac laboratory use—is of the same order of magnitude.

With leakage currents as high as these, it is clearly essential that a secure earth connection should be maintained at all times, and particularly when this equipment is to be used in the vicinity of patients with intravascular catheters. In fact, by far the majority of reported incidents with mobile X-ray units have been minor macroshock cases in which a fault on the equipment, occurring while the earth connection was broken, has allowed part of the equipment casing to become live.

No satisfactory answer has been found to this problem. Various circuits for automatically checking the integrity of the earth continuity system have been proposed but all those capable of working with the standard three-wire connections can operate only after the power has been supplied to the unit—and therefore after the danger has arisen—unless some auxiliary power supply is fitted in the equipment thus adding to its bulk and complexity. Several authorities have recommended that mobile X-ray equipments should be used only after an external auxiliary connection has been made between the frame of the X-ray equipment and some chosen earthing point. While such a routine would undoubtedly solve the earthing problem, it demands such discipline from those who are required to use this equipment in every part of the hospital, and in all kinds of circumstances, that it can never be adopted as a final solution. At the present time, the only worthwhile approach to the continued safety of mobile X-ray equipment is the institution of a regular, and frequent, preventive maintenance procedure. The weak link in the chain of safety devices is the mains connecting plug and its attachment to the trailing cable from the X-ray set. This is the component which may be dropped, twisted, forced into sockets and wrenched from them by a tug on the cable. This is the part that should be checked weekly by a competent electrician to ensure that: the three wires of the cable are securely connected to the plug terminals; the cable clamp is securely made so that no strain is placed on the internal wires of the plug by force on the cable itself; and the lengths of the wires within the plug are such that, if the cable were to be pulled out of the plug, the earth wire would be the last to be broken. The importance of a regular maintenance routine for trailing cables and their connectors cannot be overemphasized. In so many cases, accidents and near-accidents have been allowed to happen simply because of lack of attention to this simple component.

One other approach to the problem of earthing mobile units is currently being explored. A straightforward duplication of the earth connection in the power supply wiring, demanding the use of four-pin connectors would be clearly uneconomic compared with the use of the British Standard connectors now in widespread use throughout the United Kingdom. However, it does seem possible to use a double connection within the plug itself to a single earth pin which could be inserted in the standard socket. A sketch of the proposed system is shown in Fig. 22.2 and it can be seen that this allows the protective earth wire to be duplicated both at the X-ray unit and at the plug. A design exercise on a plug which would contain this feature and yet be compatible with the BS 1363 socket outlet is now in hand. Such an arrangement would undoubtedly make a significant improvement in the

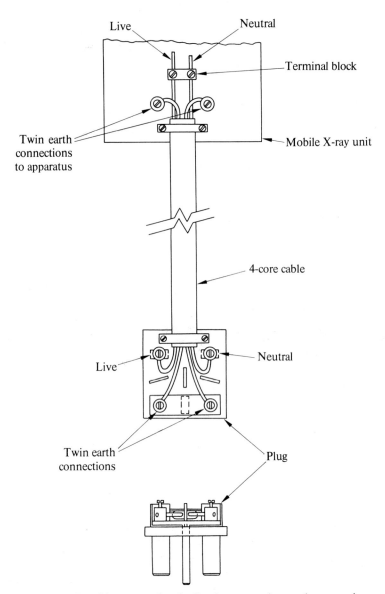

FIGURE 22.2. Possible system for duplicating protective earth connection to mobile X-ray units.

safety of mobile X-ray units but would not remove the need for frequent and regular maintenance.

MECHANICAL HAZARDS

The fact that an X-ray installation can offer mechanical hazards to patients and staff is often overlooked in the shadow of the dangers of radiation and high voltages. X-ray equipment often involves the suspension of heavy weights and the operation of complex mechanisms that call for the same high standards to be applied to the mechanical design as to the electrical circuits.

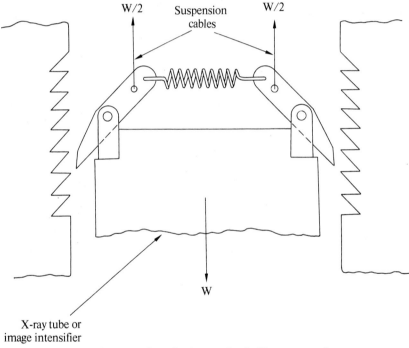

FIGURE 22.3. Incorporation of safety catches in X-ray suspension systems.

The simple stand in which the tube is supported from a column or from the ceiling is either power-operated or fitted with such good counter-balancing that it is difficult to remember that an oil-filled tube shield and diaphragm box may weigh as much as 60 kg and that this weight is generally held immediately over the patient. Such a suspension is designed to have large factors of safety but it is difficult to ensure that wear and corrosion during the life of the equipment will not defeat them, particularly if maintenance is lacking, and a second line of defence is invariably provided. This usually takes the form of either safety catches of the type shown in Fig. 22.3 or a duplicated suspension system. This must, of course, duplicate the suspension *completely*. So often have 'duplicated' suspensions been seen in which two sets of cables are finally attached by the same bolt. The tilting table is another component in which the movement mechanism has to be duplicated or otherwise constructed in a fail-safe way. In the table, particularly, great attention has to be paid to the enclosure of chains, lead-screws, etc., in order to prevent the trapping of the fingers of patients and staff.

The introduction of power-assisted movements brings certain hazards with it, in addition to its overwhelming benefits. In the complex of equipment comprising the modern screening room, collision is an ever-present possibility and, although collision guards can be fitted, they cannot be designed to cope with every contingency and it is a generally accepted rule that power-operated movements must be controlled by 'dead-man's handle' switches so that the operator's attention is demanded when movement is to be made. This principle is particularly important in remotely controlled equipment.

Temperature is not always appreciated as a possible mechanical hazard to both patients and staff. Apart from the obvious requirement for adequate ventilation of the control gear where significant quantities of heat may be dissipated, there are two locations in particular to which attention must be given. The first of these is the light-beam diaphragm box. This is often difficult to ventilate and can easily become hot enough to cause a nasty burn. For this reason a time switch is usually fitted in light beam circuits to limit the time for which the light can be switched on.

The second important location is, of course, the X-ray tube itself. The rotating-anode tube in its oil-filled enclosure has been developed over many years both to improve the dissipation of heat and to allow safe working at higher temperatures. The tube housing is not usually regarded as a touchable part and most radiologists will be well aware of how hot it becomes. They may well forget, though, that the oil-filled tube shield is a sealed vessel and that steps have to be taken to limit high pressures resulting from expansion of the oil. It is standard practice to include a bellows in the oil chamber to allow for expansion of the oil and, in the United Kingdom, it is a requirement that movement of this bellows should eventually operate a switch which opens the HT circuit to the X-ray tube. This device can offer protection for the tube housing against abuse of an X-ray set by too-frequent exposure but, because of the time taken for heat to transfer from the anode to the oil, it is too slow acting to be a safeguard against continuous excitation of the tube resulting from a fault in the control mechanism. Great attention is, of course, given to the reliability of the exposure control circuits and, indeed, they are often duplicated. To cope with the possibility of failure the Siemens company have produced their 'Loadix' device [12]. This consists of a sensor built into the tube shield which examines the under-side of the anode disc. The colour of the disc is electronically monitored and used as an indication of the anode temperature. The signals can be used either to give warnings to the operator or to de-energize the X-ray tube automatically.

With the exception of the 'Loadix' all the mechanical safety features which have been discussed are merely normal good engineering practice yet it is noteworthy that more accidents and near-accidents result from mechanical failure than from either electrical or radiation defects. Undoubtedly many of these failures have resulted from inadequate or even non-existent maintenance. Failure to lubricate can allow rapid wear in wire suspension ropes and corrosion of supposedly secure fastenings. Inadequate periodic inspection fails to reveal the broken strands of the wire ropes, structural damage and the loose, or missing nuts. In this area, more than any other, the safety of equipment which is intended to have a life of the order of ten years is dependent on regular and thorough maintenance.

SOME INCIDENTS WITH X-RAY EQUIPMENT

The need for the stringent protective measures discussed previously may be illustrated by discussion of some of the X-ray equipment fault incidents which have been investigated by the Scientific and Technical Branch of the Department of Health and Social Security during the past two years.

1   A radiographer received a severe electric shock while making a radiographic exposure on a mobile X-ray unit. The radiographer was holding the metal handswitch of the X-ray unit and was also in contact with the patient, who also suffered a mild electric shock, and possibly with other earthed objects. Investigation showed that two faults were present. The exposure counter had been incorrectly wired between line and earth instead of between line and neutral so that when an exposure was made a current of some 30 mA flowed through the casing and earth path of the unit. This fault had obviously been present since the machine was manufactured but had not been previously detected as the fault current had always flowed harmlessly to earth but, on this occasion, the earth wire had become detached from its terminal in the mains connecting plug and the electric current found its way to earth through the body of the unfortunate radiographer.

This is only one of many incidents associated with a detached earth wire in the mains connecting plug of a mobile X-ray unit. These have usually been associated with poor clamping of the cable to the plug housing and emphasize the need for regular and frequent attention to be given to these plugs.

2   A dental nurse received an electric shock as a result of damage to the all-insulated handswitch of a dental X-ray unit. The switch had been dropped to the floor causing the insert to fall out of the switch body. Live terminals at mains potential were exposed and were touched by the nurse when she picked up the switch. This design of handswitch has subsequently been modified to improve the security of fixing the insert within the body but the incident illustrates the hard usage to which handswitches are subjected and the desirability of operating them at low potentials.

3   A radiographer suffered a fatal electric shock while moving a mobile X-ray unit. This unit was equipped with an extension cable which was stored on the machine. The mains supply cable was connected to the extension reel through a metal-bodied connector mounted on a small metal box. The earthing tag, normally fitted to this connector, was missing and both the body of the connector and the box on which it was mounted were therefore isolated from earth. The insulating liner, normally fitted between the body of the connector and the cable terminations, was also missing. Investigation showed that the body of the connector and the metal box on which it was mounted had become 'live' either from an overlong assembly screw touching one of the mains terminals or a cable strand touching the metal body. It is assumed that the radiographer touched these live parts with her knee while firmly gripping the handles of the unit which were correctly connected to earth. Connectors of the type involved in this incident have been removed from service.

4   A radiographer was exposed to X-radiation, at low dose-rate, for a period of between ten and fifteen hours because of a fault in the radiographic contactor which allowed some auxiliary contacts to be made permanently. In this condition primary voltage was applied, via the surge-limiting resistors, to the HT generator whenever the set was switched on and, as the filament was continuously supplied with a stand-by voltage, X-rays were emitted continuously. This fault condition bypassed the normal indications of exposure and the fault was only discovered because of fogging of X-ray films.

5   An anaesthetized patient narrowly escaped injury when an X-ray tube burst during a series of radiographs. One end cap of the tube shield was blown off, hot oil was ejected from the shield and a small fire occurred.

Fortunately, no one was injured in what could have been a serious accident. Gradual deterioration in the control circuits of the generator had allowed a situation to arise in which the unit failed to terminate exposure while 'prepare' was selected. When making single exposures this fault was only manifest in excess blackening of the film—a fault which had been complained of by the radiographers but not recognized as an equipment defect. However, in the circumstances of the explosion, a rapid series of radiographs was being made and the generator was held permanently in the 'prepare' condition. The resulting continuous exposure caused the anode of the X-ray tube to be raised to very high temperature and it is thought that this excessive temperature caused the glass of the X-ray tube to crack allowing oil onto the anode surface. The oil then dissociated to form hydrogen and carbon and the rapid hydrogen formation burst the tube and the tube-shield. Although the tube-shield was fitted with a pressure switch, it should be appreciated that the rate of transfer of heat from the anode to the bulk of surrounding oil was too slow to allow this switch to be an effective safeguard for this kind of accident. In circumstances such as these, where two protective systems have failed at the same time, a direct measurement of anode temperature (as by the 'Loadix' device previously mentioned) or the integration of power supplied to the tube would provide a useful back-up.

6 A patient narrowly escaped injury when the suspension cable of an image intensifier fractured during an examination. The equipment was fitted with safety catches similar in principle to those illustrated in Fig. 22.3 and, although these did not hold the image intensifier, they arrested its motion so that it fell only slowly until finally it reached the stops on the guide rails. Laboratory investigation of the broken cables suggested that the failure was almost certainly due to metal fatigue accelerated by the continued bending of the cable over the support pulley and the complete lack of lubrication.

A similar failure of two wire ropes supporting a counterweight allowed a serial changer to fall onto the abdomen of a patient, fortunately without causing serious injury. This was also attributed to metal fatigue. It was estimated that the cables had been subjected to some 300 000 stress reversals before the failure occurred. Comparative tests on wire ropes, which were carried out as part of the investigation into this incident, demonstrated that adequate lubrication can improve the endurance of a wire rope by factors ranging from two to sixteen.

It is likely that both of these incidents, and others associated with similar suspension failures, could have been avoided by regular and effective maintenance. It is essential for all suspension systems, and wire ropes in particular, to be regularly inspected for fraying or other signs of damage and to be well lubricated.

7 A ceiling-mounted tube stand was forced off its suspension rails when struck by the end of a screening table which had been tilted into the vertical position. The tube stand fell to the floor without causing any injuries. The table was of an old design and did not incorporate the 'dead-man's handle' switch on the table movement. The table had been switched into the 'tilt' condition and allowed to run on until it struck the tube-stand.

8 A floor-to-ceiling tube column fell across a trolley, narrowly missing a patient. The tube column was adjustable, to allow for varying ceiling heights, and, after adjustment to the correct height, had been fastened by friction clamping. The unit had been in service for thirteen years and the friction clamping had become loose with the passage of time and had not been corrected by adequate maintenance. Eventually the clamp slipped allowing the top bearing to leave its track. More recent models of this unit rely on bolting the two parts of the tube column directly together after adjustment of length.

9 A nurse holding a baby undergoing a barium examination suffered leg injuries when the table was tilted downwards onto her knees. Efforts are still being made to provide shields and/or contact switches to these and other similar types of tables but it is impossible to design safety measures for every contingency. The most important factor in preventing accidents of this kind is the appreciation by all members of the staff of the possible dangers involved, and their continued vigilance.

This vigilance, together with a rigorous programme of regular inspection and maintenance, is the best guarantee that X-ray equipment, designed according to the principles described in this chapter, can be kept safe for both patients and staff during ten or more years of life in the X-ray department.

REFERENCES

1 DALZIELL C.F. (1972) Electric shock hazard. *IEEE Spectrum* 9, 41–50.
2 STARMER C.F. & WHALEN R.E. (1973) Current density and electrically induced ventricular fibrillation. *Medical Instrumentation* 7, 3–6.
3 RAFTERY E.B., GREEN H.L. & YAKOUB M.H. (1975) Disturbances of heart rhythm produced by 50 Hz leakage currents in human subjects. *Cardiovascular Res.* 9, 263–265.
4 STARMER C.F. (1973) New safe limit? *IEEE Spectrum* 10, 16.
5 WATSON A.B., WRIGHT J.S. & LOUGHMAN J. (1973) Electrical thresholds for ventricular fibrillation in man. *Med. J. Aust.*, 1, 1179–1182.
6 Hospital Technical Memorandum No. 8. *Safety code for electro-medical apparatus.* HMSO, London (1963, revised 1969).

7 ALBISSER A.M., PARSON I.D. & PASK B.A. (1973) A survey of the grounding systems in several large hospitals. *Medical Instrumentation* 7, 297–302.
8 Regulations for the Electrical Equipment of Buildings (14th Edition). The Institution of Electrical Engineers, London (1966).
9 *Health Devices* 3, 151–158 (1974).
10 ROWE G.G. & ZARNSTORFF W.C. (1965) Ventricular fibrillation during selective angiocardiography *J. Am. med. Ass.* 192, 105–108.
11 BS 4284: *Fibrelight cables and fittings for surgical equipment*. British Standards Institution, London (1972).
12 FRIEDEL R. & HABERBRECKER K. (1973) Increased output of rotating-anode X-ray tubes. *Electromedica* 41, 198–201.

CHAPTER 23: NOTES ON RADIOLOGICAL EMERGENCIES[1]

G. ANSELL

MAJOR EMERGENCIES IN THE X-RAY DEPARTMENT

An alarm system should be present in each X-Ray room so that, in the event of an emergency, the radiographer can summon medical assistance without leaving the patient. Primary treatment should be directed to life-saving procedures, and a rapid assessment should be made to determine which systems are predominantly at risk, e.g. cardiovascular, respiratory or central nervous systems.

RAPID SYSTEM GUIDE

Cause	System	Complication	Treatment	

Contrast media
Anaesthesia
  local
  general
Cardiac catheterization
Pneumography
Coronary disease
Electric shock etc.
	C V S	Cardiac arrest → asystole / ventricular fibrillation	Cardiac massage etc.
			Defibrillation
				O X Y G E N

		Hypotension, syncope	Horizontal posture Vasopressor drugs etc.	
		Pulmonary oedema	Sit up if possible aminophylline I.V. Lasix I.V. ? venesection, ? positive pressure ventilation ? morphine & atropine I.V.	
	R S	Respiratory arrest or obstruction	Maintain airway → natural / artificial Pulmonary ventilation ? nikethamide I.V.	I N
Local anaesthetic	C N S	Toxic convulsions	Valium I.V.	A L L
Contrast media		Coma, cerebral oedema	corticosteroid I.V. 20 per cent mannitol I.V.	
Anaphylactoid Contrast media and other drugs		Hypotensive collapse	corticosteroid I.V. (see also CVS)	C A S E S
		Bronchospasm	adrenaline S.C. aminophylline I.V. corticosteroid I.V.	
		Angioneurotic oedema Iodism	adrenaline S.C. antihistamine I.V. corticosteroid I.V. ? laryngostomy	
Pneumography		Gas embolism	Left lateral decubitus Head down (see also CVS)	

[1] 'Notes on Radiological Emergencies' is also available in a form suitable for hanging in the X-ray room as a spiral-bound collection of coloured cards with a flip index. ISBN 0 632 00401 0.

Chapter 23

CARDIAC ARREST

Diagnosis	Treatment
1 Sudden coma	1 Raise legs
2 Absent pulse	2 Thump praecordium at second intervals. If no response within *15 seconds* of onset
3 Dilated pupils	
4 General appearance (gasping, pallor, cyanosis, convulsions, etc.)	3 *External cardiac massage* — PULMONARY VENTILATION
	4 Send for assistance, ECG and defibrillator

External Cardiac Massage

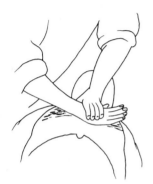

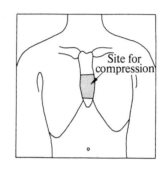

Infants Midsternal compression with thumb 100 times per minute

1 Patient supine on a *hard* surface.
2 Restrict pressure to lower sternum; sternum should move 3–4 cm.
3 60 compressions per minute; allow sternum to come up between compressions.
4 Concurrent pulmonary ventilation but do not inflate lungs at the same time that the sternum is depressed: Alternate 1 chest inflation with 6 sternal compressions.
5 Check response: pulse, pupils, etc. If not satisfactory, review technique immediately. If still not effective, proceed to direct cardiac massage (q.v.).

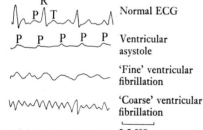

(Courtesy *British Journal of Radiology*)
ECG differentiation of ventricular asystole and ventricular fibrillation.
Note: Electrical activity does not exclude 'cardiac arrest'.

I.V. DRUGS DURING CARDIAC MASSAGE
(massage arm or wash through with drip)

ASYSTOLE
Calcium chloride (1 per cent) 5–10 ml (infants 1–2 ml).
or
* Adrenaline (epinephrine) (1/1000) 0.5–1 ml (infants 0.1–0.2 ml).
(improve cardiac tone; may induce ventricular fibrillation or change 'fine' fibrillation to 'coarse'. Can be repeated at 3–4 minute intervals.)

HYPOTENSION
* Aramine 0.05–0.5 ml (can be repeated at 10 minute intervals).

ACIDOSIS
Sodium bicarbonate (8.4 per cent) 50–150 ml.

* See section on drug interactions page 485.

External Defibrillation

1 Cardiac massage, oxygenation and adrenaline to produce 'coarse' fibrillation.
2 Precaution against accidental electrocution of operator or assistants.
3 Electrodes covered with ECG jelly, at apex and base of heart.
4 Initial shock 100 joules DC. Increase if necessary up to 400 joules, or multiple shocks (children 50–200 joules).
5 Continue cardiac massage and pulmonary ventilation as required.
(*Pacing*: 50–150 v 80–100 beats per min. Children 20–100 v.)

Direct Cardiac Massage

1 Positive pressure ventilation.
2 Incision left 4th, 5th or 6th intercostal space, from one inch lateral to border of sternum extending well back into axilla.
Avoid internal mammary artery.
3 Retract ribs.
4 Incise pericardium anterior to phrenic nerve.
5 Compress heart between palmar surfaces of *base* of fingers NOT tips of fingers.
6 Drugs for asystole (see opposite page) into ventricle.

INTERNAL DEFIBRILLATION

1 Cardiac massage, oxygenation and adrenaline to produce 'coarse' fibrillation.
2 Precautions against electrocution.
3 Electrodes covered with gauze soaked in Ringer solution or saline, on each side of heart.
4 Initial shock 20 joules DC (children 5 joules). Increase if necessary or multiple shocks.
5 Continue cardiac massage and pulmonary ventilation as required.

DEFIBRILLATION BY DRUGS

1 Only if electrical defibrillator not available, or fails.
2 Lignocaine (2 per cent) 2.5–5 ml I.V. (lignocaine = lidocaine).
3 If heart does not restart, calcium chloride (1 per cent) 5–10 ml I.V.

Cardiogenic Shock

Infusion of isoprenaline (isoproterenol) 1–5 mg in 500 ml 5 per cent dextrose solution. Steroids. Mechanical ventilation if necessary.

Cerebral Oedema

I.V. mannitol; corticosteroids; ? Lasix; ? hypothermia; ? mechanical ventilation.

Chapter 23

RESPIRATORY FAILURE

Vomiting in unconscious patient: Tilt table head down. Patient on left side if possible. Aspirate pharynx.

Endotracheal intubation should only be attempted by a person experienced in the procedure.

Oxygen Therapy

Oxygen is indicated if the patient is cyanosed or shocked. If the patient is breathing spontaneously, oxygen can be administered by face mask. If respiration has ceased, start immediate treatment with expired air resuscitation (opposite page) and, as soon as possible change to resuscitation bellows or bag with oropharyngeal airway. (Enrich with oxygen as indicated.)

WITH EXTERNAL CARDIAC MASSAGE, DO NOT INFLATE CHEST DURING STERNAL COMPRESSION. ALTERNATE 1 CHEST INFLATION WITH 6 STERNAL COMPRESSIONS.

Rib-traction method of artificial respiration (where circumstances preclude standard methods). Stand at head of patient. Hook fingers under costal margins. Pull rib-cage upward and outward for inspiration. Compress rib-cage with palms of hands for expiration. Maintain airway.

Laryngeal Oedema With Obstruction

Do not attempt endotracheal intubation unless you are experienced in the procedure. If not immediately successful, perform emergency laryngotomy or tracheotomy.

Temporary laryngotomy (*not* for children). Head extended. Steady skin between cricoid and thyroid cartilages with thumb and index finger. *Transverse* incision 2 cm long, exposing cricothyroid membrane. Incise cricothyroid membrane. Insert laryngotomy tube. If tube not immediately available, insert scalpel handle and rotate to provide an adequate airway. If prolonged artificial airway required, proceed to formal tracheotomy as soon as crisis is over and close laryngotomy.

Emergency tracheotomy. Head extended, left index finger on cricoid. Short midline *vertical* incision and, by sharp dissection expose trachea to palpating finger. Enter trachea well below the cricoid cartilage. Incise at least two rings. Insert tracheotomy tube.

Transtracheal oxygenation (possible alternative to laryngotomy or tracheotomy). Head extended, large gauge needle attached to syringe inserted through cricothyroid membrane or into upper trachea. Check position by aspirating air. Disconnect syringe and administer oxygen through needle at rate 4–5 litres per min. (Guard against gas embolism.) May require assisted respiration by rib-traction method.

Persistent apnoea. Nikethamide, I.M. or I.V. may be indicated as supplementary measure. (NOT after cardiac arrest.)

Respiratory Arrest

EXPIRED AIR RESUSCITATION

1. Clear airway by swabbing with finger or by catheter suction.
2. Keep head tilted backward.
3. Check that the chest is rising with each inflation.

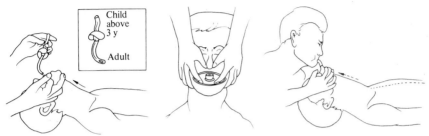

MOUTH-TO-AIRWAY METHOD

1. Take position facing top of victim's head. Tilt his head back, hold his tongue forward with fingers and insert airway *over* tongue. Insert long end for adults, short end for children. (With young children and infants use *only* the smaller paediatric-size airway or direct mouth-to-mouth breathing.)

2. Pinch victim's nose with your thumbs, press flange of airway firmly over his lips with index fingers. With other fingers hold victim's chin upward and toward yourself.

Never let chin sag.

3. Take deep breath and blow into mouth-piece. When victim's chest rises, take your mouth off to let him exhale, then blow in next breath. Blow rapidly at first, then once *every three or four seconds*. Continue until victim breathes naturally.

NOTE—*For an adult blow forcefully; for a child blow gently; for an infant, use puffs.*

THE MOUTH-TO-MOUTH METHOD OF RESUSCITATION

If the airway is not immediately available, or cannot be inserted because of a tightly closed jaw, start direct mouth-to-mouth breathing immediately.

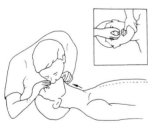

DIRECT MOUTH-TO-MOUTH METHOD IN ADULTS

Insert thumb of your left hand between victim's teeth. Hold the jaw upward so that the head is tilted backward. Close nostrils with your right hand. Take a deep breath and place your mouth tightly over victim's mouth and your own thumb. Blow forcefully enough to make his chest move. When the chest moves, take your mouth off to let him exhale passively. Repeat inflations about once every 3 to 4 seconds.

DIRECT MOUTH-TO-MOUTH METHOD IN CHILDREN (OR IN ADULTS WITH TIGHT JAW)

Grasp the angles of the child's jaw at the ear lobes with both hands and lift up forcibly so that the head is tilted backwards. Push child's lower lip toward the chin with your thumbs. NEVER LET THE CHIN SAG. Take a breath and place your mouth tightly over child's mouth (in a small child cover both mouth and nose). Blow in until his chest moves. When the chest moves take your mouth off and let him exhale passively. Repeat inflations about once every 2 to 3 seconds. In infants, use puffs

GASTROINTESTINAL TRACT

Barium impaction in colon. Administer lactulose (Duphalac) orally to soften faeces.

Gastrografin is hypertonic and may cause severe diarrhoea. In infants and debilitated adults, hypovolaemia and collapse may occur, requiring intravenous fluids. Gastrografin enema may also cause hypovolaemia requiring I.V. fluids.

Inhalation of Gastrografin may cause pulmonary oedema: oxygen, ? steroids, ? positive pressure ventilation.

Inhalation of barium. Spill-over of small amount usually does not cause major symptoms: encourage postural coughing but avoid vomiting. Inhalation of large amounts of barium may cause bronchial blockage and, if mixed with gastric contents may cause Mendelson's Syndrome: encourage postural coughing. If severe, consider bronchoscopy, antibiotics and ? steroids for late pulmonary reaction.

Barium enema perforation. The patient may appear deceptively well in the early stages. The main risk is from sudden collapse and death 10-12 hours after perforation due to septic shock and hypovolaemia.

1 Hyperhydration.
2 Colostomy and peritoneal toilet: avoid major surgery.
3 Prophylactic antibiotics (systemic and oral non-absorbable).
4 Watch for signs of shock. Monitor by central venous pressure, BP and haematocrit.
Treat by steroids, I.V. fluids, ? digoxin, ?? isoprenaline, depending on clinical indications.

Barium embolism. Very rare, usually fatal. If venous intravasation detected during fluoroscopy, tilt table feet down, lie patient on right side. Oxygen for dyspnoea.

Water intoxication due to enemas in megacolon. Patient drowsy, vomiting, cold clammy skin, BP normal, abundant dilute urine, convulsions, coma. Treat by intravenous infusion hypertonic sodium chloride. If pulmonary oedema occurs: venesection.

BRONCHOGRAPHY

In asthmatics, instillation of contrast medium may occasionally be followed by bronchial block due to bronchospasm with dyspnoea and cyanosis: oxygen, slow intravenous injection of aminophylline.

Accidental spillage of Dionosil (propyliodone) into soft tissues of the neck, during transcricoid bronchography, may cause few immediate symptoms. However, a delayed chemical inflammatory response often occurs after a few hours or days, causing dysphagia and severe neck pain. Keep under observation. If reaction occurs, treat by systemic steroids with antibiotic cover.

Allergic reactions to contrast medium: see section on iodine contrast media. Page 479

LOCAL ANAESTHETICS

The great majority of toxic reactions to local anaesthetics are due to OVERDOSAGE or excessively rapid absorption causing convulsions, status epilepticus and death. Direct cardiovascular and respiratory depression may also occur: treat as indicated in Table. In the early stage, paraesthesia, faintness, trembling or apprehension, may provide warning symptoms of an impending toxic reaction: interrupt or abandon procedure.

Rate of absorption of local anaesthetic is increased at higher concentrations.

Local anaesthetic lozenges contribute to the total dose of anaesthetic absorbed.

LOCAL ANAESTHETIC REACTIONS
PROPHYLAXIS

1 Premedication: Valium 2–5 mg orally one hour prior to examination. Atropine. 1 mg I.M. 30 minutes prior to examination.
2 Measure out dose of anaesthetic to be used before commencing, and DO NOT EXCEED MAXIMUM DOSE.
   Up to 0.2 g lignocaine (5 ml 4 per cent) for ADULT; less for frail patients and children.
3 Dilute to 1 per cent for intratracheal use.
4 Patient recumbent if possible.

TREATMENT

Toxic convulsions	Respiratory depression	Cardiovascular depression
Oxygen Valium I.V. (10 mg in 2 ml) Use smallest effective dose but increase if necessary. (Infants 0.25 mg/kg)	Maintain airway Oxygen Artificial ventilation	Posture, oxygen *Aramine 0.2–1 ml I.M. Cardiac massage

* See section on drug interactions page 485.

NOTE

1 Lignocaine = Lidocaine.
2 Citanest (prilocaine) is claimed to have only 60 per cent of the toxicity of lignocaine and the maximum dose is therefore somewhat larger. However, prilocaine may occasionally cause mild cyanosis due to methaemoglobinaemia.

True ALLERGIC reactions to local anaesthetics are very rare: treat as for contrast media reactions page 479.

Chapter 23
IODINE CONTRAST MEDIA
GENERAL CONSIDERATIONS
UROGRAPHY

Hypersensitivity reactions are unpredictable, but the risk is increased in: patients with a definite previous reaction to contrast media (30 per cent risk of reaction recurring); bronchial asthma (risk of bronchospasm probably increased with methylglucamine media); general history of allergy or hypersensitivity to other drugs; rapid injections and large doses of contrast media; high concentration sodium media (Conray 420) may cause severe arm pain.
Reactions in the older age group tend to be more severe, particularly in patients with cardiovascular disease.
Congestive failure may be exacerbated by large-dose urography.
Avoid dehydration in: infants; renal failure; diabetics with renal disease; myelomatosis.
(Note: diuretics, vomiting, etc. may predispose to dehydration.)
Urography should not be undertaken in myelomatosis unless there are compelling reasons.

CHOLANGIOGRAPHY

General principles as for urography.
Risk increased with hepatic or renal impairment (particularly with myelomatosis).
Patients undergoing oral or intravenous cholangiography should ensure good fluid intake.
Intravenous cholangiography is undesirable immediately after oral cholecystography.
Intravenous cholangiographic media should be administered very slowly, preferably by infusion over 30 minutes.
Waldeström IgM macroglobulinaemia is an absolute contraindication to intravenous cholangiography.

PROPHYLAXIS

There is no completely reliable method of preventing reactions. Preliminary intravenous antihistamine is unlikely to be of value in preventing *major* reactions, and may itself cause side effects.
Antihistamines should never be mixed with contrast media.
Where prophylaxis is considered advisable, intravenous steroids (e.g. 1-2 ml Efcortesol) can be given one hour before the contrast medium. Full precautions should be available to treat any reaction (I.V. drip, etc.). Follow-up steroids may be required at four-hourly intervals.

ROUTINE FOR UROGRAPHY
1 Enquire for history of allergy or drug idiosyncrasy.
2 Test dose 1 ml I.V.
3 Inject slowly (unless otherwise indicated). Do not re-inject blood in syringe.
4 ALL PATIENTS UNDER CONTINUOUS OBSERVATION FOR AT LEAST 20 MINUTES.

TREATMENT OF MINOR REACTIONS
1 Check compression not too tight (release if necessary).
2 Urticaria: if troublesome, may require *adrenaline (epinephrine) 0.5 ml (S.C.), *or* *antihistamine (oral or I.V.).
3 Extravenous injection: apply cold pack.
4 Vomiting: if persistent or severe: metoclopramide (Maxolon), (10 mg) 2 ml (I.M. or I.V.)
5 Abdominal colic: atropine 0.6 mg (I.M. or I.V.)

EXTRAVENOUS INJECTION
Avoid use of hyaluronidase. If swelling does not subside in 24 hours, systemic steroid therapy with antibiotic cover should be considered.

RARE MAJOR REACTIONS
Renal failure
If a moderate or severe reaction occurs following large dose urography in renal failure, treat immediate reaction as indicated in table. Institute dialysis to remove excess contrast medium.
Deterioration of renal function with oliguria or anuria may occasionally follow large dose urography in renal failure, or aortography, particularly if the patient has been inadvertently dehydrated: I.V. mannitol, high dose Lasix, or dialysis may be required. (Consult nephrologist for most appropriate course of action.)

Accidental overdose of contrast medium in infants may cause: pulmonary oedema; cardiac arrest; acidosis, hypertonic dehydration; convulsions. Institute life-saving procedures if required, administer steroids empirically, combat acidosis, treat convulsions with Valium (diazepam). If renal function is inadequate, dialysis should be considered to remove excess contrast medium.
Excessive dosage in adults may cause toxic convulsions requiring Valium.

TREATMENT OF MAJOR REACTIONS

Take immediate life-saving procedures (see table).

IN ALL SEVERE REACTIONS, WHATEVER THE SYMPTOMATOLOGY, GIVE INTRAVENOUS STEROIDS (e.g. 1-2 ml Efcortesol) AS SOON AS POSSIBLE. If no response to symptomatic therapy (see table), repeat steroids or increase dose of steroids. Repeated doses of steroids may be required at four-hourly intervals to prevent relapse.

As soon as patients condition permits, set up I.V. drip to administer drugs, etc.

SYMPTOMATIC TREATMENT FOR MAJOR CONTRAST MEDIA REACTIONS

Reaction	Oxygen	Steroids (I.V.)	*Adrenaline (S.C. or I.M.)	Aminophylline (I.V.)	Antihistamine (I.V.)	Additional measures which may be required
Hypotension	+	+	?			Posture, aramine I.M.
Cardiac arrest						Cardiac massage etc.
Respiratory failure	+	+	?			Maintain airway Pulmonary ventilation (? Nikethamide I.V.)
Bronchospasm	+	+	+	+	?	
Angioneurotic oedema	+	+	+		+	May require laryngotomy or tracheotomy
Pulmonary oedema	+	+		+		Lasix (I.V.), ? venesection Sit up if possible ? morphine and atropine I.V., ? Digoxin ? Positive and pressure ventilation
Toxic convulsions	+	+				Valium I.V. (? artificial ventilation)
Cerebral oedema	+	+				20% Mannitol I.V.
Hypertensive crisis (phaeochromocytoma)	+	+				Rogitine I.V. ? Propanolol I.V.

Steroid (100-200 mg) Efcortisol 1-2 ml (I.V.).
*Adrenaline (Epinephrine) 0.5 ml (S.C. or I.M.).
Aminophylline 0.25 g in 10 ml (I.V. slowly).
*Antihistamine: Piriton (10 mg) 1 ml (I.V.).
*Aramine 0.2-1 ml. (I.M.)
Nikethamide 1.4 ml (I.M. or I.V.).
Valium (10 mg) in 2 ml (slow I.V.) (Infants 0.25 mg/kg).

For other doses etc see 'Drug Information', p. 482.

S.C. = Subcutaneous (hypodermic). I.M. = Intramuscular. I.V. = Intravenous.

* See 'Drug interactions', p. 485.

1 Reactions may sometimes mimic myocardial infarction with transitory ECG changes, arrhythmias, conduction defects etc.
2 Prolonged hypotension may require plasma expanders in addition to steroids. Monitor by BP, haematocrit, and central venous pressure. (? noradrenaline or isoprenaline depending on clinical indications, ? digoxin.)
3 In 'allergic type' reactions, *adrenaline (epinephrine) should NEVER be given intravenously, due to the risk of causing ventricular fibrillation. (The only exception is when adrenaline is indicated in the treatment of cardiac arrest (q.v.).)
4 Aminophylline requires very slow injection to avoid hypotension.
5 Toxic convulsions requiring Valium (diazepam) should be distinguished from convulsions secondary to cardiac arrest or hypotension, which usually respond to treatment of the primary cause.
6 When the reaction has subsided, radiographs of the renal tract may show a delayed pyelogram even with small doses, and avert the necessity of a repeat examination.

CNS

Cerebral Angiography

Accidental injection of concentrated contrast medium may cause cerebral damage. As soon as suspected, commence systemic steroid therapy prophylactically. If cerebral oedema develops, continue with steroids in increased dose and also administer 20 per cent mannitol I.V.

Deterioration of cerebral function with increased intracranial pressure following routine cerebral angiography: treat as above.

Cerebral arterial spasm: intracarotid phenoxybenzamine may be of value in certain cases (see *B.M.J.* 1, 382, 1971).

Cerebral embolism following arteriography: immediate carefully controlled systemic vasopressor therapy with noradrenaline may restore neurological function. (see *Radiology* 110, 383, 1974).

SPINAL CORD DAMAGE
MYELOGRAPHY

Do not inject Myodil (Pantopaque) if CSF is bloodstained.

Late meningeal reaction: exclude infection; administer systemic steroids and antibiotic cover.

Water-soluble contrast media may produce cord irritation and muscle spasms: sit patient up; control muscle spasms by Valium I.V. In severe cases consider treatment as below.

Accidental subarachnoid injection of intravascular iodine contrast media during vertebral angiography or translumbar aortography, may cause fatal ascending myoclonic spasms.

1 Raise head to gravitate contrast medium to lumbar region.
2 Control convulsions by Valium, or anaesthesia with muscle relaxants, as necessary.
3 Wash out CSF with normal saline by cisternal and lumbar punctures, or by lumbar puncture with removal of successive 10 ml aliquots of CSF and replacement by normal saline.
4 Administer systemic steroids and prophylactic antibiotics.

Paraplegia following aortography. Treat immediately as for accidental subarachnoid injection. (Wash out CSF with saline, etc.)

LIPIODOL

Allergic reactions: treat as for contrast media p. 479.

Lipiodol embolism may be pulmonary or cerebral.

If venous intravasation noted during fluoroscopy, tilt table feet down, lie patient on right side.

If severe symptoms develop, treat as for fat embolism with heparin and systemic steroids.

Oxygen for dyspnoea.

ANGIOGRAPHY

Oral contraceptives may predispose to thrombotic complications.

ARTERIAL OBSTRUCTION

1. If needle still in artery: papaverine (I.A.) 40 mg in 20 ml saline.
2. Keep affected limb cool, warm other limbs.
3. Consider heparin, LMW dextran, fibrinolysis.
4. Urgent embolectomy.

Tolazoline hydrochloride (Priscol) has also been recommended to relieve arterial or venous spasm during catheterization. 20–50 mg injected through the catheter (infants 1 mg per kilo body weight).

AIR EMBOLISM

EFFECTS

Right heart: acute cor pulmonale.
Left heart: Coronary or cerebral ischaemia.
Vertebral veins: Cerebral ischaemia.

PROPHYLAXIS

Stethoscope strapped over praecordium during pneumography to detect characteristic churning sound.
Any change in patient's condition should be considered as possible air embolism.

TREATMENT

1. Lie on LEFT side: lower head.
2. Oxygen.
3. If no satisfactory response try external cardiac massage.
4. If there is cardiac arrest: thoracotomy, aspirate air from ventricle, direct cardiac massage.
5. Prolonged recumbency to avert recurrence.
6. Hyperbaric oxygen may be of value.

ELECTRIC SHOCK

1. Disconnect supply. If this is not possible use insulating material to move patient.
2. Prolonged artificial ventilation may be required before patient shows signs of recovery.
3. Cardiac massage, defibrillation, etc., if required.
4. Exclude fractures, dislocations, etc.
5. In less severe cases: clinical examination, ECG, reassurance and rest.

DRUG INFORMATION

Adult Doses Unless Otherwise Specified

I.A. = intra-arterial. I.C. = intracardiac. I.M. = intramuscular. I.V. = intravenous. S.C. = subcutaneous.

* See section on drug interactions page 485.

Drug	Route	Dose	Indication	Comments
*Adrenaline (epinephrine) 1/1000	S.C. I.M.	0.5 ml	Allergic type reaction, asthma etc.	Acts rapidly. 0.5 ml stat. then 0.1 ml per minute till reaction controlled.
	I.V. (I.C.)	0.5–1 ml	ONLY IN CARDIAC ARREST. To treat asystole.	Improves cardiac tone. May induce ventricular fibrillation or change 'fine' fibrillation to 'coarse'.
Aminophylline	I.V.	0.25 g in 10 ml	Bronchospasm. Cardiac asthma. Pulmonary oedema.	Inject very slowly. Vasodilatation may cause hypotension or collapse.
*Antihistamine (Piriton) (Chlor-trimeton) (chlorpheniramine)	I.V.	10 mg in 1 ml	Allergic reactions.	May cause drowsiness etc. Do not mix with contrast media.
*Aramine (metaraminol)	I.M.	0.2–1 ml	Hypotension.	Vasoconstrictor. Wait 10 minutes before repeating injection.
	I.V.	0.05–0.5 ml	Profound collapse.	
Atropine	I.V. I.M.	0.6 mg in 1 ml	Cardiac asthma. Pulmonary oedema. Also pre-op.	Decreases vagal inhibition etc. Care in glaucoma.
Calcium Chloride (1 per cent)	I.V. (I.C.)	5–10 ml	Asystole in cardiac arrest.	Cardiotonic action similar to adrenaline.
Digoxin	(I.V.)	0.5 mg in 2 ml	Acute cardiac failure especially with auricular fibrillation. ? Septic shock.	Contraindicated if patient already on digitalis. Slow injection, preferably with ECG control.
Glyceryl Trinitrate	Oral	0.5 mg Tabs	Anginal attack.	Dissolve in mouth.
Isoprenaline (isoproterenol)	I.V.	1–5 mg in 500 ml 5 per cent dextrose	Hypotensive shock with vasoconstriction.	Slow infusion. Monitor by BP and central venous pressure. Avoid if heart rate above 120. May cause arrhythmias. Cardiotonic and vasodilator action. Also bronchodilator.
Lasix (frusemide) (furosemide)	I.V. I.M.	20 mg in 2 ml	Pulmonary oedema. ? cerebral oedema. ? oliguria.	Slow injection. Separate syringe. Very rapid diuresis. Catheterize unconscious patients. Large dose may be indicated in oliguria.

DRUG INFORMATION (cont.)

Drug	Route	Dose	Indication	Comments
Lignocaine 2 per cent (lidocaine) (Xylocard)	I.V. (I.C.)	2.5–5 ml	Arrhythmias. Ventricular fibrillation.	Use preparation without preservatives. Follow by calcium chloride if heart does not respond to massage. See section on local anaesthetics (page 477).
Mannitol 20 per cent	I.V.	1.5–2 g/kg in 30–60 minutes	Cerebral oedema.	Contraindicated in over-hydration. May aggravate congestive failure, catheterize unconscious patients. (See literature for use in oliguria.)
Metoclopramide (Maxolon)	I.V. I.M.	10 mg in 2 ml	Persistent vomiting.	Very rarely causes dystonic reaction. Care if patient receiving phenothiazines.
*Morphine Sulph.	I.V.	10 mg	Cardiac asthma. Coronary infarct.	
Nikethamide (Coramine) (Nikorin)	I.M. (I.V.)	1–4 ml	Respiratory failure.	May be contraindicated in lignocaine poisoning due to convulsive effect.
*Noradrenaline (levarterenol)	I.V.	1–4 mg in 500 ml 5 per cent dextrose soln.	Persistent hypotension.	Causes marked vasoconstriction. Control by BP etc.
Papaverine	I.A.	40 mg in 1 ml	Arterial spasm.	Dilute to 20 ml with saline
Propranolol	I.V.	Up to 5 mg in 5 ml	Cardiac arrhythmias (e.g. phaeochromocytoma).	0.5 mg per min precede by atropine contraindicated in asthma, see literature.
Rogitine (Regitine) (phentolamine)	I.V.	5 mg in 1 ml	Hypertensive crisis due to phaeochromocytoma etc.	Control by BP.
Sodium Bicarbonate (8.4 per cent)	I.V.	50–150 ml	Acidosis in cardiac arrest etc.	Give by slow injection as early as possible. (Irritant.)
Steroids (Efcortesol) (hydrocortisone sodium phosphate)	I.V.	1–2 ml	Contrast media reactions. Status asthmaticus. Profound collapse especially in patients who have been treated with steroids.	May require repetition or increased dose. (If not available use alternative corticosteroid preparation.)
*Valium (diazepam)	I.V.	10 mg in 2 ml (Infants 0.25 mg/kg)	Toxic convulsions.	Slow injection. Use minimum dose. May cause hypotension or respiratory arrest.

Chapter 23

EMERGENCY EQUIPMENT FOR THE X-RAY DEPARTMENT

(A) Minimal Requirements

ALARM SYSTEM.
OXYGEN, FACE MASK, REBREATHING BAG, GUEDEL AIRWAYS, MOUTH GAG.
RESUSCITATION AIRWAYS (SAFAR or BROOK).
SYRINGES* & NEEDLES*.
SCALPELS*, SCISSORS*, HAEMOSTATS*, DISSECTING FORCEPS*.
SPHYGMOMANOMETER. STETHOSCOPE.
AMPOULES: Adrenaline, Aminophylline, Antihistamine, Aramine, Metoclopramide, Nikethamide, Normal Saline, Steroid, Valium.
TABLETS: Antihistamine, Glyceryl Trinitrate.
LIST OF EQUIPMENT: DRUG INFORMATION FILE.

(B) Supplementary List

RESUSCITATION BELLOWS. SUCTION APPARATUS & CATHETERS*. LARYNGOTOMY SET*, LARYNGOSCOPE*. ENDOTRACHEAL TUBES*. TRANSFUSION SET*.
AMPOULES: Atropine, Calcium Chloride (1 per cent), Digoxin, Isoprenaline, Lasix, Lignocaine (2 per cent), Morphine, Noradrenaline, Papaverine, Practolol, Rogitine.
BOTTLES: Normal Saline, Sodium Bicarbonate (8.4 per cent), Mannitol (20 per cent), Plasma.

(C) Equipment Available From Outside Department

ANAESTHETIC APPARATUS. ECG MACHINE.
DEFRIBILLATOR & ELECTRODES*.
THORACOTOMY DRUM*. BRONCHOSCOPE*.

* Ready Sterilized Equipment.

DRUG INTERACTIONS

Complex reactions may occasionally occur in patients who have received certain antidepressive or antihypertensive drugs within the previous two weeks.
The following drug interactions are relevant to emergency treatment in the X-Ray Department.

ANTIDEPRESSIVE DRUGS

TRICYCLIC GROUP INTERACTIONS

Sympathomimetic drugs. There may be increased sensitivity to adrenaline (epinephrine), noradrenaline (levarterenol), ephedrine and probably metaraminol (Aramine). With normal doses of these drugs, hypertensive crises may occur causing headache, cerebrovascular accidents, cardiac arrhythmias, pulmonary oedema, etc.
    Treatment: Hypertensive crisis—phentolamine (Rogitine)
                Cardiac arrhythmias—practolol.
Other drugs. The actions of antihistamines or diazepam (Valium) may be potentiated. Reactions may possibly occur with pethidine (see below).

MONOAMINE OXIDASE INHIBITOR GROUP INTERACTIONS

Sympathomimetic drugs. There may be sensitization to ephedrine and metaraminol (Aramine) causing hypertensive crises as above, but there does not appear to be significant sensitization to adrenaline (epinephrine) or noradrenaline (levarterenol), so that in these patients, noradrenaline is probably preferable to metaraminol for the treatment of shock.
Pethidine and morphine. There may be an immediate reaction with a variable syndrome (excitation, rigidity, coma, hypotension or hypertension, impaired respiration, hyperpyrexia, shock, etc.).
    Treatment: Intravenous corticosteroids and, depending on the clinical picture, one of the following may be required (in very small dosage): chlorpromazine, nalorphine or a vasopressor.
Other drugs. There may be potentiation of the actions of antihistamines, barbiturates, diazepam (Valium) and insulin. Although hyperexcitation may occur with cocaine, reactions do not, as yet, appear to have been reported with lignocaine.

GENERAL PRECAUTIONS

When therapy with interacting drugs is required small trial doses (1/10th normal dose) should be used initially and further dosage should be controlled by the clinical response.

ANTIHYPERTENSIVE DRUGS

Reserpine and similar long-acting antihypertensive drugs may cause profound hypotension during anaesthesia and require the use of larger than normal doses of vasoconstrictor drugs. These patients may be resistant to metaraminol, but they may be hypersensitive (denervation reaction) to noradrenaline. This drug should therefore be commenced at 1/10th of its normal dose and subsequent dosage controlled by the clinical response.

Tricyclic group drugs: (e.g.) amitriptyline (Triptizol, Saronten, Laroxyl, Triptafen D.A.), imipramine (Tofranil), iprindole (Prondol), nortriptyline (Aventyl, Allegron), opipramol (Insidon), protriptyline (Concordin), trimipramine (Surmontil).
Monoamineoxidase inhibitor group drugs: (e.g.) iproniazid (Marsilid), isocarboxizid (Marplan), mebanazine (Actomol), nialamide (Niamid), pargyline (Eutonyl), phenelzine (Nardil, Perfenil), pheniprazine (Cavodil), phenoxypropazine (Drazine), pivhydrazine (Tersavid), tranylcypromine (Parnate, Parstelin).

INDEX

Abdomen
  compression, contrast media reaction 9
  pain, in intravenous cholangiography 269
    during urography 12
  pseudo-acute, endoscopy 363
Abortion, and hysterosalpingography 321
Abrodil in myelography 199
Acetrizoate, excretion urography 1
Acetylcholinesterase,
  inhibition, in oral cholecystography 265
  partial, contrast media 27
Acid-base balance, and anaesthetic techniques 386
Acidosis, management 472
Adrenal
  infarction, and phlebography 234
  insufficiency, and adrenal phlebography 234
    during pneumoencephalography 184
  phlebography 233-4
Adrenaline (epinephrine)
  in contrast media reactions 32
  emergency use 482
  in renal phlebography 234
Age, and contrast media reactions 5, 6
Air embolism
  during angiography 53
  in cardiopulmonary angiography 82-3
  in gas myelography 206
  management 397-8, 481
  in phlebography 217
  in pneumoencephalography 190
  in renal arteriography 124
  Sonicaid diagnosis 398
  after ventriculography 193
  retroperitoneal pneumography, deaths 369
  see also gas embolism
Air insufflation, retroperitoneal emphysema 356
Albumin, macro-aggregated, adverse reactions 443
Alcaptonuria, false positive test after intravenous urography 25
Alcohol-dehydrogenase, partial inhibition, contrast media 27
Aldosteronism, primary, and adrenal phlebography 233
Alimentary tract
  barium appendicitis 335
    poisoning 335
    venous intravasation 352-4
  barium enema 339-60
    perforation, colonic 340-3, 345-50, 360
      vaginal 343-4, 345, 354-5
    portal gas vein 357
    retroperitoneal emphysema 355-6
    tannic acid 358-9
    water intoxication 357
  barium granuloma 351-3
  barium meal 333-6
    obstruction 334-5
    stasis 334
  barium peritonitis 350-1
  complications 333-68
  duodenography, hypotonic 336
  endoscopy 360-4
    Gastrografin 336-9
    water-soluble media 336-9
Alkaptonuria, false positive test 273

487

Althesin, and intracranial pressure changes 392
Aminophylline
  in contrast media reactions 32
  emergency use 482
Amitriptyline, drug interactions 485
Amniography 323–32
  complications, prevention 328
    fetal 324–7
    increase of circulating antibodies 326
  indications 323–4
  maternal, complications 327–8
  meconium peritonitis 327
  placental injury 326
Amplatz catheters 100
Anaesthesia
  condition of patient 382–5
  general, cerebral angiography, complication 164–9
    burst suppression 167
    electrical silence 167
    encephalograms after 165, 166
    tonsillar herniations 165
    tracheal obstruction 166
    see also Anaesthesia, techniques
  local, in bronchography 278–85
    cardiovascular depression 279
    death 278
    dose factors 280–2
    idiosyncrasy reactions 280
    overdosage 278
    reactions, prevention 282–4
    treatment 284–5
    respiratory depression 279
    toxic effects 278
  intrathecal injection, in myelography 200
  reactions, prophylaxis and treatment 477
  problems 382–403
  prolongation 387
  radiographic procedure 397–9
  techniques 385–97
    acid-base balance 386
    babies 390–1
    cardiovascular factors 392–5
    cerebral factors 390
    drug interaction 387
    hypercarbia, and hypocarbia 396
    intracranial pressure changes 391–2
    narcotic overdose 395
    premedication 386

renal function 396–7
respiratory factors 389–90
following trauma 383–4
in unconscious patients 384–5
Aneurysm, false
  during angiography 60
  during transaxillary catheterization 68
Angiocardiography 86–98
  anaesthetic difficulties 384
  arrhythmias 90–2
  cardiac catheterization 92–3
  congenital abnormalities 97
  causing death 87–8
  haemodynamic effects 86–7
  malfunction of apparatus 97–8
  neurological complications 92–3
  pulmonary complications 94–7
Angiogram, pull-out, thrombus formation 44
Angiography
  anaesthetic difficulties 384
  cardiopulmonary see Cardiopulmonary angiography
  cerebral, see Cerebral angiography
  complications, artery dissection 54–6
    bleeding 53–4
    catheter impaction 61
    catheter knotting 61
    classification 42
    embolism 52–3
    false aneurysm 60, 61
    femoral thrombosis 44–50
    gastrointestinal 60
    guide wire breakage 61
    haematoma 53–4
    infections 61
    renal see Renal
    spinal cord see Spinal cord
    thrombosis prevention 44, 49
    thrombus formation 44
    transfemoral catheterization 43
  pharmacologic adjuvants 144–7
  presedation 48
  radiation dose 421
Angioneurotic oedema, management 479
Anoxia, cerebral, during cerebral angiography 171
Antibiotics
  nephrotoxicity, and urography 24
  in renal failure in nonrenal angiography 114

Anticoagulant therapy
  needle biopsy 373
  and phlebography 216
Antidepressive drugs, interactions 485
Antihistamines
  emergency use 482, 485
  and iodipamide interaction 274
  prophylaxis, and contrast media reactions 29
Antihypertensive drugs, interactions 485
Anti-scatter grid 418
Aortic root, perforation, in transseptal left heart investigation 84
Aortography
  spinal cord injury 145
  translumbar, complications 64-5
    dissection 64
    haematoma 65
    injection into aortic branch 65
    neurological damage 65
    due to puncture 64
Apnoea, and diagnostic angiology 144
Appendicitis, barium 335
Arachnoid adhesions
  in myelography 202
  Conray and Dimer X 209
Arachnoiditis, following myelography 202
Aramine, emergency use 482
Arm
  pain, caused by contrast media 12
  phlebography 231-2
Arrhythmias
  anaesthetic technique 393
  in angiocardiography 90-2
  asystole 91
  atrial fibrillation 92
  heart block 92
  in phlebography 217
  in pneumoencephalography 186
  in ventricular fibrillation 91
Arterial
  embolism, systemic, in angiocardiography 95
  spasm, use of tolazoline 50
    treatment 481
Arterio-biliary fistula, needle biopsy, liver 374
Arterio-venous fistula, needle biopsy, kidney 374
Arteriography
  arterial obstruction, management 481

coronary, *see* Coronary arteriography
  death 43
  femoral, complications, 62-3
    dissection 63
Artery
  of Adamkiewicz, in aortography 146
  dissection, during angiography 54-6
  see also names of specific arteries
Arthritis, infective, arthrography 373
Arthrography 372-3
Ascites, contraindication to splenoportography 258
Aspiration pneumonia, peroral endoscopy 361
Aspirin
  in thrombosis prevention, angiography 47
  in uricosuric action of cholecystographic media 266
Asystole
  during angiocardiography 91
  in cerebral angiography 158
  management 472
Asystolic arrest in coronary arteriography 102
Athero-embolism, during angiography 53
Atheroma, in thrombus formation, angiography 46
Atheromatous plaques, damage during femoral arteriography 62
Atrial fibrillation, in angiocardiography 92
Atrium, right, perforation, in transseptal left heart investigation 84
Atropine
  emergency use 482
  premedication in cerebral angiography 175

Back pain, in myelography, Conray and Dimer X 208
Bacteraemia, percutaneous cholangiography 250, 254
Bacterial endocarditis, in cardiopulmonary angiography 83
Balloon catheter
  barium, enema perforations 348
  intravasation 352
  catheters, flow directed, causing ischaemia 94
Barbiturates, and intracranial pressure changes 392

Barium
  appendicitis 335
  contaminated 335
    colonic obstruction 360
    ECG changes 359
    equipment, contamination 360
    perforation 339–50
      balloon catheter 348
      carcinoma 345, 360
      colonic 340–3, 345–50
      through colostomy 349
      infants 349
      instrumentation 349
      management 476
      rectal wall abrasion 349
      vaginal 343–4, 354–5
    portal vein gas 357
    reflux, into small bowel 360
    retroperitoneal emphysema 355–6
    silicone foam 360
    with tannic acid complications 358
    water intoxication 357–8
  embolism 352
    management 476
  granuloma 351–2
  meal, impaction 335
    management 476
    obstruction 334–5
      cleared by lactulose 335
  perforation, management 476
  peritonitis 350–1
    hyperhydration 351
  poisoning 335
  sulphate, in bronchography 294
Barolith 335
Basilar artery, role in consciousness 160
Bence-Jones protein, precipitation caused by iodipamide 273
Beta-adrenergic blocking drugs, action 393
Beta-glucuronidase, partial inhibition, contrast media 27
Bile
  duct, rupture, in ERCP 256
  leakage, percutaneous cholangiography 245
  peritonitis, percutaneous cholangiography 247
Biliary cirrhosis, primary, spleno-portography 257
Biligrafin, in intravenous cholangiography 269

Biligram, in intravenous cholangiography 269
Bilivistan, in intravenous cholangiography 269
Bilopaque, in oral cholecystography 264
Biloptin, in oral cholecystography 264
Biopsy, see Bronchial brush: Percutaneous needle: Transcatheter
Bleeding
  during angiography 54
    inadequate technique 53–4
  during femoral arteriography 63
  intra-abdominal, in pancreatic phlebography 235
  from puncture site, phlebography 218
  due to splenic puncture 257
  into subarachnoid space, myelography 200
  see also Haemorrhage
Blood/bile fistulae, percutaneous cholangiography 249
Blood–brain barrier
  in cerebral angiography 154
  and contrast media toxicity 136–9
    alteration of tight junctions 137–8
Blood
  coagulation defects 17
  culture, preliminary, in percutaneous cholangiography 241
Bone infarction during phlebography 229
Brachial
  palsy, in angiocardiography 93
  plexus injury during transaxillary catheterization 68–9
Brain ischaemia, focal, treatment by vasopressor drugs 93
Bronchial
  brush biopsy 374
  tree, intubation, in ERCP 255
Bronchography 278–300
  anaesthetic problems 385
  anaesthesia, local, see also Anaesthesia, local, bronchography
  bronchospasm, management 476
  contrast media 285–95
    alteration of lung function 286–90
    barium sulphate 294
    Dionosil 291–2
    Gastrografin 294
    Hytrast 293–4
    Lipiodol 290
    Pulmidol, withdrawal 295

Bronchography (*cont.*)
   tantalum powder 294
   Visciodol 291
  local anaesthesia 278–85
  reactions following, various contrast
    media 290
  technique, complications 295–8
   contrast media in soft tissue 295
   haematoma 295
   surgical emphysema 295
  in young children 289
   halothane anaesthesia 289
   pulmonary collapse 289
Bronchospasm
  and anaesthetic techniques 387, 389
  caused by bronchography 288
  caused by contrast media 8
   aminophylline 32
   management 479
  in intravenous cholangiography 269
  management 476, 479
Bronchus, right main, intubation, risks 388
Bunamiodyl, in oral cholecystography 264
Burns, from light-beam diaphragm box 464
Burst suppression, during anaesthesia, carotid angiography 167
Bursting
  pressures, colon 345
  vein, during phlebography 223

Calcification, needle track, ventriculography 191
Calcium chloride, emergency use 482
Carbon dioxide
  increase in blood Pa,CO_2 372
  use in pneumography 369
Cardiac
  arrest, contrast media reactions,
    treatment 479
   management 472
   in phlebography 217
   treatment 394
  catheterization in angiocardiography 92–3
  equipment 459–60
  complications, of phlebography 217–18
  disorders, contrast media reaction 9
  infarction-like syndrome, oral cholangiography with pethidine 269
  intramural injection, during ventricular puncture 85
  massage, direct, management 473
   external, management 472
  puncture in aortography 64
  tamponade 397
   during angiocardiography 90
   complicating ventricular puncture 84
Cardiogenic shock, management 473
Cardiopulmonary angiography 76–110
  accidental introduction of air or infection 82–3
  angiocardiography 86–98
  contrast medium 80
  catheter problems 77–82
  coronary arteriography 98–107
  direct left ventricular puncture 84
  electrical hazard 83
  Rashkind balloon septostomy 85–6
  transseptal left heart investigation 84
Cardiovascular
  collapse during ventriculography 191–2
  depression, caused by local anaesthesia, bronchography 279
  factors, in anaesthetic techniques 392–5
Carotid angiography
  carotid sinus puncture 171
  left-sided, carotid compression after 172, 173
  right-sided, left-sided hemiparesis 162
  subintimal dissection 170
  tracings 155–8
  transitory hemiparesis 158
  wrong injections 376–9
Carotid sinus, puncture 171
Catheters
  Amplatz 100
  breakage, Rashkind 86
  buckling, in cardiopulmonary angiography 81
  in cardiopulmonary angiography, choice 77
  clotting 80–1
  in cerebral angiography complications 176
  Ducor pigtail 90
  flow-directed balloon-tip, causing ischaemia 94

Catheters (cont.)
  hypothrombogenic, use in angiography 48
  impaction, during angiography 61
  Judkins 100
  knotting during angiography 61, 81
  manipulation, causing arrhythmias 91
    in cardiopulmonary angiography 77–8
  misplacement, in cardiopulmonary angiography 78
  obstruction, cardiopulmonary angiography 82
    in cerebral angiography 172, 173
  recoil, in cardiopulmonary angiography 79
  side hole position, cardiopulmonary angiography 82
  Sones 99, 100
  tip, breakage, in cardiopulmonary angiography 82
    wedging, in pulmonary ischaemia 94
  trapping, in ventricular wall 90
Catheterization
  arterial, in vertebral angiography 153
  cardiac, in angiocardiography 92–3
  equipment 459–60
  transfemorol, angiographic approach, complications 43
  transaxillary, brachial plexus injury 68–9
    complications 66
    embolism 69
    haematoma 68
    thrombosis 67
  translumbar, in aortography 66
Cavography
  inferior vena 231
  superior vena cava 231–2
Central nervous system
  complications, in angiocardiography 92–3
  isotope hazards 440
  see also Neurological complications
Cerebral
  angiography 151–82
    anaesthesia, complications 164–9
    asystole 158
    carotid tracings 155–8
    cause of death 151–2
    contrast media 175–6
    classification of reactions 152
    contrast media, choice 152
    convulsions 168
    emergencies, management 480
    mechanical factors, complications 169
    pre-existing disease 151–2, 176–7
    technical factors, complications 169, 176
    vasomotor reflexes 154
    see also Vertebral angiography
  artery, posterior, role in consciousness 160
  cavities, development, ventriculography 191
  embolism, in angiocardiography 93
    management 480
    oil, in lymphography 307
    treatment by vasopressor drugs 93
  oedema, and contrast media 400
    management 473, 479
    and water intoxication 357
  reactions to contrast media 400
Cerebrospinal fluid, effect of Myodil 205
Chest radiography, in pregnancy 428
Chlorpromazine, in hypertension 184
Cholangiography
  infusion 270
    contrast media 270
      transport maximum 270
  intravenous 269–74
    antihistamines 274
    contrast media 269
      reactions 269
    hepatic impairment 271
    malarial relapse 274
    and oral, interaction between media 271
  meglumine iodoxamate 271
  percutaneous transhepatic 239, 240–54
    bacteraemia 250
    bile, leakage 245
      peritonitis 247
    blood/bile fistula 249
    coeliac plexus injection 244
    complications 241–54
    contraindications 241, 252
    decompression procedures 248
    fine needle 253
    general considerations 241
    haemorrhage 249
    intrapericardial injection 244

Cholangiography (*cont.*)
    intrathoracic injection of opaque
        medium 244
    pain 246
    precautions 241
    puncture, extrahepatic structures 242
    pyrexia 251
    septicaemia 250
    shock 244, 250
    subphrenic abscess 247
Cholangiopancreatography, endoscopic
      retrograde 243, 254-7
    acute pancreatitis 255
    aspiration pneumonitis 257
    duodenal wall damage 257
    general indications 239
    intubation, bronchial tree 255
    perforation, oesophagus 255
    risks 254
    rupture, bile duct 256
        oesophageal varices 255
    septicaemia 256
Cholangitis, in retrograde cholangio-
      pancreatography 362
Cholebrin, in oral cholecystography 264
Cholecystitis, gangrenous, retrograde
      cholangiopancreatography 362
Cholecystogram, caused by contrast media
    24
Cholecystography, oral 264-77
    contrast media 264
        accidental overdose 269
        deposition in stomach 269
        side effects 264-5
        uricosuric action 266
            use of aspirin 266
    drug interactions 268
    gout, precipitation 267
    jaundice 266
    liver function, impaired 265
    metabolic factors 268-9
    renal function, impaired 265
    effects on thyroid function 267
Cholesterol emboli in arteriography 123
Cholestyramine, and abnormal gall
      bladder 268
Cholografin, in intravenous cholangio-
      graphy 269
Cholovue, *see* Meglumine iodoxamate
Chromosome aberrations, after ultrasound
    448
Chylothorax in aortography 64

Chymopapain, anaphylactic reactions 376
Cine angio-cardiography, radiation dose
    421
Cinefluorography, high radiation exposure
    413
Citanest, toxicity 282
Coagulation defects, caused by contrast
      media 17
Coeliac plexus injection in percutaneous
      cholangiography 244
Collapse during pneumoencephalography
    184
    see also Hypotensive collapse
Colon
    haemorrhage, during colonoscopy 364
    perforation, barium enema, 340-3,
        345-50
        bursting pressures 345
        extraperitoneal 345
        hydrostatic pressure 347
        intraperitoneal 345
    colonoscopy 363-4
Colonoscope, incarceration 364
Colonoscopy 363-4
    perforation 363-4
Colostomy, barium enema, 349
Coma in vertebral angiography 160, 162
Conray
    myelography, complications 207-9
    in positive contrast ventriculography
        193
        reactions 194
Consciousness, impaired, in vertebral
      angiography 158, 160, 162
Contrast media
    angiography, neurotoxic actions 135
    in cardiopulmonary angiography 80
    cerebral angiography, blood-brain
        barrier 154
        choice 152
        dose schedule 175
    in coronary arteriography 100-1
    epidural injection 231
    causing gangrene 221-6
    nephrotoxicity 117-19
    neurotoxicity 135-47
        evolution 136
        gravitational effect 141
        local sludging 136
        myelomalacia 136-41
        effects of norepinephrine 143
        spinal convulsions 136

Contrast media (*cont.*)
  reactions 31, 399–400
    cerebral reactions 400
    hypertonicity 399
    major 478–9
    spinal cord damage 400
    treatment 31
    see also specific headings
  skin necrosis 221–6
  toxicity, blood–brain barrier 136
    alteration of tight junctions 137–8
  treatment of major reactions 478–9
  urography, accidental overdose 21
    adrenaline 32
    antihistamine prophylaxis 29
    idiosyncrasy reactions 2–17
      abdominal compression 9
      administration route 19
      aetiology 25–8
      allergic hypothesis 25
      allergy, history 18
      arm pain 12
      bronchospasm 8
      cardiac disorders 9
      classification 2, 3
      clinical characteristics 6
      convulsions 14
      death 3
      dose 20
      dose, incidence 21
      drug interactions 27
      ECG changes 10
      effects of age 5, 6
      enzyme inhibition 27
      fear 30
      haematological 16
      histamine release 26
      hyperthyroidism 15
      hypotension 8
      incidence 2, 4
      injection rate 20
      intermediate 3, 7
      iodide effects 14
      methylglucamine 19
      minor 3
      mucocutaneous 6
      predisposing factors 18–24
      pretesting 29
      prevention 28
      previous reactions 18
      protein-binding 27
      pulmonary oedema 11
      renal function 22
      rigors 11
      salivary gland enlargement 14
      severe 3, 7
      soft tissue changes 12
      steroids, intravenous 32
      steroid prophylaxis 30
      syncope 9
      tetani 14
      treatment 28, 30, 31
      urinalysis 25
      vomiting 8
    retrograde 33–5
Convulsions
  in cerebral angiography 168
  and contrast media 400
    reactions 14
    management 479
  during pneumoencephalography 185
  following ventriculography 194–5
Coronary
  artery embolism, in coronary arteriography 101
  arteriography 98–107
    catheters 99–100
    complications 101
      prevention 104
    contrast medium 101–1
    death 102
  ECG changes 101
  infarction, in oral cholecystography 265
  orifice, obstruction, in cardiopulmonary angiography 78
  sinus, rupture, causing death 95
  techniques 98–9
Cotton fibre embolism
  during angiography 53, 378
  in renal arteriography 124
Cranial nerve palsies, following myelography 202
Cyanosis
  and Prilocaine 282
  and Visciodol 291

Death
  cause, and X-ray examination 151–2
    angiocardiography 87–8
    catheter in coronary sinus 95
    and cerebral angiography 152–4
    in coronary arteriography 102–4

Index 495

Death (*cont.*)
    haemopericardium after ventricular puncture 87
    infusion cholangiography 273
    IV urography 3
    air embolism 369
    following intraosseous phlebography 230
    iron hydroxide precipitate 442
    local anaesthesia in bronchography 278
    Myodil ventriculography 194
    oral cholangiography 269
    pneumoencephalography 189-90
    tannic acid with barium enemas 358
Decompression procedures, percutaneous cholangiography 248
Defibrillation
    drugs 473
    external, method 473
    internal, management 473
Dehydration
    and high-dose urography 23
    hypernatraemic, after hypertonic phosphate enemas 358
    in renal failure, in nonrenal angiography 113
Dental radiography
    radiation dose 428-9
    finger injuries 408
Dermatitis
    reaction to lymphography 302
    caused by radiation 407, 408
Dextran, in thrombosis prevention, angiography 47
Diabetes
    insipidus, nephrogenic, in angiography 116
    mellitus, and high dose urography 23
    and intravenous cholangiography 272
    in renal failure, in nonrenal angiography 113
Diaginol Viscous, in hysterosalpingography 317
Dialysis, contrast media removal, in renal failure 32
Diamentor 413
Diarrhoea
    Gastrografin 476
    in intravenous cholangiography 269
    in oral cholecystography 264

Diatrizoate
    excretion urography 1
    in Gastrografin, alimentary tract diagnosis 336 *et seq.*
    in hysterosalpingography 317
    neurotoxicity 136
    precipitation 338
Diazepam
    in angiography presedation 48
    convulsion control in pneumoencephalography 185
    prevention, lignocaine-induced 283
    emergency use 483
    in myoclonic spasms 208
Diclonine hydrochloride, in bronchography 282
Digitalis
    intoxication in coronary arteriography 104
    medication, and contrast media, interaction 28
Digoxin, emergency use 482
Dimer X
    in hysterosalpingography 318
    myelography, complications 207-9
    in positive contrast ventriculography 193, 195
Dionosil, bronchography 292-3
Dipyridamole, in thrombosis prevention, angiography 47
Discitis, during discography 375
Discography 375-6
Disease, pre-existing, and cerebral angiography 151-2, 176-7
Dissection
    during aortography, translumbar 63
    coronary artery, in arteriography 101
    during femoral arteriography 62
    renal artery, in arteriography 124
    subintimal, in carotid angiography 170
Diuretics, preceding urography, reaction 23
Dizziness in oral cholecystography 264
Doppler apparatus, ultrasound 450
Dosemeter, radiation 407
Drug interactions 485
    and anaesthetic techniques 387
    antidepressive drugs 485
    antihypertensive drugs 485
    monoamine oxidase inhibitor drugs 485
    in oral cholecystography 268
    potentiation by contrast media 27

Dubin–Johnson and intravenous
    cholangiography 273
Ducor pigtail catheter 90
Duodenal wall, damage, in ERCP
    257
Dysuria, in oral cholecystography 264

Electric shock 465
    effects 452
    management 481
Electrical hazards 452–63
    cardiac catheterization 459–60
    double insulation 454
    earthing 454, 456, 465
        detached earth wire 466
    endoscopy 460
    equipment, general radiography 455–8
    isolating transformer 454, 455
    leakage currents 454, 457
    mobile X-ray equipment 460
    recommendations for electro-medical
        equipment 453
    shock 452
        macroshock 452
        microshock 452
    ventricular fibrillation 452
Electrocardiogram
    monitoring 153
    after vertebral angiography 162
Electrocardiographic changes
    in barium enema 359
    in coronary arteriography 101
    in infusion urography 10
Electroencephalogram
    after carotid angiography 162–4
    after general anaesthesia 165, 166
    monitoring 153
Electroencephalograms, after vertebral
    angiography 161
    postural hypertension 184, 185
Emboli, in renal arteriography 121–3
    cholesterol 123
Embolism
    air, see Air embolism
    in angiography 52–4
    cerebral, see Cerebral embolism
    fat, see Fat embolism
    gas, see Gas embolism
    peripheral, in 52
    pulmonary, see Pulmonary embolism
    during transaxillary catheterization 69

Emergencies
    equipment, X-ray department 484
    radiological 471–86
    rapid system guide 471
Emphysema
    mediastinal 356
    muscle, arthrography 372
    retroperitoneal, air insufflation 356
    surgical, in bronchography 295
End-hole jets, catheter, causing
    perforations 90
Endografin in hysterosalpingography 317
Endoscopy 360–4
    equipment 460
    peroral 361–3
        aspiration pneumonia 361
        biliary complications 362
        pancreatic complications 362
        perforation 361–2
Enzyme
    induction, and oral cholecystography
        268
    inhibition, contrast media 27
        iodipamide 274
Epidermal inclusion cyst, needle biopsy
    374
Epileptic seizures, see Convulsions
ERCP, see Cholangiopancreatography,
    endoscopic retrograde
Erythema, caused by radiation 409
Ethiodol, see Lipiodol
Ethylenediamine-tetracetic acid (EDTA),
    contrast media stabilization 1
Excretion urography, contrast media
    1–33
    see also Contrast media, urography
Extravasation
    contrast 231
    phlebography 216–17, 220, 221, 228
    skin necrosis 221–6
    Lipiodol, lymphography 303
        dermal backflow 303
    see also Myocardial extravasation
Eyes, radiation dose 422

Facial nerve, paralysis, caused by contrast
    media 15
Fat embolism
    complicating intraosseous phlebography
        230
    following trauma 383

Fetus
　anomalies, after ultrasound 448
　complications, amniography 324–7
　　injection into vertebral canal 326
　　skin sloughing 326
　death, in intrauterine transfusion 330–1
　heart monitoring, ultrasound 450
　in intrauterine transfusion 328–31
　　risks 330–1
Fever, following pneumoencephalography 188
Fibregastroscopy, perforation 361
Fibrescope
　infection 363
　sterilization 362–3
Fibrinogen degradation products, increased in pancreatic cancer 249
Fibromuscular dysplasia, in renal artery dissection 125
Fibrosis, caused by Thorotrast 429
Fluorescein cineangiography, causing spinal injury 136
Fluoroscopy, radiation in 420
　automatic brightness control 420
Fluroxene, and intracranial pressure changes 392
Foraminal crowding in pneumoencephalography 186
Foreign body
　granuloma, in hysterosalpingography 320
　retrieval 174
Fractures following myelography, Conray and Dimer X 209

Gall bladder
　abnormal, and cholestyramine 268
　puncture, bile peritonitis 247
　　external biliary fistula 248
　　in percutaneous cholangiography 242
　　bile peritonitis 247
Gangrene, caused by contrast media extravasation 221–6
Gas
　embolism, percutaneous needle biopsy 373
　during retrograde pneumography 369–72
　　treatment 370–2

　myelography 206
　portal vein, in barium enema 357
Gastric casts 336
Gastrografin
　alimentary-tract diagnosis 336–9
　　cathartic action 337
　　causing hypovolaemia 337
　　meconium ileus 337
　　precipitation 338
　diarrhoea, management 476
　hypersensitivity 339
　inhalation 338
　　accidental 294, 476
Gastrointestinal complications during angiography 61
Genetic effects of radiation 411
Glass particles, causing micro-emboli 378
Glove powder embolism
　during angiography 378
　renal arteriography 124
Glucose-6-phosphate dehydrogenase, partial inhibition, contrast media 27
Glucuronyltransferase, and oral cholecystography 268
Glyceryl trinitrate, emergency use 482
Gonads, male, radiation dose 425
Gout
　contrast media reaction 23
　precipitation, oral cholecystography 267
Granuloma, lipoid, following Myodil myelography 204
Great vessels, perforation, during angiocardiography 88
Guide wires breakage
　in cardiopulmonary angiography 81
　cerebral angiography 174
　retrieval 174

Haemarthrosis, during phlebography 229
Haematobilia, percutaneous cholangiography 250
Haematoma
　during angiography 53
　　inadequate technique 53
　in bronchography 295
　during femoral arteriography 62
　intramural, endoscopy 362

Haematoma (*cont.*)
   mesentery 364
   subdural, following pneumoencephalography 189
   during transaxillary catheterization 68
   during translumbar aortography 66
Haemodynamic effects of angiocardiography 86–7
Haemoglobin–oxygen equilibrium, change during cardio-angiography 17
Haemoglobinuria
   in angiography 116
   caused by contrast media 17
Haemolysis
   in angiography 116
   caused by contrast media 16
   caused by Iodipamide 274
Haemopericardium, death after ventricular puncture 84
Haemoptysis
   in lymphography 307
   percutaneous needle biopsy 373
Haemorrhage
   anaesthetic technique 392
   colonic, during colonoscopy 364
   endobronchial, percutaneous needle biopsy 373
   in hysterosalpingography 318
   percutaneous cholangiography 249
   see also Bleeding
Haemothorax
   complicating direct left ventricular puncture 84
   following rib penetration, phlebography 231
Haloperidol, in pneumoencephalography, extrapyramidal effects 185
Halothane
   and intracranial pressure change 391
   causing pulmonary collapse, children 289
Headache
   following myelography 202
      Conray and Dimer X 208
   in oral cholecystography 264
   in pneumoencephalography 185, 188
   following ventriculography 194
Heart
   block, in angiocardiography 92
   perforation, during angiocardiography 88
   in transseptal left heart investigation 84
Hemiparesis
   left-sided, after right-sided carotid angiography 162
   transitory, in carotid angiography 158
Heparin
   use following phlebography 216
   in embolism, lymphography 314
Heparinization, systemic, in angiography 49
Hepatic
   function, and contrast media reactions 22
   injection in percutaneous cholangiography 261
   in spleno-portography 262
   oil embolism, in lymphography 312
   see also Liver
Hepatotoxic reaction following iodipamide 272
Histamine release, and contrast media reactions 26
Hydrocephalus, obstructive
   in myelography 202
   in Myodil ventriculography 193
Hydrocortisone
   in adrenal collapse, pneumoencephalography 184
   emergency use 483
Hyperbilirubinaemia, and oral cholecystography 268
Hypercarbia, anaesthetic management 396
Hyperpyrexia, malignant 389
Hypersensitivity
   to Gastrografin 339
   to Dionosil, bronchography 293
   to Hytrast 294
   to Myodil 203
   to local anaesthetics, bronchography 280
   to propyliodone 339
Hypertension
   management in pneumoencephalography 184
   drug interaction 485
   pulmonary, after lung scanning 443
   in phlebography 217
   in pulmonary arteriography 95
Hypertensive crisis
   following adrenal phlebography 234
   management 479

Index

Hyperthyroidism, urographic contrast media reaction 15
Hypertonic phosphate enemas, and hypernatraemic dehydration 358
Hyperuricaemia in renal failure, in non-renal angiography 113
Hypnosis, and contrast media reactions 30
Hypocarbia, anaesthetic management 396
Hypotension
  contrast media reaction 8
    treatment 479
  management 472
  and myelography, Conray and Dimer X 208
  postural, in encephalogram chair, prevention and management 184, 185
Hypotensive collapse
  in intravenous cholangiography 269
  in oral cholecystography 265
Hypothermic perfusion, in *ex vivo* renal arteriography 129
Hypovolaemia, caused by Gastrografin 337
  cathartic effect 337
Hysterosalpingography 317–22
  abortion 318
  chemical and inflammatory reaction 320
  contrast media 317
  embolisms 320
  haemorrhage 318
  intravasation 318
  trauma to uterus and cervix 318
  tubal pregnancy, rupture 321
Hytrast
  in bronchography 293
    hypersensitivity reactions 294

Ileus
  during angiography 61
  gastrografin 337–8
Iliac phlebography 231
  iliac vein, perforation during 231
Imipramine, drug interactions 485
Immunoglobulin M paraproteinaemia, and infusion cholangiography 273
Infection
  during angiography 61

in cardiopulmonary angiography 83
in lymphography 302
pelvic, in hysterosalpingography 317
Inferior vena cavography 231
  perforation during 231
Injection materials, preparation 378
Internal jugular phlebography 235
Intra-abdominal bleeding in pancreatic phlebography 235
Intracranial pressure
  changes, anaesthetic techniques 391
  ventriculography 191
Intraosseous phlebography, lower limb 226–31
Intrapericardial injection
  in percutaneous cholangiography
  in spleno-portography 261
Intraperitoneal transfusion 329–30
Intrathoracic injection in percutaneous cholangiography 244
Intrauterine transfusion, prevention of complications 328–31
Intravasation, venous
  barium 352
  in hysterosalpingography 320
Intravenous cholangiography, effects on blood 273–4
Intubation, before carotid angiography 165
Iocetamic acid, in oral cholecystography 264
Iodamide, excretion urography 1
Iodide, and contrast media reactions 15
Iodine
  contrast media, treatment of major reactions 478–9
  free, in urographic contrast media, reaction 15
  levels, blood, in oral cholecystography 267
  protein-bound, serum, elevation due to Myodil 204
  sensitivity, treatment 309
Iodipamide
  causing blood coagulation 274
  enzyme inhibition 274
  causing haemolysis 274
  causing hepatotoxic reaction 272
  in intravenous cholangiography 269
  causing renal failure 272–3
  uricosuric action 273
  see also Methylglucamine iodipamide

Iodism
   in bronchography 291
   contrast media reaction 29
Iodized oil, *see* Lipiodol
Iodomethamate, excretion urography 1
Iodopyracet, excretion urography 1
Ioglycamate
   causing renal failure 273
   in intravenous cholangiography 269
Ion radiography, radiation dose 426
Iopanoic acid, in oral cholecystography 264
Iophendylate, *see* Myodil
Iophenoxic acid, elevated blood iodine levels 267
Iopydol and iopydone, in bronchography 293
Ipodate, in oral cholecystography 264
Iothalamate
   excretion urography 1
   neurotoxicity 136
Iozamate, in hysterosalpingography 318
Iprindole, drug interactions 485
Iproniazid, drug interactions 485
Iron hydroxide precipitate, radioactive, adverse reactions 442
Irradiation, lymphography after 312–14
Ischaemia, pulmonary, in angiocardiography 94
Isocarboxizid, drug interactions 485
Isopaque Cerebral, in cerebral angiography 152, 175
Isoprenaline, emergency use 482
Isotope techniques 439–46
   CNS hazard 440
Isotopes
   complications 439–41
   contamination 440
   pharmaceuticals 441
   precautions 445
   unsealed sources 440

Jaundice
   contraindication in oral cholecystography 266
   intravenous cholangiography, prolonged infusion 271
Joint effusion, during arthrography 372
Judkins catheter 100

Ketamine, and intracranial pressure changes 392
Kidneys
   diseased, angiography risks 126
   in *ex-vivo* renal arteriography 128–9
   see also Nephrotic: Nephrotoxicity: Renal
Klippel–Trenaunay syndrome, phlebogram 222

Lactic dehydrogenase, serum, elevation, in renal infarction 122
Largactil in hypertension 184
Laryngeal oedema, management 474
Lasix, emergency use 482
LD_{50}, of contrast media 1, 20–1
Lethal dose, *see* LD_{50}
Lethidrone, in respiratory depression 184
Leukaemia, caused by radiation 411
Levalorphan tartrate, use in respiratory depression 395
Lidocaine, in bronchography, toxic effects 278
Lignocaine
   bronchography, reactions 280
      prevention 283
      toxic effects 278
      treatment 284
   convulsions, prevention 283
      treatment 284
   dosage, bronchography 281
      endotracheal administration 281
      half-life 281
   emergency use 483
Lipiodol
   in bronchography 290
   embolism, management 480
   lungs in lymphography, management 310–12
   lymphatico-venous anastomoses 310
Livedo reticularis, in cholesterol emboli 124
Liver
   abscess, barium intravasation 352
   disease, and lignocaine in bronchography 281
   failure, in tannic acid barium enemas 358–9
   function, impaired, and oral cholecystography 265

Liver (*cont.*)
   and intravenous cholangiography 271
   effect of tannic acid barium enemas 359
   tests, abnormal in oral cholecystography 268
   pain, and intravenous cholangiography 272
Loeffler syndrome, and Hytrast 294
Lorazepam, in angiography presedation 48
Low back pain, following myelography 202, 208
Lung
   function, alteration by bronchography 286
      in lymphography 304–7
         haemoptysis 307
         impaired 305
         pulmonary oil embolism 304
         thrombophlebitis 306
   injection, in spleno-portography 261
   lymphography, complications, treatment 309–14
Lymphangiography, anaesthetic problems 385
Lymphatico-venous anastomoses, in lymphography 310
Lymphocyte transformation test, and allergic mechanism, contrast media 25
Lymphography 301–16
   allergic reactions 302
   cancer spread 308
   complication, treatment 308–15
   extravasation 303
   infection 302
   lungs in 304–7
      complications, treatment 309–14
   neurological complications 307–8
   effect on nodes 308
   oedema 303–4
   pain 302–3
Lysozyme, partial inhibition, contrast media 27

Macroshock, electrical 453, 452
Malarial relapse following intravenous cholangiography 274
Malignancy, caused by radiation 411

Mammography, radiation dose 424
   beryllium window 425
   molybdenum target 425
Mannitol, emergency use 483
   to reduce intracranial pressure 165, 186
Mebanazine, drug interactions 485
Mechanical hazards 463–8
Meconium
   ileus, treatment by Gastrografin 337
   peritonitis, in amniography 327
Meglumine
   iocarmate, in positive contrast ventriculography 193, 195
   iodoxamate, in intravenous cholangiography 271
   iothalamate, in positive-contrast ventriculography 193
      reactions 194
Membranes, premature rupture, in amniography 328
Mendelson's syndrome
   Gastrografin inhalation 338
   and stomach aspiration 383
Meningeal reactions, ventriculography 195
Meningitis
   aseptic, after intrathecal isotope administration 444
   myelography 200
   septic, following pneumoencephalography 188
   and ventriculography, Dimer-X 195
Methaemoglobinaemia
   and prilocaine 282
   and Visciodol 291
Methoxamine, in postural hypotension 184
Methoxyflurane
   and intracranial pressure changes 391
   causing renal changes 396
   in renal failure in nonrenal angiography 113
Methylglucamine
   side effects 19
   diatrizoate, *see* Diatrizoate
   iocarmate, in hysterosalpingography 318
Methylprednisolone 193
Metoclopropamide, emergency use 483
Metrizamide
   myelography 209–11
   in ventriculography, complications 195

Metrizoate
  in cerebral angiography 152, 175
  excretion urography 1
Mobile X-ray equipment 460-1
Monoamine oxidase inhibitor drugs,
    interactions 485
Morphine
  reaction in pneumoencephalography
    184
  sulphate, emergency use 483
Murphy's law 376-9
Muscle emphysema, during arthrography
    372
Myelographic block, in myelography 202
Myelography 199-214
  Abrodil 199
  bleeding from puncture 200
  Conray, complications 207-9
  Dimer X, complications 207-9
  gas 206
  local anaesthetic, intrathecal injection
    200
  management of emergencies 480
  meningitis 200
  metrizamide 209-11
  Myodil 199-206
    cerebral vasospasm 205
    cerebrospinal fluid effects 205
    complications 201-6
    elevation of serum protein-bound
      iodine 204-5
    embolism 201
    headache 205
    impotence 205
    lipoid granuloma 204
    meningeal reaction 202
    urine retention 205
    visual loss 205
  sodium diatrizoate, causing myoclonic
    spasms 207
  sodium methiodol 207
  water-soluble 207-11
Myeloma, multiple, in renal failure in
    nonrenal angiography 113
Myelomalacia, contrast media toxicity
    136-41
Myelomatosis
  and intravenous cholangiography 273
  and urography 24
Myocardium
  extravasation, during angiocardiography
    88

  in cardiopulmonary angiography 77
  infarction, in coronary arteriography
    102, 104
  perforation 397
Myoclonic spasms, following myelography,
    Conray and Dimer-X 208-9
Myodil
  embolism 201
  in ventriculography 194
  extradural, in myelography 200
  hypersensitivity to 203
  intrathecal, reactions 201-6
  in myelography 199-206
  in positive contrast ventriculography
    193-4
  complications 193
  death 194
  reactions, treatment 204
  subdural, in myelography 200
Myoglobinaemia, following sodium
    diatrizoate myelography 207

Nalorphine hydrobromide in respiratory
    depression 184
Naloxone in respiratory depression 396
Narcotic overdose 395
Neck stiffness following pneumoencephalo-
    graphy 188
Necrosis, skin, caused by contrast medium
    extravasation 221-6
Needle biopsy, see Percutaneous needle
    biopsy
Neomycin
  paralysis, retrograde pyelography 34
  precipitation of sodium alginate 35
Neoplasia, and Thorotrast 429
  see also Tumour
Nephrogenic diabetes insipidus, in
    angiography 116
Nephrograms
  dense, in acute renal failure 112
  persistent, in oral cholecystography
    266
Nephropathy, antibiotic, and urography
    24
Nephrosis, osmotic, contrast media
    reaction 23
Nephrotic syndrome in angiography
    115
Nephrotoxic drugs, and renal failure in
    nonrenal angiography 113

Nephrotoxicity, contrast media 117–19
Nerve palsies, caused by Thorotrast 429
Neuroleptanalgesia in pneumoencephalography 185
Neurological complications
 in angiocardiography 92–3
 in lymphography 307–8
 in translumbar aortography 66
Neurotoxic actions, of contrast medium in angiography 135
Nialimide, drug interactions 485
Nikethamide, emergency use 483
Nitrous oxide
 explosive hazard 372
 and intracranial pressure changes 392
Noradrenaline, emergency use 483
l-Noradrenaline, in postural hypotension 184
Norepinephrine, in contrast media neurotoxicity 143
Nortriptyline, drug interactions 485
Nystagmus, following myelography 202

Obturation arteriography 120–1
Oedema, following lymphography 303–4
 see also Angioneurotic oedema: Cerebral oedema: Pulmonary oedema
Oesophagus
 perforation in ERCP 255
  fibregastroscopy 361–2
 varices, rupture in ERCP 255
Operidine, reaction in pneumoencephalography 184
Opipramol, drug interactions 485
Orabilix, in oral cholecystography 264
Oragrafin, in oral cholecystography 264
Oral contraceptives
 and impairment of cholangiographic media 272
 and thrombus formation, in angiography 46
Orbital phlebography 235
Osmotic nephrosis, contrast media reaction 23
Osteomyelitis, during phlebography 229
Oxford fig leaf 414, 415
Oxygen therapy 474

Pain
 back, in myelography 202, 208
 in hysterosalpingography 317–18
 in lymphography 302–3
 in percutaneous cholangiography 246
  fine-needle 253
  intraperitoneal injection 246
  subcapsular injection 246
 pericardial, after ventricular puncture 84
 in sialography 374
 following spleno-portography 259–61
 see also Abdomen, pain: Arm, pain: Pericardial pain
Palsy, brachial, in angiocardiography 93
 cranial nerve, following myelography 202
 see also Paralysis
Pancreatic
 damage, retrograde pancreatography 362
 phlebography 234–5
Pancreatitis, acute, in ERCP 255, 362
 in aortography 64
Pancreatography, retrograde, pancreatic damage 362
Pancytopenia, caused by diatrizoate 15
Pantopaque, see Myodil
Papaverine, emergency use 483
Paralysis
 facial nerve, contrast media reaction 15
 neomycin, in retrograde pyelography 33
Paraplegia, post-traumatic, cord circulation 147
 see also Spinal cord complications: Contrast media neurotoxicity
Pargyline, drug interactions 485
Parotid gland enlargement, contrast media reaction 14
Patent Blue Violet
 allergic reactions 302
 complications, treatment 308–9
 in lymphography 301
Pelvis
 female, irradiation 423
 infection in hysterosalpingography 317
Pentazocine, causing respiratory depression 396
Pentothal anaesthesia, potentiation by contrast media 27
Percutaneous needle biopsy 373–4

Perforation
  barium, management 476
  cardiac, in angiocardiography 88–90
  colonoscopy 363–4
    extraperitoneal 363
    intraperitoneal 363
  fibregastroscopy 361–2
  great vessels, in angiocardiography 88–90
  myocardium 397
  right ventricle, in angiocardiography 98
Pericardial
  pain, after ventricular puncture 84
  reaction in direct left ventricular puncture 84
Peripheral artery, pressure injection into, during cardiopulmonary angiography 79
Peritonitis, barium 350–1
  hypertension 351
Pethidine
  in oral cholangiography, cardiac infarction-like syndrome 269
  reaction in pneumoencephalography 184
Phaechromocytoma, and adrenal phlebography 234
  management 479
Pharmacoangiography, risks 128
Phenelzine, drug interactions 485
Phenergan, in hypertension 184
Pheniprazine, drug interactions 485
Phenoxypropazine, drug interactions 485
Phenylephrine in pulmonary hypertension 217
Phlebography 215–38
  air embolism 217
  in anticoagulated patients 216
  arm 231–2
  ascending 218–26
  bleeding from puncture site 218
  complications, cardiac 217–18
    local 216–17
  contrast media 215
  extravasation of contrast 216–17, 220, 228
  skin necrosis 221–6
  iliac 231
  intraosseous, lower limb 226–31
  lower limb 218–26
  pulmonary embolism 218
  pulmonary hypertension 217

  retrograde 233–5
    specific sites 233–5
  causing thrombosis 216
Photofluorography, radiation dose 427
Pivhydrazine, drug interactions 485
Placenta in amniography, injury 326
  site 329
Plaster of paris, administration, accidental 336
Pneumoencephalography
  adrenal insufficiency and collapse 184–5
  anaesthetic problems 385
  complications 183–91
    in children 190
  death 189–90
  fever 188
  foraminal crowding 186–7
  headache 188
  meningitis 188
  minor complications 185
  neuroleptanalgesia 185
  post-examination complications 187–91
  postural hypotension, management and prevention 184, 185
  reaction to premedication drugs 184–5
  respiratory depression 184
  response, and inflammatory reactions, differentiation 188, 189
  subdural haematoma 189
  vomiting 188
Pneumography, retrograde
  deaths 369
  gas embolism 369–72
    treatment 370–2
Pneumomediastinum
  arthrography 373
  in pneumoencephalography 190–1
Pneumonia, aspiration 361
  caused by Hytrast 293
Pneumonitis, aspiration, in ERCP 257
Pneumoperitoneum
  contraindication to splenoportography 258
  gas embolism 372
  complicating retroperitoneal emphysema 356
Pneumothorax
  in aortography 64
  in percutaneous cholangiography 244
  percutaneous needle biopsy 373

Pneumothorax (*cont.*)
  following rib penetration, phlebography 231
  complicating ventricular puncture 84
Portal vein gas, in barium enema 357
Practolol, emergency use 483
Pregnancy
  chest X-ray, 428
  radiation, dose 407, 410-11
    indication for abortion 411
    DHSS letter and form 433-6
    radiation doses 436-8
  tubal, rupture, hysterosalpingography 321
Premedication, and anaesthetic techniques 386
Prepontine veins, rupture, in internal jugular phlebography 235
Prilocaine, toxicity 282
Promethazine hydrochloride, in hypertension 184
Propyldocetrizoate, in bronchography, contraindications 295
Propyliodone
  in bronchography 292-3
  hypersensitivity 339
  into soft tissues, management 476
Prostigmine, and water intoxication 357
Protein, false positive test, after intravenous urography 25
Protein-binding, contrast media 27
Protriptyline, drug interactions 485
Pulmidol, in bronchography, contraindications 295
Pulmonary
  arteriography in pulmonary hypertension death due to 95
  collapse, bronchography, children 289
  complications in angiocardiography 94-7
  embolism in angiocardiography 94
    barium 352
    in cardiopulmonary angiography 80
    oil in hysterosalpingography 318
    following lymphography 304
    during phlebography 217, 230
    in right ventricular perforation 89
  hypertension, *see* Hypertension, pulmonary
  oedema, anaesthetic technique 389-90
    in angiocardiography 94
    caused by diatrizoate 338

in infusion urography 11
management 479
Puncture
  of carotid sinus, during carotid angiography 171
  splenic, in splenoportography 257
  ventricular, direct left complication 84
  of vertebral artery, during cerebral angiography 171
Pyelogram, as sign of blood/bile fistula 249
  neomycin paralysis 34
  renal failure 33
Pyrexia
  percutaneous cholangiography 251
  following pneumoencephalography 188

Radiation problems 303-38
  angiography 421
  beam, collimation 416-17
  dental radiography 428-9
  doses, from diagnostic procedures 412-15
    patients 413-15
    personnel 412-13
  reduction 415-19
    anti-scatter devices, air gap and grid 418
    film screen combination 418
    kilovoltage, input exit dose ratio 419
  regional 436-8
  dosemeter 407
  eye 422
  filtration, added 417
  fluoroscopy 420-1
    automatic brightness control 420
  general effects 411
  genetic effects 411-12
  gonads, male 425
  hazards 398
    to others 412
    potential, personnel 408-9
  ion radiography 426
  leakage 416
  mammography 424
  maximum permissible dose 406-8
    ancillary workers 407-8
    indication for abortion 441
    patients 408
    pregnancy 407, 410-41

Radiation problems (*cont.*)
  public 408
  radiation workers 406–7
  monitoring 407
  off-focus, reduction 416
  pelvis, female
  pregnancy, DHSS letter and form 433–6
  small format radiography 427
  target angle 418
  thyroid, 422
  tomography 427
  units 404–6
    rad (r) 405
    g/rads 405
    R × cm² 405
    rém 405
    Röntgen (R)
  videotape 428
  xerography 426
Radioactive tracers, *see* Isotopes
Radiopharmaceuticals, adverse reactions 441–4
Rashes, following carotid arteriography 377, 378
  see also Dermatitis: Skin
Rashkind
  balloon septostomy, complications 86
  catheter breakage 86
Rectal wall, abrasion, barium enema 349
Red cells, clumping, caused by contrast media 16
Renal
  angiography, use of drugs 128
  arteriography, contrast media nephrotoxicity 117–19
    ex-vivo 128–9
    obturation 120–1
    over injection 119–20
  renal complications 117–29
  artery, dissection, in arteriography 124
    embolism, during angiography 53
  complications, of angiography 111–33
    in infants and children, 116
    rare 115
    related to technique 116
    of selective renal arteriography 117–29
  failure, contrast media reactions, dialysis 32
    and intravenous cholangiography 272
    following iodipamide 272
  in nonrenal angiography 111–15
  in oral cholecystography 265
  prevention, in nonrenal angiography 115
  in retrograde pyelography 33
  in urography, management 478
  function, and anaesthetic techniques 396–7
    and contrast media reactions 22
    deterioration 23
    impaired, in oral cholecystography 265
  infarction, elevated LDH, in renal arteriography 122
    in renal phlebography 234
  phlebography 234
  scintigraphy 120
  see also Kidney: Nephrotic: Nephrotoxicity
Renography, isotope 120
Renovascular hypertension, and renal artery dissection 125
Respiratory
  depression, caused by local anaesthesia
    in bronchography 279
    in pneumoencephalography 184
  factors, and anaesthetic technique 389
  failure, management 474, 479
Resuscitation
  expired air 475
  mouth-to-mouth 475
Retinal embolism in hysterosalpingography 318
Retrograde
  cholangiopancreatography, endoscopic, *see* Cholangiopancreatography
  phlebography 233–5
  urography, contrast media 33–41
Retroperitoneal
  emphysema, barium enema complication 355–6
  infection, pneumography 372
Rigors during urography 11
Rogitine, emergency use 483

Salivary gland enlargement, contrast media reaction 14
Salpix, in hysterosalpingography 317
Scintigraphy, renal 120
Sedation
  and anaesthetic technique 388

Sedation (*cont.*)
  excessive 373
Septicaemia
  in ERCP 256–7
  in percutaneous cholangiography 250
  during phlebography 229
Serenace, in pneumoencephalography, extrapyramidal effects 185
Septostomy, Rashkind balloon, *see* Rashkind balloon septostomy
Serum complement, activation, by contrast media 27
Shock
  in angiography, spinal cord risks 145
  cardiogenic, management 473
  electric 465
    effects 452
    due to faulty foot switch 83
    management 481
  endotoxic, in percutaneous cholangiography 244, 250
S.I. units 430–1
  gray (Gy) 430
Sialography 374–5
Sickle cell crises, caused by contrast media 17
Skin
  blue, after Patent Blue Violet 301
  radiation hazards 407
    dental X-rays 408
  rashes in intravenous cholangiography 269
  sloughing, fetal amniography 326
Sodium
  acetrizoate, causing spinal injury 136
    in hysterosalpingography 317
  alginate, precipitation by neomycin 35
  bicarbonate, emergency use 483
  diatrizoate myelography, causing myoclonic spasms 207
  methiodol myelography 207
Soft tissue changes, contrast media reaction 12
Sones catheter 99, 100
Sonicaid in diagnosis of air embolism 398
Spasm, axillary artery, during catheterization 67
Spinal cord
  complications, of angiography 134–50
    contrast media toxicity 133–47
  damage, during angiography 57–60
    incidence, presentation 57
    prevention 57
    treatment 57, 147–8
  complicating aortography 145
  caused by contrast media 400
    management 480
Spitz Holter valve, care in use 398
Spleen
  avulsion, colonoscopy 364
  in spleno-portography, *see* Splenoportography
Spleno-portography 257–62
  bleeding due to splenic puncture 257
  contraindications 258
  injection into extra-splenic organs 261
    into lung 261, 262
  pain following 259–61
  in primary biliary cirrhosis 257
  spleen, damage 258
    puncture 257
    rupture 258
Steroids
  emergency use 483
  intravenous, in constrast media reactions 32
  in lymphography complications 314
  prophylaxis, and contrast media reactions 30
Stomach, perforation fibregastroscopy 361–2
Subarachnoid space, accidental contrast media injection 400
  management 480
Subphrenic abscess in percutaneous cholangiography 247
Sulphonilamide, with Visciodol 291
Sulphonyl-ureas, and impairment of cholangiographic contrast media 272
Sulphur colloid, radioactive adverse reactions 442
Superior vena cavography 231–2
Switch, foot, faulty, shock due to 83
Syncope, contrast media reaction 9
Syringes, faulty 378–9

Tamm-Horsfall protein, precipitation by contrast medium 23
Tannic acid, in barium enemas, complications 358–9
Tantalum powder, bronchography contraindications 294

Telepaque
  accidental overdose 269
  in oral cholecystography 264
Teridax, elevated blood iodine levels 267
Tetani, contrast media reactions 14
Tetany, in intravenous cholangiography 269
Thorotrast
  local fibrosis 429
  causing neoplasia 429
  nerve palsies 429
  radiation hazards 429–30
Thrombophlebitis, in lymphography 306
Thrombosis
  angiography, incidence 46
    prevention 47
  complicating phlebography 216
  during transaxillary catheterization 67–8
  femoral, angiographic complication 44
Thrombus, formation, angiographic complication 44–6
Thyroid
  function, and oral cholecystography 267
  hyperplasia, and amniography 327
  radiation to 422
Thyrotoxicosis, and oral cholecystography 267
Tolazoline, in relaxation of arterial spasm, angiography 50
Tomography, radiation dose 426–7
  middle ear 422
Tonsillar herniation, in cerebral angiography 165
Tourniquet, too-tight, in phlebography 223, 226
Tracheal obstruction, during anaesthesia 166
Tracheotomy, emergency 474
Transcatheter biopsy 374
Transseptal
  left heart investigation, complications 84
  needle breakage in cardiopulmonary angiography 81
Transtracheal oxygenation 474
Tranylcypromine, drug interactions 485
Trauma, anaesthetic problems 383
Trichlorethylene, and intracranial pressure changes 391
Tricyclic group drugs, interactions 485

Trimipramine, drug interactions 485
Tubular necrosis, and oral cholecystography 266
Tumour spread, needle biopsy 373
  see also Neoplasia
Tyropanoate, in oral cholecystography 264

Ultrasound, diagnostic 447–51
  cavitation 449
  chromosome aberrations 448
  continuous 450
  effect on fetus 448
  fetal anomalies 448
  fetal heart monitoring 450
  threshold intensity 449
Unconsciousness
  anaesthetic problems 383–5
  in vertebral angiography 159, 160, 162
Urethrocystography, tissue toxicity, from anaesthesia and contrast media 34
Urinalysis, after intravenous urography 25
Urine
  blue, after Patent Blue Violet 301
  enzyme excretion, in contrast media nephrotoxicity 118
  retention, after myelography 205
  in myelography, Conray and Dimer-X 208
Urografin 76, in hysterosalpingography 317
Urograms, repeat high-dose, and antibiotic nephropathy 24
Urography
  contrast media 1–41
  see also Contrast media, urography
  excretion, contrast media, *see* Contrast media, urography
  retrograde, contrast media, *see* Retrograde urography
  routine, treatment of reactions 478
    accidental overdosage 478
    extravenous injections 478
    renal failure 478
Urticaria
  in oral cholecystography 264
  reaction to contrast medium 163

Vagina, perforation, barium 343–4, 345, 354–5

Valium, *see* Diazepam
Vasomotor reflexes, cerebral angiography 154
Vasopressor drugs
  in cerebral embolism 93
  in focal brain ischaemia 93
  in contrast medium neurotoxicity 143
Vasospasm, cerebral, following Myodil myelography 205
Vein
  bursting, during phlebography 223
  perforation, arm phlebography 232
  in retrograde phlebography 233
  see also specific names of veins
Venous spasm 232
  treatment 481
Ventricular
  asystole, management 472
  fibrillation, during angiocardiography 91
    cardiopulmonary 83
    in coronary arteriography 102
    management 472
  puncture, direct left, complications 84
Ventricle, right, perforation, in angiocardiography 98
Ventriculography 191-6
  anaesthetic problems 385
  cerebral cavities, development 191
  intracranial pressure 191
  meningitis, 195
  minor complications 194
  needle puncture sequelae 191
  positive contrast 193-4
    complications 193-4
    contrast media 193
  water-soluble media 194-6
Vertebral
  angiography, coma 160, 162
    impaired consciousness 158, 160, 162
    tracings 156
    unconsciousness 159, 160, 162
    see also Cerebral angiography
  artery, obstruction 172, 173
    puncture 171
Video-disc, instant replay, radiation 428
Videotape, radiation dose 428
Visciodol, in bronchography 291
Visual
  disturbances, encephalograms 160
  loss, after myelography 205
Vitamin K deficiency, in obstructive jaundice 249
Vomiting
  caused by contrast media 8
  in oral cholecystography 264
  following pneumoencephalography 185, 188
  following ventriculography 194

Waldenström's disease, and infusion cholangiography 273
Water intoxication
  in barium enema 357-8
  management 476
Wire breakage during angiography 62
Wound infection, in cardiopulmonary angiography 83

X-ray
  equipment, hazards, examples 464-8
    see also Electrical hazards: Mechanical hazards
  tube, anode temperature 465
    burst 466
    continuous excitation 465
    Loadix device 465
Xerography, radiation dose 426
Xylocaine, in bronchography, toxic effects 278

THIS BOOK IS DUE ON
Books not returned on time are
Lending Code. A renewal may be
consult Lending Code.

14 DAY	
OCT 16 1985	
OCT 18 1985	
14 DAY	
FEB 2 6 1988	
RETURNED	
FEB 1 8 1988	

RETURNED
MAY 1 5 1984

14 DAY
JUN 1 8 1984
Renewen July 3

RETURNED
JUL 5 1984

14 DAY
AUG 1 9 1984
RETURNED
AUG 1 5 1984